Integration of Cloud Computing and IoT

This book presents cutting-edge research, advanced techniques, and practical applications of the integration of Cloud Computing and Internet of Things (IoT). It explores the practical challenges associated with the development and deployment of Cloud Computing and IoT-based solutions, including security, privacy, and interoperability issues. It includes various case studies, including smart cities, network security, healthcare, smart urban sensing, agriculture, mining, railways, and crime investigations. It offers in-depth insights, developments, and practical applications, showcasing helpful tools and techniques in IoT and Cloud Computing.

Key Features

- Provides a comprehensive and accessible resource that covers a broad range of topics on the integration of Cloud Computing and IoT.
- Shows how Cloud Computing and IoT are applied to various challenging situations.
- Covers in-depth inference of Cloud Computing and IoT to real-world issues, which will assist the industrial sector in growth.
- Includes discussion of how robust privacy solutions are important to enable effective integration between Cloud Computing and IoT-based applications.
- Offers case studies and real-life examples from various sectors, including healthcare, agriculture, mining, smart transport, and underwater communication.

The book is a comprehensive guide for professionals, researchers, and students interested in the latest Cloud Computing and IoT developments.

Emerging Technologies: Research and Practical Applications

Series Editors:

Aryan Chaudhary
Chief Scientific Advisor, Bio Tech Sphere Research, India

Raman Chadha
Chandigarh University, Punjab

About the Series: The series offers a comprehensive exploration of frameworks and models that harness information and knowledge from diverse perspectives. It delves into the application of these insights to address global issues in industries, the environment, and communities. The focus is on knowledge transfer systems and innovative techniques that facilitate effective implementation. This amalgamation of intelligent systems and various applications necessitates collaboration across disciplines like science, technology, business, and humanities.

Covering a wide range of cutting-edge topics, the series includes books on Artificial Intelligence, Big Data Analytics, Cloud Computing Technologies, Design Automation, Digital Signal Processing, IoTs, Internet of Medical Things, Machine Learning, Natural Language Processing, Robotics and Automation, Signal Processing, Healthcare applications, Convolutional neural organization, Medical Imaging, Voice Biomarker (Voice Tech in Health), Image Recognition, and more. The series adopts a gradual approach, starting from the basics and progressing to more intricate subjects.

Comprising Handbooks, Monographs, and edited volumes, this series serves as an invaluable resource for those curious about the world's most popular technologies of the present and future. Whether newcomers or seasoned professionals, readers will find valuable insights into emerging advancements.

List of titles:

Artificial Intelligence and Society 5.0
Issues, Opportunities, and Challenges
Edited by Vikas Khullar, Vrajesh Sharma, Mohit Angurala, Nipun Chhabra

Advances on Mathematical Modeling and Optimization with its Applications
Edited by Gunjan Mukherjee, Biswadip Basu Mallik, Rahul Kar, and
Aryan Chaudhary

Practical Book on Cybersecurity and Forensics
Advanced Tools and Techniques
Edited by Keshav Kaushik, Mariya Ouaissa, Aryan Chaudhary

Integration of Cloud Computing and IoT
Trends, Case Studies and Applications
Edited by Prabh Deep Singh and Mohit Angurala

For more information about the series, please visit: https://www.routledge.com/Emerging-Technologies/book-series/CRCCHETRPA

Integration of Cloud Computing and IoT
Trends, Case Studies and Applications

Edited by
Prabh Deep Singh and
Mohit Angurala

CRC Press
Taylor & Francis Group
Boca Raton London New York

CRC Press is an imprint of the
Taylor & Francis Group, an **informa** business

A CHAPMAN & HALL BOOK

Designed cover image: metamorworks, Shutterstock contributor

First edition published 2025
by CRC Press
2385 NW Executive Center Drive, Suite 320, Boca Raton FL 33431

and by CRC Press
4 Park Square, Milton Park, Abingdon, Oxon, OX14 4RN

CRC Press is an imprint of Taylor & Francis Group, LLC

ISBN: 978-1-032-64741-8 (hbk)
ISBN: 978-1-032-65668-7 (pbk)
ISBN: 978-1-032-65669-4 (ebk)

DOI: 10.1201/9781032656694

Typeset in Times
by KnowledgeWorks Global Ltd.

Contents

Preface

Integration of Cloud Computing and IoT: Trends, Case Studies and Applications explores the transformative landscape of Cloud Computing and the Internet of Things (IoT) across various domains. As editors, our intent is to present a holistic view of this dynamic integration, offering readers insights into the evolving trends, real-world applications, and the profound impact it has on diverse sectors.

The foundational chapters lay the groundwork by elucidating fundamental concepts and architecture design in Cloud Computing. These serve as a primer for readers, ensuring a solid understanding before delving into the specialized applications explored in subsequent sections. The book starts with the basics of Cloud Computing and IoT, exploring their integration in agriculture, healthcare, transport, railways, and smart systems. It highlights advancements in wildlife conservation, underwater communication, and healthcare, emphasizing the role of AI and machine learning. The book later focuses on education, showcasing the transformation of traditional learning through these technologies.

Fog Computing emerges as a key theme, with dedicated chapters providing a comprehensive review of recent trends, challenges, and case studies. The future of Cloud Computing is examined, with a special focus on the paradigm shift facilitated by Fog Computing. As we progress, the book investigates the profound impact of Cloud and IoT integration in the education industry, underscoring its role in shaping the future of learning and knowledge acquisition. The exploration concludes with a deep dive into sustainable marketing and retailing, revealing how Cloud Computing and IoT are not just technological enablers but also catalysts for sustainable business practices.

This book is not just a compilation of disparate chapters; it is a collaborative effort that seeks to provide readers with a comprehensive understanding of the integration of Cloud and IoT. The contributors, experts in their respective fields, bring diverse perspectives and experiences to the table, creating a mosaic of insights that reflects the multifaceted nature of this technological convergence. Whether you are a researcher, educator, industry professional, or technology enthusiast, we invite you to immerse yourself in the pages that follow, where each chapter serves as a window into the transformative possibilities of Cloud and IoT integration. As the digital landscape evolves, we hope this book serves as a guiding light, illuminating the path toward a future where Cloud and IoT seamlessly intertwine to shape the way we live, work, and innovate.

Contributors

Jyoti Agarwal
Graphic Era (Deemed to be University)
Dehradun, India

Shivani Aggarwal
Chandigarh University
Mohali, India

Tisha Aggarwal
Graphic Era (Deemed to be University)
Dehradun, India

Shikha Agnihotri
Institute of Management Studies
Ghaziabad, India

Mohit Angurala
Department of Computer Science,
    Guru Nanak Dev University College
Pathankot, Punjab, India

Yojna Arora
Sharda University
Greater Noida, India

Oroos Arshi
University of Petroleum and
    Energy Studies
Dehradun, India

Ramesh Bharti
Jagan Nath University
Jaipur, India

Maneet Kaur Bohmrah
G.N.D.U
Amritsar, India

Aryan Chaudhary
Bio-Tech Sphere Research
Ghaziabad, Uttar Pradesh, India

Ekta
Lovely Professional University
Phagwara, India

Duragaprasad Gangodkar
Graphic Era (Deemed to be University)
Dehradun, India

Abhiraj Gautam
Graphic Era (Deemed to be University)
Dehradun, India

Subhajit Ghosh
IMS Engineering College
Ghaziabad, India

Upma Jain
Graphic Era (Deemed to be University)
Dehradun, India

Arvinder Kaur
Chitkara University
Punjab, India

Harjot Kaur
G.N.D.U
Amritsar, India

Kanwaldeep Kaur
Global Group of Institutes
Amritsar, India

Wuriti Kavya
Geethanjali College of Engineering
    and Technology
Hyderabad, India

Arnav Kotiyal
Graphic Era (Deemed to be University)
Dehradun, India

Vinod Kumar
Maharishi Markandeshwar (Deemed
  to be University)
Haryana, India

Mohit Lalit
Chandigarh University
Punjab, India

Yogesh Lohumi
Graphic Era (Deemed to be University)
Dehradun, India

Shivani Malhan
Chitkara University
Punjab, India

Rekha Mewafarosh
Shri Ram Institute of Management
  and Technology
Kashipur, India

Tanusha Mittal
Graphic Era (Deemed to be University)
Dehradun, India

Ambika N
St. Francis College
Bangalore, India

Prerna
S.G.R.R. University
Dehradun, India

Nidhi Punj
Chitkara University
Punjab, India

Yogita Yashveer Raghav
K R Mangalam University
Gurugram, India

Priyanka Ranga
Maharishi Markandeshwar (Deemed
  to be University)
Haryana, India

Sanjay Sharma
S.G.R.R. University
Dehradun, India

Anuj Kumar Singh
Chandigarh University
Mohali, India

Dhawan Singh
Amity University, Madhya Pradesh
Gwalior, India

Harmeet Singh
Sant Baba Bhag Singh University
Jalandhar, India

Prabh Deep Singh
Department of Computer Science and
  Engineering Graphic Era (Deemed
  to be University)
Dehradun, India

Tejinder Deep Singh
Jagan Nath University
Jaipur, India

Tejinder Deep Singh
Global Group of Institutes
Amritsar, India

Sneha
Graphic Era (Deemed to be University)
Dehradun, India

Prakash Srivastava
Graphic Era (Deemed to be University)
Dehradun, India

Anita Tanwar
Chitkara University
Punjab, India

Swathi Tejah
Yalla Chaitanya Bharathi Institute
  of Technology
Hyderabad, India

Vikas Tripathi
Graphic Era (Deemed to be University)
Dehradun, India

Varsha
Lovely Professional University
Phagwara, India

Vaibhav Vyas
K R Mangalam University
Gurugram, India

1 Fundamental Concepts of Cloud Computing

Oroos Arshi and Aryan Chaudhary

1.1 INTRODUCTION

An introduction to the fascinating world of cloud computing, a ground-breaking technical idea that has revolutionized modern computing. This enlightening chapter, titled "Fundamental Concepts of Cloud Computing," explores the basic concepts, frameworks, and arrangements that underpin this trailblazing area. As we study the complexity of cloud computing, we will look at all its elements, from its historical history to its complex design, deployment methods, and security considerations. Throughout five in-depth sections, this chapter examines cloud computing's foundations in great detail. We begin our adventure in the initial stage, where we explore the fundamentals of cloud computing. In this section, we'll look at the definition's main ideas as well as what makes them unique and how they've changed over time. We will weigh its positive effects on the technical environment against any potential negative effects to paint a complete picture of its impact on our world. The detailed design of cloud computing is explained in the second section of the chapter. As we explore the many cloud architecture worlds, we'll look into the components that make up this ethereal environment. The chapter delves into the intricate operation of cloud storage systems, illuminating the technological advancements that make it possible for straightforward data management and accessibility. Every aspect of the operation of cloud computing and the substantial security risks it presents will be thoroughly investigated. The final portion of our discussion focuses on the many deployment models that make up the cloud computing ecosystem. We'll go over the options available to businesses seeking the ideal cloud environment for their needs, from the public cloud's extensive accessibility to the private cloud's precise management and security. The fourth section thoroughly discusses the difficulties brought on by the cloud server models for the platform as a service, or PaaS, infrastructure as a service (IaaS), and software as a service (also known as SaaS). We'll look at the advantages and disadvantages of each model as we go, giving readers the information they need to make decisions that are right for them. The final section provides a summary of the significant findings from our analysis. Here, we provide an overview for readers to deepen their awareness of this dynamic field by summarizing the major insights gained from our investigation of the fundamental concepts behind cloud computing.

DOI: 10.1201/9781032656694-1

1.2 DEFINITION OF CLOUD COMPUTING AND ITS CHARACTERISTICS

A cloud is an internet connection or an interconnected system. It is a network infrastructure that stores, monitors, and retrieves data online through remote servers rather than local storage. Files, documents, images, videos, audio recordings, and various kinds of media can all be included in the data.

The following operations are made possible by cloud computing:

- The development of new services and applications
- Data recovery, archiving, and backup Providing on-demand software, running blogs and websites, and doing data analysis
- The ability to stream audio and video

Small and big IT firms both employ traditional approaches to provide IT infrastructure. As a consequence, any IT company, must have a Server Room at a bare minimum. An email server, interpersonal interaction, antivirus programs, modems, routers, toggles, the state of Queensland State Police (Query Per Second specifies the total number of requests or inquiries processed by the server), a computer programming system, quick internet, and maintenance specialists should all be located in the same server room. To create an IT infrastructure, we will need to invest a substantial sum of money. All of these concerns are addressed by the freely available internet computing system, which helps reduce the overall cost of IT infrastructure. Figure 1.1 illustrates the clear distinction between the Cloud Computing Era and the Pre-Cloud Era, highlighting the transformative impact of cloud technology on various aspects of computing.

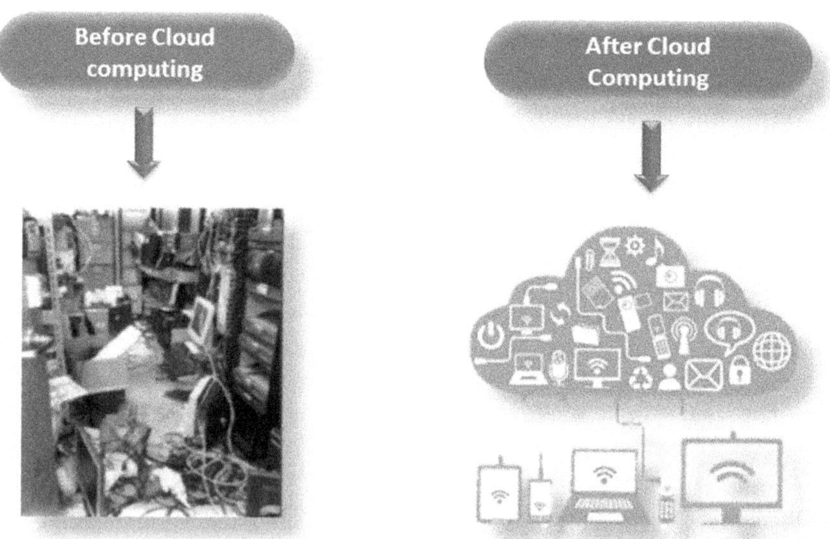

FIGURE 1.1 Contrasting the Cloud Computing Era with the Pre-Cloud Era.

1.2.1 ESSENTIALS OF CLOUD COMPUTING

Those listed below are some basic characteristics of cloud computing:

1. **Quickness:** The cloud is based on a distributed computing platform. It allocates resources among clients and functions quickly.
2. **Dependability and high availability:** Because the infrastructure is unlikely to collapse, server availability and reliability are high.
3. **Extensive Scalability:** Cloud computing enables massive amounts of" ondemand" provision of resources without the requirement for high-traffic engineers.
4. **Multiple Sharing:** Many persons and apps can work more successfully and save costs by collaborating on shared infrastructure with the help of cloud computing.
5. **Repair and upkeep:** Cloud computing programs are simple to maintain because they can be accessed from various locations and are not required to be installed on each person's laptop. As a result, the price dropped.
6. **Low cost:** The cost of computing via the cloud will be reduced because the IT company doesn't have to create its data center and will just pay for the hardware and software that are used.
7. **Pay-per-use services:** Customers can utilize APIs or interfaces for application programming for using cloud resources and pay for the service based on how much they use it.

1.3 HISTORY OF CLOUD COMPUTING

The following section will look through the history of cloud computing. There is also a history of client-server technologies, distributed technology, and cloud computing [2].

The client-server design was used in the days before computers, with every bit of client data and management being held on the server's end. To access data, one user must initially connect to the computer; only after that will the user be given permission. However, it has several shortcomings. The growth of client-server computing gave rise to the concept of distributed computation. Every machine is connected via this kind of computing, enabling customers to share resources as required. There are still some drawbacks. As a result, cloud computing has emerged as a means of avoiding the limitations of a system that is distributed. Figure 1.2 provides a visual representation of the historical evolution of cloud computing, showcasing its key milestones and transformative developments over time. This concise graphic offers a comprehensive overview of the journey that has shaped the modern cloud computing landscape.

- John MacCarty delivered a speech at MIT in 1961 titled "Computing Should be Offered as a Utilities, Comparable to Consuming and Electricity." John MacCharty thought it was a brilliant idea. However, individuals at the time were cautious about employing this technology. They considered their

History of Cloud Computing

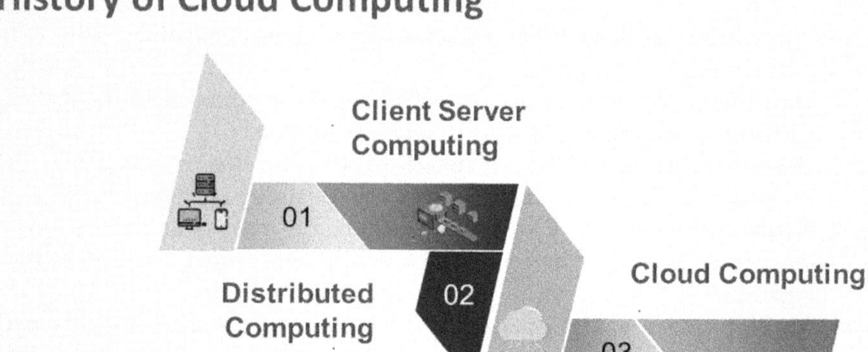

FIGURE 1.2   History of cloud computing.

current technology to be sufficient. Little will be investigated in this field, given the unfavorable criticism regarding this computer idea. However, as time passes and technology advances to keep up with it, the idea eventually becomes a reality. Salesforce.com initially launched this in 1999.

- The Public Cloud Computational boom started due to this startup's initiative to release a corporate program online.
- Amazon Web Services, also known as AWS, was established in 2002 to provide processing and storage through the internet. Amazon will make the Flexible Computing Cloud Professional Service available to everyone in 2006. Following that, Google Play introduced the development of commercial cloud computing applications in 2009. As soon as other companies realized the advantages of cloud computing, they began providing their cloud-based services. As a result, other businesses like Alibaba, which are Oracle, IBM, and HP, quickly followed with their iterations of cloud services after Microsoft released Microsoft Azure in 2009. The importance and popularity of cloud computing have increased recently.

Advantages

1. It is easier to receive backup in the cloud.
2. It allows us to simply and rapidly recover saved information from everywhere and at any time.
3. It can be used on mobile devices to retrieve information.
4. It is simple to use and lowers both hardware and software costs.
5. Its key benefit is database integrity.

Disadvantages

1. It is necessary to have a reliable internet connection A reliable internet connection is required.
2. The user only has limited authority over the data.

1.4 CLOUD COMPUTING DEVELOPMENT

Outsourcing computing resources is akin to using the cloud. This idea first gained traction throughout the 1950s. Five technologies were necessary to develop cloud computing to its present position. Examples include utilitarian computing, distributed systems or peripherals, service-based design, and virtualization (web 2.0) [3]. Figure 1.3 provides a visual timeline depicting the progressive evolution of cloud computing, showcasing its pivotal milestones and advancements.

a. **Distributed System:** Users are provided with an overview of a distributed system, which is made up of several different separate systems. In contexts where resources are shared, distributed computing tries to use them as effectively and efficiently as possible. Scalability, concurrency, continual accessibility, variations, and failure independence are attributes of distributed systems. The network's primary design flaw was the requirement for all systems to be situated in one specific area. As a result of distributed computing, three innovative computing paradigms—platform computational science, clustering computer science, and machines on grids—have arisen to address this issue.

b. **Mainframe computers:** Introduced in 1951, mainframes are powerful and dependable computing machines. These processes manage vast volumes of data, including huge input-output approaches. Nowadays, computers are still used for large-scale processing tasks like online transactions. These systems have very low downtime due to their strong fault tolerance. The system now has a greater processing capacity thanks to distributed computing. But the price for these was outrageous. Cluster computing is a low-cost substitute for traditional computer technology.

c. **Cluster computing:** Mainframe computers were superseded by cluster computing in the 1980s. A high-bandwidth network linked each of the

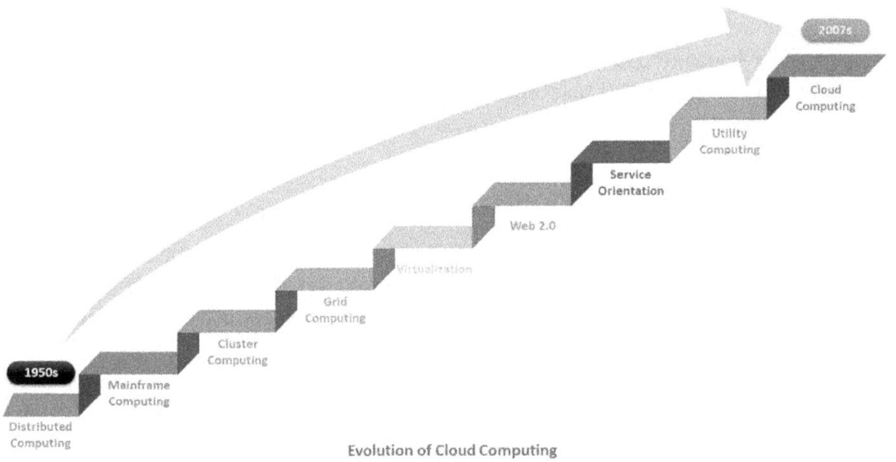

FIGURE 1.3 The evolutionary journey of cloud computing.

group's gadgets to one another. These are considerably cheaper than mainframe systems. These can carry out complex calculations. More nodes can also be added to the current group quickly if more are required. The cost issue was only partially resolved as a result, and the geographic issue continued. To solve this problem, the idea of computing on a grid was developed.

d. **Grid computing:** In the 1990s, the idea of computing on an organized basis first emerged. It demonstrates how various systems were set up in various geographically distinct locations and connected via the internet. The grid had a variety of nodes since these systems belonged to different organizations. Although it resolved some problems, others appeared as the space between the nodes grew. The fundamental problem, along with other network-related issues, was an absence of high-bandwidth connectivity. Because of this, cloud computing is often described as the "successor to grid computing."

e. **Virtualization:** Virtualization first appeared about 40 years ago. It refers to the procedure of creating a layer of virtualization on hardware to ensure a user can run several instances on the already in-use hardware at once. A crucial component of cloud computing technology, it is the foundation upon which vital computer services such as VMware VCloud and Amazon EC2 are constructed. One of the types of virtualization that is most frequently utilized is hardware virtualization.

f. **Web 2.0:** Cloud computing businesses communicate with their customers using the Web 2.0 interface. Online 2.0 has produced interactive and dynamic web pages. Additionally, it improves how adaptable websites are. Google, Facebook, and Twitter Maps are a few examples of Web 2.0 applications. Communication through the internet is unquestionably only possible as a result of this technology. In 2004, it was a popular choice.

g. **Service orientation:** In the form of service-oriented cloud computing. It enables the development of scalable, adaptable, and inexpensive solutions. This computer model was composed of two essential components (Quality of Service) and QoS SaaS (Software as a Service), which includes SLAs (Service Level Agreements), were two among them.

h. **Utility technology:** An IT paradigm that offers a pay-per-use provisioning method for computer resources as well as additional essential services like conservation, facilities, and so forth.

1.5   ADVANTAGES AND DISADVANTAGES

As is well known, every organization has migrated its activities to the cloud to hasten its growth [5]. Figure 1.4 visually portrays the manifold advantages of cloud computing, emphasizing its transformative benefits for businesses and individuals alike.

Advantages

The advantages of cloud computing are listed below:

1. **Make a copy of the data and restore it:** Backup and recovery are much simpler once data is in the cloud.

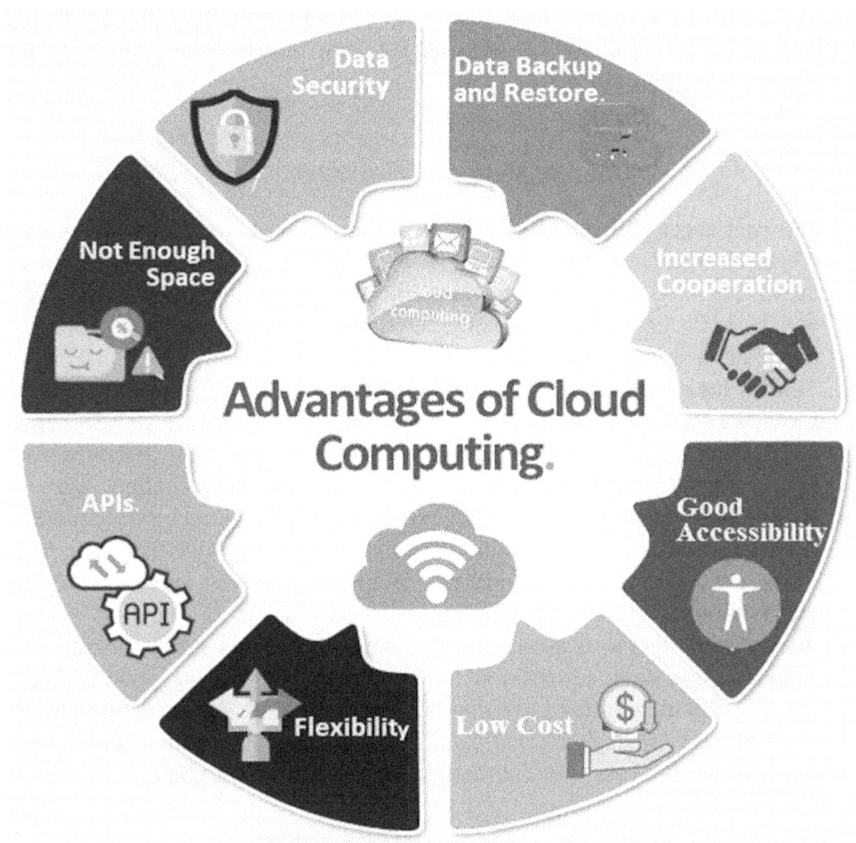

FIGURE 1.4 Advantages of cloud computing.

2. **Better collaboration:** Teams can share information more quickly and easily thanks to cloud applications as they are given access to shared cloud storage.
3. **Excellent accessibility:** By utilizing the cloud and an internet connection, we can rapidly and easily retrieve data that is stored anywhere, at any time. An Internet cloud system boosts organizational productivity and efficiency by guaranteeing our knowledge is available.
4. **Reduced maintenance costs:** Businesses can save money on software and hardware maintenance thanks to cloud computing.
5. **Flexibility:** We now have access to everything stored in the cloud using a mobile device, thanks to cloud computing.
6. **Application Programming Interfaces:** Cloud computing customers can access cloud-based services using programmatic interactions, or APIs, and are then charged depending on how frequently they use those services.
7. **Ample storage is available:** We have enough space on the internet for centralized storage of all of our important documents, images, audio files, and films.

Data security is one of the main advantages of cloud computing. According to the cloud's numerous state-of-the-art security features, information may be managed and maintained appropriately.

Disadvantages

The disadvantages of cloud computing are listed below:

1. **Internet accessibility:** As you likely already know, cloud computing keeps all data online, making it accessible from anywhere. This includes photos, audio files, and video files. If the connection to the internet is unreliable, you cannot access this information. Despite [6], we don't have any alternative ways to access cloud-based data.
2. **Vendor lock-in:** The main drawback of cloud computing is vendor lock-in. Moving a company's products or services from one supplier to another could be challenging. Because there are so many vendors and many different platforms, transitioning from a single cloud to another can be challenging.
3. **Restricted Authority:** As soon as we are aware, the cloud's infrastructure is entirely owned, managed, and watched over by a service provider. As a result, cloud users have less influence over how the services provided within the framework operate and function.
4. **Security:** While cloud assistance providers use the best safety measures to store receptive data, you must be conscious that before using cloud computing, you will be granting a supplier of cloud computing services accessibility to all of the company's critical data. There is a risk that criminals will steal sensitive details from your company while they are being sent to the cloud.

1.6 ARCHITECTURE OF CLOUD COMPUTING

It is generally accepted that large and small businesses use computing technology to store information within the cloud and give access to it wherever and whenever they are connected to the internet [7]. Figure 1.5 offers an architectural overview of cloud computing, detailing its core components and their interconnections within the system.

Event-driven architecture and service-based design are combined in cloud computing architecture

- Front End
- Back End

The cloud computing architecture is depicted in the diagram below:

Front End: The client uses the front end. The programs and client-side interfaces needed to use cloud computing services make up the software. Lean and bulky people, cell phones, and portable electronics make up the front end, along with internet servers (such as Firefox, Chrome, Microsoft's Internet Explorer, etc.).

Back end: The service provider uses the back end to provide the service. It is in charge of overseeing all of the assets needed to deliver cloud computing services. A ton of data is kept, and there are security precautions.

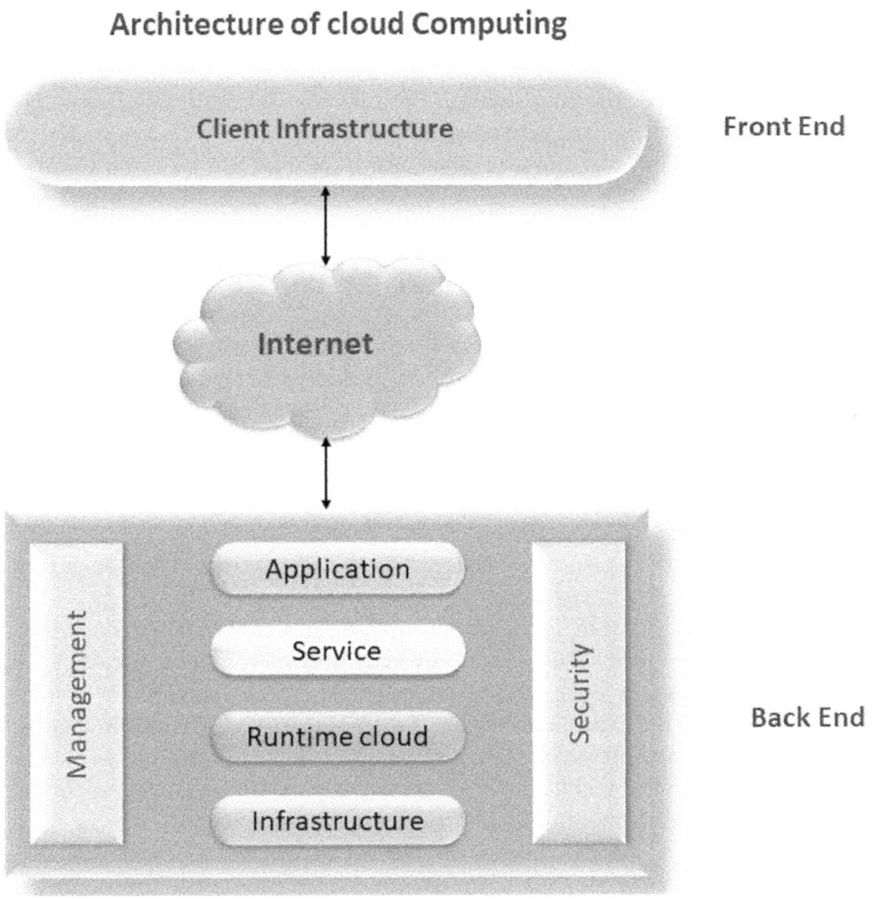

FIGURE 1.5 Architecture of cloud computing.

1.7 CLOUD COMPUTING INFRASTRUCTURE

One of the most popular technologies today, cloud computing has emerged as a revolutionary business trend for businesses of all sizes. It oversees a sizable and intricate infrastructure set up to offer cloud services and resources to consumers. The back end of the cloud architecture includes the hardware and software elements that collectively constitute the cloud infrastructure, such as storage devices, networking components, management applications, deployment software, virtualization software, etc. The entire cloud computing system is run by back-end cloud infrastructure. Cloud computing is a method of providing clients with on-demand services no matter where they are, at any point they require them, and under any conditions. The trigger that turns on every aspect of the cloud computing system is the cloud infrastructure. Cloud infrastructure enhances our capacity to provide customers with services equivalent to those provided by physical infrastructure. Private clouds, public cloud services, and hybrid cloud systems can all be used since they are more economical, adaptable, and scalable [8].

1.7.1 COMPONENTS OF THE CLOUD INFRASTRUCTURE

The computational needs of the cloud computing paradigm are supported by a variety of cloud infrastructure components. The fundamental elements of cloud architecture include more than just networks, computers, software, and storage. The three main categories of infrastructure for clouds, i.e.

1. Computing
2. Networking
3. Storage

The most important thing to keep in mind is that cloud infrastructure must comply with several basic infrastructure criteria, including flexibility, reliability, openness, and intelligent monitoring [9]. Figure 1.6 presents an insightful view of the infrastructure underpinning cloud computing, illustrating the foundational elements and their networked structure.

Following is a diagram that shows the cloud infrastructure.

- **The hypervisor:** commonly referred to as a hypervisor or low-level program, is required to enable.
- **Virtualization:** Multiple clients may utilize cloud resources by using virtualization. The term "hypervisor" is widely utilized to refer to the software that manages and looks after cloud resources and services, which is additionally referred to by the acronyms The VMM (Virtual Machine Manager).
- **Management Software:** Management software facilitates infrastructure configuration and maintenance. Cloud management software allows for the optimization and monitoring of services, apps, data, and resources.

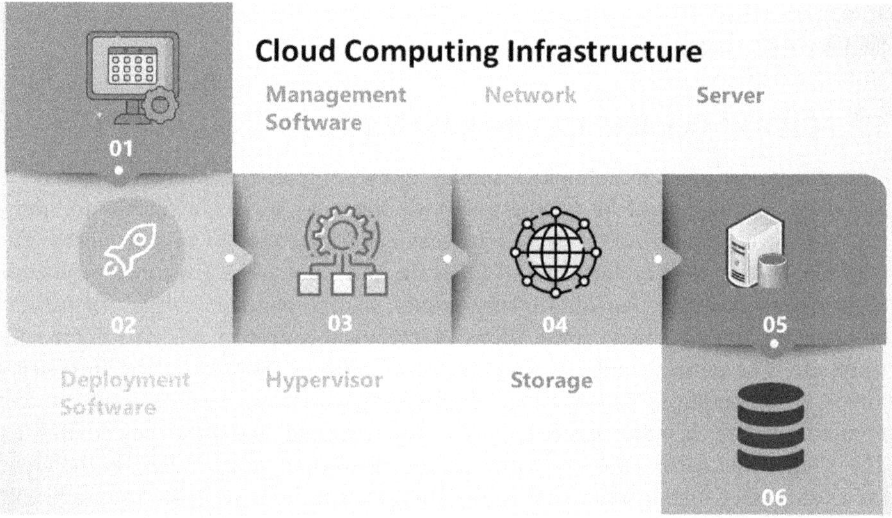

FIGURE 1.6   Unveiling the infrastructure of cloud computing.

- **Deployment Software:** With the help of deployment tools, the cloud service is integrated and deployed. As a result, it is frequently made easier to create a virtual computer environment.
- **The network:** The network that brings cloud services online and is one of the essential components of the cloud infrastructure. Transferring resources and information both internally and outside requires the usage of a network.
- **Server:** The computer or system that powers the cloud's infrastructure is in charge of managing and supplying cloud services to an assortment of partners and offerings, as well as keeping an eye on security and carrying out other responsibilities.
- **Storage:** The management of data and its storage space are referred to as storage and are made available to different enterprises. As it maintains several storage copies, it provides the option of retrieving an alternative resource in the case that one fails.

Additionally, virtualization is regarded as a crucial element of cloud infrastructure. Because customers communicate with their cloud services via a GUI (Graphical User Interface), which isolates the data that is accessible storage and computational capacity from the real hardware.

1.8 CLOUD MANAGEMENT IN CLOUD COMPUTING

Cloud computing management entails the upkeep and oversight of on-demand services and resources, whether hybrid, private, or public. Some of its capabilities include distributing load as well as efficiency, safeguarding, backups, capacity, implementation, and other elements. To do this, a cloud management team needs full access to all cloud resource capabilities. Combining different software packages and technologies results in an integrated cloud management approach and process [10].

Since a private cloud system is exclusively utilized by one business, it can be internally or externally managed by that business. Public cloud services are delivered over an accessible network that is frequently used. In this method, the IT infrastructure is owned by a private company, and consumers can purchase or rent processing capacity as needed. Hybrid cloud computing environments include private and public cloud services provided by diverse vendors. Most companies store their data on their cloud servers to safeguard their clients' privacy. Public cloud computing services, which are less expensive, handle less crucial data. The term "hybrid cloud servers" refers to those that combine both public and private cloud [11].

1.8.1 CLOUD MANAGEMENT IS NECESSARY

Large organizations now select the cloud as their main information storage option. A minor error or outage can cost an organization a lot of money and inconvenience. Specific team members are in charge of designing, managing, and maintaining a cloud computing service. They ensure that everything goes as planned and that any issues that may arise are resolved.

1.8.2 Cloud Management Platform

An IT system's information technology (IT) system can access data from every area of the system thanks to a software system called a cloud administration platform, which has a robust and extensive set of APIs. The Cloud Management Protocol (CMP) allows an IT organization to offer a standardized approach to IT security and management that can be used throughout the organization's entire cloud environment.

1.8.3 Cloud Computing Tasks

Figure 1.7 visually delineates a spectrum of cloud management tasks, offering a comprehensive view of the responsibilities involved in effectively overseeing cloud resources and services

- **Auditing System Backups:** It is required to routinely audit the backups to ensure the reconstruction of randomly selected files from various users. This could be handled by the company or the cloud service provider.
- **Flow of data:** The managers are responsible for drawing a diagram of the data flow that shows how the information is supposed to move throughout the business.
- **Vendor Lock-In:** If the business chooses to switch service providers, managers must be allowed to transfer data from one server to another.
- **Understanding the provider's security procedures:** Managers need to be aware of the provider's security procedures, particularly those relating to multitenant use, processing of electronic payments, hiring practices, and encryption policy.
- **Monitoring the Capacity, Preparation, and Scalability Capabilities:** The management needs to be aware of the cloud provider's capacity for scaling and whether the current cloud provider can meet the organization's future demand.

Cloud Computing Tasks

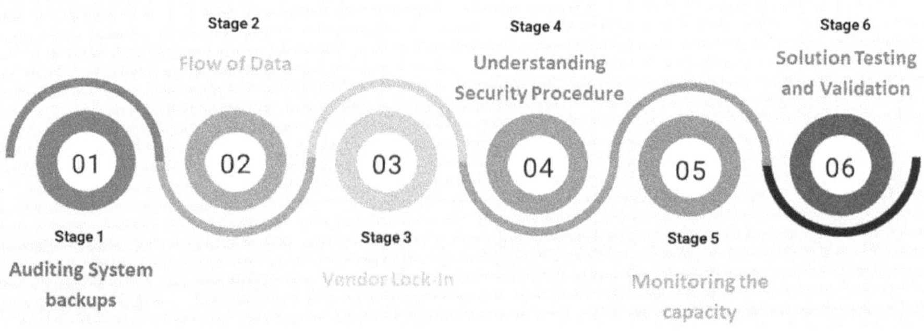

FIGURE 1.7 Visualizing cloud management tasks.

- **Monitoring the audit log:** The management regularly audits the logs to search for recurring errors.
- **Solution Evaluation and Validation**: It's crucial to test cloud services and validate the results to make sure the solutions are error-free.

1.9   STORAGE SYSTEM IN THE CLOUD

Broadly, alternative internet services are referred to as cloud computing. It has many resources, such as servers, databases, networking, programs, instruments for data storage, etc. The platform, infrastructure, and applications it contains are surrounded by servers, laptops, personal computers, phones, and tablets [12].

1.9.1   FEATURES OF CLOUD COMPUTING

The main characteristics of cloud computing are listed below.

1. There are more resources available.
2. The ease of maintenance is one of the key benefits of using cloud computing.
3. Cloud computing has large network access as one of its features. It is automatically operated.
4. Security is one of the essential components which cloud computing enables you to protect across all networks.

1.9.2   STORAGE SYSTEM IN THE CLOUD

The three main types of cloud storage are as follows:

1. Block-Based Storage System
2. File-Based Storage System
3. Object-Based Storage System

Let's discuss each one independently in the order listed below.

1. Block-Based Storage System
   - The storing method utilized in hard disks is block-based. Your operating system, like Windows or Linux, can see your hard drive. It consequently recognizes a disk on which you can create a volume that you can subsequently partition and format.
   - As an illustration, if the entire system has a storage capacity of 1000 GB, we can split it into local drives C and D, which are each 800 GB and 200 GB in size.

     As you may remember, a drive would represent a block-based storage system on your computer, from which you might form volumes and partitions.

2. File-Based Storage System
   - In this instance, connecting is accomplished through the use of a Network Interface Card (NIC). A network connection is required to

use any network-attached storage server (NAS). NAS units employ file-based storage platforms.

- With a different disk within, each storage server is a unique computer. It already has a file system set up, has partitions that are formatted, and will share that file system over a network. The drive's networking location can be found here.
- The user is not required to format the drive or create partitions with this method, as the previous one. This is currently used in file-based storage systems. Thus, the operating system identifies a system of files with a local drive letter. [13]

3. Object-Based Storage System

- Through the use of a web browser, customers can do this to upload products and store these inside an object container for storage. This makes use of the additional APIs (such as GET, PUT, POST, which stands for SELECT and DELETE), as well as the HTTP Protocols.
- After connecting to a website, you might need to download text, images, or other content from it. For that reason, it qualifies as an HTTP GET request. Utilize PUT and POST requests when evaluating products.
- The container also lacks hierarchical components. A storage system that makes use of objects maintains all files at the same level.

Advantages

1. **Scalability** - Enables both performance enhancement and capacity and storage extension.
2. **Flexibility** - Scaling and data modification are allowed by the rules.
3. **Simpler Data Migrations** - Disruption data transfers are no longer required because they can add and remove new and old data as needed [14].

Disadvantages

To operate effectively, data centers require access to electricity and a dependable internet connection.

1.10   CLOUD COMPUTING TECHNOLOGIES

The following is a list of cloud computing technologies:

a. Virtualization
b. Service-Oriented Architecture (SOA)
c. Grid Computing
d. Utility Computing

a. **Virtualization**: To execute various applications and operating systems on the same server, a virtualized environment must be created through the process of virtualization. A computerized setting can be a thing, including only one instance or multiple instances of multiple operating systems, storage platforms, network-based application servers, and environments.

Cloud virtualization is a concept that broadens the applications of virtual computers. A virtual machine is a piece of software or a computer program that may perform tasks for the user, such as running applications or other programs when needed, while also acting as a physical machine [15].

Types of Virtualizations

Below is a list of various virtualization types:

1. Hardware virtualization
2. Server virtualization
3. Storage virtualization
4. Operating system virtualization
5. Data Virtualization

b. **Service-Oriented Architecture (SOA):** To adapt to changing business needs, organizations can quickly adopt cloud-based computing solutions thanks to SOA. It is possible to function with or without cloud computing. Numerous benefits of SOA implementation include great scalability, platform independence, and ease of maintenance [16].

Service-Oriented Architecture Applications

- The use of SOA can be seen in the following situations:
- The healthcare sector makes use of it.
- Numerous mobile games and applications are made with it.
- Situational awareness systems are deployed in the Air Force using SOA infrastructure.
- The two main roles in SOA are service consumers and service providers.
    Figure 1.8 elucidates the intricacies of an SOA within cloud computing, highlighting the modular and interconnected nature of cloud services.

c. **Grid computing:** Grid computing is often referred to as distributed computing. It is a sort of processing design that combines a significant amount of computational power from various sources to focus on a single goal. Simultaneous nodes connect the grid to form a computer cluster in grid computing. These computer clusters come in a variety of sizes and with different operating systems.
    In grid computing, the following three kinds of machines are employed:
1. The entire network is controlled by a group of servers called a control node.
2. Provider: A provider is a piece of hardware that adds features to a network's capacity pool.
3. A user is a device that makes use of network resources.
    Grid computing is mostly utilized in back-end infrastructures, marketing research, and ATMs.

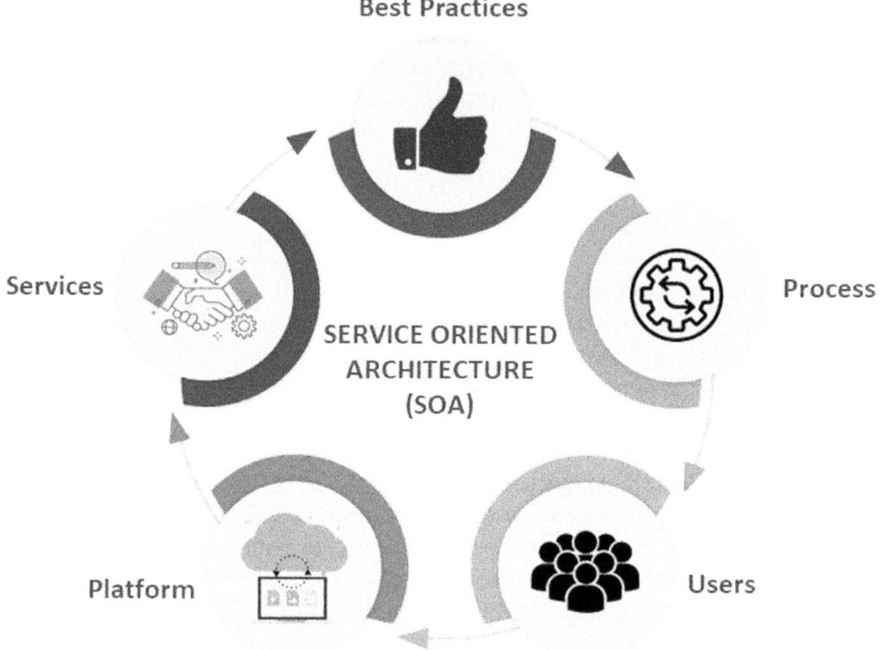

FIGURE 1.8 Depicting a service-oriented cloud architecture.

d. **Utility Computing**: Utility computing is now the strategy used to supply IT services that are most in demand. It provides an infrastructure with immediate computing features (computation, preservation, and programming services via API) and a pay-per-use business model. While minimizing related costs, it maximizes resource efficiency. The benefits of utility computing include decreased IT expenditures, increased flexibility, and simplicity of administration [17].

Large firms like Google and Amazon developed application utility services and compute storage infrastructure. Figure 1.9 showcases the concept of utility computing, illustrating how resources are allocated and billed based on actual usage, enhancing cost-efficiency in cloud environments.

1.11   CLOUD COMPUTING VS. GRID COMPUTING

Cloud Computing: Using the Cloud Client-Server Architecture, cloud computing allows the pay-per-use distribution of computer resources such as storage, databases, servers, and software across the cloud (internet).

Cloud computing is a very popular option for businesses because of its advantages, which include cost savings, better productivity, efficiency, and reliability as well as backups of information, disaster recovery, and security. Figure 1.10 provides a diagrammatic representation of cloud

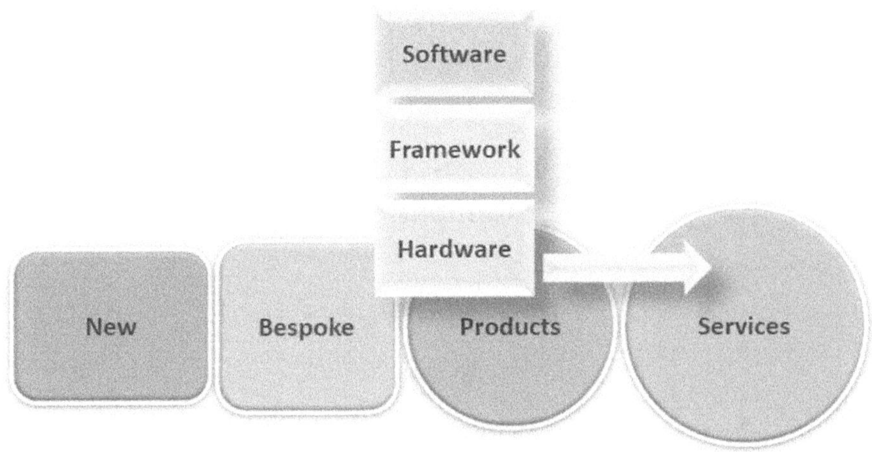

FIGURE 1.9 Visualizing utility computing in cloud services.

FIGURE 1.10 Diagrammatic representation of cloud computing.

FIGURE 1.11   Visualizing utility computing in cloud services.

computing, offering a visual overview of its key components and their interconnections.

Grid Computing: "Distributed computing" is another term for grid computing. It provides customers access to several computing resources, including workstations, servers, PCs, and storage components.

The main advantages of cloud computing include quicker turnaround times and higher user productivity because resources are transparently available. Figure 1.11 illustrates the concept of utility computing, demonstrating how computing resources are provisioned and charged based on actual usage, enhancing cost-efficiency in cloud environments.

Let's examine the differences between grid computing and the cloud explained in Table 1.1

1.12   WORKING OF CLOUD

Assuming you are an executive at a sizable business. Making sure that all of your employees possess the software and equipment they need to do their tasks is one of your specific responsibilities. Purchasing desktops for everyone would be insufficient. You also need to purchase software and permissions and then make them available to your team as needed. If you hire a new worker, you must either purchase additional software or make sure the licenses you already own for the program permit the addition of a second user. Because of the stress, you must spend an excessive quantity of money.

For CEOs like you, there can be another choice. Instead of installing a full suite of programs on each system, you only need to open one application. Through that

TABLE 1.1

Difference between Cloud Computing and Grid Computing

Cloud Computing	Grid Computing
In cloud computing, client-server computer architecture is utilized.	Grid computing takes advantage of a distributed computing architecture.
Extremely scalable.	Scaling is normal.
Cloud computing is more flexible than grid computing.	Grid computing is less adaptable than cloud computing.
The cloud's administration system is centralized.	Grid is a distributed management system that is in use.
Infrastructure providers own the cloud servers in cloud computing.	In grid computing, the organization owns and maintains the grids.
Services like IaaS, PaaS, and SaaS are used in cloud computing.	Grid computing makes use of distributed pervasive, decentralized information and distributed computing systems.
The cloud is service-oriented computing.	Application-focused grid computing.

application, the staff members will have access to a Web-based service that will house all the tools a user needs to do his or her duties. On server sites owned by another company, everything will run, including email, word processors, and complex data analysis software. Cloud computing is what it is, and it possesses the power to completely change the computer industry [18]. Figure 1.12 provides a comprehensive visual representation of the core workings and components of cloud computing, offering insights into its fundamental architecture and operations.

The workload is significantly modified in a cloud computing system. Programs are no longer exclusively executed on local machines. On the other hand, cloud computing can quickly and automatically manage that huge task. Users are making

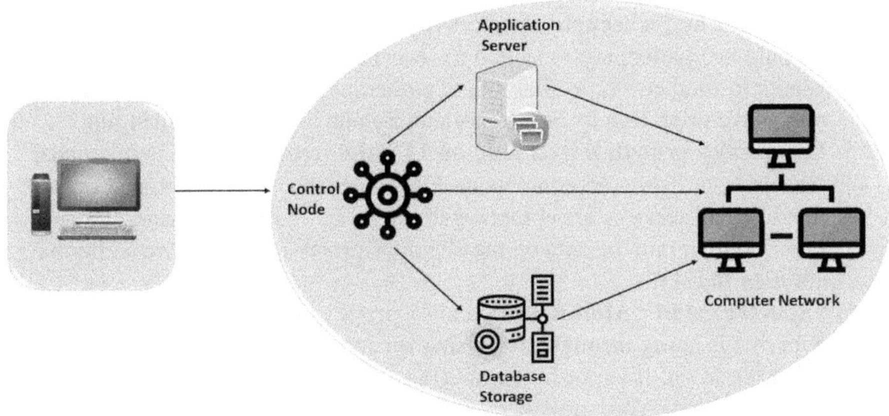

FIGURE 1.12 The mechanisms of cloud computing.

fewer demands for software and hardware. Cloud-based networking software, which might be as basic as a Web browser, is all that is needed on the user's personal computer to execute the system. The remaining tasks are handled by the cloud network.

1.13 SECURITY RISKS OF CLOUD COMPUTING

Cloud computing offers many advantages, including improved collaboration, increased accessibility, mobility, capacity for storage, etc. Cloud computing, however, poses security risks as well.

The most frequent security threats related to cloud computing are listed below.

1. **Data loss:** Data loss is an especially frequent security concern with cloud computing. The phrase "data leak" is occasionally used to describe it. Data loss happens when someone, a program, or an application actively tampers with, eliminates, or renders unusable data. Data loss occurs in the cloud computing environment even when the confidential information is in the hands of an external organization, when the data owner is unable to access some or all of the data, when the storage device is malfunctioning, and when technology is never updated [19].

2. **Hacked Interfaces and Insecure APIs:** APIs that aren't secure and hacked interfaces to protect the interface and application programming languages (APIs) that are used by external users, which cloud computing completely depends on, security measures must be taken. APIs are the most practical way to reach the majority of cloud services. The public can access only a small number of cloud computing services. Due to the fact that these services are open to the public, hackers may try to harm or compromise them.

3. **Data breach:** A "data breach" occurs when private information is accessed, retrieved, or copied without consent by a third party. This is caused by hacking.

4. **One of the major security dangers in cloud computing is vendor lock-in:** It could be challenging to switch the services of an organization from one vendor to another. The availability of numerous providers' diverse platforms may make migrating from one particular cloud to another challenging.

5. **Complexity growth puts a load on IT employees:** IT personnel is under stress due to increased complexity. The migration, integration, and operation of cloud services are complicated for the IT staff. To manage, incorporate, and maintain the data on the cloud, IT professionals require additional abilities and skills.

6. **"Spectre" and "Meltdown":** These programs can monitor and steal data that is going through processing on computers as a result of Spectre and Meltdown. It supports computing on desktop PCs, mobile devices, and the cloud. Your password can be stored in the database of other open applications, along with other private data like emails, images, and work papers.

7. **DoS (denial-of-service) attacks:** Denial-of-service (DoS) cyberattacks occur whenever the server is unable to handle the volume of information the system is getting. DoS assaults frequently target the web servers of significant institutions, including financial institutions, media outlets, and the government. To process the data and recover the lost data, DoS attackers charge a high price in terms of both time and money.
8. **Theft of a user's account:** The hijacking of accounts is a serious security problem in cloud computing. It is the method through which cybercriminals seize control of a user's or the organization's cloud account (which may include a bank account, username, password, and social media account). The hackers utilize the account that was taken to commit crimes [20].

1.14 CLOUD DEPLOYMENT MODELS

With cloud computing, we can access a pool of shared internet computing resources (servers, storage areas, software, etc.). Simply ask for more resources as needed. Clouds make it easy to launch resources quickly. It's possible to release services that are no longer needed. With this approach, you only have to pay for what you expect to utilize. Your cloud provider is in charge of all maintenance [21].

- **Cloud Computing Models:** The cloud deployment model functions as a simulated computer system with a changing deployment architecture, depending on the amount of data that it wishes to keep and who gets access to the infrastructure.
- **Cloud computing deployment models:** The type of cloud environment depends on ownership and flexibility, access, the nature of the cloud, and the intended use and is determined by the cloud deployment strategy. The exact place and person in charge of the computer servers you're utilizing are identified by a cloud deployment technique. The architecture of the cloud's architecture is discussed, along with what you may change, how you can use services, and whether you must construct everything from scratch. Various cloud deployment techniques connect your users with computer systems. These descriptions of various cloud computing deployment approaches can be found below.

1.15 TYPES OF CLOUD

Figure 1.13 categorizes and visually presents the various types of clouds in the context of cloud computing, showcasing distinctions between public, private, and hybrid cloud deployments.

1. Public Cloud
2. Private Cloud
3. Hybrid Cloud
4. Community Cloud

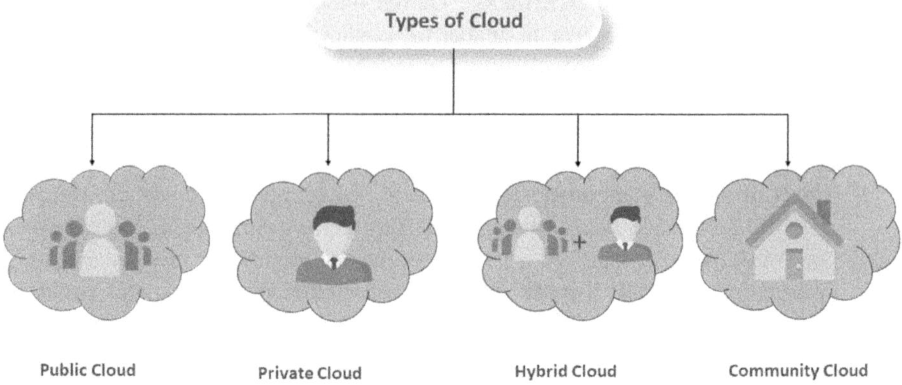

FIGURE 1.13   Exploring the different types of cloud deployments.

1. Public Cloud

Utilizing a pay-per-use business model, anyone can store data in the public cloud and gain online access to it. The cloud services company (CSP) manages and controls the computing power in a public cloud. Figure 1.14 delves into the specifics of public cloud computing, offering a visual breakdown of its characteristics, advantages, and key use cases.

Examples include Amazon's EC2 elasticity-based computing cloud, Google App Engine, IBM SmartCloud Enterprise, and Windows Azure Services Platform.

- To utilize the public cloud's shared platform, you need an internet connection.
- A third-party organization or cloud service provider maintained the public cloud, which utilized a pay-per-use model.
- On the public cloud, numerous users are concurrently using the same storage.

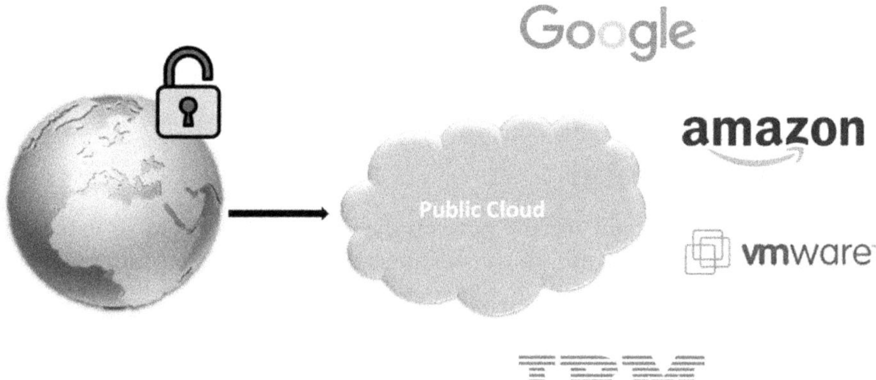

FIGURE 1.14   Unveiling the public cloud paradigm.

- Public clouds are owned, run, and managed by companies, educational organizations, governmental entities, or a combination of these, as well as by people.
  - The public cloud includes services like Google Cloud, the Elastic Compute Cloud from Amazon (EC2), Azure by Microsoft, IBM's Blue Cloud, Solar Cloud, and Amazon Variable Compute Cloud (EC2).

Advantages of Public Cloud

1. Low cost: Public clouds are more affordable than private or hybrid clouds since many individuals share the same resources.
2. Location Disconnected: Public computing services based on the internet are not based on a specific place.
3. Save time: Users can save time when installing and constructing servers, enabling connectivity, deploying new products, and other tasks by using the public cloud, where the provider of cloud services the data centers where data is stored.
4. Quick and easy setup: Businesses can quickly and easily buy public cloud services online, deploy them remotely, and configure them through the cloud service provider.

Disadvantages of Public Cloud

1. Limited Safety Resources are less secure because they are distributed publicly on public clouds.
2. Application: The public cloud's performance depends on how rapidly someone can connect to the internet.
3. Less flexible: The private cloud can be customized more than the public cloud.

1. **Private Cloud**

   An institutional cloud or business cloud is another name for a privately owned cloud. A private cloud, as opposed to a public cloud, offers computer support for a small number of customers and a private internal network (inside the company). Through firewalls and internal hosting, private online computing facilities provide a high level of privacy and protection for data. Additionally, it guarantees that vital or secret data won't be accessible to third-party vendors [22]. Figure 1.15 provides an in-depth exploration of private cloud computing, visually presenting its unique attributes, benefits, and common applications in enterprise environments.

   Examples of private clouds are Elantra, Microsoft, HP Data Centers, Ubuntu, and HP.

Advantages of Private Clouds

1. **Greater control:** Users have more control over their computer hardware and resources because fewer people can use private clouds versus public ones.

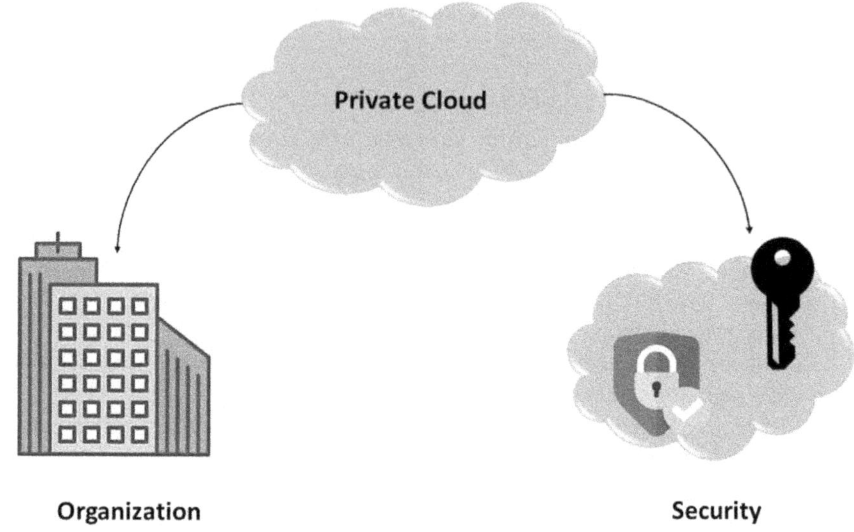

FIGURE 1.15   Navigating the world of private cloud computing.

2. **Privacy:** Cloud computing's security and privacy features are a key plus. In contrast to public clouds, private clouds offer higher security.
3. **Improved performance:** Private clouds deliver improved performance with quicker speeds and greater storage space.

Disadvantages of Private Clouds
1. **Expensive:** Pricing is higher than that of public clouds due to the high installation and upkeep costs for hardware assets.
2. **Limited functioning area:** Due to the private cloud's internal-only accessibility.
3. **Private clouds:** Only able to expand to a certain number of the resources they host internally.
4. **Requires skilled professionals:** individuals are needed for cloud service administration and operation.

3. Hybrid Cloud
A hybrid cloud system allows public and private clouds to coexist. A hybrid cloud is created by combining public and private clouds. A unified, computerized, and well-managed computing environment is what these two clouds (Public and Private) aim to achieve most when combined. In a hybrid cloud, crucial tasks are performed by a private cloud, while optional tasks are taken care of by a public cloud. Hybrid clouds are most frequently used in the healthcare, financial, and educational sectors [23]. Figure 1.16 illustrates the concept and configuration

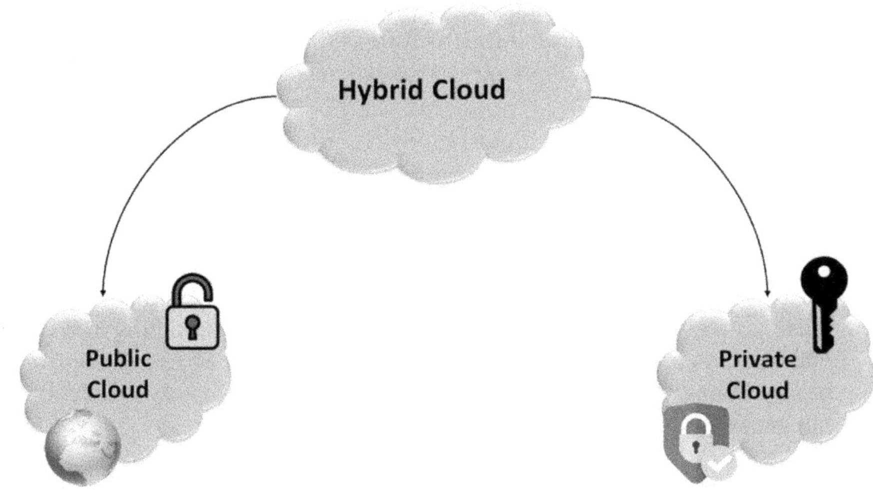

FIGURE 1.16 Unveiling the hybrid cloud paradigm.

of hybrid cloud environments, showcasing the integration of both public and private cloud resources for enhanced flexibility and efficiency.

The leading providers of hybrid clouds include organizations like Amazon, Google, Microsoft, Cisco, and NetApp

Advantages of Hybrid Clouds

The advantages of the hybrid cloud are as follows:

1. **Secure and adaptable:** Combining a private cloud with a general-purpose cloud provides flexible and secure resources.
2. **Price:** Hybrid clouds are less expensive than private clouds. It enables companies to reduce expenses for both application support and infrastructure.
3. **A wise investment:** It offers the benefits of both public and private clouds. A hybrid cloud may be able to meet any organization's architectural, memory, or storage needs.
4. **Security:** Hybrid data centers are secure since the secure cloud handles critical functions.
5. **Effective risk management:** It is possible for businesses to use hybrid clouds.

Disadvantages of Hybrid Clouds

1. **Networking issues:** Since the hybrid cloud uses both private and public clouds, networking becomes more difficult.
2. **Infrastructure Support:** The main problem with a hybrid cloud is the infrastructure's adaptability. Because a private cloud is managed by the

business and a public cloud is not, they can run across multiple stacks because of the dual-level infrastructure.

3. **Reliability:** Cloud computing service providers are in charge of the service's reliability.

4. Community Cloud

A cloud architecture that enables a collection of many organizations to use systems and services without exchanging data is illustrated by the community cloud. A variety of locally based businesses, a third party, or a combination of them own, manage, and operate it [24]. Figure 1.17 provides a concise overview of the community cloud model, highlighting its collaborative approach to cloud computing among specific user groups or organizations. Our Indian government agency, for instance, may share computational resources in the cloud to manage data.

Advantages of Hybrid Computing

The following are some advantages of Community Cloud:

1. **Efficient in terms of price:** Community clouds are economical since the entire cloud is shared by many enterprises or a community.
2. **Scalable and flexible:** The collaborative cloud is flexible and scalable since it integrates with every user. Users can make changes to the documents to suit their needs.
3. **Security:** The community cloud is safer than the private cloud, which is safer than the public cloud.
4. **Shared infrastructure:** Community clouds allow us to share cloud resources, infrastructures, as well as additional capabilities between different organizations.

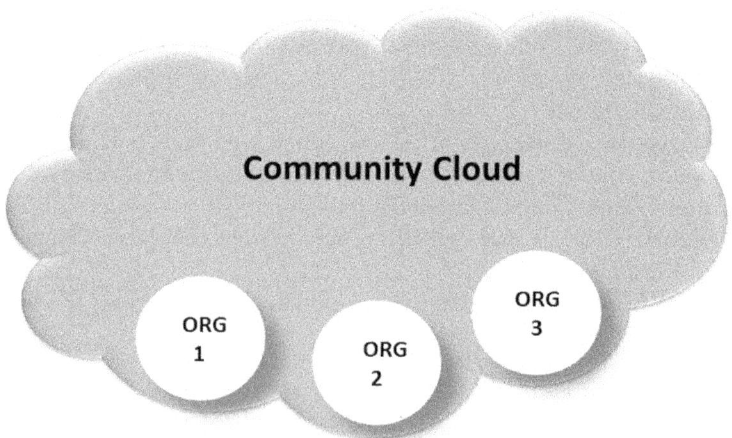

FIGURE 1.17   Unveiling the community cloud paradigm.

Disadvantages of Hybrid Cloud

The Community Cloud has the following drawbacks:

1. Not appropriate for all sectors: Not every organization should use community clouds.
2. Slow uploads: Sluggish uptake of data
3. Limited capacity: Each community member contributes to the fixed quantity of data storage and bandwidth.
4. Expensive: Costlier than public clouds community clouds.
5. Specialized management: It's challenging for organizations to divide up duties.

Similarities between Public and Private Cloud

- Virtualization: technologies for virtualization are employed by both public and private clouds to build virtual machines that make the most of real hardware and resources.
- Automation: Both public and private clouds utilize automation technologies for resource management and allocation. This increases effectiveness and simplifies corporate procedures.
- High Availability: Both public and private clouds have multiple data centers, independent systems, and failover protocols in place to provide high reliability and availability while minimizing downtime.
- Scalability: Both the private cloud and the public cloud are designed to be scalable, allowing users to swiftly add or remove resources to accommodate fluctuating demands.
- Resource Pooling: Both public and private clouds make use of resource pooling, which allows several users or applications to share resources like processing power, networking, and storage.
- SOA is a technology that both public and private clouds employ to provide resources as functions that can be accessed and used by numerous applications or users. Table 1.2 shows the difference between Public and Private Clouds, and Table 1.3 shows the difference among Public, Private, and Hybrid Clouds.

1.16   CLOUD SERVICE MODELS

Cloud computing is the technique of storing, managing, and processing data via a network of remote computers located on the internet as opposed to a physical server or a personal computer. These companies—often referred to as cloud providers—impose fees for cloud computing solutions based on usage. Clusters and grids are the foundation of cloud computing [25]. Figure 1.18 categorizes and visually presents the various cloud service models, including Infrastructure as a Service (IaaS), Platform as a Service (PaaS), and Software as a Service (SaaS), to clarify their distinctions and applications within cloud computing.

TABLE 1.2
Difference between Public and Private Cloud

Public Cloud	Private Cloud
Public access to the infrastructure of cloud computing service providers is provided over the internet. It serves a large number of customers or enterprises.	With private firms, service providers trade cloud computing infrastructure online. It benefits a single company.
Under a multi-tenancy approach, data from several enterprises is kept in a single, common environment. Security, permission, and regulations are used to control data sharing.	Under a single tenancy, information about a single enterprise is retained.
Considering that their clientele is global, cloud service providers offer every type of hardware and service imaginable. Different persons and organizations may require different services and tools, so the services offered must be adaptable.	Specialized tools and services customized to the needs of the business are offered by a private cloud.
It is stored on the service provider's server.	It is preserved on the website of the business or service provider.
It has access to the general internet.	Only connections via private networks are supported.
Scalability is quite great, while reliability is average.	Although scalability is limited, reliability is quite good.
The cloud is utilized by the users and is run by the cloud service provider.	Management and employees are all under one firm.
Compared to a private cloud, it is cheaper.	In comparison to the public cloud, costs are higher.
Security is important and is based on the service provider.	It provides a very high level of security.
Performance that is low to mediocre.	Outstanding performance.
It is done using shared servers.	There are presently devoted servers.
Both AWS and Google AppEngine are examples.	Microsoft KVM, Red Hat, VMware, and HP are a few examples.

The following are the three different cloud service model types:

1. Infrastructure as a Service (IaaS)
2. Platform as a Service (PaaS)
3. Software as a Service (SaaS)

1. Infrastructure as a Service (IaaS)

- One phrase frequently used to refer to IaaS is "hardware as a service" (HaaS). It functions as one of the levels of the cloud computing platform. Numerous IT system elements, such as servers, networking, storage, processing capacity, virtual machines, and other resources, are available for customers to contract out. Customers use these online resources using a cost-per-use business model.

TABLE 1.3

Differences between Private, Public, and Hybrid Clouds

Factors	Public	Private	Hybrid
Resources	The same resources are used by lots of consumers	The use of shared resources is permitted by one organization	Public and private clouds are mixed together in response to demand
Tenancy	The public cloud is where many different organizations' data is stored	Data from a single organization is stored in a public cloud	The public cloud, wherever data is kept, offers security
Operated by	Unaffiliated service providers	Particular organizations	Possibly combining both
Expense	Less expensive	More expensive	It could be more costly or less expensive depending on the demands of the organization and its specific requirements
	Expense your consumption has	A variety of pricing models, such as pay-as-you-go public clouds, could be used with fixed-price private clouds	There are other price models, such as consumption- and subscription-based ones
Scalability and flexibility	It is scalable and more flexible	It is reliable and predictable	The cloud offers versatility by allowing enterprises to mix public and private cloud services
Availability	Public at large (through the internet or a combination of both)	Restricted to a certain organization	A combination of both public and private

- Traditional hosting services involve renting computer hardware and IT infrastructure for a certain amount of time. No matter how much planning or time was utilized, the consumer got charged for it. Customers can change their setup to meet changing requirements while only paying for the services they use [26] thanks to the IaaS computing platform layer.
- IaaS comes in three varieties—private, cloud, and hybrid. On the client's property, hardware and software for a private cloud are developed. While the hybrid cloud combines the best elements of both the public and private clouds, the public cloud is run out of the company's data center, which has developed the platform for cloud computing. An IaaS provider provides the services listed below:
  1. Computing as a service is the first step. This includes processing units and virtual.

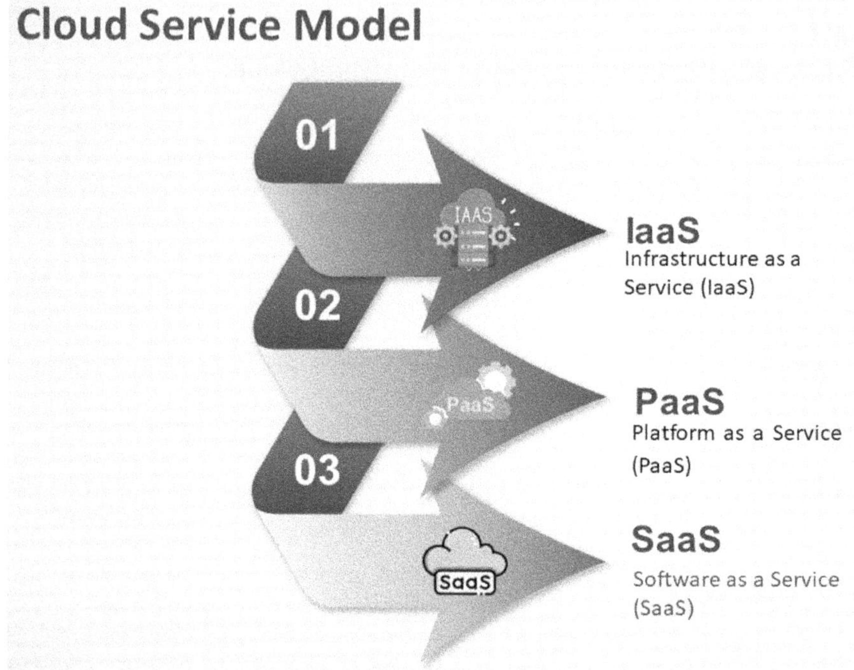

FIGURE 1.18 Cloud service models.

2. Main memory for systems of computers that are made accessible to end users.
3. Storage: The IaaS supplier offers back-end storage to safeguard files.
4. Network: Through the use of network as a service (NaaS), virtual computers can connect to networks of switches, routers, and bridges.
5. Load Balancer: These devices provide infrastructure-layer load balancing functionality.

Figure 1.19 illustrates the key features and components of IaaS, emphasizing its role in providing scalable and virtualized infrastructure resources to users.

The Advantages of IaaS Cloud Computing

The advantages of the IaaS computing layer are as follows:

1. **Shared infrastructure:** Using shared infrastructure, IaaS allows several users to utilize a single physical infrastructure.
2. **Web-based resource access:** IaaS enables IT users to access resources online.

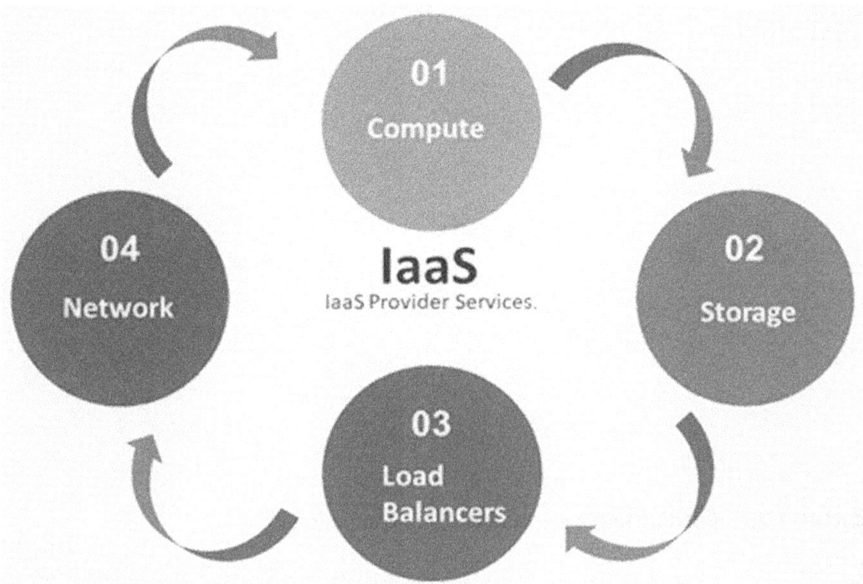

FIGURE 1.19   IaaS (infrastructure as a service).

3. **Pay as per-use model:** Service delivery that is pay-as-you-go. IaaS providers offer services on a pay-per-use basis. The resources that users have used must be paid for.
4. **Focus on the core business:** Pay more attention to the company's core business than to its IT infrastructure. IaaS providers pay more attention to the organization's core business.
5. **Demand-driven scalability:** One of the key advantages of IaaS is its flexibility to expand as needed. IaaS users are relieved of the burden of upgrading software and troubleshooting hardware-related issues.

The Disadvantage of IaaS Cloud Computing

1. **Security:** The majority of IaaS providers are unable to offer total security.
2. **Software upgrades and maintenance:** Not every firm using IaaS services keeps their software updated and maintained.
3. **Interoperability issues:** Users may encounter vendor lock-in concerns because moving virtual computers between IaaS providers is difficult.

IaaS Computing Layer Points of Importance

The IaaS cloud computing system offers more, and every resource is predictable depending on demand, even though it cannot replace the traditional hosting strategy. Even with the IaaS cloud computing platform, there may still be a need for an internal IT department. The IaaS system will need to

PaaS Providers

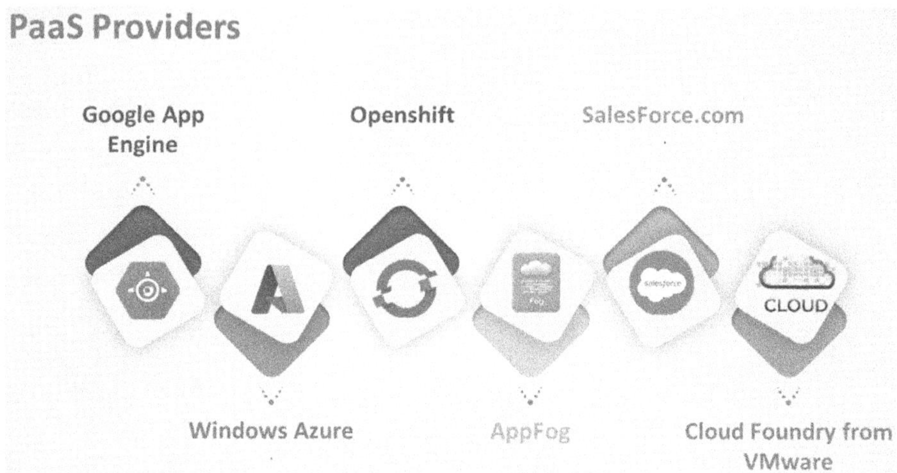

FIGURE 1.20 Leading IaaS providers in cloud computing.

be monitored or managed. IT pay costs may not be significantly reduced even while other IT expenses can be reduced. The IaaS cloud computing infrastructure vendor may have access to your data. So, partner with trustworthy organizations or businesses. For safety, review the policies and procedures [27]. Figure 1.20 outlines leading IaaS providers, offering a visual overview of key industry players and their services in cloud computing. Table 1.4 shows the difference between IaaS Supplier, IaaS Solutions, and their details.

2. Platform as a Service (PaaS)
- PaaS, which stands for platform as a service, provides a runtime setting. Programmers can create, test, use, and deploy web apps more easily. These programs are accessible to anybody with a web connection and are provided by cloud service companies on a cost-per-use basis.
- End users are spared the duty of infrastructure maintenance thanks to PaaS, which handles back-end scalability on the part of the cloud service provider.
- PaaS is made up of a platform (middleware, tools for growth and development, systems for maintaining databases, intelligence for business, and more) and architecture (servers, storage, and network) to support the life cycle of web applications [28].
- Examples include Google App Engine, Azure, Force.com, and Joyent.

PaaS suppliers offer resources, including databases, programming languages, and frameworks for applications. Figure 1.21 visually represents the

TABLE 1.4

Difference between IaaS Supplier, Solutions, and Their Details

IaaS Supplier	IaaS Solution	Details
Flexible Amazon Web Services	MapReduce, Route 53, EC2, virtual private clouds, etc., are a few examples.	Amazon, the company that developed the first cloud computing platform, offers load balancing, cloud observing, and auto-scaling services.
Rackspace	Websites, servers, and storage in the cloud, etc.	The company that provides the cloud computing infrastructure lays a lot of emphasis on commercial hosting services.
Technology SifyIaaS	The cloud computing platform for Sify is run on convergent infrastructure from SifyHP.	The three types of cloud services offered by the provider are IaaS, PaaS, and SaaS.
Netmagic Services	Netmagic Cloud IaaS.	In addition to a virtual data center in the US, Netmagic operates from data centers in Mumbai, Chennai, India, and Bangalore. Services are currently planned to be expanded to Asia's west.
Communications Reliance	Internet data center reliance.	With data centers in the cities of Mumbai, Bangalore, Hyderabad, and Chennai, RIDC delivers both conventional hosting and cloud services. IaaS and SaaS are two of RIDC's cloud offerings.
Tata Communications' InstaCompute	InstaCompute is the name of the IaaS offering from Tata Communications.	InstaCompute operates data facilities in both Singapore and Hyderabad, where the business also has operations.

array of PaaS provider services, showcasing the diverse offerings that empower developers to build and deploy applications in the cloud.

1. Programming languages: PaaS companies give developers access to a variety of programming languages so they may build applications. A few popular programming languages that PaaS providers support are Java, PHP, Ruby, Perl, and Go.
2. Application framework: PaaS providers provide application frameworks that simplify the process of developing applications. PaaS companies offer a variety of well-known application frameworks, including Node.js, the Drupal platform, and Joomla, the platform used by WordPress, Springtime, Playing, Rack, and Zend.

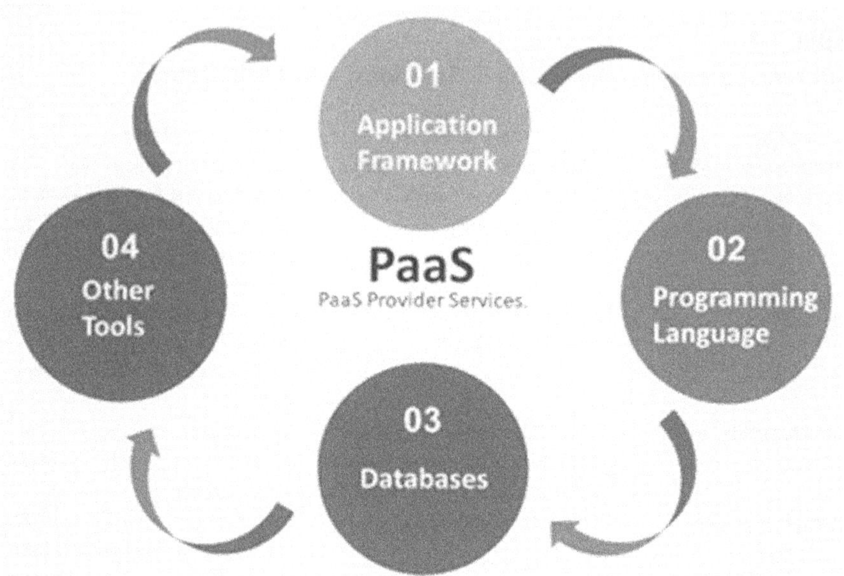

FIGURE 1.21   Exploring PaaS provider services.

3. Databases To interface with the apps, PaaS providers offer a variety of files, including ClearDB, the MySQL database, MongoDB, and Redis.
4. Additional devices: The PaaS providers supply extra devices needed for app development, testing, and deployment.

Advantages of PaaS

1. **Initial Condensed Construction:** Because of PaaS, developers may focus on building and inventing instead of managing infrastructure, which is one benefit of PaaS.
2. **Less risk:** Initial hardware and software investments are unnecessary. Developers only need a desktop or laptop and a web connection to start designing applications.
3. **Integrated business tools:** Several PaaS providers also offer pre-built business functions to save users time and allow them to concentrate only on their projects.
4. **Easily approachable:** PaaS businesses generally provide online forums so developers can learn from others, share experiences, and obtain support.
5. **Scalability:** Despite the apps themselves changing, the number of users for installed applications can rise from one to a thousand [29].

Disadvantages of PaaS

1. **Vendor lock-in:** Since one must create applications consistent with the platform provided by the PaaS vendor, moving apps to other PaaS operators would be challenging.

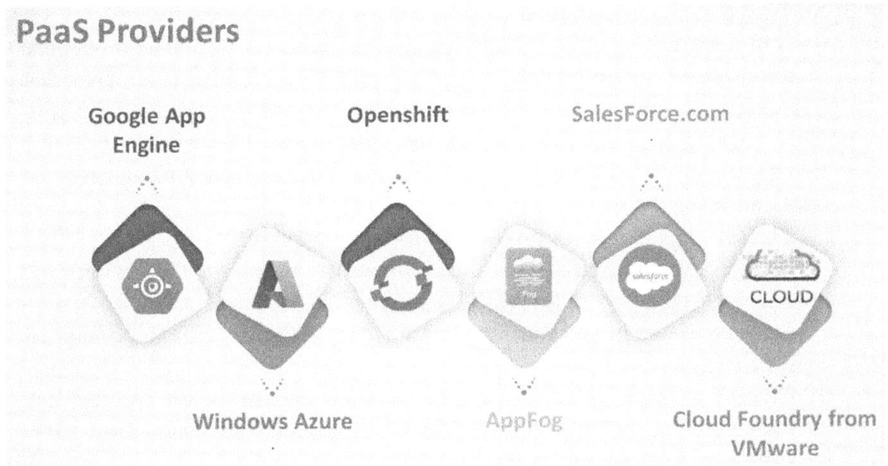

FIGURE 1.22   Leading PaaS providers in the cloud.

2. **Data Protection:** Regardless of its importance, corporate data is private therefore storing it somewhere other than the company's walls could compromise its privacy.

3. **Integration with the system's other applications:** While some applications may be local, others may be cloud-based. Therefore, if we were to merge local data with cloud data, there is a chance that complexity would rise.

  • **Popular PaaS vendors**

Figure 1.22 highlights a selection of key PaaS providers, offering a visual overview of notable industry players and their PaaS offerings. Table 1.5 shows the difference between the Service and Providers.

TABLE 1.5

Difference Between Providers and Services

Providers	Services
(GAE) Google App Engine	Cloud Storage Client Library, Logservice, URL Fetch, and App Identity
On Windows Azure	There is data storage, IoT, security, and computation
Microsoft Azure	RedHat Openshift
Salesforce.com	Chat, sales cloud, and CRM capabilities, as well as simpler setup and scalability
AppFog	SkyDrive, Google Docs, and JustCloud
loud Foundry from VMware	Data, messaging, and other services are provided

3. Software as Service (SaaS)

"On-demand software" or SaaS is another name for it. The features used in this type of software delivery are hosted by a cloud service provider. These services are accessible by users without downloading additional software because they are provided online [30]. Figure 1.23 visually presents the spectrum of SaaS provider services, showcasing the diverse range of applications and solutions available to users in the cloud.

The services that SaaS providers provide comprise:

1. Business services: To assist the business in getting off the ground, the SaaS provider provides various business services. ERP, CRM, payment processing, and marketing are a few instances of SaaS company services.
2. Document management: Document management is the production, management, and tracking of electronic documents using third-party software (SaaS providers).
3. Social networks: As is common knowledge, many people use social networking websites. SaaS is, therefore, used by social media service providers to manage user information.

Advantages of SaaS

1. SaaS is easy to purchase: Businesses can have business skills for a cheap cost that is less expensive than licensed software because SaaS pricing depends on an occasional or yearly price subscription. SaaS businesses

FIGURE 1.23   Exploring SaaS provider offerings.

often charge a subscription fee for the use of the apps, most frequently on a monthly or annual basis. Traditionally, the software has been sold as licensed software with an upfront charge (and usually an additional, optional monthly maintenance fee).

2. Several-to-One: SaaS services are delivered via one-to-many approaches, which enable several users to share a single copy of the program.

3. Less hardware is needed for SaaS: Businesses are not required to invest in additional hardware because the program is remotely hosted.

4. SaaS needs less maintenance: Software as a service relieves businesses of the burden of installation, configuration, and continuous maintenance. SaaS frequently has less initial setup expense than business software. SaaS providers base the cost of their programs on use variables such as user count. SaaS provides automatic upgrades and is, therefore, easy to keep track of.

5. It's not required to use a specific model of hardware or software: All users will have access to the same version of the software, which they usually access using a web browser. SaaS minimizes the cost of IT support by contracting out support and upkeep of hardware and software to the IaaS provider [31].

6. API integration: Due to the widespread use of APIs, SaaS services are easy to link to other programs or services.

7. Easily accessible: SaaS services can be accessible online directly from the service provider, negating the need for client-side software installation.

Disadvantages of Cloud Computing

1. **Security:** Given that data is stored on the cloud, some customers might worry about it. However, deploying locally does not offer more security than using the cloud [32].

2. **A delayed response:** Because data and applications are stored in the cloud at various locations far from the end-user, there is a risk of a higher latency when talking with the application compared to local deployment. The SaaS approach is therefore unsuitable for systems that require demanding response times of just milliseconds.

3. **Total dependency on the internet:** The majority of SaaS applications require a connection to the web to run.

4. **Changing providers of SaaS:** It can be challenging to change SaaS providers since it entails transferring extremely large data files over the internet, transfiguring them, and then combining them with a new SaaS.

Figure 1.24 presents the spectrum of SaaS provider services, showcasing the diverse range of applications and solutions available to users in the cloud. The services offered by several well-known SaaS businesses are included in the Table 1.6.

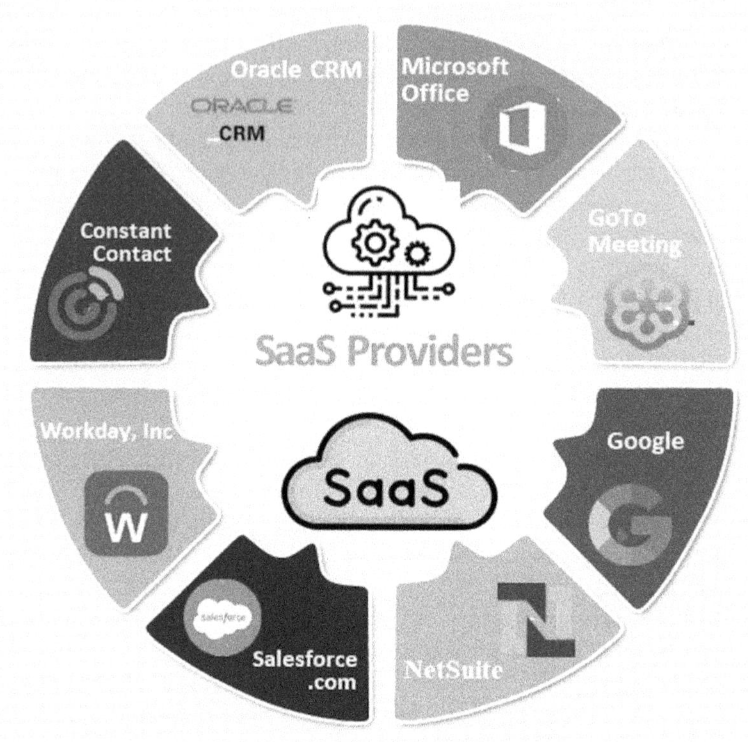

FIGURE 1.24 Diverse offerings from SaaS providers.

TABLE 1.6

Difference between Providers and Services

Providers	Services
salseforce.com	Options for on-demand CRM
Microsoft Office 365	Digital workspace
Oracle CRM	CRM software
IncPayroll, financial, and human capital management. Workday,	
Google Apps, Gmail,	Google Calendar, Docs, and Sites
NetSuite ERP	Applications for professional service automation (PSA), accounting, CRM, and online retail
GoToMeeting	Software for video chats and conferences over the internet email marketing, internet surveys, and event promotion through Constant Contact

REFERENCES

1. Jain, A., Mahajan, N. (2017). Introduction to Cloud Computing. In: The Cloud DBA-Oracle. Apress, Berkeley, CA. https://doi.org/10.1007/978-1-4842-2635-3_1
2. Stanoevska-Slabeva, K., Wozniak, T. (2010). Cloud Basics – An Introduction to Cloud Computing. In: Stanoevska-Slabeva, K., Wozniak, T., Ristol, S. (eds) Grid and Cloud Computing. Springer, Berlin, Heidelberg. https://doi.org/10.1007/978-3-642-05193-7_4
3. Buyya, R. (Jan.-June 2013). Introduction to the IEEE transactions on cloud computing. IEEE Transactions on Cloud Computing, 1(1): 3–21. https://doi.org/10.1109/TCC.2013.13
4. Vaquero, L.M., Rodero-Merino, L., Caceres, J., Lindner, M. (2009). A break in the clouds: Toward a cloud definition. ACM SIGCOMM Computer Communication Review, 39(1): 50–55.
5. Wikipedia. (2010). Cloud computing. Retrieved from http://en.wikipedia.org/wiki/Cloud_computing
6. Jiyi, W., Lingdi, P., Xuezeng, P. (2009). Cloud computing: Concept and platform. Telecommunications Science, 12: 23–30.
7. Strickland, J. (2010). How cloud storage works[OL]. http://communication.howstuffworks.com/cloud-storage.htm.
8. Storage Networking Industry Association. Cloud Storage Reference Model, Jun. 2009.
9. Layton, J.B. (2013). Cloud storage concepts and challenges [OL]. http://www.linux-mag.com/cache/7617/1.html,2010.
10. Garrison, G., Kim, S., Wakefield, R.L. (2012). Success factors for deploying cloud computing. Communications of the ACM, 55: 62–68.
11. Herhalt, J., Cochrane, K. (2012). Exploring the cloud: A global study of governments' adoption of cloud. Alassafi MO, AlGhamdi R, Alshdadi A, Al Abdulwahid A, Bakhsh ST. Determining factors pertaining to cloud security adoption framework in government organizations: an exploratory study. IEEE Access. 2019 Sep 19; 7:136822-35.
12. Sales force, —CRM‖. http://www.salesforce.com/
13. Venters, W., Whitley, E.A. (2012). A critical review of cloud computing: Researching desires and realities. Journal of Information Technology, 27: 179–197.
14. Yang, H., Tate, M. (2012). A descriptive literature review and classification of cloud computing research. Communications of the Association for Information Systems, 35–60.
15. Marston, S., Li, Z., Bandyopadhyay, S., Zhang, J., Ghalsasi, A. (2011). Cloud computing — The business perspective. Decision Support System, 51: 176–189.
16. Nguyen, T.D., Nguyen, D.T., Cao, T.H. (2014). Acceptance and Use of Information System: E-Learning Based on Cloud Computing in Vietnam. In: Information and Communication Technology-EurAsia Conference. Springer, Berlin, Heidelberg, pp. 139–149. https://doi.org/10.1007/978-3-642-55032-4_14
17. Ahmad, N., Hoda, N., Alahmari, F. (2020) Developing a cloud-based mobile learning adoption model to promote sustainable education. Sustainability, 12: 3126. https://doi.org/10.3390/su12083126
18. Korepin, V.N., Dorozhkin, E.M., Mikhaylova, A.V., Davydova, N.N. (2020). Digital economy and digital logistics as new area of study in higher education. International Journal of Emerging Technologies in Learning, 15(13): 137–154. https://doi.org/10.3991/ijet.v15i13.14885
19. Papadakis, S., Kalogiannakis, M., Sifaki, E., Vidakis, N. (2017). Access Moodle Using Smart Mobile Phones. A Case Study in a Greek University. In Interactivity, Game Creation, Design, Learning, and Innovation. Springer, Cham, pp. 376–385. https://doi.org/10.1007/978-3-319-76908-0_36

20. Talan, T. (2020). The effect of mobile learning on learning performance: A meta-analysis study. Educational Sciences: Theory and Practice, 20(1): 79–103. http://dx.doi.org/10.12738/jestp.2020.1.006

21. Papadakis, S., Kalogiannakis, M., Sifaki, E., Vidakis, N. (2018). Evaluating Moodle use via smart Mobile phones. A case study in a Greek university. EAI Endorsed Transactions on Creative Technologies, 5(16): 1–9. https://doi.org/10.4108/eai.10-4-2018.156382

22. Masud, M.A.H., Huang, X. (2012). An e-learning system architecture based on cloud computing. System, 10(11): 255–259.

23. Mell, P., Grance, T. (2011). The NIST Definition of Cloud Computing. National Institute of Standards and Technology.

24. Shirzad, M., Ahmadipour, M., Hoseinpanah, A., Rahimi, H. (2012). E-learning based on cloud computing. In 2012 International Conference on Cloud Computing Technologies, Applications and Management (ICCCTAM). IEEE, pp. 214–218. https://doi.org/10.1109/iccctam.2012.6488101

25. Chandra, D.G., Borah, M.D. (2012). Cost benefit analysis of cloud computing in education. In 2012 International Conference on Computing, Communication and Applications. IEEE, pp. 1–6. https://doi.org/10.1109/iccca.2012.6179142

26. Shahzad, N., Ismail, A., Golamdin, A.G. (2016). Opportunity and challenges using the cloud computing in the case of Malaysian higher education Institutions. International Journal of Management Science and Information Technology, 20(20): 1–18.

27. Siddiqui, S.T., Alam, S., Khan, Z.A., Gupta, A. (2019). Cloud-Based e-Learning: Using Cloud Computing Platform for an Effective e-Learning. In Smart Innovations in Communication and Computational Sciences. Springer, Singapore, pp. 335–346. https://doi.org/10.1007/978-981-13-2414-7_31

28. Moravčík, M., Segeč, P., Uramová, Y., Kontšek, M. (2017). Teaching cloud computing in cloud computing. In Conference: 2017 15th International Conference on Emerging eLearning Technologies and Applications (ICETA). IEEE, pp. 1–6. https://doi.org/10.1109/iceta.2017.8102512

29. Naveed, Q.N., Ahmad, N. (2019). Critical success factors (CSFs) for cloud-based ELearning. International Journal of Emerging Technologies in Learning, 14(01): 140–149. https://doi.org/10.3991/ijet.v14i01.9170

30. Liu, Z.-Q., Dorozhkin, E., Davydova, N., Sadovnikova, N. (2020). Effectiveness of the partial implementation of a cloud-based knowledge management system. International Journal of Emerging Technologies in Learning, 15(13): 155–171. https://doi.org/10.3991/ijet.v15i13.14919

31. Joshi, N., Shah, S. (2019). A Comprehensive Survey of Services Provided by Prevalent Cloud Computing Environments. In Smart Intelligent Computing and Applications. Springer, Singapore, pp. 413–424. https://doi.org/10.1007/978-981-13-1921-1_41

32. Colman-Meixner, C., Develder, C., Tornatore, M., Mukherjee, B. (2016). A survey on resiliency techniques in cloud computing infrastructures and applications. IEEE Communications Surveys & Tutorials, 18(3): 2244–2281. https://doi.org/10.1109/comst.2016.2531104

33. Giannakis, M., Spanaki, K., Dubey, R (2019). A cloud-based supply chain management system: Effects on supply chain responsiveness. Journal of Enterprise Information Management, 32(4): 585–607. https://doi.org/10.1108/jeim-05-2018-0106

34. Nazir, R., Ahmed, Z., Ahmad, Z., Shaikh, N.N., Laghari, A.A., Kumar, K. (2020). Cloud computing applications: A review. EAI Endorsed Transactions on Cloud Systems, 6(17): 5. https://doi.org/10.4108/eai.22-5-2020.164667

35. Li, Y., Xia, S., Cao, B., Liu, Q. (2019). Lyapunov optimization-based trade-off policy for mobile cloud offloading in heterogeneous wireless networks. IEEE Transactions on Cloud Computing, 1–3. https://doi.org/10.1109/tcc.2019.2938504

36. Galletta, A., Carnevale, L., Celesti, A., Fazio, M., Villari, M. (2017). A cloud-based system for improving retention marketing loyalty programs in industry 4.0: A study on big data storage implications. IEEE Access, 6(c): 5485–5492. https://doi.org/10.1109/access.2017.2776400

37. Blackboard (2018). Blackboard product help. https://help.blackboard.com/

38. Blackboard (2018). Blackboard learn. https://www.blackboard.com/teaching-learning/learning-management/blackboard-learn. https://doi.org/10.13021/g8460x

39. Hussein, L.A., Hilmi, M.F. (2020). Cloud computing based E-learning in Malaysian universities. International Journal of Emerging Technologies in Learning, 15(08): 4–21. https://doi.org/10.3991/ijet.v15i08.11798

40. Chipps, J., Kerr, J., Brysiewicz, P., Walters, F. (2015). A survey of university students' perceptions of learning management systems in a low-resource setting using a technology acceptance model. CIN: Computers, Informatics, Nursing, 33(2): 71–77. https://doi.org/10.1097/cin.0000000000000123

2 Fundamental Concepts of IoT

Oroos Arshi and Aryan Chaudhary

2.1 INTRODUCTION

IoT stands for the Internet of Things. It refers to the connectivity of actual physical items that have connectivity, apps, and sensors built into them so they can connect and exchange data. Cars and appliances are two examples of these products. This technology provides the way for the creation of autonomous and more efficient systems by enabling the collection and sharing of data from a vast network of devices. The networking of physical objects with electronics built into their design to communicate and detect their interactions with each other or the world around them is another way to describe IoT. In the upcoming years, IoT-based technology will offer a wider range of services, fundamentally changing how people conduct their daily lives. IoT is well-established in several industries, such as smart houses, smart cities, genetic counseling, agriculture, and power [1].

The IoT describes a network of networked computers that are built into common items to facilitate data sharing. Currently, the internet is connected to around 9 billion "things" (both actual objects). In the not-too-distant future, it is predicted that there will be an astounding 20 billion people on the planet [1]. People commonly inquire about the definition of the IoT, which is a subject that is quite popular in today's industry. Simply expressed, the term "Internet of Things" refers to a group of items that may gather, store, and transmit information to other devices or services through the internet. These products include those that are enabled by software, hardware, firmware, and electronics. Another method of defining IoT is through an example. Imagine yourself wearing a watch that keeps track of precise biological information, such as your pulse, oxygen saturation, stress levels, and the distance you have covered. What is the way it works? Due to the special hardware, firmware, and sensors on the smartwatch, which enable real-time data collecting, an application or a piece of software analyses the data for the user.

2.2 IMPORTANCE OF INTERNET OF THINGS

Before talking about the importance of the IoT, it's crucial to understand that people do accept and value technology. Any gadget that enables communication with other objects or the acquisition of information soon acquires popularity. Isn't it practical to use an app on your phone to plan your washing machine for use when you go to the store? Wouldn't it be nice to be able to have smart devices on your body that could teach you how to run more efficiently? The IoT, which additionally provides a lot of data, has made many tasks easier for humans to perform. When people have

DOI: 10.1201/9781032656694-2

access to data, they feel like they have more influence over every aspect of their environment. These technologies are now more well-liked as a result of the creation of product prototypes. If a business owner has a great idea for an IoT device, they can look for a company that creates prototypes and develops the product on a budget [2]. The IoT device will succeed if there is a market or a problem that has to be solved. The business owner presents his concept, and the design and production company handles the rest by putting the machinery or product together with assistance from a knowledgeable engineering team. Let's look more closely at the usefulness of IoT.

2.2.1 AMAZING IoT BENEFITS FOR DAILY LIVING

2.2.1.1 Enhanced Standard of Living

- Modern life is significantly impacted by smartphones. People utilize these devices for almost all tasks, such as making phone calls and sending emails.
- The standard of life can be enhanced through communication. IoT devices additionally make that possible.
- If you rely on a smart device to constantly remind you of your visit or if you can easily download your medical report, you might be able to spend a larger percentage of your time on other activities.
- When you save time, your productivity increases. Your mind is relieved of unnecessary worries, making it easier to focus on more important duties.
- There are currently a lot of innovative IoT devices that emphasize well-being, which is another advantage of this technology.

2.2.1.2 Automated Processes

- The lights are turned on. Releasing the garage door. Changing the thermostat. Cleaning clothes. Prescribing a medication. Acquiring food. Keeping track of the calories you burn during exercise. What traits do all of these have in common?
- These actions can be carried out automatically. You can automate and manage these tasks with IoT devices. By installing sensors, you may program the lights in your home to turn on remotely or turn themselves on at midnight when someone enters the room.
- You might be able to open the garage door from your car so you don't have to spend time getting out to do it on your own when you get home.
- Since the wearable on the wrist will gather that information for you, you may exercise without thinking about how many calories you've burned. This is a key advantage of the IoT for wearable devices.

2.2.1.3 Data-Driven Conclusions

- Well-known author Sean Covey once said, "We are free to select our paths, but we are unable to choose the repercussions that come with them."
- The foundation of life is decision-making. IoT is important to many people because it allows for the use of obtained data to inform decisions.
- If it might help your crops grow more effectively, would you measure the soil's nutrients and humidity? Most likely, the answer is yes. If you

implement sensors or smart technology, your crops might reach whole new heights.

- Of course, data by itself is meaningless. The knowledge acquired will help you identify potential problems as well as make decisions in light of them.

2.2.1.4 Real-Time Monitoring

- These gadgets will have the capacity to monitor the standard of domestic items at the moment. You can save time and work by deciding whether or not you should upgrade your items by being aware of their condition rather than having to keep evaluating their state.
- According to Markets and Markets, "the IoT-based Asset Tracking and Monitoring market is projected to grow worth USD 3.9 billion in 2022 to a worth of USD 6.6 billion by 2027." Businesses in a range of industries want to be able to monitor resources in real-time. They can use this capability to examine the performance, availability, internal hazards, and prospective threats for the asset under observation [3].

2.3 IoT OVERVIEW

2.3.1 Evolution of the Internet of Things in the Present Past and Future

- The first connected network, or ARPANET, served as the forerunner of the current internet. The history of the IoT began with ARPANET. In Figure 2.1, we can see a timeline diagram of the evolution of the IoT.
- In 1982, David Nichols, a computer science doctorate student at Carnegie Mellon University, asked if there were any cold beverage bottles in the Coke vending machine in the department. He was interested in knowing in advance so that he wouldn't go to the dispenser, which was off campus from his college, only to find that there wasn't a cold drink available [4].

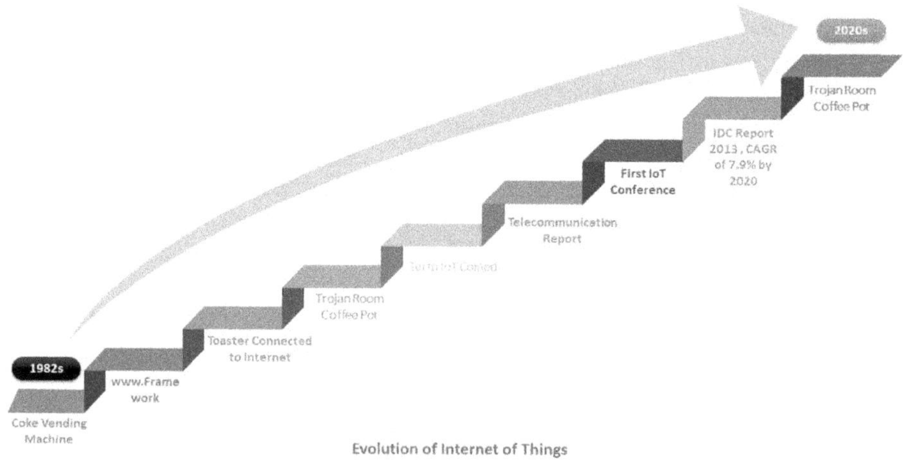

FIGURE 2.1 Timeline of IoT evolution.

- He received assistance with this study from undergraduates Mike Kazar and Ivor Durham, as well as researching engineer John Zsarnay from the institution. They developed a code that could tell whether Coca-Cola was cold or not if it was in the vending machine. Anyone accessing the university's ARPANET may verify the state of the Coke machines right now.
- When Tim Berners-Lee proposed the design of the "world open web," the internet was established in 1989 [5].
- In 1990, John Romkey developed a first-of-its-kind toaster that was able to be operated via the internet. The microwave was connected to the PC with a cable before Wi-Fi was a thing. This toaster is the first "thing" that the IoT has been declared to contain [5].
- Academics seem to prefer caffeine, whether it is consumed hot or cold. The Trojan Room Coffee Pot was created in 1993 in the computer lab at the University of Cambridge by Quentin M.N. Kamdar, V. Sharma, and S. Nayak. [5]. The pot's contents were damaged and relayed to the building's server three times each minute. Browsers that supported image display made it possible for users to view these pictures online.
- The term "the World Wide Web" or "IoT" was first used in 1999 by Kevin Ashton, who is presently the CEO of Auto-ID Labs. This expansion has been extremely beneficial to IoT. It was the title of a speech he delivered about how RFID may be used to connect items to customers at Procter & Gamble, the company where he was working at the time [6].
- The term "Internet of Things" was first used in journals like The Guardian and Scientific America in 2003–2004. Around this time, RFID implementation started in Walmart and the US Department of Defense [6].
- IoT's implications were highlighted by the International Telecommunication Union of the United Nations in a 2005 report. The IoT was designed to assist in building a brand-new dynamic network of connections [7].
- The first IoT conference took place in Zurich in March 2008. To encourage information sharing, researchers and professionals from business and academia were gathered. The US National Intelligence Council designated the IoT as one of six significant civil technologies that year.
- In 2008 and 2009, when more objects were connected to the internet than people, the IoT was completely born, according to a 2011 white paper by Cisco Internet Business Solution Group (CIBSG). The ratio of commodities to people increased from around 0.8 in 2003 to 1.84 in 2010, according to calculations by CIBSG.
- Cisco released several educational tools on the topic along with the white paper, and it ran marketing campaigns to entice customers who were interested in using IoT. Following that, IBM and Ericsson joined the contest.
- IoT was discussed in Gartner's 2011 Hype cycle, which monitored the advancement of emerging technologies.
- According to IDC's predictions, the worldwide IoT industry will grow with a CAGR of 7.9% and reach USD 8.9 trillion by 2020 [7]. Figure 2.1 depicts the evolutionary timeline of IoT, highlighting key milestones and advancements in a concise visual format.

2.4   COMPONENT OF INTERNET OF THINGS

Below are the explanations behind these. Figure 2.2 illustrates the core components integral to the IoT ecosystem.

1. **Items or Devices:** These have sensors and actuators fitted. Sensors gather environmental data and deliver it to a gateway as actuators act (as directed after data processing).
2. **Gateway:** Data from the sensors is transmitted to the Gateway, after which it passes through some sort of pre-processing. It also serves to raise the security of the network and the information that is being transmitted.
3. **Cloud:** The collected data is subsequently transmitted to the cloud. The cloud is a collection of computers that are always online.
4. **Analytics:** Processing is finished once the data has been received in the cloud. A range of techniques, including machine learning, are employed for proper data analysis.

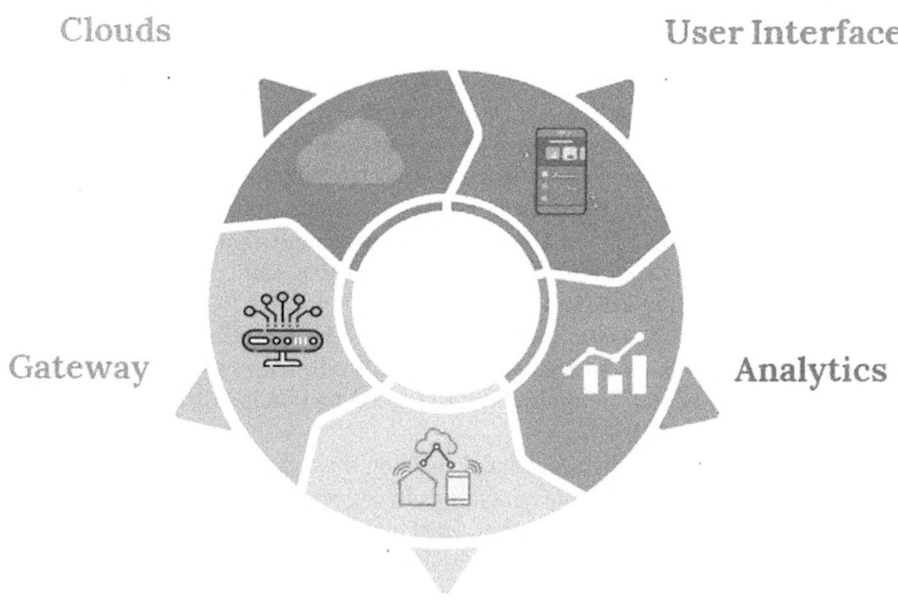

FIGURE 2.2   Key components of the IoT system.

2.5   SENSORS AND ACTUATORS IN IoT

Sensors are utilized in the majority of IoT device architectures. Objects, equipment, and other items can all be detected with sensors. A sensor is a device that responds to a certain measurement by generating a usable output.

The sensor gathers anatomical information and converts the data into an electrical signal that may be analyzed to determine the presence of a particular physical quantity (for example, electrically, mechanically, or visually). The output of a sensor is a signal with diverse characteristics, such as resistance, capacitance, impedance, etc. that can be interpreted by humans [8]. Figure 2.3 showcases the essential hardware elements driving the functionality of IoT systems.

Transducer

- Transducers change data from one physical element to another by transferring one kind of electricity into another.
- It could be used in a variety of systems as an actuator.

Characteristics of Sensors:

1. Static
2. Dynamic

- **Accuracy:** Accuracy is a term used to describe a measuring tool's ability to deliver an outcome that is roughly comparable to the true worth of the quantity being measured. Errors are documented. We employ both comparison and absolute error metrics, assessing the output's dependability against a more developed, older technology. Absolute error is the calculated value less the real value.

$$\text{Relative error} = \text{Measured value}/\text{True value}.$$

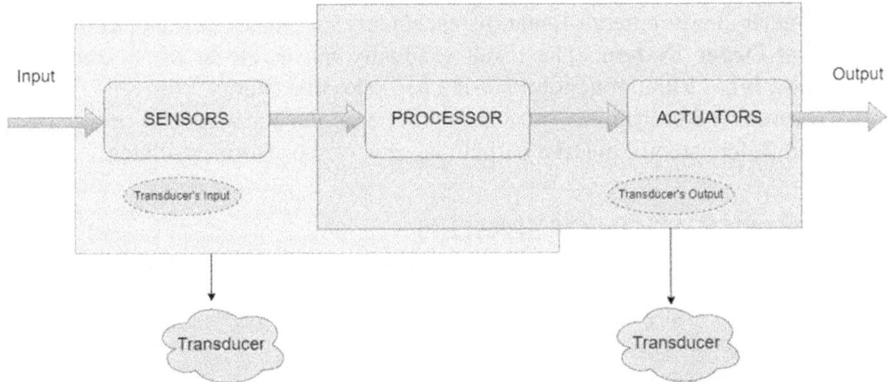

FIGURE 2.3   IoT hardware components overview.

- **Range:** Gives the physical quantity's highest and lowest values, based on what the sensor is capable of tracking. Outside of these ideals, there can be no sense or way to react. For instance, an RTD can measure temperatures between –200°C and 800°C.
- **Resolution:** Resolution is a crucial consideration when choosing sensors. The precision improves as the resolution increases. The threshold is the point at which new material cannot accumulate. Change the data in any way that the sensor is capable of detecting.
- **Precision:** A measuring device must consistently produce the same measurement when used for measuring the same quantity by the same set of guidelines. Future measurements will, therefore, be constant, as opposed to getting closer to the real figure. It has to deal with the variance of a set of measurements. Even if it is crucial for precision, it is not enough on its own.
- **Sensitivity:** Sensitivity indicates how responsive a system is to those little changes in its responses caused by modifications in its input parameters. The angle at which the slope of a sensor's output features curve can be used to calculate it. The instrument's reading will alter with even the slightest quantity fluctuation.
- **Linearity:** Deviation from a substantially straight line in the sensor curve data. The calibration curve determines how linear everything is overall. The static measurement diagram displays the relationship between the output and input amplitudes.
- **Drift:** The variation in the sensor's measurement from a particular reading when that value is maintained for an extended length of time.
- **Repeatability:** The variance in measurements made consecutively under the same circumstances. The measurements must be taken over a brief enough period to prevent significant long-term drift.

2.5.1 Systems' Characteristics: Dynamic Features

- **Zero-Order System:** The response to the input signal at the output is instantaneous. It doesn't have anything that can hold energy, such as potentiometer measurements, linear displacements, and rotational displacements.
- **First-Order System:** The result gradually moves closer to its ultimate value. It has a component for storing and releasing energy.
- **Second-Order System:** Intricate output reaction for a second-order system. Before stabilizing, the output response of the sensor oscillates.

2.6 CLASSIFICATION OF SENSORS

1. Active and Passive
2. Scalar and Vector
3. Analog and Digital

1. **Passive Sensors:** Active sensors can recognize input, whereas passive sensors cannot. Thermometers, water level sensors, soil moisture sensors, and acceleration sensors are a few examples [9].

2. **Active Sensors:** Input can be detected by active sensors on their own. Examples of altimeter sensors include laser, sounder, and radar ones.
3. **Analog Sensor:** The response or output of the sensor is continuously determined by one or more of its input parameters. Examples include an analog hall effect, an ongoing analog temperature sensor, as well as a temperature sensor.
4. **Digital sensor:** Only binary responses are generated by digital sensors. Design can overcome the drawbacks of analog sensors. It also features additional bit conversion hardware in addition to the analog sensor. Examples include electronic temperature gauges (DS1620) and passive infrared (PIR) sensors.

2.6.1 Types of Sensors

Electrical Sensor: Contact or non-contact proximity electrical sensors are also possible. Using the component, simple touch sensors complete an electrical circuit. Electrical proximity sensors use capacitance or induction as detection mechanisms for items other than metals. These sensors are non-contact and operate on these electrical principles.

Lighter Sensor: One of the most important sensors, often known as photo sensors, is the light sensor. An LDR is the most fundamental type of light sensor currently accessible. The characteristic of LDR is that the resistance has an inverse relationship with the level of ambient light; as the level of light increases, the resistance decreases, and vice versa.

Touch Sensor: Using a pen or a finger, a touch sensor can detect touches from various items.

The word "detection" is a reference to anything.
They are separated into two groups:

1. Resistive Type
2. Capacitive Type

Capacitive touch sensors make up the vast majority of touch sensing available today because they have a superior information-to-noise ratio and are more accurate.

Speed Sensor: The speed sensor is a tool that gauges the speed of a moving object or automobile. UDAR, ground speed radar, and wind velocity sensors are a few examples.

Speed Sensor: Temperature sensors are devices that monitor temperature over time using an electrical signal to represent it. A voltage measurement of these electrical impulses will be employed, which has a direct inverse relationship to temperature data.

Ultrasonic Sensor: Ultrasonic sensors operate similarly to SONARs and RADARs, which analyze radio or sound wave echoes to determine a target's characteristics by emitting high-frequency sound waves.

PIR Sensor: A PIR sensor, also called a pyroelectric sensor, is a technological tool that tracks and detects infrared (IR) light coming from anything in its area of vision. A passive infrared sensor is referred to as PIR. Movement and motion of people detection are its main uses.

2.7   ACTUATORS IN IoT

An IoT device is made up of a physical object (or "thing"), a control system (or "brain"), sensors, actuators, or networks. A device or function that moves or controls a machine's mechanism is known as an actuator. The device's sensors collect data about its surroundings, after which control signals are generated for the actuators in line with the actions that must be conducted [9].

A tool that functions identically to a servo motor that rotates is an actuator. They can either be rotatory actuators or typical linear actuators, meaning they can move or spin in a straight line or at an angle. In IoT applications, we can use servo motors to simply rotate the energy source by an incline of 90 degrees, 180 degrees, etc., to fulfill our needs. In Figure 2.4, IoT devices operate by collecting and exchanging data, and the integration of actuators enhances functionality by enabling physical actions in response to the data, illustrating the dynamic interplay between sensing and actuation in the IoT ecosystem.

The control system alters the environment through the actuator. A control signal and an energy supply are both required. In reaction to a control signal, it converts an energy source into a mechanical operation. It comes in the several types described below depending on the type of energy it uses [10].

2.8   TYPES OF ACTUATORS

1. **Hydraulic Actuators:** Using hydraulic force, a hydraulic actuator performs a mechanical action. They are driven by a cylinder or liquid motor. The physical activity is transformed into rotary, simple, or oscillating motion depending on what the IoT device needs. Hydraulic motors have traditionally been seen in construction machinery due to the force they could provide.

Advantages

- A hydraulic actuator has a high force and speed output capacity.
- Used to elevate or lower cars in car carriers; also used for welding and clamping.

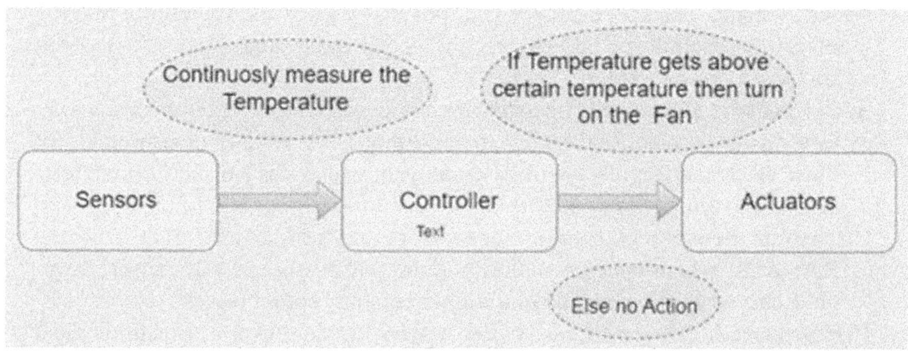

FIGURE 2.4   Working of IoT devices with actuator integration.

Disadvantages

- Performance can suffer, and cleaning issues can arise as a result of hydraulic fluid leaks.
- It is expensive.
- There is a requirement for heat exchangers, noise-canceling technologies, and high-maintenance systems.

2. **Air-Powered Actuators:** Using a pneumatic actuator, you can create linear or rotary motion out of pressure or highly compressed gas energy. The use of air-compressed sensing in robotics that resembles the fingertips of humans is one example [11].

Advantages

- They start and stop very rapidly, are dependable, have a long operational life, and require little maintenance.
- They are utilized in high-temperature environments because using air is safer than using chemicals since they are a more affordable alternative.

Disadvantages

- Pressure loss could reduce its effectiveness.
- The compressor that pumps air needs to be running constantly.
- Air requires upkeep and is susceptible to contamination.

3. **Electric Motor Actuators:** Electric actuators are pushed by electric motors, which convert electrical current into hydraulic thrust. Electric actuators are fueled by electricity. An electric actuation is equivalent to a solenoid-powered electric bell.

Advantages

- It is helpful in various industries since it can automate industrial valves.
- It is quieter to operate and safer because there are no fluid leaks.
- It has the best positioning accuracy and control, and it can be programmed.

Disadvantages

- It is expensive.
- It's significantly impacted by the environment.

Various other actuators are:

1. **Thermal Actuator:** Power sources over actuators are magnetic or thermal actuators. These actuators make use of magnetic shape-memory alloys or alloys with shape-memory characteristics. A piezo motor is one type of SMA-based thermal/magnetic controller.
2. **Mechanical Actuators:** A mechanical actuator can move any object by converting rotating motion into linear motion. To function, it needs gears, rails, pulleys, shackles, and several other parts. By way of illustration, a crankshaft.

2.9   CONNECTIVITY PROTOCOL

The Edge Things: A gateway is a significant object that is found in the edge zone together with other things, including sensors, actuators, tools, and other things. Controlling interactions between objects and services in the cloud while also fostering communication between them is the gateway's main purpose. Edge computing, which processes data as close as feasible to its source at the network's edge, is where the term "edge" originates. The edge can be a factory floor, a power grid, a smart city, a smart construction, a drilling rig, a turbine for wind energy, a dairy farm, a train, a plane, or a car. The crucial component that gives edge processing its importance is activating the data processor and making judgments that are most closely related to real-time [12].

2.9.1   FIELD PROTOCOLS

A mechanical actuator can move any object by converting rotating motion into linear motion. To function, it needs gears, rails, pulleys, shackles, and several other parts. By way of illustration, a crankshaft.

Bluetooth: For applications using the IoT, Bluetooth is standard. Its power usage has been significantly reduced by design. The minimal requirements include the Bluetooth 4.2 core specification, 2.4GHz (ISM) rate, 50–150 m (Smart/BLE) range, and 1 Mbps (Smart/BLE) data rate [12].

ZigBee: It has a sizable installed base, much like Bluetooth, even if historically, it may have been used more commonly in industrial settings. Two of the current ZigBee profiles, ZigBee PRO and ZigBee Remote Control (RF4CE), are based on the IEEE 802.15.4 protocol. This well-liked 2.4GHz wireless networking technique is used for applications that call for the transmission of extremely rare information over a constrained area and within a 100-meter spectrum, such as inside a home or an additional structure.

Wi-Fi: Given the extensive use of Wi-Fi in LANs in residential areas, many developers typically choose this kind as a preferred alternative. Massive volumes of data may be handled, and it provides rapid data transport.

NFC: It is a type of wireless technology that enables quick and secure communication in both directions between electronic devices. Users can connect devices, access electronic materials, and make contactless payments using this technology, which is particularly well suited to smartphones. It effectively expands the range of sensitive card technology, enabling devices to connect at a distance of less than 4 cm [13].

2.10   IoT GATEWAY

In the IoT, a gateway acts as a bridge between several communication channels to create connections between controllers (sensors/devices) and the cloud. The usage of gateways allows for the establishment of communications between gadgets or between devices and the cloud. A gateway can be created using any common hardware or software. It supports IoT communication, establishes a connection across

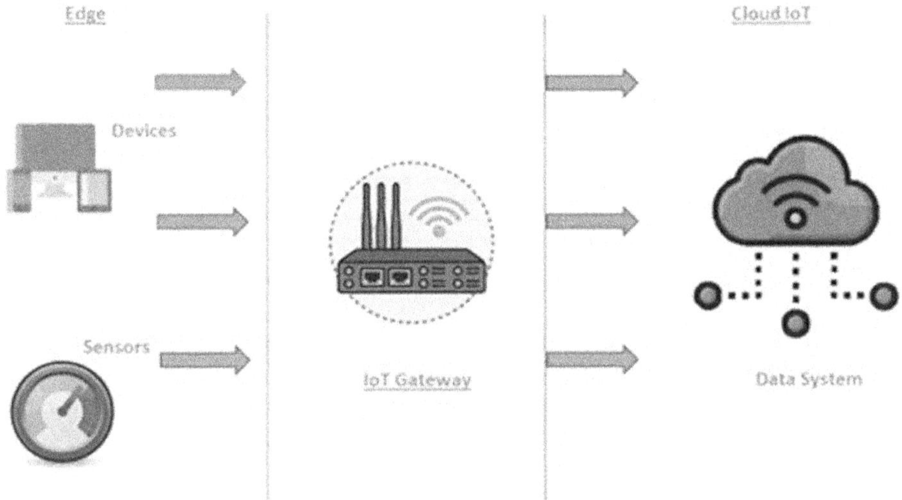

FIGURE 2.5   Internet of things gateways connectivity between sensors.

the network of sensors and the internet, and performs several other tasks as well. IoT gateways interpret protocols, gather data, perform local analysis on it, and filter it before sending it off to the cloud. Additionally, by interpreting protocols and storing data locally, they operate devices independently and add to device security [14].

The illustration below Figure 2.5 demonstrates how the IoT gateways facilitate connectivity between sensors and the IoT.

Direct connection with the cloud or the internet will not be energy-efficient for IoT devices since they have low power consumption (battery power) or are energy-constrained. As a result, they initially establish a connection with the Gateway utilizing low-power devices and short-range wireless technologies like the ZigBee Bluetooth, etc. Long-distance networks like Wi-Fi and cellular can also be used to connect them. The Gateway transmits the data to the Web or cloud by transforming the data into a common protocol, such as MQTT, using an Ethernet, Wi-Fi, mobile phone, or satellite connection. Additionally, gateways are frequently mains-powered, in contrast with sensor nodes, which are powered by batteries. There are many gateway devices in use [15].

2.10.1   Key Features of an IoT Gateway

- Building a bridge of connection.
- Gives you more security.
- Aggregates data is performed.
- Data preprocessing and filtering.
- provides a cache or buffer with local storage.
- Data processing at the edge.
- Ability to control the whole device.
- Device analysis.

- Increasing the functionality.
- Inspecting protocol.

2.10.2 THE WORKINGS OF AN IoT GATEWAY

1. A sensor network sends data, which is received.
2. On unprocessed information, before the process, filtering, and cleaning operations are performed.
3. Protocols that transport communication standards to the cloud for data.

IoT Gateways are an essential part of IoT architecture since they generate connections for communication and carry various additional duties as indicated above. A gateway for the IoT is, therefore, one of the most important components to take into account when thinking about an IoT ecosystem.

2.11 CLOUD PROTOCOL

All IoT solutions, including those that function primarily at the edge, must interact with other IoT solutions built on the cloud or with cloud-based services. For communication, the following cloud standard must be used:

MQTT: Messaging Queue Transmission Transport MQTT was developed by IBM in 1999 and standardized by OASIS in 2013. MQTT, a well-liked protocol for machine-to-machine communications, is frequently supported and used by embedded electronics. The goal is to provide an embedded connection between the applications, networks, middleware, and communication of the other side [16].

Advanced Messaging Queuing Protocol (AMQP): This protocol was created particularly for the banking sector. It uses TCP and has a publish/subscribe architecture similar to MQTT. The intermediary is different because its two primary components are exchanges and queues. Publisher communications are forwarded to queues by the exchange depending on predefined roles and criteria after being received by the publisher. Queues are essentially subscription-based topics for which consumers receive sensory input as it becomes available in the queue.

Constrained Application Protocol (CoAP): To offer a straightforward RESTful (HTTP) interface, the IETF-constrained Resource Environment working group created this session layer protocol. Between HTTP clients and servers, Representational State Transfer, or REST, is the primary method of communication. For lightweight applications, such as IoT, REST may, on the other hand, result in severe overhead and energy waste. Document transfer is made possible using CoAP. CoAP was created to help low-power sensors utilize RESTful services while still staying under their power limits. Unlike HTTP, which often uses TCP, this protocol is developed over UDP and contains a simple approach to ensure dependability.

HTTP: HTTP is a "connectionless" protocol. While utilizing the HTTP bridge, devices that are connected to Cloud IoT Core stop exchanging data. Instead, they converse and ask questions. The IoT will continue to communicate using the existing standards for internet services. We occasionally employ this form of communication even though it has already been used before when duration and capacity were important.

2.12 IoT ARCHITECTURE

The IoT architecture is made up of several different elements, such as sensors, protocols, actuators, cloud-based services, and layers. In addition to hardware like sensors and gadgets, the IoT architecture layers are unique in that they use gateways and protocols to monitor a system's consistency [17].

2.12.1 LAYERS OF IoT ARCHITECTURE

The complexity underlying IoT system architectures could prove a significant barrier to IoT adoption. Interestingly, by using the numerous layers of the framework of IoT solutions, anyone can learn the basics of IoT. The four various models that outline how IoT solutions are created include

- Three-layer architecture
- Five-layer architecture

2.12.2 LAYER ARCHITECTURE

The three-layer design is the most straightforward illustration of the topology of an IoT network. However, the three-layer model was unable to take into account more significant aspects of IoT network operation. The five-layer design builds on the existing three-layer architecture to give a broad overview of how IoT solutions ought to operate.

IoT solutions for both programs will likely require distinct designs given the variety of use cases for various industries. However, the three-layer IoT architecture is the first paradigm for describing the structure of IoT applications. Despite the various use cases, IoT solutions have always included some constant layers. The three-layer framework for IoT applications has been the most widely used framework for defining IoT app development. The following is a list of the elements that make up the three-layer IoT architecture. Figure 2.6 illustrates a three-layer IoT architecture, showcasing the distinct layers of perception, network, and application that collectively enable the seamless functioning and communication of IoT devices in a structured framework.

2.12.2.1 Perception Layer

The core architecture of the IoT includes a perception layer that contains the sensors utilized by IoT networks and applications. It serves as the key building block for IoT capability by granting access to information that can be acquired from various

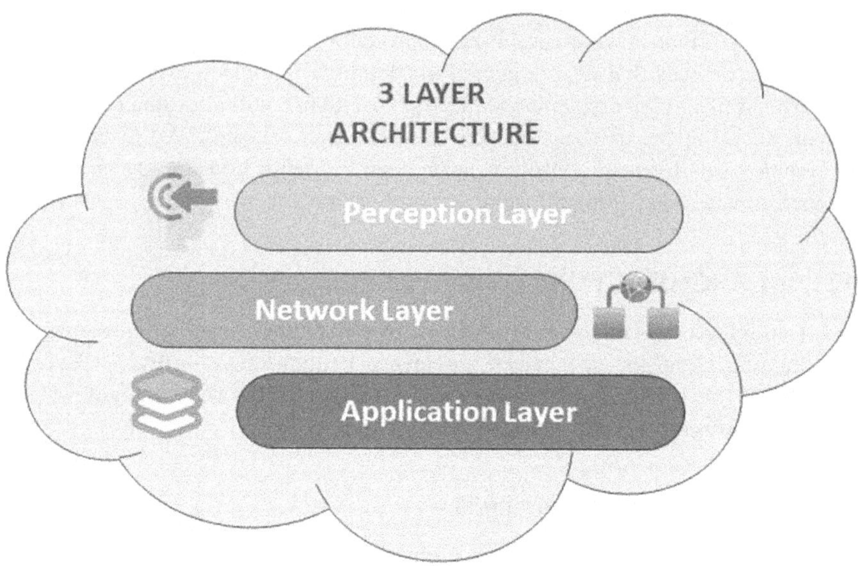

FIGURE 2.6 Three-layer architecture of IoT system.

sensors on connected devices. The sensory layer also includes actuators, which respond to changes in their environment.

2.12.2.2 Network Layer

The network layer, which covers data transmission through the IoT network, is another crucial component of the IoT ecosystem. The network layer of the IoT architecture is crucial for creating connectivity between different devices and sending data to the required backend services.

2.12.2.3 Application Layer

The application layer, which is the third component in the three-layer architecture for IoT solutions, serves as the user-facing layer for an IoT solution, such as the app for phones used to control IoT devices in smart homes.

2.12.3 FIVE-LAYER ARCHITECTURE

The five-layer architecture's introduction also serves to explain how the IoT is structured. Although the three-layer IoT architecture helps to explain how IoT solutions work, it does not focus on the implications of using data. Figure 2.7 delineates a comprehensive five-layer IoT architecture, providing a layered framework for efficient connectivity, data processing, and application deployment.

2.12.3.1 Perception Layer

The perception layer is the link between the three-layer architecture and the five-layer IoT system design model. In IoT networks, it serves as a stand-in for the actual hardware elements that "perceive" the data necessary for processing. IoT sensors utilized

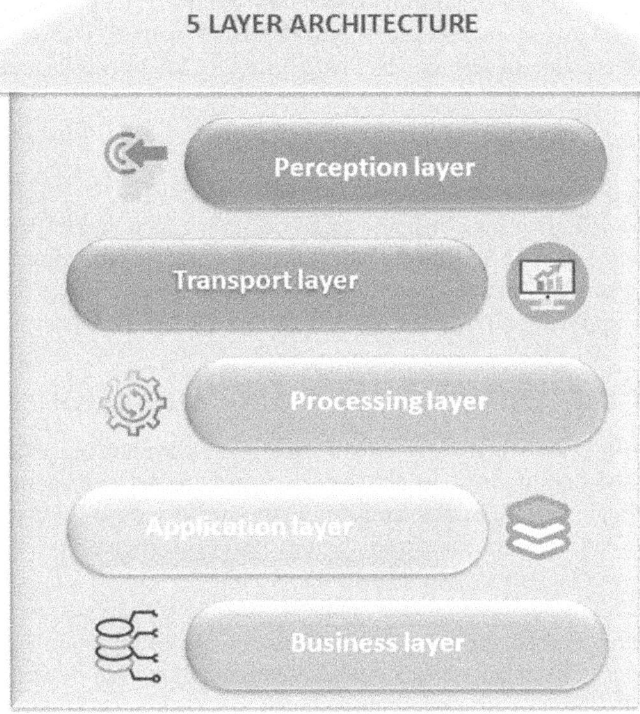

FIGURE 2.7 Five-layered IoT architecture framework.

in several applications, such as medical tracking systems, self-driving automobiles, security features, and lighting systems, serve as an example of an IoT perception layer.

2.12.3.2 Transport Layer

In the five-layer model, the transport layer is the third layer of the IoT architecture. It is in charge of delivering the obtained data to the cloud- or edge-based processing centers. The transport layer uses internet gateways to convey data from the perception layer, or sensors, to the processing layer. IoT network administrators employ cellular and Wi-Fi connections to move data over the transport layer. BLE and Low-Power Wide-Area Networks may also be significant highlights in the transport layers of IoT architectures. The architecture of a solution for the IoT may combine several transport protocols, depending on the requirements of the use case [18].

2.12.3.3 Application Layer

The third layer of the architecture, the application layer, controls how IoT solutions communicate with users. It is a layer that each of the three parts of the architecture share. Most processing tasks in IoT designs can be completed without the assistance of a human. The responses to the question "What is the framework of IoT architecture?" would prove inadequate if the application layer weren't present.

Human input is required to tell the server whether a set of rules has been followed. Using the application layer, administrators may control IoT devices and define the rules that should govern how the IoT network should operate. The application layer influences the creation of service-level bargaining for IoT networks and systems.

2.12.4 BUSINESS LAYER

A potential addition to the five-layer IoT design is the business layer. It concentrates on converting IoT data into business analytics that can power efficient decision-making techniques. The business layer relies heavily on reporting and real-time dashboards to provide business intelligence. With the aid of further connectors, IoT data from the application layer can be further improved at the business layer.

2.13 INTERNET OF THINGS COMMUNICATION MODELS

Communications between device-to-device: The communication model shows connections and communication between two or more devices happening directly instead of going through a middle application server. As seen in Figure 2.8, although these devices can communicate over various networks, including the World Wide Web or IP networks, they commonly use Bluetooth, Z-Wave, or ZigBee [19].

These device-to-device networks make it possible for devices to communicate with one another and perform the intended function. This communication architecture is commonly used by applications like home automation systems, which frequently use small data packets to communicate between gadgets with only modest data rate requirements.

Residential IoT items like lightbulbs, light switches, cooling and heating systems, and door locks frequently exchange brief messages with one another in a home automation scenario (such as a locked door status message or a turn-on light command). Since the core device-to-device protocol for communication is incompatible, the user is frequently compelled to select a variety of gadgets that share a common protocol [19].

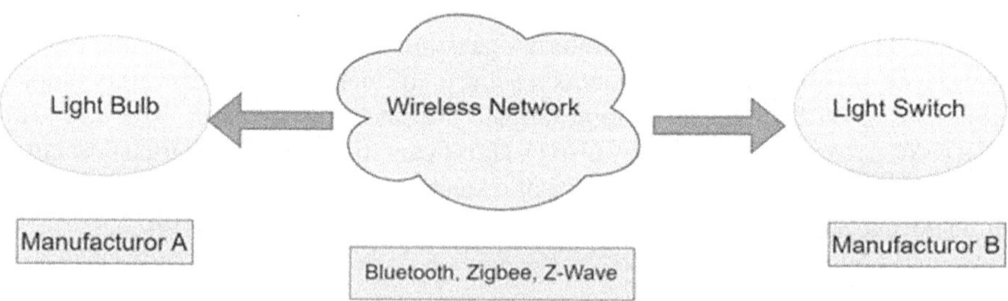

FIGURE 2.8 Communication between device-to-device.

2.13.1 Communications between Devices and the Cloud

The device-to-cloud communication protocol connects an internet cloud service, such as an application distributor of services, directly to an IoT device to exchange data and manage message flow. Using Wi-Fi or traditional wired Ethernet connections, this method frequently connects a gadget to the IP (internet protocol) network, which then communicates with the cloud service. Figure 2.9 illustrates the communication flow between devices and the cloud, portraying the crucial link in IoT data exchange.

The Samsung Smart TV and Nest Labs Learning Thermostat are two popular consumer IoT products that make use of this communication method. With the help of the Nest Learning Thermostat, information is transmitted to a cloud database, which can be examined to figure out how much energy is used at home. The user also has remote access to their temperature via a mobile device or web interface thanks to this cloud connection, which also enables thermostat software updates. Like other internet-connected speech recognition technologies, Samsung SmartTV technology sends data about user viewing to Samsung for analysis and links to the internet to enable interactive speech recognition features. The device-to-cloud approach, in these circumstances, helps the end user by enhancing the device's functionality beyond its built-in features.

2.13.2 Device-to-gateway Communication Model

An application layer gateways (ALG) service is used in the device-to-gateway concept to connect an IoT device to a cloud-based platform. This merely implies that an on-premises gateway device has installed software that serves as a conduit across the device's hardware and cloud services and offers security as well as additional features like communications or protocol translation. Figure 2.10 depicts the communication dynamics between devices and gateways, illustrating the pivotal link in IoT connectivity

In consumer electronics, this paradigm is applied in a variety of contexts. The local gateway device is frequently a smartphone with an app that enables the transmission of data to a cloud service and communication with a device. Wearable fitness trackers are one well-known example of a product that commonly uses this concept. Due to their inability to establish an instant connection to a cloud service,

FIGURE 2.9 Communication between the device and cloud.

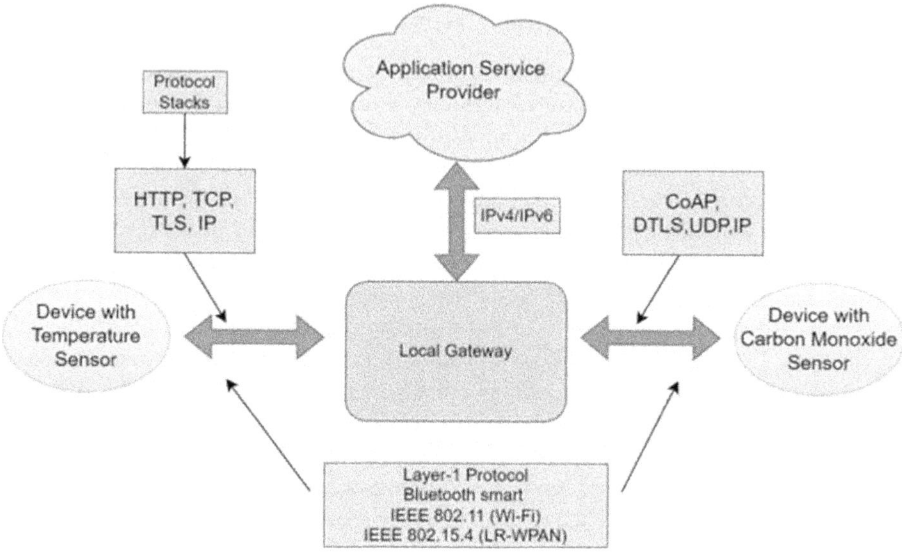

FIGURE 2.10 Device-to-gateway communications.

these gadgets frequently rely on cellular device software to serve as an independent gateway that links the exercising gadget to the cloud.

Another example of this device-to-gateway structure is the inclusion of "hub" devices in programs for home automation. The local incompatibility of various IoT devices with cloud services can be bridged by these devices. Z-Wave and ZigBee transceivers have been added to the Smart Things hub, a standalone gateway device, enabling it to communicate with both kinds of gadgets. The smart device's cloud is then connected after that.

Customers who use the service can access the gadgets online and via a smartphone app.

The following technological advantages of the device-to-gateway technique are covered in the *Internet Engineering Task Force Journal* article:

This communication method is used when interoperability between devices that do not use IP is required for smart objects. When using this method to integrate IPv6-only devices, a gateway is necessary for older IPv4-only gadgets and services.

To put it differently, this communications paradigm is generally used to integrate new electronic devices into an older system that already has devices that are not already compatible with them. Because the application layer gateways technology and system must be developed, this approach has the disadvantage of making the entire system more complex and expensive [20].

2.14 IoT SECURITY CHALLENGES

IoT faces various kinds of security challenges, some of which are shown in Figure 2.11.

FIGURE 2.11   IoT security challenges.

2.14.1   Using Outdated or Hazardous Components

Due to the inability to purchase software components from reliable supply networks, this issue arises. Unpatched third-party software, for instance, might enable device hacking.

2.14.2   Insecure Data Transfer and Storing

Inappropriate encryption or access management during data transport could allow unauthorized access to sensitive data.

2.14.3   Privacy Protection Is Insufficient

Users' data might be exposed to third parties if stored on unprotected devices or in unprotected connected ecosystems.

2.14.4   Passwords that Are Illogical or Poor

If the passwords you use are short, weak, or easily guessed, your IoT device's security is in jeopardy. This also applies to hard-coded passwords that the manufacturer implemented but may be visible to the public.

2.14.5   Establishing Dangerous Network Services

Users are warned against installing useless or risky network services on their devices. The data's privacy could be in jeopardy while it is being stored.

2.14.6 No Physical Barriers Exist

IoT device owners could neglect to keep their devices physically secure, and manufacturers might overlook the design element. The result is tangible interference.

2.14.7 Inadequate Updating Mechanisms

Another security concern that unsecured software update systems pose is the possibility of allowing hackers to access private information via an unencrypted connection.

2.15 APPLICATIONS OF IoT IN DIFFERENT AREAS

Most commonly used programs are already Intelligent but cannot talk to one another. They can develop a wide range of inventive applications by being given the freedom to do so and by exchanging crucial information. The quality of our lives will undoubtedly increase thanks to these new, autonomous apps [21]. Some of the applications discussed in this portion, as depicted in Figure 2.12, are related to IoT applications.

2.15.1 Smart Homes and Buildings Using IoT

In today's quick-paced world, a smart home is increasingly essential. A smart home can communicate with many household items by connecting them to the internet. Several home appliances, such as cooling and heating, as well as doors, windows, and lighting, and the refrigerator and washing machine can also be operated manually, as seen in

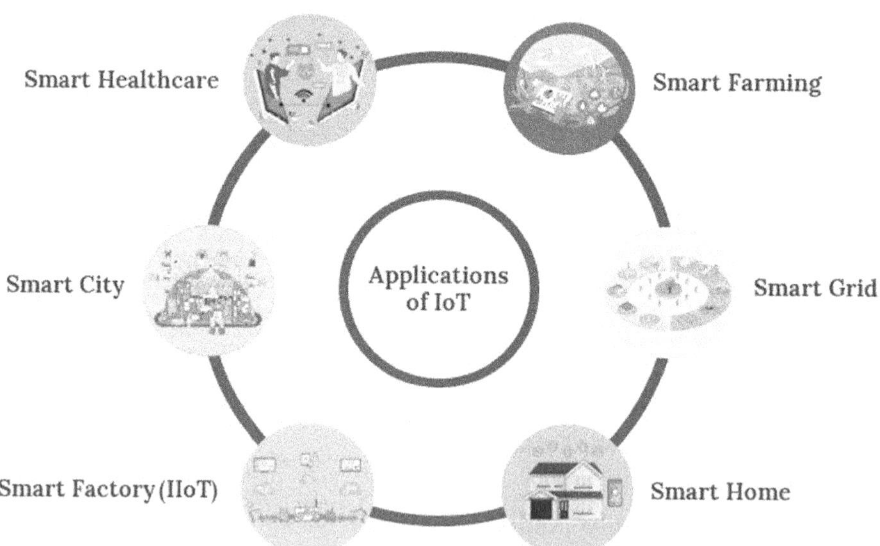

FIGURE 2.12   Application areas of IoT.

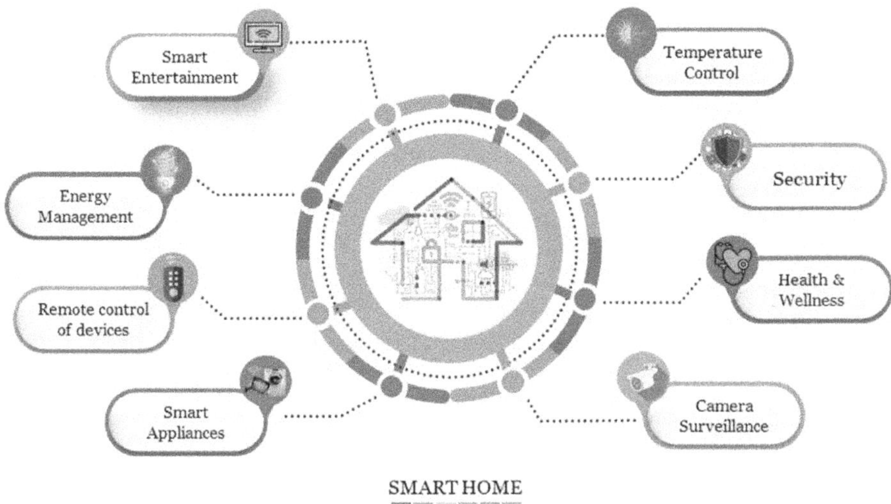

FIGURE 2.13   Smart homes and building using IoT.

Figure 2.13. Wireless sensors and IoT integration into complex building management of energy solutions are possible through networks. Using a laptop or a cell phone will allow us to gain access to the building's energy control and management system [22].

2.15.2 SMART FARMING

Intelligent agricultural systems can use networked sensors to monitor a range of crop field variables, such as sunlight, temperature, humidity, and precipitation prediction, including soil moisture, as shown in Figure 2.14. Automation of irrigation systems is made possible by IoT. Numerous advantages of smart farming include improved crop management, increased output control, monitoring of weather patterns, and improved product quality and volume. The automation of corporate processes enables all of these advantages. Additional advantages include improved internal control of operations and reduced manufacturing hazards.

2.15.3 SMART HEALTHCARE

IoT monitoring technology is capable of helping regularly check on the physiological status of hospitalized patients, who need to be constantly monitored. When sensors are used in the field of health technology, accurate physiological data can be collected, evaluated, and saved through a gateway via the cloud. The data are automatically delivered to caretakers so they can further evaluate and analyze them [23]. Instead of having a doctor visit the patient to check on their vital signs, an ongoing, automated flow of data is used instead. As seen in Figure 2.15, this method improves the level of care by including continuous care, reducing the cost of medication, and gathering and analyzing data.

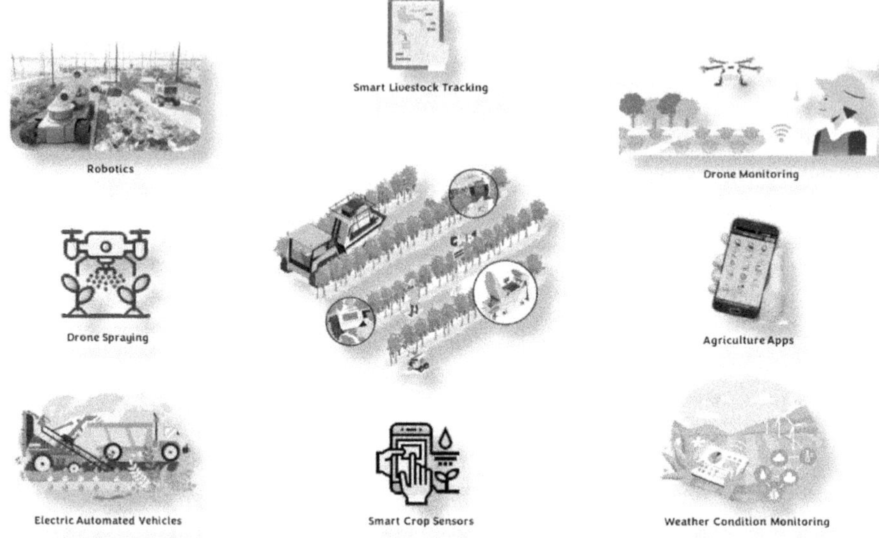

FIGURE 2.14　Smart farming using IoT.

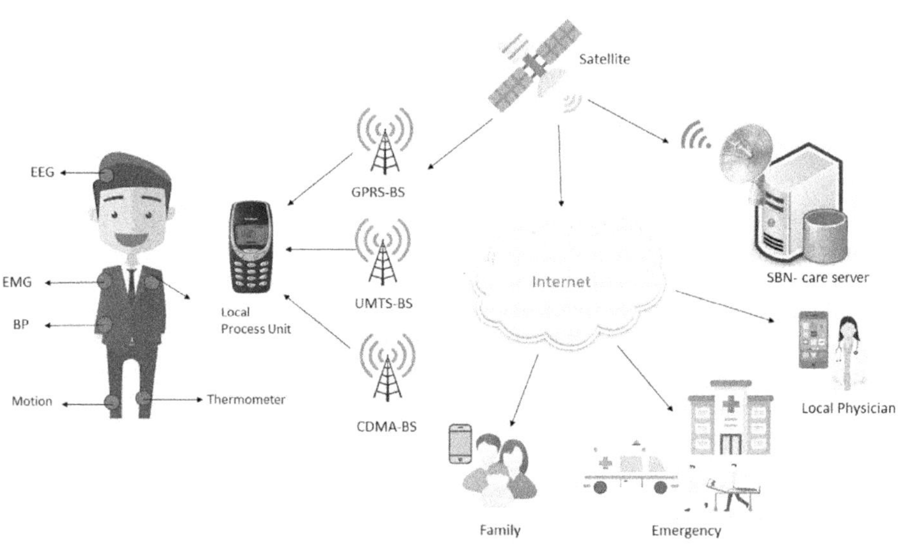

FIGURE 2.15　Smart healthcare using IoT.

2.15.4 SMART CITY

Before deploying the IoT, smart cities must be fully prepared at all levels, with backing from government officials as well as residents. IoT can improve cities in several ways, such as by boosting public transport, reducing traffic, and protecting citizen safety and health. A smart city's design is seen in Figure 2.16.

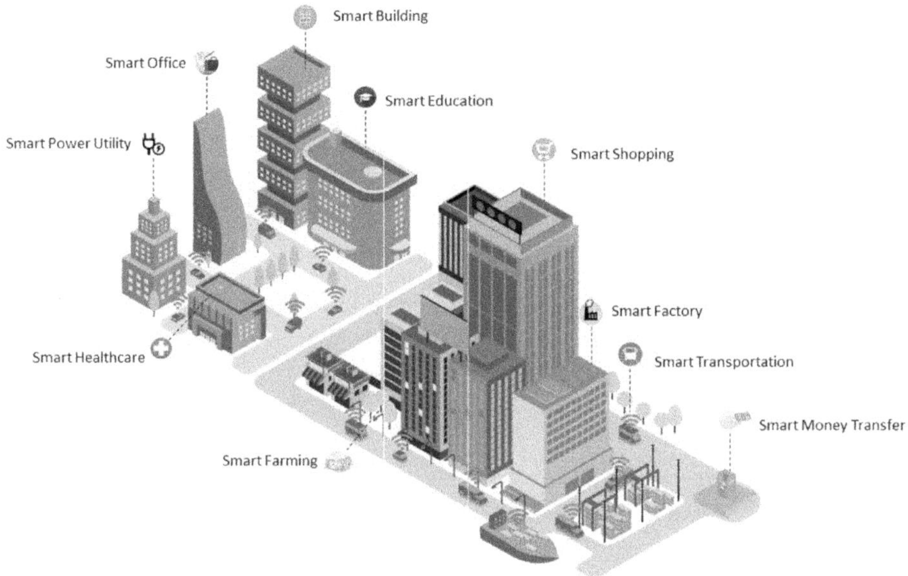

FIGURE 2.16 Smart City using IoT.

2.15.5 Smart Grid and Smart Energy

Customers and suppliers can communicate with each other in real-time via a smart grid that combines the use of ICTs with the electrical network. As an outcome, the environment where energy flow takes place becomes more dynamic, improving the effectiveness and long-term profitability of the provision of power [24]. The fundamental building elements of knowledge include communications technology, which will include digital communications, power flow monitoring, and sensing. Electrical systems for transmission of data, smart meters in homes that display information on energy use, and coordination and control created a highly collaborative, responsive electrical grid using integrated automation systems to collect and analyze a variety of data. Figure 2.17 displays the Smart grid using IoT.

2.16 ADVANTAGES AND DISADVANTAGES OF IoT

The advantages and disadvantages of IoT applications are listed in this section.

2.16.1 Advantages of IoT Application

Security: You can monitor and manage your house using mobile devices. It might enhance someone's security.

Keep in Touch: You can always be online with your family. It is possible to maintain an electronic connection despite being physically apart.

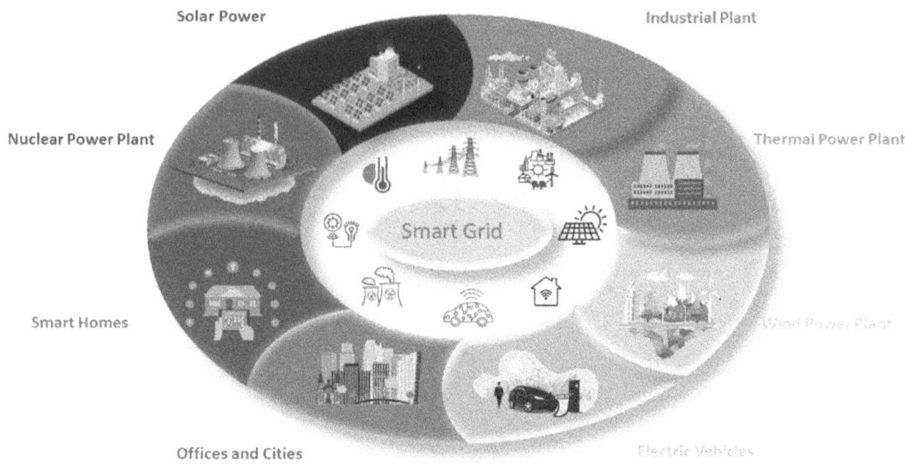

FIGURE 2.17   Smart grid using IoT.

Proper Use of Electricity and Energy: You can quickly maintain and repair your home appliances if you can speak with them. If appliances are self-sufficient, electricity utilization can be done efficiently.

Best Medical Care and Administration: During a doctor's visit, continuous tracking of patients is possible, and it also gives them the ability to make decisions and provide therapy in the event of an emergency.

Affordable Business Procedures: The IoT can be utilized to effectively track the number of business operations, such as transportation and location, security, asset monitoring, as well as inventory control, person tracking of orders, client management, personalized marketing, and sales operations, etc.

2.16.2   Disadvantages of IoT Application

Security Issues: Hackers could access the network and possibly steal data.

Increasing Indolence: People are getting used to click-based work, which makes them lazy about employing research in everyday activities or engaging in any kind of physical activity.

Unemployment: Unskilled workers, for example, may face serious risks of losing their jobs.

2.17   FUTURE TRENDS AND IoT CHALLENGES

2.17.1   Emerging IoT Technologies

The advantages of IoT technologies for businesses have significantly improved the lives of both owners and users. For instance, leveraging cutting-edge IoT technologies can help construct resilient supply networks. Businesses could use IoT apps concurrently to optimize asset usage and maintenance. Customers can benefit from IoT's advantages,

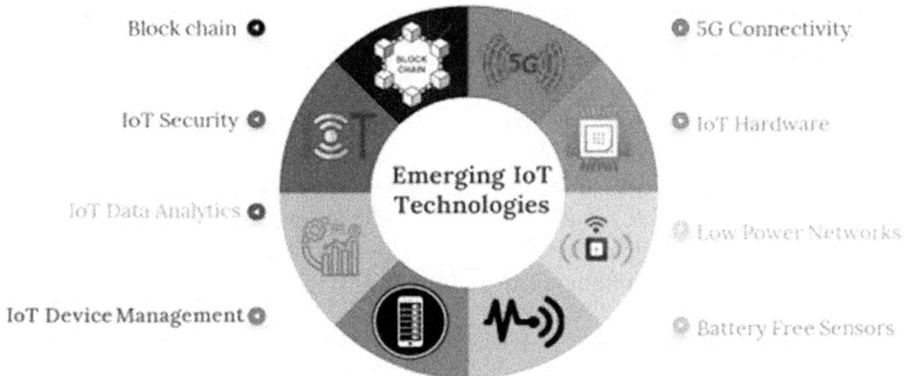

FIGURE 2.18 Emerging technologies of the Internet of Things.

which make using internet-connected devices easier. IoT is still growing, though, and future developments will greatly impact how it develops. The major technological advancements that will have the greatest influence on the IoT by 2023 are listed below in Figure 2.18 [26]. Figure 2.18 illustrates the emerging technologies of Internet of Things.

2.17.2 BLOCKCHAIN

When leveraging IoT networks, user data security is a significant issue. Because IoT networks are so large, there are more security vulnerabilities when connected devices share sensitive user data. In response to the inquiry, "What are the most recent developments in IoT?" It would develop blockchain technology. As it can guarantee the privacy of information, blockchain is rapidly becoming a fundamental technology supporting IoT. The blockchain network's permanent archives can help preserve user data and transactional details. The transparency of the distributed ledger in the context of the blockchain system may increase IoT security. Furthermore, the decentralization feature of blockchain would provide uninterrupted communication between different network nodes. Blockchain technology will, therefore, be a crucial technology that will dominate IoT in 2023 [7].

2.17.3 5G NETWORK

In response to the query "What are the most recent advances in IoT?" 5G would be the next entry on the list. Wireless connectivity is one of the requirements for constructing IoT networks and ensuring their performance. 5G offers the opportunity to fully leverage the IoT in addition to being a wireless technology advancement.

Stronger connectivity will be required for dependable and high-performance IoT devices. 5G wireless communication may have benefits such as decreased latency, extensive coverage, processing of data in real-time, and network slicing. The major advantage of 5G connection is that it can provide dependable networks in remote areas, bridging the gap between urban and rural areas [8].

2.17.4 IoT Security

One of the important improvements in IoT for 2023 will be IoT security. IoT security for networks is a top priority for businesses and developers. Security technologies will capture the interest of the IoT market in 2023 thanks to new technological advances for defending IoT devices and platforms from both online and offline threats [9]. Future security tools for the IoT would also address modern issues like IoT network device impersonation. IoT security measures might also be used to thwart denial-of-sleep attacks, which could deplete the power supplies of IoT devices. Using security technologies to encrypt communications might assist in preventing attacks [27].

2.17.5 Summary

The IoT chapter in this book offers a concise yet in-depth overview of the subject. The definition and significance of the IoT are discussed, as well as its use in a variety of sectors, including smart homes, healthcare, agriculture, manufacturing, and energy management. The architecture of the IoT, which includes layers like perception, network, and application, as well as its fundamental components and history, are all described in this chapter. Security concerns and communication methods such as Device-to-Device, Device-to-Cloud, as well as Device-to-Gateway are discussed. Future technologies such as IoT and the link between 5G and IoT are indicated as trends. This chapter provides a thorough explanation, making it an invaluable guide for comprehending the quickly evolving IoT environment.

REFERENCES

1. C. Perera, C. Harold Liu, and S. Jayawardena, "The emerging internet of things market place from an industrial perspective: A survey," IEEE Transactions on Emerging Topics in Computing, vol. 3, no. 4, pp. 585–598, 31 Jan 2015.
2. A. Al-Fuqaha, M. Guizani, M. Mohammadi, M. Aledhari, and M. Ayyash, "Internet of things: A survey on enabling technologies, protocols, and applications," IEEE Communications Surveys & Tutorials, vol. 17, pp. 2347–2376, 2015.
3. A. Luigi, I. Antonio, and M. Giacomo, "The internet of things: A survey," Science Direct Journal of Computer Networks, vol. 54, pp. 2787–2805, 2010.
4. W. Miao, L. Ting, L. Fei, S. ling, and D. Hui, "Research on the architecture of the Internet of things," IEEE International Conference on Advanced Computer Theory and Engineering (ICACTE), Sichuan Province, China, pp. 484–487, 2010.
5. M.N. Kamdar, V. Sharma, and S. Nayak, "A survey paper on RFID technology, its applications and classification of security/privacy attacks and solutions," IRACST - International Journal of Computer Science and Information Technology & Security (IJCSITS), vol. 6, no. 4, pp. 2249–9555, July-August 2016.
6. Edge computing - http://searchdatacenter.techtarget.com/definition/edge-computing
7. IoT protocols - https://www.rs-online.com/designspark/eleven-internet-of-things-iot-protocols-you-need-to-know-about
8. The architecture of IoT gateways - https://dzone.com/articles/iot-gateways-and-architecture
9. Internet of things protocols and standards - http://www.cse.wustl.edu/~jain/cse570-15/ftp/iot_prot/
10. R. Abdmeziem, and D. Tandjaoui, "Internet of Things: Concept, Building Blocks, Applications and Challenges," Computers and Society, Cornell University.

11. E.S.A. Ahmed, and Z. Kamal, "Internet of things applications, challenges and related future technologies," World Scientific News, vol. 67, no. 2, pp. 126–48, 2017.

12. D. Niewolny, How the internet of things is revolutionizing healthcare - https://cache.freescale.com/files/corporate/doc/white_paper/IOTREVHEALCARWP.pdf

13. E.A. Nuaimi et al., "Applications of big data to smart cities," Journal of Internet Services and Applications, 1–15 6, Article number: 25 (2015).

14. Smart grid enabling energy efficiency and low-carbon transition - https://www.gov.uk/government/uploads/system/uploads/attachment_data/file/321852/Policy_Factsheet_-_Smart_Grid_Final__BCG_.pdf

15. R.R. Mohassel et al., "A survey on advanced metering infrastructure," International Journal of Electrical Power & Energy Systems, vol. 63, pp. 473–484, 2014.

16. T. Mufti, N. Sami, S.S. Sohail, "A review paper on internet of things(IOT)," Indian Journal Of Applied Research, vol. 9, no. 8, August 2019.

17. V. Sharma, R. Tiwari, "A review paper on "IOT" & It's smart applications," 2016.

18. "What is the internet of things, or IoT? A simple explanation," IOT for all, [Online]. Available: https://www.iotforall.com/what-is-internet-of-things

19. J.A. Stankovic, "Research directions for the internet of things," Internet of Things Journal, IEEE, vols. 1, pp. 3–9, 2014, doi: 10.1109/JIOT.2014.2312291

20. B. Ray, "Four types of IoT wireless networks," IOTa, 24 March 2020. [Online]. Available: https://www.iotacommunications.com/blog/types-of-iot-networks/

21. A. Al-Fuqaha, M. Guizani, M. Mohammadi, M. Aledhari, and M. Ayyash, "Internet of things: A survey on enabling technologies, protocols, and applications," IEEE Communications Surveys & Tutorials, vol. 17, no. 4, pp. 2347–2376, 2015.

22. C.-L. Zhong, Z. Zhu, and R.-G. Huang, "Study on the IOT architecture and gateway technology," [Online]. Available: https://www.semanticscholar.org/paper/Study-on-the-IOT-Architectureand-Gateway-Zhong-Zhu/262f051026165e84eb53a068cd75866b6af6c038/figure/0

23. C.-L. Zhong, Z. Zhu, and R.-G. Huang, "Study on the IOT architecture and gateway technology," in 14th International Symposium on Distributed Computing and Applications for Business Engineering and Science (DCABES), Guiyang, China, 2015, pp. 196–199.

24. A. El Hakim, "Internet of things (IoT)," System Architecture and Technologies, White Paper, 2018, doi: 10.13140/RG.2.2.17046.19521

25. K. Mundle, "Home smart IoT home: Domesticating the internet of things," [Online]. Available: https://www.toptal.com/designers/interactive/smart-home-domestic-internet-ofthings#:~:text=IoT%20home%20automation%20is%20the,controlled%20by%20a%20mobile%20app.

26. K.K. Patel, S. Patel, P.G. Scholar, and C. Salázar, "Internet of things-IOT: Definition, characteristics, architecture, enabling technologies, application & future challenges," International Journal of Engineering Science and Computing, vol. 6, no. 5, pp. 6122–6131, 2016.

27. R. Abdmeziem, and D. Tandjaoui, "Internet of things: Concept, building blocks," Applications and Challenges, Computers and Society, Cornell University Scientific News, vol. 67, no. 2, pp. 126–148, 2017.

3 Sustainable Marketing and Retailing

A Cloud Computing and Internet of Things (IoT) Approach

Shivani Malhan, Anita Tanwar, Rekha Mewafarosh, Shikha Agnihotri, and Divya Gupta

3.1 INTRODUCTION

Cloud Computing is a platform and can be considered a type of application as well. It describes the various applications that can be made accessible through the internet. These applications of the cloud use data centers that are very large, and servers that are very powerful to host applications related to the host web and services related to the web. A cloud application can be accessed by any person who has a suitable internet connection and a standard browser.

The Internet of Things (IoT) is a paradigm that has gained focus in the current scenario as it helps in communication between the devices which are electronic in nature and sensors. With the help of IoT, a connection can be enhanced between smart devices through the internet, which can further help provide innovative solutions for private and public institutions as a whole.

IoT is able to provide innovation through smart systems, intelligent devices, frameworks, and sensors. A lot of research has been done on the potential of IoT and its effectiveness (Zhang et al, 2021). IoT is used in almost all the sectors, but in this chapter, we will be discussing the role of IoT in the retail industry. Through this chapter, the technological and societal aspects of IoT will be discussed.

In the retail sector, it is important that the customer is provided a convenient environment to buy the products. The retail framework must be such that it requires less time and effort.

IoT is extremely imperative in retail as it helps in providing a seamless experience to the customer through the SMARTMART application, through which the customer can easily find the product of his choice as the products are RFID enabled (Verma et al., 2019).

Recently, Amazon has started with Amazon Go, where the customer will get a cashless and seamless experience. All the products are RFID enabled, so when the customer collects all the products in the shopping cart, the products are automatically

DOI: 10.1201/9781032656694-3

billed. Then the bill is received by the customer through the application on his/her mobile phone.

Through IoT, the retailer is able to understand the needs and preferences of the customer. It also helps to identify the products that are out of stock so that they can be replenished in real-time. If the customer does not see the products on the shelves, he thinks that the products are out of stock, and he might go to another shop to buy the products. This can cause dissatisfaction to the customer. On-shelf availability is essential for the retailer as it urges the customer to buy the products and increases sales and profitability.

3.2 LITERATURE REVIEW

3.2.1 SUSTAINABILITY

Sustainability is stated as sustainable development. The definition of sustainable development was provided by Brundtland Commission Report (1987), and it was stated that it is "meeting the needs and requirements of the present without compromising the ability of future generations to meet their own needs".

3.2.2 SUSTAINABLE MARKETING

Environmental damage is caused by businesses due to manufacturing, processing, discarding, and polluting (Saha and Darnton, 2005: 117). It is very difficult to find a sustainable future if we are degrading the ecological balance through growth in population and high consumptive lifestyles. The role of sustainable marketing is immense and it needs to be harnessed properly as it can help in motivating the customer to recycle, reuse, decrease wastage and save energy. Sustainable marketing is attained through three existing marketing disciplines, which include green marketing, social marketing, and critical marketing. The components of sustainable marketing are given in Figure 3.1.

3.2.3 SOCIAL MARKETING (SOCIAL LAYER OF SUSTAINABLE MARKETING)

Green marketing when combined with social media can help in developing sstainable marketing. Social marketing is the application of marketing tools, techniques, and concepts to achieve certain behavioral goals that are good for the public or society. The main aim of social marketing includes changing the behavior related to health, related to society, and related to sustainability. In the latest case, Britain's first cycle city has been introduced through the Bristol Cycle City project, where less car usage is introduced through social marketing. Social marketing managers are promoting the social goods in a persuasive manner. Commercial organizations should be involved in social initiatives through partnerships with NGOs, which they have been doing in the past as well (Kotler and Lee, 2008).

3.2.4 CRITICAL MARKETING (ECONOMIC LAYER OF SUSTAINABLE MARKETING)

Critical marketing involves green innovation, a closed-loop supply chain, and adjustment of strategies of marketing.

3.2.5 FRAMEWORK OF SUSTAINABLE MARKETING

FIGURE 3.1   Framework of sustainable marketing.

3.2.6 INTERNET OF THINGS

IoT helps in enhancing man-to-machine and machine-to-man interactions by connecting various devices to the World Wide Web. The applications of IoT are not only limited to the field of retail, but it has widespread applications in the healthcare industry, automation of homes, automation of various vehicles, automation of factory lines, medical, etc.

3.2.7 COMPONENTS OF INTERNET OF THINGS

There are four components of IoT through which the data is processed. This is described in Figure 3.2.

3.2.7.1 Sensors

Sensors are used to collect the data from various sources of the surrounding areas. The data collected may vary with different complexities and it might be collected

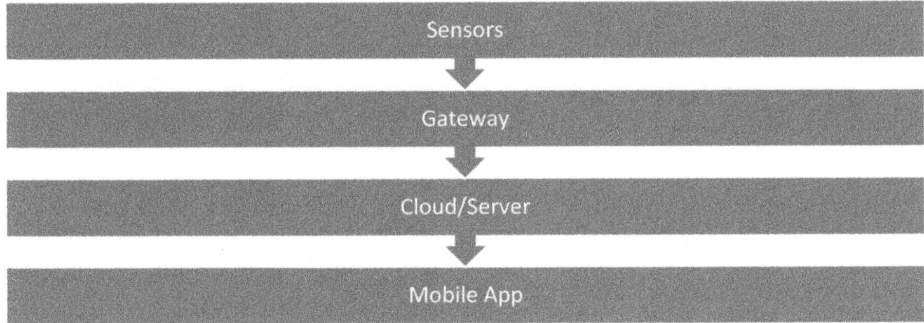

FIGURE 3.2 Internal working of Internet of Things.

from various sources. It can be as simple as a sensor that senses the outside temperature to a sensor that collects the data through a full video feed.

3.2.7.2 Connectivity

The data which is collected by the sensor is sent to the cloud infrastructure through a medium as a medium is required to transport the data. The various modes of communication and transport include Wi-Fi, WAN, networks which are cellular in nature, Bluetooth, networks related to satellites, etc.

3.2.7.3 Data Processing

After the data has been collected in the cloud infrastructure, the data is then processed through various software. All the things with the networked sensors are sent to the data stores and then to the analytical engines like machine learning, servers, etc. Then feedback and control are sent to the things. Data processing is explained in Figure 3.3.

The data is sent to the user in the form of an update on the phone, message, or an email alert. The user can easily check their home camera and see the people who have visited his/her home and get an update if the temperature of the refrigerator is not working properly.

3.3 RESULTS AND DISCUSSION

3.3.1 Role of IoT and Cloud Computing in Improving Sustainable Marketing

IoT and cloud computing are very important in order to understand customer requirements and insights and increase the campaigns related to marketing. When companies keep an eye on the smart devices of their customers and the data they view on them, they come to know the needs and preferences of their customers. This also helps the marketers understand the products and services the customer wants. The needs of the customers can be predicted through IoT and cloud computing, and sales and profitability can increase over a period of time. The customer experience should

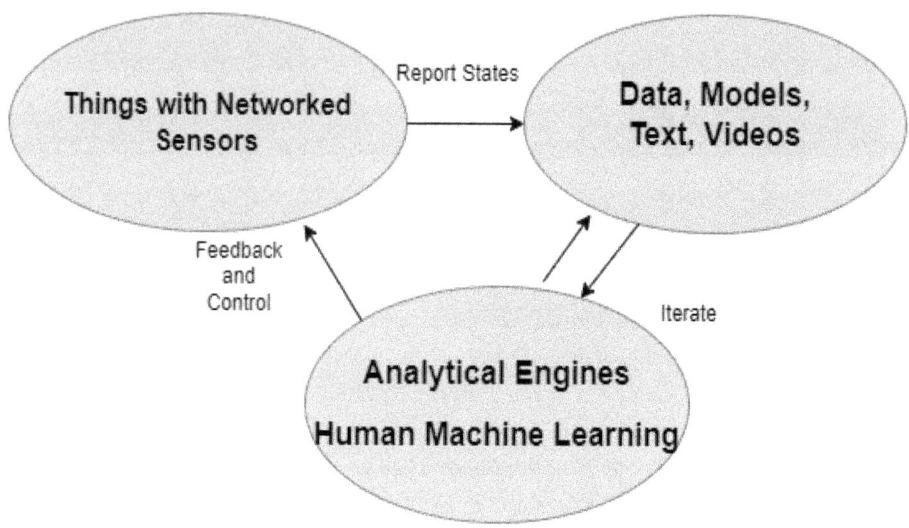

FIGURE 3.3 Data processing.

not be sacrificed in order to increase the sales of the product and services. Collecting the data manually can take a lot of time and effort, but marketing automation done through IoT and cloud computing considerably decreases the time. IoT is doing a very important job in marketing automation as previously, the main task of the marketing manager was to collect customer data, customer feedback, and understand the customer by taking interviews. But now the job of the marketing manager has been eased out. The data collection and database creation is already done through IoT, but the marketing manager has to analyze the data already collected by the IoT database (Rong et al., 2016). This information is provided to the company, and the company officials make the necessary customizations. Through IoT and cloud computing the human-to-human interactions have been decreased considerably as IoT is able to manage all the things through automation

3.3.2 Role of IoT and Cloud Computing in Improving Retailing

3.3.2.1 Next Stage in Retail Technology Development and Implementation

The previous decade has been considered the electronic retail era, but the next decade will be the decade of technological development in the traditional retail sector. This is because most companies have started using IoT to provide excellent performance to their customers. They want their customers not to stand in lines, take the help of the salesforce, take receipts after billing, etc. Nowadays, products are RFID enabled, and with the help of IoT, the customers will be able to get information related to the product by just picking up the product.

Although the retailers previously used RFID technology, now it is used through improved technology, improved sensors, fast internet connections, and easy communication through mobile devices. This has taken place with the help of RFID.

Companies provide personalized experiences through RFID. IoT has become the need of the hour because it provides a connection between physical as well as digital objects and allows bidirectional interaction with the customers inside and outside the store (Gregory, 2015).

IoT helps the retailer as it provides an opportunity to link each customer with the store and track various activities it performs while purchasing the store's product. IoT makes the retailer's work more efficient.

1. IoT as a factor of efficiency of retail operations

    Through IoT, a lot of data related to the customer has been used by the retailer for analyzing the requirements and needs of the customer in the future. Many firms in Europe have captured around 20–30% of revenue on the basis of Big data gathered from the customer.

    The data management of customers and security issues related to the organization are a threat to the organization. Due to the increase in technology, there can be chances that the data can be hacked.

2. Implementation of the IoT in retailing

    Due to IoT many business processes have been streamlined. The disruptive technologies that have changed and transformed the business include Machine learning, AI, autonomous vehicles, and Artificial Intelligence (World Economic Forum).

    The implementation of IoT is explained in Figure 3.4. Modern technologies used in retail include the virtual product, zone experience, zone virtual

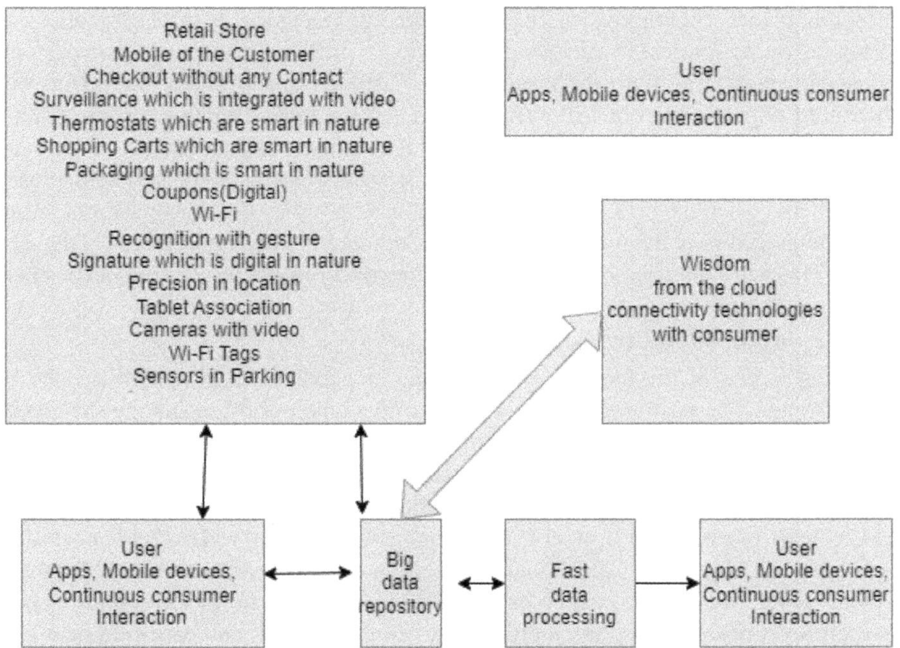

FIGURE 3.4   Implementation of IoT.

simulation, physical product, and prototypic zone. This figure describes all the technologies that are used in retail and their connection with each other. It also emphasizes that IoT improves continuous interaction with the customer and thereby enhances the customer experience.

3.3.2.2   Internet of Things in Retail

IoT helps reduce the energy consumption used in various business processes. It helps manage waste, improve the quality of products, and delivers the product to the customer in a hassle-free manner. With the help of IoT the efficiency of the organization is increased as it helps improve the internal processes. It also helps in increasing the experience of the customer as it provides an ecosystem that is digital in nature and provides the goods and services to the customer in a hassle-free environment (Brynjolfsson et al., 2009; Gregory, 2015). It improves the experience of all its stakeholders, including customers, retailers, and suppliers (McFarlane and Sheffi, 2003; Patil, 2016).

3.3.3   INNOVATION IN RETAIL INDUSTRY

3.3.3.1   IoT Store Layout and Customer Tracking

The IoT helps optimize store layout. Sensors can track the movement and behavior of customers inside the store. Heat sensors are used by companies like Hugo Boss to track the movement of the customer, and based on that, changes are being made by placing the products of premium pricing at crowded places. Also, the products are placed in places where they can be easily picked up (Gregory, 2015). Companies can check customer interest in various products by calculating how much time customers spend interacting with the product. The customer and product interaction was measured with the help of IoT. With the help of the radio frequency identification technique, all the products are given a digital identity. The reading time was measured through RFID. Also, Light Dependent Resistor sensors are placed behind each product, and when the product is lifted, a light is received by the sensor. Through this, it can be judged how many times a particular product is liked and picked by the customer. This helps managers to access and prevent the products from going out of stock.

3.3.3.2   Automated Checkout

Checkout is one of the most time-consuming and tedious tasks which is done by the customer. The customer has to wait for a very long period of time at the checkout counter. IoT will solve this problem by scanning all the RFID-enabled products from the shopping cart and charging the sales to the customer through their paying account via their mobile phones. This will help the customers to leave without any hassle. It will save a lot of time for the company. Also, it will reduce labor costs as it will replace cashier staff. The labor cost might be reduced to 75% by 2025. Disney is already using MagicBand, which uses RFID tags to check in for attractions, access to their hotel rooms, food, and merchandise, as well as cash and card-free payment. Disney also keeps track of the activity of the customers and how they are using certain Disney services.

3.3.3.2.1 IoT-Based Store Mapping

One of the customers' biggest frustrations is that they cannot find the specific product in the retail store. They have to search for the given product and sometimes get fed up with searching and give up. Also, they might search for the product from different other stores. Through IoT, companies put RFID tags on their products so they can be easily mapped. This increases the overall shopping experience and customer loyalty. SmartMart IoT-based mobile phone application helps find the product easily as it is able to track and locate products through its indoor positioning system. The SmartMart app is linked with the RFID tags and, thus, helps with easy navigation. When the grocery list is uploaded, this application even helps in finding the shortest path to reach the product as it will take the shortest time.

3.3.3.3 IoT Shelf Availability

In retail, it's very important for the companies to display all the products on the shelves. On-shelf availability is the availability of the product on the shelf according to the customer's desire. On-shelf availability is essential as it promotes sales. No product should be out of stock at a given point in time, as it creates customer dissatisfaction and reduces sales. The customer might buy the product from another retailer or might not buy the product at all. On average, 4% percent of the sales decreased due to the unavailability of the out-of-stock products. The products are not on the shelves mainly for two reasons: Firstly, the store manager is not able to forecast product demand, and secondly, the products are kept in the backrooms of the store.

In manual shift audit, the employee checks the shelves and identifies the out-of-stock products, whereas in the point of-sale model, the number of products on each shelve is identified. Both methods are fine, but the method used by IoT is completely different.

3.3.4 Real World Use Cases of IoT in Retail

3.3.4.1 Amazon Go

This smart store has automatic checkpoints and no human intervention. It was started in 2016 and is based in the US. Amazon- Go works because of machine learning, IoT sensors, deep learning, and methodologies related to computer vision. It uses the latest innovations related to artificial intelligence, IoT, integrated systems of payment, QR code recognition, GPS, etc. Amazon has around 100 cameras, and Amazon Go can easily identify and remember each thing without fitting chips to every product.

3.3.4.2 Zippin

Zippin is a budding start-up in San Francisco. It is a retail store with checkout points, which are automated in nature. It minimizes the time and effort and increases comfort. Also, it uses smart cameras and AI-enabled software to track the customers in their store.

3.3.4.3 Watasale

Watasale was established in 2018, and it was one of these computerized retail stores in India. The creators were well-versed in AI and deep learning and provided problem-free services to the customer in terms of customer experience. The smart retail store enhances the customer experience of shopping without waiting in line. The customers will receive the bill on their smartphones at the end of their shopping journey. This bill can be settled with a debit card or mobile wallet.

3.3.5 ROLE OF IoT IN ENABLING SMART RETAILING EXPERIENCES

IoT works with issues which are customer centric as well with internal tasks related to the organizations. It helps in personalization, tracking of inventory, personalized recommendations as well as monitoring. Real-time inventory information is provided in order to understand the customer demands and upgrade inventories. This helps in increasing the level of productivity, speeding up work and avoids threat. The role of IoT is explained in Figure 3.5.

IoT engages the customer throughout the lifecycle of the product. Through IoT various objects which are smart in nature can interact with smart phones in order to interpret data which is biometric. A satisfactory IoT design is required for life cycle support so that productivity and secure handling of information can be done.

Retailers use IoT for various purposes like managing the inventory, managing the hardware, tracking various products and engaging with the customer.

The model given below focuses on support of the customer, satisfaction related to purchase and customer experience.

With the help of the IoT platform sensory measurements of the natural conditions are taken.These are further utilized by the Building Management system to change controls to work in biological conditions at the retail store. An appropriate relation between business performance and sensors should be established for the same. This is explained in Table 3.1.

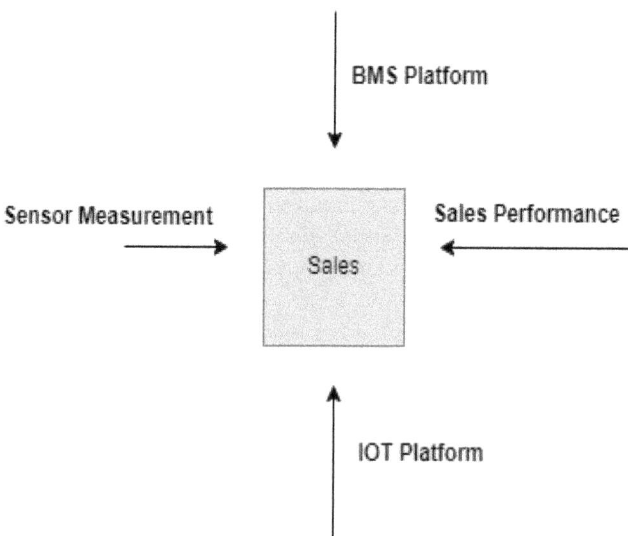

FIGURE 3.5   The representation of IDEF0 model of sales activity in smart retailing.

TABLE 3.1

Demand-Side and Supply-Side of IoT Smart Objects

Demand		Supply	
High Throughput	Camera Networks	High Density	Passive RFID tags
Medium Throughput	Smartphones	Medium density	Bluetooth, Wifi
Low Throughput	Smartcard chips	Low density	GPS services

Role of IoT in Sustainability in the firm

Role of IoT is explained in Figure 3.6. IoT has a significant role to play in improving the environmental performance by using the environmental friendly business practices which stimulates the economic performance of the organization as a whole. IoT is used for environmental monitoring, purchasing of products from various sources, selling the products from various stores, logistics, Agriculture Industry, Automative Industry and Manufacturing Industry.

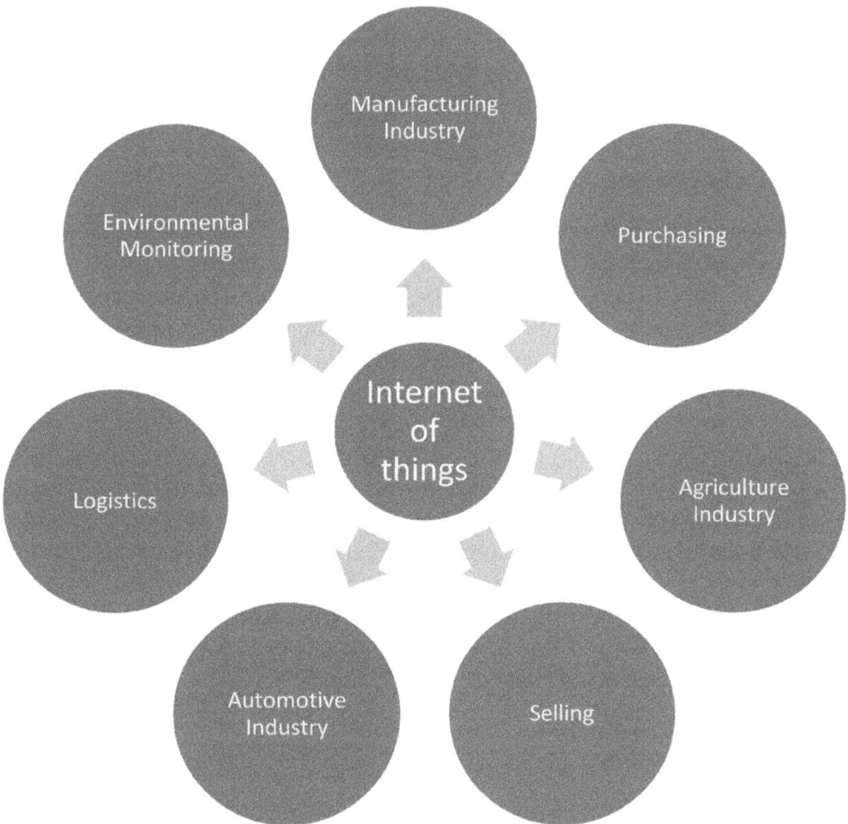

FIGURE 3.6 Role of IoT in sustainability in the firm.

3.3.6 CHALLENGES OF IoT IN RETAIL INDUSTRY

3.3.6.1 Millennials

The spending patterns of the millennials is increasing and is currently more than 1.4 trillion US dollars. This age group of people have different choices and spending patterns. So, it becomes difficult for the companies to provide data-driven and consistent customer experience to the millennials. As the population of the millennials is more, so the companies who are not able to focus on the millennials will be left behind.

3.3.6.2 Omnichannel Reach

As the online retailers have started the trend of buying a particular product from a place return it quickly and easily it is very difficult for all the retailers to match up with the trend. So, now most of the retailers have opted for omnichannel retailing. Many retailers have stated that satisfaction related to omnichannel retail is very difficult and stated that there is a lot of competition from other online and offline retailers for same day delivery.

Retailers need to coordinate their supply chains with the warehouse and then straight to the client. Nowadays, the retailers need to be updated with different designs of varied color and sizes. Retailers who are not equipped with latest technology which is used for understanding the needs and requirements of the customer will not be able to survive in the market for a longer period of time.

3.4 CONCLUSION

IoT implementation in retail has shown promising results. Cashless checkout systems using IoT-enabled payment devices have simplified transactions and reduced queues. Real-time inventory management through IoT sensors has improved stock accuracy and replenishment efficiency. Personalized marketing campaigns based on IoT data analysis have enhanced customer engagement and loyalty. Overall, IoT in retail has increased operational efficiency, customer satisfaction, and revenue for businesses.

In conclusion, the integration of IoT in retail has revolutionized the industry, offering numerous benefits to retailers and customers alike. By leveraging IoT tools and technologies, retailers can enhance operational efficiency, optimize store management, and deliver personalized experiences to customers. However, careful consideration must be given to data privacy and security issues, along with the necessary infrastructure and workforce readiness, to fully harness the potential of IoT in retail. Also, IoT in marketing has been highly imperative because IoT tracks the needs and preferences of its customers and provides insights of their purchase behavior. This helps in understanding the customer.

3.5 IMPLICATIONS

The implications of IoT in retail and marketing are significant. Retailers can gain valuable insights into consumer behavior, preferences, and buying patterns through IoT-generated data. This data-driven approach enables targeted marketing strategies, efficient inventory management, and proactive customer service. Marketers can get the data regarding the interest shown by the customers for their products and can use this to create meaningful information related to consumer buying behavior. Moreover, IoT in retail opens up new opportunities for innovation and business growth. However, there are various challenges to consider, such as privacy of data, security concerns, infrastructure which is robust and workforce which is skilled to support IoT implementation.

BIBLIOGRAPHY

Baranwal, T., & Pateriya, P. K. (2016, January). Development of IoT based smart security and monitoring devices for agriculture. In *2016 6th International Conference-Cloud System and Big Data Engineering (Confluence)* (pp. 597–602). IEEE.

Ben-Daya, M., Hassini, E., & Bahroun, Z. (2019). Internet of things and supply chain management: A literature review. *International Journal of Production Research, 57*(15–16), 4719–4742.

Brundtland Report (The World Commission on Environment and Development) (1987). *Our Common Future.* Oxford, UK: Oxford University Press.

Brynjolfsson, E., Hu, Y., & Rahman, M. S. (2009). Battle of the retail channels: How product selection and geography drive cross-channel competition. *Management Science, 55*(11), 1755–1765.

Chamorro, A., Rubio, S., & Miranda, F. J. (2009). Characteristics of research on green marketing. *Business Strategy Environ, 18,* 223–239. [Google Scholar]

Dangelico, R. M., & Vocalelli, D. (2017). "Green marketing": An analysis of definitions, strategy steps, and tools through a systematic review of the literature. *Journal of Cleaner Production, 165,* 1263–1279. [Google Scholar]

De Graaf, J., Wann, D., & Naylor, T. H. (2005). *Affluenza: The All Consuming Epidemic* (2nd ed.). San Francisco, CA: Berrett-Koehler.

Fredette, J., Marom, R., Steiner, K., & Witters, L. (2012). The promise and peril of 0hyperconnectivity for organizations and societies. *The Global Information Technology Report, 2012,* 113–119.

Gregory, J. (2015). The internet of things: Revolutionizing the retail industry. *Accenture Strategy,* 1–8.

Hasan, M., Nekmahmud, M., Yajuan, L., & Patwary, M. A. (2019). Green business value chain: A systematic review. *Sustainable Production and Consumption, 20,* 326–339. [Google Scholar]

Headey, D. D., & Hoddinott, J. (2016). Agriculture, nutrition and the green revolution in Bangladesh. *Agricultural Systems, 149,* 122–131. [Google Scholar]

Hsu, C. W., & Yeh, C. C. (2017). Understanding the factors affecting the adoption of the internet of things. *Technology Analysis & Strategic Management, 29*(9), 1089–1102. https://data-flair.training/blogs/how-iot-works

James, O. (2007) *Affluenza.* London: Vermilion.

Kassaye, W. W. (2001). Green dilemma. *Marketing Intelligence & Planning, 19,* 444–455. [Google Scholar]

Kotler, P., & Lee, N. (2008) *Social Marketing: Influencing Behaviours for Good.* New York: SAGE.

Lam, J. S. L., & Li, K. X. (2019). Green port marketing for sustainable growth and development. *Transport Policy, 84,* 73–81. [Google Scholar]

Li, Y., Liu, H., Lim, E. T., Goh, J. M., Yang, F., & Lee, M. K. (2018). Customer's reaction to cross-channel integration in omnichannel retailing: The mediating roles of retailer uncertainty, identity attractiveness, and switching costs. *Decision Support Systems, 109,* 50–60.

Martínez, M. P., Cremasco, C. P., Filho, L. G., Junior, S. S. B., Bednaski, A. V., Quevedo-Silva, F., Correa, C. M., Da Silva, D., Padgett, R. C. M. L., & Gabriel, C. P. C. (2020). Fuzzy inference system to study the behavior of the green consumer facing the perception of greenwashing. *Journal of Cleaner Production, 242*, 116064. [Google Scholar]

McFarlane, D., & Sheffi, Y. (2003). *The Impact of Automatic Identification on Supply Chain Operations* (pp. 1–27). Cambridge, UK: University of Cambridge, Department of Engineering.

Mostafa, M. M. (2007). A hierarchical analysis of the green consciousness of the Egyptian consumer. *Psychology & Marketing, 24*, 445–473. [Google Scholar]

Ng, I., Scharf, K., Pogrebna, G., & Maull, R. (2015). Contextual variety, internet-of-things and the choice of tailoring over platform: Mass customisation strategy in supply chain management. *International Journal of Production Economics, 159*, 76–87.

Ng, I. C., & Wakenshaw, S. Y. (2017). The internet-of-things: Review and research directions. *International Journal of Research in Marketing, 34*(1), 3–21.

Patil, K. (2016, December). Retail adoption of Internet of Things: Applying TAM model. In *2016 International conference on computing, analytics and security trends (CAST)* (pp. 404–409). IEEE.

Peattie, K. (2001). Towards sustainability: The third age of green marketing. *The Marketing Review, 2*, 129–146. [Google Scholar]

Polonsky, M. J. (2011). Transformative green marketing: Impediments and opportunities. *Journal of Business Research, 64*, 1311–1319. [Google Scholar]

Rong, W., Vanan, G.T., & Phillips, M. (2016). The internet of things (IoT) and transformation of the smart factory. In International Electronics Symposium (IES), 399–400.

Roy, S. K., Balaji, M. S., Quazi, A., & Quaddus, M. (2018). Predictors of customer acceptance of and resistance to smart technologies in the retail sector. *Journal of Retailing and Consumer Services, 42*, 147–160.

Saha, M., & Darnton, G. (2005). Green companies or green con-panies: Are companies really green, or are they pretending to be? *Business and Society Review, 110*(2), 117–157.

Sana, S. S. (2020). Price competition between green and non green products under corporate social responsible firm. *Journal of Retailing and Consumer Services, 55*, 102118. [Google Scholar]

Stefko, R., Frankovsky, M., & Fedorko, R. (Eds.) (2016). Presov, Slovakia: University of Prešov, pp. 433–439. [Google Scholar]

Vafaei, S. A. Görgényi-Hegyes, E., & Fekete-Farkas, M. The role of social media and marketing in building sustainability orientation. In *Management 2016: International Business and Management, Domestic Particularities and Emerging Markets in the Light of Research.*

Verma, K., Bhardwaj, S., Arya, R., Islam, U. L., Bhushan, M., Kumar, A., & Samant, P. (2019). Latest tools for data mining and machine learning.

Welford, R. (2000). *Hijacking Environmentalism.* London, UK: Earthscan. [Google Scholar]

Yang, L. T., Di Martino, B., & Zhang, Q. (2017). Internet of everything (editorial). *Mobile Information Systems, 2017*, 1–3.

Zhang, X., Rane, K., Kakaravada, I., & Shabaz, M. (2021). Research on vibration monitoring and fault diagnosis of rotating machinery based on internet of things technology. *Nonlinear Engineering, 10*(1), 245–254. https://doi.org/10.1515/nleng-2021-0019

4 Embedded IoT, Cloud, and Deep Neural Network-based Stress Prediction System Using Facial Expressions

Maneet Kaur Bohmrah and Harjot Kaur

4.1 INTRODUCTION: BACKGROUND AND DRIVING FORCES

With the current pandemic and other changes in cultural and social behaviors, stress, one of the emotions, plays a significant role in the health issues of humanity. The degree of stress might rise at times, which can lead to a variety of health issues. In order to explore various aspects of emotion recognition with reference to stress, multiple ML-based algorithms have been used in the past few years. On the other hand, deep learning-based approaches have recently become more well-liked because of their excellent performance and accuracy: Deep Neural Networks (DNN) perform better in image processing and are also used to detect stress using image data. The discipline of human emotion recognition has been intensively studied for years. Intelligent systems have been long anticipated and predicted to develop to mimic human behavior without the aid of humans. According to Duffy et al. [1], deploying a machine that would be as clever as humans and able to communicate and interact with their surrounding environment as people do, and such computational mechanism will be possible due to artificial neural networks (ANN). How the human brain functions and how it can be utilized to categorize face expressions, to some extent, as a specific emotion in visuals [2, 3].

The impact of artificial intelligence (AI) on healthcare has improved as a result of technological advancements [4]. AI has made it simpler for users to diagnose symptoms. As a result, a disease can be detected at an early stage, and, hence, treatment will be provided at the initial stage where the probability of health recovery is increased [5]. Identifying the trends in the patient's prior health data, ML-based models can identify or classify the disease instantly [6]. Additionally, proper doctor's assistance will be given to the patients to improve their lifestyle and benefit healthcare providers in understanding the typical causes of the disease [5]. Large amounts of data pertaining to stressed patients are produced and kept in open databases [6]. Thus, in order to accurately anticipate heart illnesses, a creative solution combining cloud computing and machine learning is required [4]. Early diagnosis can assist patients in altering their lifestyles to prevent disease and warn patients if there will be any major concerns [7].

ANNs are computing techniques that aim to simulate, to some extent, how the brain of humans being functions to characterize images of facial expressions which belong to the specific category of emotion [2, 3]. Self-organizing maps [8], SVM [9], CNN [10], and other variations of this technique have all been successful at classifying facial expression images (CNN) [11].

It is observed that most of the authors used the KDEF dataset [12] consisting of face expression images to classify human emotions using hybrid SVM and Multilayer Perceptron (MLP). These approaches rely on Gabor filter-based feature selection that is hard-coded. In this article, we investigated a different strategy that took into account the aforementioned method by swapping the Gabor filters with a approach that provides a better feature selection technique based on self-learning by employing a deep CNN, which is a core assertion of this paper.

numerous traditional approaches are used to extract facial features, including geometric and texture aspects such as self-organizing maps and Gabor wavelets. Although it provides good accuracy, it is not clear what features determine the performance of classification. On the other hand, the CNN and the RNN enable the automatic extraction of features and classification, and deep learning has recently proven to be a very effective and successful approach [13]. This led researchers to begin applying this method to recognize human emotions. Researchers put a lot of effort into creating deep neural network designs, and the outcomes are extremely good in this field.

One major consideration in effective recognition is the feature extraction system. It demands that features maintain invariant properties within the same class while having the most distinct qualities across other classes. However, automatic extraction approaches may directly and effectively extract the characteristics from unprocessed pictures, the manual extraction of features is quite difficult, complicated, and even time-consuming.

Performance is much better when integrated ML and Internet of Things (IoT) cloud-based applications are deployed. This design can be employed in various emerging applications in education, military, health care, and business. IoT-based cloud technologies, in particular, are widely applicable in the field of healthcare. It can be used to monitor and access patient records without geographical limitations. IoT-based healthcare systems are great for gathering important information, such as regular updates to vital health metrics and updating the intensity of medical parameters over a predictable time period. Additionally, its sensor readings connected to medical parameters are successfully employed to diagnose disease at the right moment and before grave circumstances [14]. In the process of making decisions, machine learning is crucial. It can also manage large amounts of data. Data might be produced from a variety of sources, so it's crucial to analyze the system and create techniques for managing the data [15, 16]. The previous work has mostly served to build a framework for collaboration between the healthcare and computing technology domains, particularly the IoT. However, it is not suitable to employ robust computer science principles in these procedures. Consider a situation where no medical specialist is on hand to assist the patient. Or, in a different circumstance, a medical misdiagnosis may be made by the medical expert. Utilizing machine learning in these situations appears to be a wise choice. It is helpful to have a computer expert on hand to diagnose patients' issues [17, 18]. IoT devices generate a huge amount of data, and this data belongs to real-time. Hence, machine learning algorithms are used to analyze this real-time data used for predictions and further generate relevant decisions. In this paper, a large amount of data was collected in the form of images, and further, this image data is classified into seven prescribed emotions using ML and DL models [19].

In this work, a hybridization of ML and DL approach is used to classify human emotions, and IoT- an IoT-based sensor monitors the facial expressions. If some inevitable expressions are recorded during a specified time period, then a notification to the user as well as to the concerned health expert is automatically sent by the smart device. CNN model is used for the extraction of multiple features from the input images, and different ML classifiers are used to classify the images from the dataset. In other words, to recognize human emotions efficiently, we proposed a mixed CNN-SVM approach. The main innovation and contribution of this paper is structured as follows:

- Framework proposed for stress prediction and diagnosis using deep neural network, machine learning, and cloud computing.
- Feature extraction done using deep neural network.
- Extracted features used for further classification, which is implemented by ML algorithms.
- Different classifiers were implemented, and it was observed from the experiments a combination of CNN and SVM provides better accuracy.
- Real-time data analytics done using Matlab Mobile.

The remaining part of the article is structured as follows: Section 2 discusses the literature survey, and Section 3 illustrates the methodology used in this paper. Furthermore, Section 4 presents the experimental setup and results, discussion, and comparisons described in Section 5. Finally, Section 6 draws the conclusion.

4.2 BACKGROUND AND LITERATURE REVIEW

In this article, we concentrate on the study of creating an emotion detection model that can function in a real-world setting. The previous research done on recognizing emotions is reviewed in this section, along with recent developments in the field of machine intelligence. In an article proposed by Sohail et al. [10], SVM is used to classify facial expressions. The authors suggested a technique that involves counting 15 distinct features and finding dissimilarity between these features with the features of neutral expression using the Euclidean distance method. Paul et al. [20] suggested a model with average recognition rates using an SVM for classification values of 92 and 86.33. These results show that SVMs are effective at recognizing emotions from facial expressions when used as a classifier. SVMs have been used for a variety of classification tasks, including face formalization and recognition issues [21, 22].

MLPs have also shown to be effective at classifying facial expressions. An MLP was created by Hewahi et al. [23] with an accuracy rate of 83.3 thanks to the usage of ethnic background data. The authors find that a similar approach on the MSDEF dataset performs at a performance rate of 75 when no ethnic background detail is included. Implementing the image pre-processing using global pattern averaging and an MLP classifier for face emotion recognition, Khashman [24] attained 87.78% accuracy.

CNNs are now being considered for difficulties involving face expression categorization due to their success with image classification issues. In the paper [11], the author implemented the five-layer convolutional network and SVM model. The author achieves an accuracy rate of 94.4% on the Extended Cohn-Kanade Dataset. A benchmark with 99.6% accuracy was established by Burkert et al. [25] using seven layers convolutional with 10-fold cross-validation on the CK dataset.

There are a lot of discussions on the selection of features. Most of the models covered in this study provide a high level of accuracy; it is not obvious which characteristics affect classification performance. The lips are more important in recognizing happiness, while the eyes and brows are more important for recognizing despair, based on an observation by Beaudry et al. [26]. On the other hand, the authors found that fear could benefit from a holistic approach but could not determine other emotions. Determining which face characteristics define particular emotions could be very useful when used with emotion identification models. In this work, we attempt to identify facial expression pictures by emphasizing certain points of interest such as lips, eyes, and brows. To emphasize these features in an image, CNN is used. In the further section, feature extraction techniques and classification models are discussed.

Sequential forward selection (SFS) is a feature selection method used in an IoT-based hybrid system for predicting cardiovascular disease, using a random forest to classify data. This technique gives patients nutrition and exercise recommendations depending on age and gender in relative and demonstrated 98% accuracy compared with other heuristics models [27]. In the article [28], data collected from multiple sensors processes and classified using kernel random forest achieves an accuracy of 98%. Another IoT and deep CNN framework was proposed in [29] in which information regarding blood pressure and ECG was collected using wearable sensors and attained 98.2% accuracy compared to logistic regression. Based on the ensemble, the deep learning model Logit boost classifier and feature fusion technique was deployed which predicts the risk of heart attacks [30] using data collected by wearable sensors and the patient's medical backgrounds. The proposed framework achieves an accuracy of 98.5% and automatically suggests food regimens based on the patient's health.

4.3 MATERIALS AND METHOD

In this section, we discussed the techniques used to develop a stress management system using IoT, Cloud, Machine, and deep learning approaches. In this research, a stress management system is proposed using the image data. The user simply clicks their photo by using the proposed application, and our algorithm predicts the emotion among the six emotions (anger, disgust, fear, sad, surprise, and happiness) defined in the system. The label of predicted emotion is stored, and, on the basis of this predicted label, stress is identified. If the continuous label of emotion is sad or angry, or fearful, then the probability of having stress is increased, and in this case, a notification is sent to the user. The proposed methodology can be explained in the following steps:

Steps Involved in Proposed Methodology

1. Real-time image data is collected using the camera of mobile through the Matlab Mobile application.
2. Matlab Mobile application is connected with the Mathworks Cloud, where the classification of emotions of real-time images is done.
3. Classification of image achieved by using DNN and machine learning SVM algorithm is used.
4. The proposed model is trained and tested using the facial expression dataset prior.
5. A ready-to-use model predicts the emotion of the input image, and the output label is saved.
6. On the basis of the output emotion label, the stress is identified and notified to the user.

The overall schema of the proposed methodology is explained in Figure 4.1. Matlab Mobile and Mathworks Cloud are available free of cost. For designing and

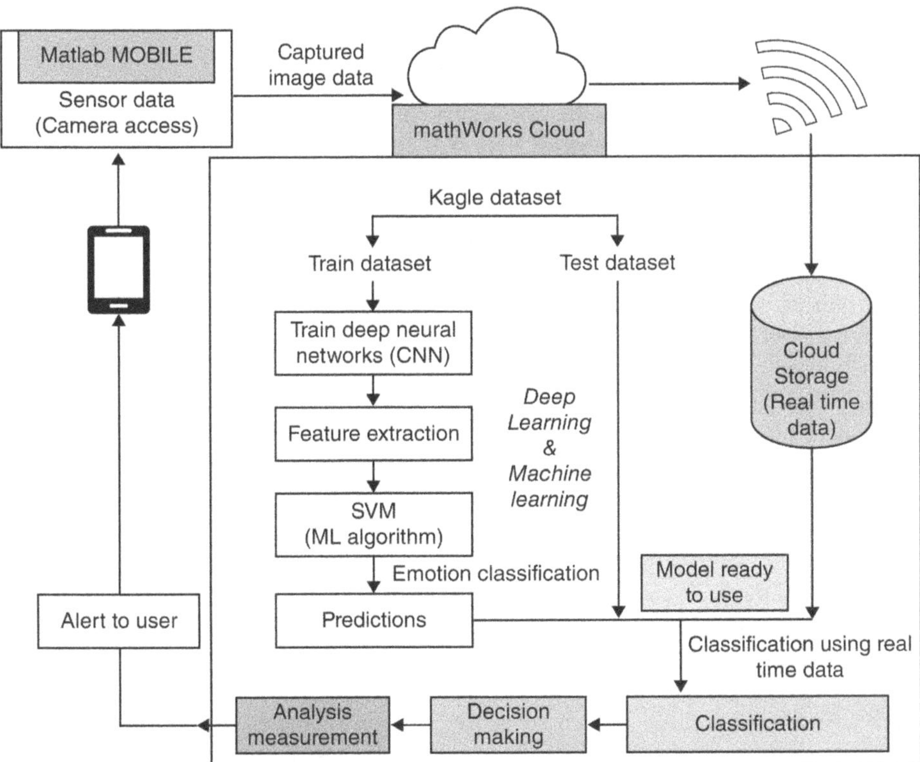

FIGURE 4.1 The complete schema of the proposed methodology.

testing purposes, we proposed this platform. Further, this work can be extended with the introduction of more sensors for testing blood pressure and heartbeat in combination with the existing system. In addition, Matlab code can be converted to the Android application using Simulink Support Packages. The further paragraphs explain the architecture of IoT and deep learning framework for stress prediction on the basis of emotion classification.

The SVM and CNN classifiers were used to develop the proposed system. In Section 3.1, we will provide a brief introduction to the SVM theory and the CNN structure in Section 3.2. The hybrid CNN- SVM trainable feature extractor model will then be given in Section 3.3, with a discussion of its advantages and disadvantages.

4.3.1 MACHINE LEARNING ALGORITHMS- SVM CLASSIFIER

Finding the ideal hyperplane that separates all of the data points in one class from those in the other is how SVM categorizes the data into different groups. The hyperplane with the biggest margin between the two classes is the best one for SVM. The slab's maximal margin is parallel to the hyperplane but has no interior data points [31]. The data points on the slab's edge, which are near the separating hyperplane, are known as support vectors. Hence, the concept of SVM where positive data points are denoted by + or first class and − for negative data points or second class.

4.3.1.1 Convolutional Neural Networks

A deep neural network with a number of convolutional layers known as a Convolutional Neural Network (CNN) [17] can be seen as being composed of an automatic classifier and feature extractor. The classifier model is used to classify the data, and the feature extractor extracts distinguishing characteristics of images mainly using two procedures from the image data: Filtering and down sampling. The best image pre-processing technique must be employed to prepare the images used for training, as the effectiveness of classifier techniques significantly depends on the effectiveness of the feature vector describing the image. Gabor filter is one of the widely used techniques in image processing because it can recognize edges. In this article, we extract features using CNN and, hence, replace Gabor filters.

Gabor filters are required to be efficiently tailored to provide true image representations and enable the identification of boundaries and, consequently, face features that are crucial for emotion identification. Convolutional Neural Networks, on the other hand, provide another established technique that has the capacity to autonomously collect the characteristics required for emotion recognition. The architecture of the CNN model is well explained in Figure 4.2.

Convolutional, MaxPooling, ReLU, and batch normalization layers make up the building blocks of the CNN emotion recognition model [19]. Batch normalization enables higher learning rates and quicker convergence. During training, batch normalization has been proven to accelerate convergence by decreasing internal co-variate shifts or the shift in how activations across networks are distributed as changes in the parameters of the network are always there [20]. This normalization is carried out for every distinct feature map in the case of convolutional layers.

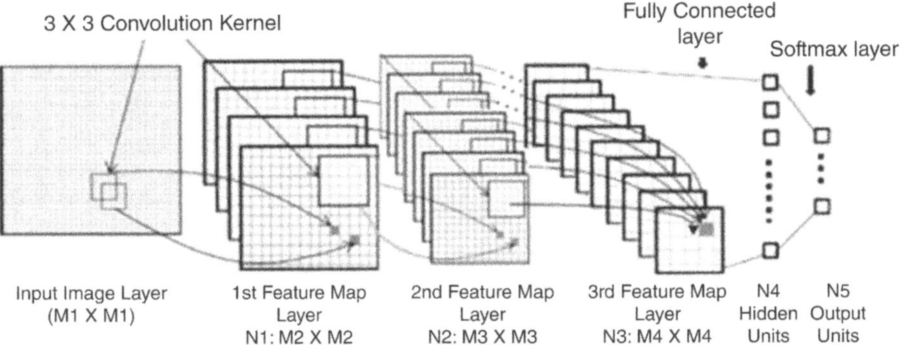

FIGURE 4.2   Basics of CNN architecture.

The three layers CNN used in this experiment and its constitution are as follows: Filters of size 3 × 3 used and a different number of filters used for each layer. A sliding window with size one and zero padding is used in all the convolutional layers. Max pooling with a stride of size two and ReLU activation function is used. Using larger filters seemed unnecessary as the image size is only 48 × 48, and larger filters often miss relevant values. The Final Softmax layer is also termed as the output layer connected with the fully connected layer and further connected with the last block, which is the output layer. The output of each layer is also normalized using batch normalization. Figure 4.2 illustrates a presentation of this approach using CNN as the feature extractor as well as the classifier.

4.3.1.2   Hybrid SVM and CNN Approach

The construction of the proposed framework is done with the extraction of features from the last output layer, called the Softmax layer, and extracted features are treated as input to the SVM model for further classification of the emotion dataset. An activation function calculates each output's probability. The weighted sum of the outputs from the preceding hidden layer with trainable weights, along with a bias term, makes up the input of the activation function. Viewing the hidden layer's output value has no meaning outside of the context of the CNN network; however, these values can be used as features by other classifiers. Figure 4.3 represents the concept of hybridization of CNN and SVM. Prior to training, the normalized images are given to the input layer, and the original CNN with the output layer is constrained throughout numerous epochs.

4.3.1.3   Advantages of Hybrid SVM-CNN Classifier

We anticipate that our innovative proposed model will outperform each individual classifier since it compensates for the limitations of CNN and SVM classifiers by combining their strengths. It is an enhanced model of the MLP because CNN acquires the same theoretical process as the MLP. In our experience, the training phase ends when the first separating hyperplane is discovered by the back-propagation algorithm, regardless of whether it is the local or global minima. The algorithm does not maintain its separation hyperplane improvement solution.

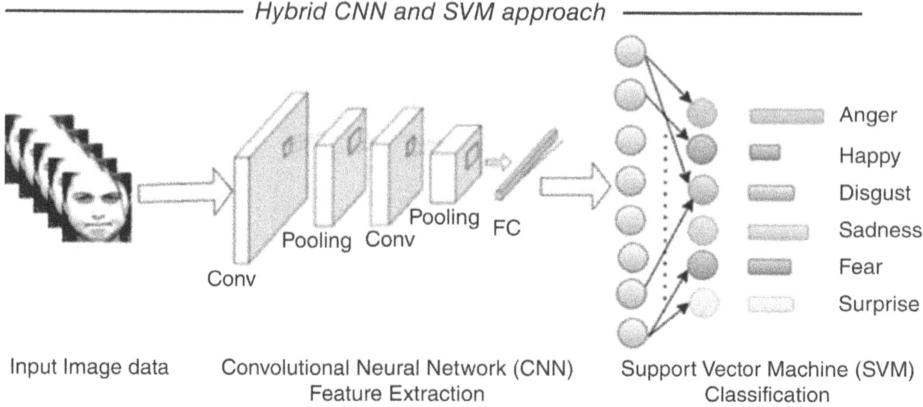

FIGURE 4.3   Proposed methodology for integration of ML and DL.

4.4   EXPERIMENTAL SETUP, RESULT, AND DISCUSSION

This research aims to create a novel framework for stress prediction based on an emotion identification that will be able to recognize human emotions in real time. On the basis of the accurate detection of emotion, stress was identified. We have created a series of emotion detection models using CNN for feature representation and SVMs as classifiers. In this paper, the FER 2013 human emotion database was used, which is available on the Kaggle repository. More than 26000 images of different emotions such as anger, disgust, fear, happiness, sad, neutral, and surprise were present in this database. We skipped the disgust category because it contains less number of images as compared to other categories. Figure 4.4 displays a few images of the human emotion database. Further, on the basis of the following below-mentioned performance metrics, the models were evaluated.

4.4.1   PERFORMANCE METRICS

The proposed model is finally evaluated based on some standard classification metrics such as accuracy, precision, recall, sensitivity, specificity, and F-Score. The following performance was used further to evaluate and compare the ML classifiers.

$$\text{a. } Accuracy = \frac{TP + TN}{TP + FP + TN + FN}$$

$$\text{b. } Precision = \frac{TP}{TP + FP}$$

$$\text{c. } Recall = \frac{TP}{TP + FN}$$

$$\text{d. } Sensitivity = \frac{TP}{TP + FN}$$

FIGURE 4.4 Sample dataset of human emotion.

e. $Specificity = \dfrac{TN}{TN + FP}$

f. $F - Score = \dfrac{2*TP}{2*TP + FP + FN}$

Where TP=True positive, TN=True Negative, FP=False Positive and FN=False Negative. It is observed from Table 4.1 that the accuracy for CNN+SVM is higher as compared to other classifiers.

TABLE 4.1

Performance Evaluation of Various ML Classifiers on Different Performance Metrics

Model	Accuracy	Precision	Recall	F1-Score
CNN	0.8977	0.9058	0.9015	0.9036
CNN-DT	0.8939	0.8955	0.8963	0.8959
CNN-NB	0.8977	0.8981	0.8983	0.8982
CNN-KNN	0.9205	0.9201	0.9225	0.9213
CNN-SVM	0.9242	0.924	0.9268	0.9254

Our proposed system, which includes three convolutional layers, is trained using Adam optimizer and use batch normalization for faster convergence. For the CNN itself, it acts as a classifier, too; the noted accuracy is 89.77%; for CNN + Decision tree it is 89.39%; CNN + Naive Bayes accuracy is 89.77%, and CNN + K-Nearest Neighbor accuracy is 92.05%. However, a combination of CNN + SVM gives an accuracy rate of 92.42%. Table 4.1 defines the performance of all the used methods with overall accuracy. Figure 4.5 defines the confusion matrix of various implemented models, and they are CNN, CNN + Decision Tree, CNN + Naive Bayes, CNN + kNN, and Figure 4.6 depicts the confusion matrix for CNN + SVM, which produced better results as compared to the other ML models.

Receiver Operating Curve, also called ROC curve, is a graph plotted between true positive and false positive rates and defines the performance of classification models. Figure 4.7 represents the ROC curve for various ML models deployed in this work. The proposed model performs better as compared to other ML classifiers shown in Figure 4.8. On the other hand, Figure 4.8 displays the output produced by this model, which classifies the emotions in images with classification or prediction accuracy percentage.

Our findings show that the proposed approach has quite potential for classification techniques due to three vital reasons: One is the key traits can be drawn out

Covid Classification Using CNN

True Class \ Predicted Class	Anger	Disgust	Fear	Happy	Sadness	Surprise		
Anger	29	11		6			63.0%	37.0%
Disgust	3	41					93.2%	6.8%
Fear			37	1			97.4%	2.6%
Happy				41			100.0%	
Sadness			1		42	2	93.3%	6.7%
Surprise		3				47	94.0%	6.0%
	90.6%	74.5%	97.4%	97.6%	87.5%	95.9%		
	9.4%	25.5%	2.6%	2.4%	12.5%	4.1%		

Predicted Class

FIGURE 4.5 Confusion matrix for (a) CNN as a classifier (b) CNN+ decision tree (c) CNN + KNN (d) CNN + naive bayes. *(Continued)*

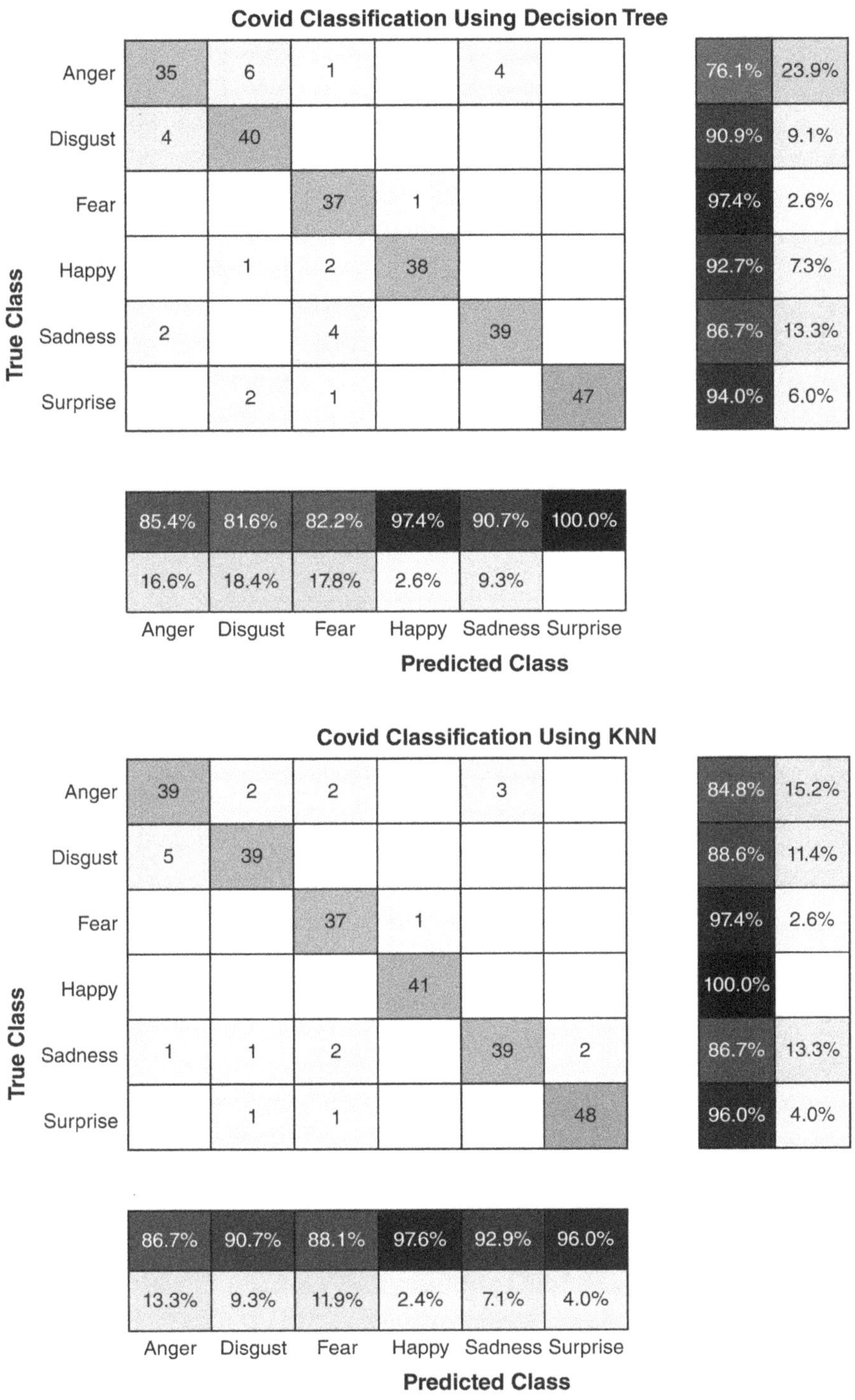

Covid Classification Using Decision Tree

	Anger	Disgust	Fear	Happy	Sadness	Surprise		
Anger	35	6	1		4		76.1%	23.9%
Disgust	4	40					90.9%	9.1%
Fear			37	1			97.4%	2.6%
Happy		1	2	38			92.7%	7.3%
Sadness	2		4		39		86.7%	13.3%
Surprise		2	1			47	94.0%	6.0%
	85.4%	81.6%	82.2%	97.4%	90.7%	100.0%		
	16.6%	18.4%	17.8%	2.6%	9.3%			
	Anger	Disgust	Fear	Happy	Sadness	Surprise		

True Class / Predicted Class

Covid Classification Using KNN

	Anger	Disgust	Fear	Happy	Sadness	Surprise		
Anger	39	2	2		3		84.8%	15.2%
Disgust	5	39					88.6%	11.4%
Fear			37	1			97.4%	2.6%
Happy				41			100.0%	
Sadness	1	1	2		39	2	86.7%	13.3%
Surprise		1	1			48	96.0%	4.0%
	86.7%	90.7%	88.1%	97.6%	92.9%	96.0%		
	13.3%	9.3%	11.9%	2.4%	7.1%	4.0%		
	Anger	Disgust	Fear	Happy	Sadness	Surprise		

True Class / Predicted Class

FIGURE 4.5 *(Continued)*

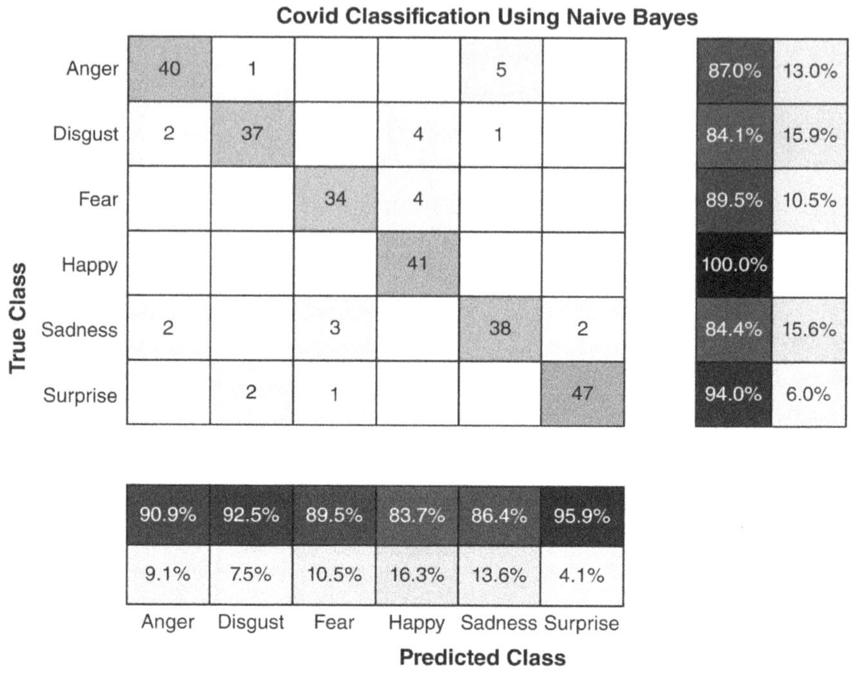

Covid Classification Using Naive Bayes

FIGURE 4.5   *(Continued)*

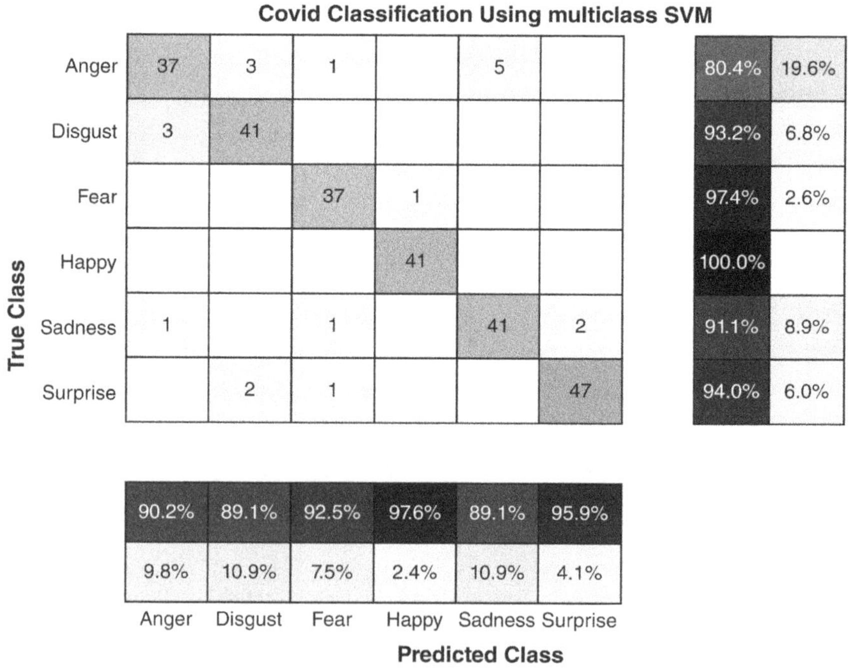

Covid Classification Using multiclass SVM

FIGURE 4.6   Confusion matrix for CNN and SVM.

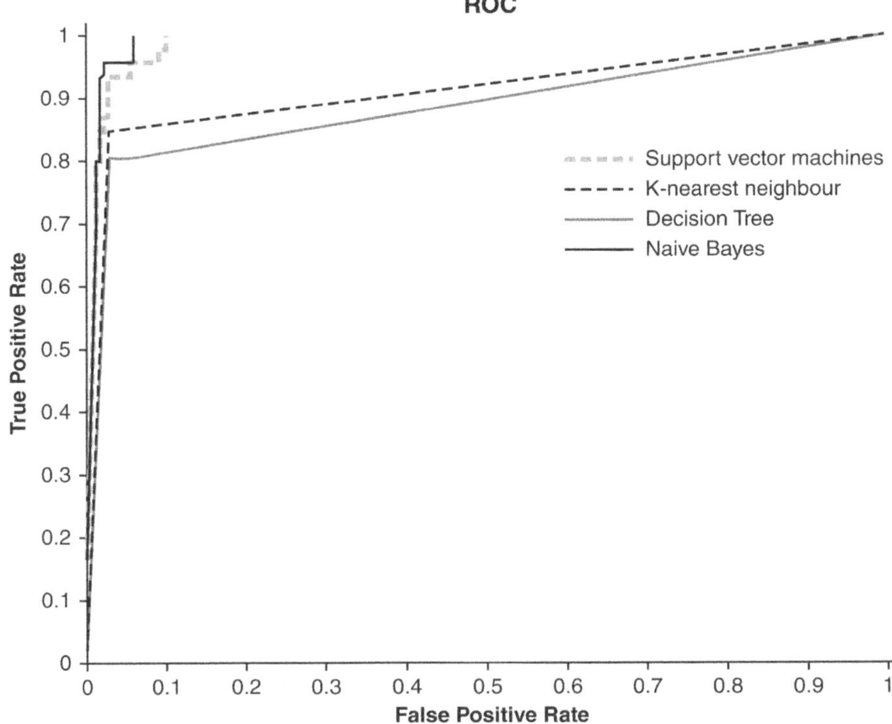

FIGURE 4.7 ROC curve.

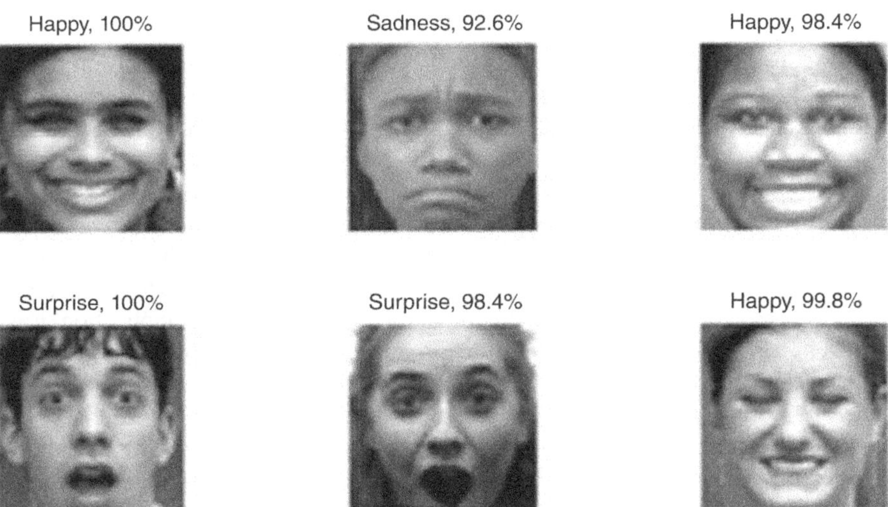

FIGURE 4.8 Emotion predicted by the model with accuracy in percentage.

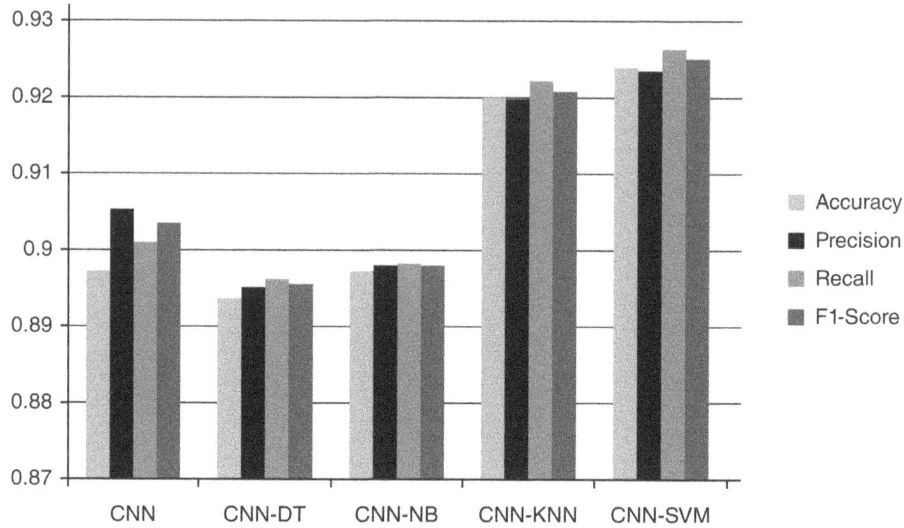

FIGURE 4.9 Comparisons of different ML classifiers used for emotions classification.

as conventional classifiers rely heavily on historical data. The second is that the proposed method merges the benefits of CNN and SVM because both the models excel in their field and are efficient; the third is the intricacy of the proposed method while making decisions slightly improves when comparing our experiments with the other classifiers shown in Figure 4.9. It is advantageous when applied in realistic applications.

4.4.2 Implementation of Matlab MOBILE and MathWorks Cloud

The model was applied in the application in accordance with the decision that the combination of CNN + SVM was the best algorithm for the stress prediction application. In this paper, MATLAB R2020b is used to perform all the experiments. Once the classification algorithm had been used, the test data set's accuracy, precision, recall, and F1-score were supplied, and they were 92.42%, 92.4%, 92.68%, and 92.54%, respectively, which is comparatively more than the other ML classifiers. Using Matlab Mobile, the DNN can be easily accessed with the mobile phones. Matlab Mobile is connected with MathWorks Cloud, which provides 5GB for the backup computations. As soon as the model was included in the application, it was ready for use. Together with some fresh data, 30% of the entire image database was used to test this model to make sure it was accurate and error-prone and was forecasting the desired target. Figure 4.10 represents the sample screenshots of the proposed system. In the future, the Matlab code will be linked with an Android application using the Simulink Support package, which converts a Simulink model to an Android application.

FIGURE 4.10 Sample screenshots of proposed stress management system using Matlab Mobile.

4.5 CONCLUSION

In this paper, a self-diagnosis stress management system is introduced in a user-friendly environment. An IoT-deep learning framework proposed to deploy the self-diagnostic mechanism in the field of healthcare. On the basis of images, emotions can be predicted and, hence, stress was detected. The proposed hybrid CNN-SVM framework, which addresses the issue of human emotion recognition, used CNN as a feature extractor and multi-class class SVM classifier as an output predictor; the suggested method's effectiveness and viability were assessed in two areas: recognition accuracy and experimental findings from the human emotion database as achieved 92.42%. The model is also tested on real-time images and classifies the images with a testing accuracy of 93.2%. The efficiency of the hybrid model can still be enhanced, improved by hyper-tuning its parameters and optimizing its design. Moreover, a fog layer can be used to reduce the computational latency.

4.6 FUTURE WORK

For future work, more efficient DNN can be used for emotion detection, as accuracy in predicting emotion is the foundation of this work. On the basis of predicted emotions, stress can be identified. A ready-to-use model designed in Matlab code can be converted into Android applications. In addition, working and installing other sensors for blood pressure, heart rate, and oxygen level can also be done in the future.

ABBREVIATIONS

ML	Machine learning
DL	Deep learning
CNN	Convolutional Neural Network
SVM	Support Vector Machine
kNN	k-Nearest Neighbor
NB	Naive Bayes
DT	Decision Tree
RNN	Recurrent Neural network
MLP	Multi-layer Perceptron

REFERENCES

1. Brian R Duffy. "Fundamental issues in social robotics." *The International Review of Information Ethics* 6 (2006): 31–36.
2. Boughrara, Hayet, Mohamed Chtourou, Chokri Ben Amar, and Liming Chen. "Facial expression recognition based on a MLP neural network using constructive training algorithm." *Multimedia Tools and Applications* 75 (2016): 709–731.
3. Ebrahimi Kahou, Samira, Vincent Michalski, Kishore Konda, Roland Memisevic, and Christopher Pal. "Recurrent neural networks for emotion recognition in video." In *Proceedings of the 2015 ACM on international conference on multimodal interaction*, pp. 467–474. 2015.

4. Yan, Yang, Jia-Wen Zhang, Guang-Yao Zang, and Jun Pu. "The primary use of artificial intelligence in cardiovascular diseases: What kind of potential role does artificial intelligence play in future medicine?" *Journal of Geriatric Cardiology: JGC* 16, no. 8 (2019): 585.

5. Kishor, Amit, and Chinmay Chakraborty. "Artificial intelligence and internet of things based healthcare 4.0 monitoring system." *Wireless Personal Communications* 127, no. 2 (2022): 1615–1631.

6. Mincholé, Ana, and Blanca Rodriguez. "Artificial intelligence for the electrocardiogram." *Nature Medicine* 25, no. 1 (2019): 22–23.

7. Shrestha, Sirish, and Partho P. Sengupta. "Imaging heart failure with artificial intelligence: Improving the realism of synthetic wisdom." *Circulation: Cardiovascular Imaging* 11, no. 4 (2018): e007723.

8. Gupta, Alka, and Madan Lal Garg. "A human emotion recognition system using supervised self-organising maps." In *2014 International conference on computing for sustainable global development (INDIACom)*, pp. 654–659. IEEE, 2014.

9. K., Sarnarawickrame, and S. Mindya. "Facial expression recognition using active shape models and support vector machines." In *2013 International Conference on advances in ICT for emerging regions (ICTer)*, pp. 51–55. IEEE, 2013.

10. Wu, Dazhong, Connor Jennings, Janis Terpenny, and Soundar Kumara. "Cloud-based machine learning for predictive analytics: Tool wear prediction in milling." In *2016 IEEE International Conference on Big Data (Big Data)*, pp. 2062–2069. IEEE, 2016.

11. Muhammad, Ghulam, SK Md Mizanur Rahman, Abdulhameed Alelaiwi, and Atif Alamri. "Smart health solution integrating IoT and cloud: A case study of voice pathology monitoring." *IEEE Communications Magazine* 55, no. 1 (2017): 69–73.

12. Lundqvist, Daniel, Anders Flykt, and Arne Öhman. "Karolinska directed emotional faces." *PsycTESTS Dataset* 91 (1998): 630.

13. Simard, Patrice Y., David Steinkraus, and John C. Platt. "Best practices for convolutional neural networks applied to visual document analysis." *Icdar* 3, (2003): 1–6.

14. Gelogo, Yvette E., Ha Jin Hwang, and Haeng-Kon Kim. "Internet of things (IoT) framework for u-healthcare system." *International Journal of Smart Home* 9, no. 11 (2015): 323–330.

15. Lee, Hyunsoo. "Framework and development of fault detection classification using IoT device and cloud environment." *Journal of Manufacturing Systems* 43 (2017): 257–270.

16. Chen, Chien-Hung, Che-Rung Lee, and Walter Chen-Hua Lu. "Smart in-car camera system using mobile cloud computing framework for deep learning." *Vehicular Communications* 10 (2017): 84–90.

17. Hossain, M. Shamim, and Ghulam Muhammad. "Cloud-assisted industrial internet of things (iiot)–enabled framework for health monitoring." *Computer Networks* 101 (2016): 192–202.

18. Sohail, Abu Sayeed Md, and Prabir Bhattacharya. "Classifying facial expressions using level set method based lip contour detection and multi-class support vector machines." *International Journal of Pattern Recognition and Artificial Intelligence* 25, no. 06 (2011): 835–862.

19. Kanade, Takeo, Jeffrey F. Cohn, and Yingli Tian. "Comprehensive database for facial expression analysis." In *Proceedings fourth IEEE international conference on automatic face and gesture recognition (cat. No. PR00580)*, pp. 46–53. IEEE, 2000.

20. Paul, Padma Polash, Md Maruf Monwar, Marina L. Gavrilova, and Patrick SP Wang. "Rotation invariant multiview face detection using skin color regressive model and support vector regression." *International Journal of Pattern Recognition and Artificial Intelligence* 24, no. 08 (2010): 1261–1280.

21. Khan, Sajid Ali, Ayyaz Hussain, Muhammad Usman, Muhammad Nazir, Naveed Riaz, and Anwar Majid Mirza. "Robust face recognition using computationally efficient features." *Journal of Intelligent & Fuzzy Systems* 27, no. 6 (2014): 3131–3143.

22. Hassner, Tal, Shai Harel, Eran Paz, and Roee Enbar. "Effective face frontalization in unconstrained images." In *Proceedings of the IEEE conference on computer vision and pattern recognition*, pp. 4295–4304. 2015.
23. Hewahi, Nabil M., and Abdul Rhman M. Baraka. "Impact of ethnic group on human emotion recognition using backpropagation neural network." *BRAIN. Broad Research in Artificial Intelligence and Neuroscience* 2, no. 4 (2011): 20–27.
24. Khashman, Adnan. "Application of an emotional neural network to facial recognition." *Neural Computing and Applications* 18 (2009): 309–320.
25. Burkert, Peter, Felix Trier, Muhammad Zeshan Afzal, Andreas Dengel, and Marcus Liwicki. "Dexpression: Deep convolutional neural network for expression recognition." *arXiv preprint arXiv:1509.05371* (2015).
26. Olivia, Beaudry, Annie Roy-Charland, Melanie Perron, Isabelle Cormier, and Roxane Tapp. "Featural processing in recognition of emotional facial expressions." *Cognition\& Emotion* 28, no. 3 (2014): 416–432.
27. Jabeen, Fouzia, Muazzam Maqsood, Mustansar Ali Ghazanfar, Farhan Aadil, Salabat Khan, Muhammad Fahad Khan, and Irfan Mehmood. "An IoT based efficient hybrid recommender system for cardiovascular disease." *Peer-to-Peer Networking and Applications* 12 (2019): 1263–1276.
28. Muzammal, Muhammad, Romana Talat, Ali Hassan Sodhro, and Sandeep Pirbhulal. "A multi-sensor data fusion enabled ensemble approach for medical data from body sensor networks." *Information Fusion* 53 (2020): 155–164.
29. Khan, Mohammad Ayoub. "An IoT framework for heart disease prediction based on MDCNN classifier." *IEEE Access* 8 (2020): 34717–34727.
30. Ali, Farman, Shaker El-Sappagh, SM Riazul Islam, Daehan Kwak, Amjad Ali, Muhammad Imran, and Kyung-Sup Kwak. "A smart healthcare monitoring system for heart disease prediction based on ensemble deep learning and feature fusion." *Information Fusion* 63 (2020): 208–222.
31. Chih-Chung, Chang. "LIBSVM: A library for support vector machines." *http://www.csie.ntu.edu.tw/cjlin/libsvm* (2001).

5 Cloud IoT in Healthcare Transforming Patient Monitoring and Data Analytics

Nidhi Punj, Anita Tanwar, Priyanka Ranga, and Vinod Kumar

5.1 INTRODUCTION

The continuous revolution of cloud computing and the Internet of Things (IoT) has made remote patient monitoring possible and practical. These interrelated technologies are extensively employed to deliver healthcare services and enable real-time patient monitoring [1]. This integration has ushered in a new epoch of potential, revolutionizing the healthcare industry by enabling seamless patient monitoring and data analytics. The IoT offers numerous advantages in streamlining and improving healthcare delivery, enabling proactive prediction of health issues, efficient diagnosis, treatment, and patient monitoring, both within healthcare facilities and remotely. Governments and decision-makers worldwide recognize the potential of technology, especially in response to the novel COVID-19 pandemic and are implementing policies to integrate IoT-based healthcare services. Understanding the role of established and emerging IoT technologies becomes crucial in supporting health systems to provide safe and effective care to patients [2]. Cloud IoT represents the synergy between IoT devices and cloud computing infrastructure, offering unprecedented opportunities for healthcare applications. IoT devices, including wearable health trackers, smart medical devices, and remote monitoring tools, continuously collect and transmit real-time patient data. These devices have become integral in tracking vital signs, physical activity, and other health-related metrics, providing a wealth of information to healthcare providers. The reach of IoT applications extends to diverse areas such as smart cities, connected cars, entertainment systems, homes, and healthcare [3, 4]. The notable accomplishments of IoT, and subsequently, seven key IoT technologies, such as big data, cloud computing, smart sensors, software, artificial intelligence, actuators, and virtual reality, were singled out for their potential to aid healthcare amidst the COVID-19 pandemic. Furthermore, an examination revealed sixteen fundamental IoT applications in the medical sector during the COVID-19 crisis, each accompanied by a concise explanation [5]. Simultaneously, cloud computing has emerged as a reliable and scalable solution for processing, storing, and analyzing vast amounts of data generated by these IoT devices. Cloud platforms offer robust computational capabilities, allowing for

DOI: 10.1201/9781032656694-5

advanced data analytics, machine learning, and predictive modeling. The foundation for a new era of data-driven healthcare is transforming the way patient monitoring and data analytics are conducted by integrating IoT devices with cloud computing. Cloud computing plays a crucial role in enhancing the efficiency of cloud computing, reducing turnaround time for patient requests, and maximizing CPU utilization while minimizing waiting times by leveraging the optimal allocation of virtualized resources. To improve the accuracy of physiological sensor-data fusion calculations in the IoT framework, an advanced PSO (Particle Swarm Optimization) approach is applied, enabling automated diagnosis of natural epilepsy and brain abnormalities based on EEG signals detected by healthcare centers. The discrete wavelet transform is also utilized to enhance signal noise elimination. Moreover, the combination of neurological disability diagnosis with particle swarm computation optimizes neural network propagation and EEG analysis. This sophisticated approach enables more precise results, particularly when handling complex signals like EEG as input data [6].

Moreover, Cloud IoT simplifies remote patient monitoring and telemedicine, bridging the geographical divide and providing healthcare services to individuals in remote or underserved areas. It also gives the advantage of virtual consultations and telehealth services that make it more accessible, enhancing healthcare accessibility and affordability. The transformative capability of Cloud IoT extends to the potential of data analytics. Healthcare providers can extract valuable insights, identify patterns, and make data-driven decisions by analyzing the vast amount of data collected from IoT devices. To enable accurate diagnoses, personalized treatment plans, and proactive disease management, predictive analytics are implemented. Additionally, facing a myriad of challenges, including but not limited to data capture, storage, search, sharing, exchange, analysis, and visualization, it is observed that healthcare is undoubtedly one of the most data-intensive industries that face challenges that are multifaceted, encompassing three main levels: Data, Process, and Management [7]. It is essential to address challenges related to data security, privacy, interoperability, and reliability, and the potential benefits of Cloud IoT in healthcare are remarkable. Personal health records are a prime target for cyber-attacks because of their sensitive nature, making them a sought-after commodity. Cryptography and encryption play a vital role in ensuring their protection, as security and privacy are critical aspects of safeguarding these records. However, strict security measures can sometimes limit the availability of patient data to authorized personnel, creating obstacles in the transition to a digital health system. To facilitate the sharing of healthcare data, a promising solution to address security and privacy concerns lies in using permissioned blockchain technology. Unlike the conventional centralized systems prevalent in most healthcare setups, a permissioned blockchain offers enhanced security by reducing vulnerabilities and mitigating potential threats [8].

5.2 IoT DEVICES AND CLOUD COMPUTING IN HEALTH CARE

5.2.1 IoT Devices in Healthcare

The utilization of IoT devices in the healthcare sector has a revolutionary transformation in the method and way of taking care of patients. These devices are an integral part of the IoT ecosystem, encompassing a wide range of smart and interconnected healthcare gadgets. The IoT devices include wearable health trackers, smart medical devices, and remote patient monitoring tools that contribute to creating a patient-centric healthcare environment.

- **Wearable Health Trackers:** in recent years, Wearable health trackers have gained immense popularity. Individuals wear these devices, and they continuously collect various health-related data points throughout the day. Fitness bands, smart watches, and health-monitoring rings are some common examples of wearable health trackers. Vital signs such as heart rate, blood pressure, respiratory rate, and body temperature are measured by these devices. In addition, they also track physical activity metrics, such as step count, distance walked, calories burned, and even sleep patterns. Valuable insights into an individual's overall health and well-being are collected by these trackers. Moreover, a study conducted keeping in mind the Covid 19, the implementation of IoT-based wearable health monitoring systems in healthcare has the capacity to bring about a transformative change. For patients in quarantine, this technology offers a cost-effective and remote monitoring solution that is particularly beneficial for such cases. Enabling healthcare providers to remotely monitor patients in real time not only lessens the strain on medical resources but also optimizes the utilization of limited healthcare facilities and personnel [9].
- **Smart Medical Devices:** To monitor and manage various health conditions and chronic diseases, smart medical devices are specifically designed. These devices are equipped with sensors and communication capabilities to provide real-time data to healthcare providers and patients [10]. Examples of smart medical devices include smart glucose meters for diabetic patients, smart inhalers for asthma management, and smart blood pressure monitors [11]. Smart medical devices empower patients to actively participate in their care and help healthcare providers make informed treatment decisions by enabling continuous monitoring and data transmission.
- **Remote Patient Monitoring Tools:** Remote patient monitoring tools play a crucial role in extending healthcare services beyond traditional clinical settings. With the usage of these devices, healthcare providers can remotely monitor the health conditions of patients, and if any abnormalities are detected, they can intervene promptly. For various medical conditions, such as cardiac monitoring for heart patients, telehealth solutions for remote consultations, and remote fetal monitoring for expectant mothers, remote monitoring tools can be used. Such tools play a critical role in improving patient outcomes, reducing hospital readmissions, and enhancing the overall quality of care. A comprehensive and patient-centric approach to medical care is a result of recent developments in remote patient monitoring. Healthcare providers can monitor patients' vital signs and health status remotely, enabling personalized care and early intervention by leveraging modern communication and sensor technologies. As this technology continues to evolve, it holds immense promise in enhancing patient outcomes and managing healthcare resources more efficiently in an increasingly complex and aging world population [12].

The continuous data collection from IoT devices in healthcare offers several advantages:

- **Real-Time Monitoring:** To monitor changes in vital signs and health metrics promptly, real-time access to patient data is provided by IoT devices [9]. This enables timely interventions and reduces the risk of complications.
- **Personalized Care:** IoT devices collect data that can be used to tailor treatment plans and interventions based on an individual's specific health needs, leading to more personalized and effective care [13].
- **Improved Patient Engagement:** Patient engagement is improved by wearable health trackers and smart medical devices that encourage patients to actively engage in their health management, fostering a sense of responsibility and empowerment [14].
- **Remote Access to Data:** To remotely access patient data, remote patient monitoring tools enable healthcare providers, which is particularly beneficial for patients in rural or underserved areas [15].

However, the widespread adoption and usage of IoT devices in healthcare raises concerns about data security and privacy. Since these devices collect sensitive and subtle health information is to protect patient privacy. IoT devices have become indispensable tools in healthcare settings, revolutionizing patient monitoring and data collection, so ensuring robust encryption and secure data storage are of paramount importance. These devices, from wearable health trackers to smart medical devices and remote patient monitoring tools, enable real-time data collection and continuous health monitoring. Healthcare providers can deliver more personalized and efficient care, leading to improved patient outcomes and enhanced overall healthcare experiences by leveraging the power of IoT. Nevertheless, to harness the full potential of IoT devices in healthcare, striking the right balance between technology and data security remains crucial.

5.3 CLOUD COMPUTING IN HEALTHCARE

Revolutionizing how healthcare services are delivered, managed, and accessed, cloud computing in healthcare has emerged as a game-changer. Cloud computing in the healthcare sector facilitates convenient access and storage of patients' medical records anytime and from anywhere, ultimately enhancing physicians' capabilities to deliver improved patient care [16]. Numerous advancements benefiting patients, healthcare providers, and administrators alike are achieved through integrating cloud computing technology in the healthcare industry.

For storing, managing, and analyzing vast amounts of health-related data, cloud computing provides a flexible and scalable infrastructure. This data can include electronic health records (EHRs), medical images, patient information, research data, and more. Healthcare organizations can reduce their reliance on costly on-premises infrastructure, leading to cost savings and improved resource utilization by moving data and applications to the cloud [17]. The significant advantage of cloud computing in healthcare is real-time accessibility [18]. With an internet connection from any location and cloud-based systems, authorized healthcare professionals can securely access patient data and medical records. It facilitates telemedicine, remote consultations, and collaborative care to enable healthcare providers to deliver timely and efficient services, especially in remote or underserved areas. Cloud computing also supports seamless data sharing and interoperability among different healthcare systems and institutions [19]. This allows the exchange of patient information and medical records between hospitals, clinics, and other healthcare entities, promoting continuity of care and reducing redundant medical procedures.

Furthermore, cloud computing enhances data security and disaster recovery in healthcare. By implementing robust security measures, including encryption, access controls, and regular backups, reputable cloud service providers ensure the protection of sensitive patient information. Cloud-based data can be easily recovered, minimizing downtime and data loss in the event of a natural disaster or hardware failure.

Despite the numerous advantages, the adoption of cloud computing in healthcare also presents challenges [20], particularly concerning data privacy [21] and regulatory compliance. Healthcare organizations must ensure strict adherence to data protection regulations as healthcare data is highly sensitive. For instance, the Health Insurance Portability and Accountability Act (HIPAA) in the United States and the General Data Protection Regulation (GDPR) in the European Union protect data. By offering a scalable, accessible, and secure infrastructure for data management, analytics, and collaboration, cloud computing has transformed the landscape of healthcare. Modernizing the healthcare industry and its potential to improve efficiency, patient care, and data-driven decision-making makes it a crucial tool. As cloud technology continues to evolve, empowering healthcare providers to deliver better and more personalized care to patients worldwide and its impact on healthcare is expected to grow.

5.4 REAL-TIME PATIENT MONITORING WITH CLOUD IoT

With Cloud IoT, Real-time patient monitoring has emerged as a transformative approach in the healthcare industry. Healthcare providers can continuously monitor patients' vital signs, health metrics, and medical conditions in real-time, both within clinical settings and remotely, by leveraging the power of Cloud IoT technologies. To ensure the integrity of real-time electrocardiogram (ECG) data. Researchers have introduced a real-time remote patient monitoring system based on the IoT. Real-time ECG information transmitted from the system to a web server utilizes the Message Queuing Telemetry Transport (MQTT) protocol. To monitor both real-time and previously recorded ECG data of the patients, healthcare professionals, such as doctors, can conveniently access the web server using a smartphone or computer [22]. Cloud IoT enables the seamless integration of IoT devices, wearable health trackers, smart medical devices, and sensors, which continuously collect and transmit patient data to cloud-based platforms that provide healthcare professionals with instant access to real-time patient information, allowing for timely interventions and proactive care.

Real-time patient monitoring with Cloud IoT enhances patient safety and improves the efficiency of healthcare services in clinical environments. For instance, if any abnormalities or critical changes in vital signs are detected, monitoring devices connected to patients can trigger immediate alerts to healthcare staff. This reduces the risk of adverse events, improves patient outcomes, and enables swift responses and rapid medical assistance.

Cloud IoT empowers remote patient monitoring and telehealth services beyond clinical settings. To transmit the collected data securely to cloud servers, patients can use wearable health trackers or smart medical devices at home. It also helps health-care providers to remotely monitor patients' health status, track treatment adherence, and assess progress without requiring physical visits to healthcare facilities. Patients with chronic conditions, elderly individuals, and those living in remote areas partic-ularly benefitted from such an approach; it also enhances accessibility to healthcare services and reduces the need for frequent hospital visits.

With Cloud IoT, real-time patient monitoring facilitates data-driven healthcare. The wealth of patient data collected by IoT devices can be subjected to advanced analytics, machine learning algorithms, and predictive modeling. This enables healthcare professionals to identify patterns, trends, and potential health risks, leading to personalized treatment plans and proactive interventions. Hu et al. [23] developed the Simultaneously Aided Diagnostic Model (SADM) with the aim of improving the efficiency of outpatient doctors while reducing their workload. The model carries out a number of tasks, such as data collection, storage, pre-processing, feature extraction, machine learning, performance testing, and immediate assisted diagnosis by giving clinicians access to reference indexes. Information, including outpatient medication records, treatment schedules, budgets, costs, treatment results, test results, and imaging data, is collected through data acquisition. In order to accu-rately classify hyperlipidemia through machine learning, the model uses Support Vector Machine and Neural Network algorithms to train the data on previous medi-cal datasets. Additionally, cloud-based infrastructures' scalability and adaptability ensure that healthcare organizations can manage the expanding volume of patient data without making substantial hardware investments. In addition to offering strong data security measures, Cloud IoT solutions ensure the confidentiality and privacy of sensitive health data. Conversely, Sandhu et al. [24] developed a revolutionary data mining approach based on tensor and Granularity Computing (GrC) to handle three significant issues: Abstraction, heterogeneity, and privacy in healthcare data. GrC addresses computer issues at several degrees of depth, concentrating on extracting important data while deleting unimportant aspects from the dataset. Their three-phase methodology is designed to efficiently mine healthcare data. To store various sorts of information, including raw text data, medical records, audio files, and vid-eos, within a Health data Tensor (HdTr), a data matrix is first established. The sec-ond stage entails using a concept hierarchy to extend the data matrix across several granularity levels. Finally, in the third phase, granules are used to process queries and generate final results, assuring efficient data retrieval and analysis.

5.5 DATA ANALYTICS IN CLOUD IoT HEALTHCARE

5.5.1 Predictive Analytics and Personalized Medicine

Data analytics is Another area where cloud computing significantly improves health-care [25]. Healthcare firms can examine massive datasets to find patterns, trends, and insights by utilizing the cloud's computational capacity. Clinical decision-making, predictive analytics, population health management, and research are all aided by this data-driven methodology. A strategic framework for cloud computing was developed by the researchers to facilitate decision-making and is based on a thorough and all-encompassing approach in Saudi Arabia [26].

5.5.2 POPULATION HEALTH MANAGEMENT

Improved health outcomes for a particular group of people within a defined population is the goal of population health management, a crucial component of contemporary healthcare. Healthcare organizations may now analyze aggregated patient data using Cloud IoT technology, which helps them identify public health trends, monitor disease outbreaks, and improve population health management techniques. The IoT has recently attracted significant attention on a global scale and is becoming more readily available for applications linked to forecasting, preventing, and monitoring infectious diseases [27].Cloud IoT permits the seamless integration of data from numerous sources, including EHRs, wearable health trackers, smart medical devices, and other IoT-enabled sensors. A comprehensive view of a specific population's health status and behavior patterns by aggregated data enables healthcare providers to make informed decisions and develop targeted interventions.

From Cloud IoT platforms, healthcare organizations can analyzing the data and also identify prevalent health issues and trends within the population. Public health officials can proactively address emerging health concerns by detecting patterns in disease prevalence, risk factors, and health behaviors. Timely containment, prevention measures, and the early identification of potential disease outbreaks allow for reducing the impact on the population and healthcare resources. Intelligent disease surveillance systems powered by the IoT promise to be a game-changer in the fight against the pandemic. By leveraging existing infrastructure, including smartphones, wearable devices, and internet connectivity, its potential lies in its capacity to analyze and leverage existing data to effectively manage and contain the pandemic's spread [28]. In addition, Cloud IoT enables the development of custom population health management plans. With the assistance of advanced analytics and advanced machine learning algorithms, healthcare providers can break down the population into different risk categories according to their health profile. This allows them to tailor interventions and preventive measures to high-risk populations, improving resource allocation and health outcomes.

Cloud IoT also permits healthcare providers to deploy health promotion programs across the entire population. Through in-person and remote communication and monitoring, providers can encourage healthy behaviors, boost adherence to treatment, and provide continuous support for people living with chronic conditions. For example, Liu and Tu [29] carried out a study and implemented an IoT wireless sensor network to form a context-sensitive health promotion system for elderly care. The system is composed of three main sub-systems: IoT-based Physiological Information, Context Awareness-based Services, and Elderly Nutrition and Health Promotions. In the app, one can choose the Aging Diet Module or the Aging Exercise Module and receive personalized recommendations and long-term health management support.

The goal of the research is to develop an IoT-based health promotion system. Cloud IoT allows for real-time monitoring and analysis of data. This allows for continuous assessment of population health indicators. Healthcare providers can monitor the impact of interventions and strategies in real time and make changes based on feedback and results.

5.6 CHALLENGES AND CONCERNS

Addressing the below-mentioned challenges is essential to ensure the successful and secure implementation of Cloud IoT in healthcare.

Sr.No.	Challenges and Concerns	Description
1.	Data Security and Privacy	Sensitive healthcare data stored in the cloud may be vulnerable to cyber threats and unauthorized access. To ensure privacy, data encryption and access controls are essential.
2.	Interoperability	Standardized efforts for seamless data exchange and communication are required to integrating data from diverse healthcare systems and IoT devices.
3.	Data Integration and Quality	It is crucial to maintain and ensure data accuracy and consistency from various sources for accurate diagnoses and treatments.
4.	Reliability and Uptime	To meet the demands of real-time patient monitoring and critical applications, Cloud IoT services must maintain high reliability and uptime.
5.	Regulatory Compliance	To ensure legal and ethical data usage, compliance with healthcare data regulations, such as HIPAA or GDPR, is necessary.
6.	Latency and Connectivity	For real-time patient monitoring and data analysis, low latency and reliable internet connectivity are vital.
7.	Data Ownership and Control	To have clear agreements with cloud vendors on data access and usage, healthcare organizations need to retain control and ownership of their data.
8.	Cost Management	It is crucial to optimize resource allocation to Manage costs related to data storage, processing, and network usage.
9.	Ethical Considerations	It is essential for responsible data handling to address ethical concerns related to data usage, consent and potential bias in algorithms.
10.	Resistance to Adoption	It may require education and proper demonstrations of benefits to encourage healthcare professionals and patients to adopt new technologies.

5.7 FUTURE DIRECTIONS: TOP OF FORM

5.7.1 EDGE COMPUTING INTEGRATION

To reduce data transmission delays and enhance real-time processing capabilities, further improving healthcare services integration edge computing with Cloud IoT can be used. As technology continues to evolve, integrating edge computing with Cloud IoT in the healthcare industry holds significant promise for improving patient care, optimizing resource utilization, and enhancing overall healthcare services. Here are some future directions and potential benefits of this integration:

- **Real-time data analytics and decision-making:** Healthcare providers can collect, process, and analyze patient data in real-time at the point of care. This enables faster decision-making, diagnosis, and treatment adjustments, leading to better patient outcomes.
- **Reduced latency and enhanced responsiveness:** The integration of edge computing with Cloud IoT can reduce data transmission delays, and network latency can also be minimized. This is particularly crucial in time-sensitive scenarios, such as remote monitoring of critical patients or transmitting data from medical wearables, where even milliseconds can make a significant difference.
- **Enhanced security and privacy:** By processing sensitive patient data locally rather than transmitting it to a centralized cloud, edge computing can help address security and privacy concerns. The attack surfaces and potential risks associated with data breaches can be reduced by implementing this approach.
- **Scalability and cost-effectiveness:** It is easier to scale the deployment across healthcare facilities through Cloud IoT allows for centralized management and updates of edge devices. As only relevant and processed data is sent to the cloud, edge computing helps reduce data transmission costs and cloud computing expenses.
- **AI-powered edge devices:** To enable real-time analysis of complex medical data, integrating artificial intelligence (AI) capabilities at the edge can be used. Edge devices can use machine learning algorithms to recognize patterns, predict patient deterioration, or identify anomalies, allowing for early intervention and proactive healthcare management.
- **Telemedicine and remote healthcare:** By supporting high-quality video conferencing, remote consultations, and virtual diagnostics, edge computing facilitates the growth of telemedicine. The seamless communication between patients and healthcare providers, regardless of geographical distance, is ensured by the combination of edge and Cloud IoT.
- **Edge-assisted robotic surgery:** Edge computing can be integrated into surgical robots, enabling precise, and low-latency control during procedures. Additional support, such as access to medical knowledge bases or collaboration with other specialists in real-time, can be provided through the cloud.
- **IoT-enabled medical devices:** The capabilities of IoT medical devices can be enhanced by edge computing by enabling local data processing and decision-making. For example, an IoT-enabled insulin pump can analyze blood glucose levels at the edge, delivering the right insulin dosage without relying on constant cloud connectivity.
- **Predictive maintenance of medical equipment:** Predictive maintenance of medical equipment is possible by combining edge computing with Cloud IoT. Monitoring machine health locally and transmitting relevant data to the

cloud for long-term analysis, ensuring equipment reliability and reducing downtime by edge devices.

- **Regulatory compliance and data governance:** Healthcare organizations must comply with strict regulations concerning data privacy and security. A careful approach is required to ensure compliance with regional and international laws, such as HIPAA, GDPR, and others, to integrate edge computing and Cloud IoT.

Healthcare holds immense potential for transforming the industry by enhancing patient care, optimizing operations, and revolutionizing medical services through the integration of edge computing with Cloud IoT. We can expect to witness the realization of these future directions, leading to improved healthcare outcomes for patients worldwide as technology advances and becomes more accessible.

5.8   AI-DRIVEN DECISION SUPPORT

Cloud IoT systems are empowered through advancements in AI that provide intelligent decision support to clinicians, aiding in diagnosis, treatment planning, and predicting patient outcomes. As we look into the future, advancements in AI will have the potential to significantly empower Cloud IoT systems. It will also enable them to provide intelligent decision support to clinicians and revolutionize healthcare. These advancements will result in more precise and personalized patient care, enhanced diagnosis, treatment planning, and accurate prediction of patient outcomes. Here are some potential future directions in this domain:

- **Integration of AI and Cloud IoT Platforms:** Future developments will focus on seamlessly integrating AI capabilities into Cloud IoT platforms. Real-time data processing and analysis of massive volumes of patient data collected from various connected medical devices, wearables, and sensors are possible through such integrations.
- **Deep Learning and Neural Networks:** AI's ability to learn complex patterns from diverse datasets, allowing for more accurate and efficient medical diagnosis and predictive analytics, will be enhanced by advancements in deep learning techniques and neural networks.
- **Explainable AI (XAI):** There will be a growing need for transparency and interpretability in their decision-making processes as AI algorithms become increasingly sophisticated. XAI will be essential in healthcare to provide clinicians with understandable explanations for AI-generated recommendations, ensuring they can trust and validate the outcomes.
- **Federated Learning:** Privacy concerns and data security will continue to be paramount in healthcare. AI models, to be trained across multiple devices or cloud nodes without centralizing the data, will gain traction through a technique known as Federated learning. It will enable the creation of robust AI models while keeping sensitive patient data secure.
- **Digital Twins and Personalized Models:** Based on their health data, AI-enabled digital twins of patients under different treatment scenarios will be created to simulate individual health conditions and predict potential

outcomes. Based on an individual's unique characteristics, these personalized models will help clinicians make informed decisions.

- **Natural Language Processing (NLP) and Voice Interfaces:** To interact with Cloud IoT systems more naturally, AI-driven NLP and voice interfaces will assist clinicians. It also helps in making data entry, query, and retrieval processes more efficient and user-friendly.
- **Real-time Monitoring and Alerts:** In real-time, AI-powered systems will continuously monitor patient vitals and other health metrics. The system will generate alerts for immediate intervention in case of any anomalies or potential health risks, thereby reducing response times and preventing adverse outcomes.
- **AI for Drug Discovery and Treatment:** Drug discovery will accelerate by analyzing vast molecular data and predicting potential drug interactions and side effects through Cloud IoT systems equipped with advanced AI. Personalized treatment plans can be developed based on an individual's genomic data and can be designed by AI.
- **Longitudinal Data Analysis:** Patient data longitudinally over extended periods will be analyzed through AI. It also enables the identification of long-term trends and early signs of chronic conditions or disease progression.
- **Telemedicine and Remote Diagnostics:** A crucial role is played by AI-driven Cloud IoT systems in expanding telemedicine services. To enhance healthcare, remote diagnostics, continuous monitoring, and virtual consultations, especially in underserved areas.
- **Ethical AI Use and Regulation:** The development of ethical frameworks and regulations to govern its use will be essential as AI becomes more pervasive in healthcare. It also ensures patient privacy, data security, and unbiased decision-making will be critical.

Healthcare will pave the way for a new era of intelligent decision support for clinicians through the convergence of AI and Cloud IoT systems. To process vast amounts of data and extract meaningful insights, healthcare professionals will be better equipped to make accurate diagnoses, devise personalized treatment plans, and predict patient outcomes with unprecedented precision by harnessing the power of AI. This transformative potential holds the promise of significantly improving patient care and revolutionizing the healthcare industry as a whole.

5.9 BLOCKCHAIN FOR DATA INTEGRITY

Enhanced data integrity, ensuring transparency and immutability of health records, and enabling secure data sharing among authorized parties are some benefits of implementing blockchain technology. Future Directions of Advancements in Blockchain for Data Integrity in Healthcare:

- **Interoperability and Standardization:** Efforts will be made to standardize data formats and protocols to ensure seamless interoperability between different blockchain networks and healthcare systems as blockchain technology becomes more prevalent in healthcare. This will also lead to better patient care and medical research and enable secure and efficient data sharing across healthcare providers, researchers, and other authorized parties.

- **Scalability and Performance:** When dealing with large-scale healthcare data, current blockchain systems face challenges related to scalability and performance. To handle the growing number of medical data produced through multiple healthcare applications, future developments will concentrate on improving the throughput and transaction processing speed of blockchain networks.
- **Privacy and Consent Management:** Although blockchain's decentralized nature inherently offers data security, privacy concerns still exist, especially in healthcare. While ensuring data availability to authorized entities, future developments will aim to incorporate advanced privacy-preserving techniques such as zero-knowledge proofs and differential privacy to protect sensitive patient information.
- **Smart Contracts for Healthcare Automation:** Smart contracts, as we know, are self-executing contracts with predefined conditions. They have enormous potential to automate a variety of healthcare procedures, including payment settlements, clinical trial agreements, and insurance claims. The creation of reliable and secure smart contracts will result in simplified administrative processes, lower expenses, and boost overall effectiveness.
- **Integration with IoT and Wearable Devices:** For healthcare monitoring and diagnostics, the IoT and wearable devices are becoming an integral part. It will not only enhance data integrity and secure data transmission but also facilitate secure data sharing among healthcare providers, patients, and researchers.
- **Tokenization of Healthcare Assets:** The concept of tokenization consists of representing physical or digital assets on a blockchain using tokens. This idea can be applied in the healthcare industry to maintain and handle patient health information, medical devices, pharmaceutical supplies, and even data from healthcare research. The traceability, origin, and transparency of these assets will all be improved through tokenization.
- **Cross-Border Health Data Exchange:** Cross-border health data exchange can be possible through blockchain by creating a secure and standardized platform for sharing medical information across international boundaries. This will be predominately valuable in emergencies or crises and when patients seek medical care abroad.
- **Data Analytics and Research Advancements:** Blockchain's immutability and transparency can significantly improve data reliability for medical research and analytics. Advanced analytics tools and advancements in integrating blockchain will lead to more accurate and insightful research outcomes.
- **Regulatory Frameworks and Compliance:** Regulatory frameworks will evolve to address the unique challenges and opportunities it presents as blockchain technology gains traction in healthcare. For the use of blockchain in healthcare, government agencies will work toward defining guidelines while ensuring compliance with data protection and privacy regulations.

- **Decentralized Autonomous Organizations (DAOs) in Healthcare:** Without a central authority, DAOs are organizations governed by smart contracts on a blockchain. In healthcare, DAOs could foster collaborative research, decision-making, and resource allocation while ensuring transparency and accountability.

Blockchain technology in healthcare appears to have a bright and promising future. Data integrity, transparency, and security will be substantially enhanced as technology develops and is more widely used, which will ultimately result in better patient care, medical research, and industry-wide improvements.

5.10 FUTURE SCOPE OF STUDY

The healthcare sector has witnessed significant advancements in recent years with integrating Cloud IoT technologies. The transformative changes in patient monitoring and data analytics, revolutionizing healthcare delivery and patient outcomes, become possible with the convergence of cloud computing and IoT. The future scope of study in this domain holds immense potential for further advancements and innovations as the field continues to evolve [30].

- **Predictive Analytics for Disease Management:** collection and aggregation of diverse patient data, which can serve as a foundation for predictive analytics, is achievable through the integration of Cloud IoT. In the future, the creation of algorithms and machine learning models can help analyze historical patient data to predict disease progression, identify potential complications, and recommend preventative measures. This could result in more efficient resource allocation, reduced hospital readmissions, and improved overall patient outcomes.
- **Remote Patient Monitoring and Telemedicine:** The growth of telemedicine by enabling remote consultations and virtual healthcare services is already enabled through Cloud IoT. The future scope of study could emphasize refining the telemedicine experience through seamless integration of IoT devices and cloud infrastructure. New ways might be explored to enhance the quality of remote patient examinations, enable real-time diagnostic capabilities, and create immersive virtual healthcare environments.
- **Health Data Interoperability and Integration:** Seamless interoperability and integration of health data are crucial to fully realize the potential of Cloud IoT in healthcare. To facilitate data exchange between different IoT devices, EHR systems, and cloud platforms, future research could delve into the development of standardized data formats, protocols, and application programming interfaces (APIs). This interoperability would enable a holistic view of patient health and support informed decision-making by healthcare professionals.
- **Ethical and Regulatory Considerations:** As Cloud IoT in healthcare evolves, researchers must address the ethical and regulatory challenges

that arise. Future studies might explore and develop frameworks for ethical data usage, patient consent mechanisms for data sharing, and compliance with evolving data protection regulations. These deliberations are essential to build trust between patients, healthcare providers, and technology stakeholders.

- **Integration of Augmented Reality (AR) and Virtual Reality (VR):** Medical training, surgical procedures, and patient education can be enhanced with the fusion of CloudIoT with augmented and virtual reality. Future research could investigate how AR and VR technologies can be seamlessly integrated into the IoT ecosystem, enabling medical professionals to visualize patient data in immersive 3D environments and perform remote surgeries with greater precision.

Expansive and brimming with possibilities, the future scope of study in the realm of Cloud IoT in healthcare is very bright. As it offers avenues to improve patient care, revolutionize medical practices, and drive advancements in data analytics, the transformative potential of this convergence is vast. As researchers delve deeper into these areas, they will play a pivotal role in shaping the future of healthcare, where Cloud IoT becomes an integral part of patient monitoring, treatment, and overall wellness management.

5.11 CONCLUSION

In healthcare, the integration of Cloud IoT technology has revolutionized the industry, offering unprecedented opportunities for improved patient care, operational efficiency, and medical advancements. Cloud IoT has demonstrated its potential throughout this transformative journey to enhance data collection, analysis, and communication in healthcare settings, resulting in better outcomes for both patients and healthcare providers.

One of the significant advantages of Cloud IoT in healthcare is not only its ability to seamlessly connect various medical devices and sensors, but also generating real-time data from patients and medical equipment. Healthcare professionals have empowered with this data-driven approach, valuable insights into patient conditions, enabling timely interventions and personalized treatment plans. Moreover, Cloud IoT continuously monitors and has allowed for early detection of anomalies, leading to proactive care and reduced hospital readmissions.

Furthermore, have significantly improved the efficiency of healthcare operations has significantly improved as a result of the scalability and flexibility of Cloud IoT. Healthcare organizations can manage vast amounts of information. By storing and processing data in the cloud, without the burden of maintaining complex on-premises infrastructures, as a result, both healthcare institutions and patients receive benefits in the form of cost savings, streamlined processes, and improved resource allocation.

Regardless of these tremendous advancements, adopting Cloud IoT in healthcare has not been without challenges. As any data breach could have severe consequences for patients and healthcare providers alike, ensuring the security and privacy of sensitive patient data remains a top priority. Therefore, continuous efforts to maintain trust in Cloud IoT solutions and to implement robust security measures, such as encryption, authentication, and access controls, are imperative to safeguard patient information.

As we look to the future, the potential and prospective of Cloud IoT in healthcare appear limitless. Advancements in AI, machine learning, and data analytics will further intensify the benefits of Cloud IoT, enabling predictive and personalized medicine. Healthcare systems can ensure the immutability and transparency of patient records by integrating blockchain for data integrity and privacy, data sharing among authorized parties, and enhancing trust.

Furthermore, the divergence of Cloud IoT with other emergent technologies, such as 5G and edge computing, will lead to enhanced connectivity, reduced latency, and increased responsiveness, enabling real-time monitoring and decision-making in critical healthcare scenarios.

In conclusion, Cloud IoT has ushered in a new era of data-driven, patient-centric healthcare. It has had a transformative impact because of its ability to connect devices, collect and analyze data, and improve operational efficiency in the healthcare industry. While embracing further advancements by addressing security and privacy concerns, Cloud IoT will continue to revolutionize healthcare, empowering healthcare professionals, and improving the lives of patients worldwide. With the collaboration of healthcare providers and technology innovators, the future of Cloud IoT in healthcare holds the promise of a healthier, more connected, and data-driven world.

REFERENCES

1. Sharma, G., & Kalra, S. (2019). A lightweight user authentication scheme for cloud-IoT based healthcare services. *Iranian Journal of Science and Technology, Transactions of Electrical Engineering, 43*, 619–636.
2. Kelly, J. T., Campbell, K. L., Gong, E., & Scuffham, P. (2020). The internet of things: Impact and implications for health care delivery. *Journal of Medical Internet Research, 22*(11), e20135.
3. Nasajpour, M., Pouriyeh, S., Parizi, R. M., Dorodchi, M., Valero, M., & Arabnia, H. R. (2020). Internet of things for current COVID-19 and future pandemics: An exploratory study. *Journal of Healthcare Informatics Research, 4*, 325–364.
4. Liu, Y., Dong, B., Guo, B., Yang, J., & Peng, W. (2015). Combination of cloud computing and internet of things (IOT) in medical monitoring systems. *International Journal of Hybrid Information Technology, 8*(12), 367–376.
5. Javaid, M., & Khan, I. H. (2021). Internet of things (IoT) enabled healthcare helps to take the challenges of COVID-19 pandemic. *Journal of oral Biology and Craniofacial Research, 11*(2), 209–214.
6. Goyal, A., Kaushik, S., & Khan, R. (2021). IoT based cloud network for smart health care using optimization algorithm. *Informatics in Medicine Unlocked, 27*, 100792.
7. Firouzi, F., Rahmani, A. M., Mankodiya, K., Badaroglu, M., Merrett, G. V., Wong, P., & Farahani, B. (2018). Internet-of-things and big data for smarter healthcare: From device to architecture, applications and analytics. *Future Generation Computer Systems, 78*, 583–586.
8. Ali, A., Rahim, H. A., Pasha, M. F., Dowsley, R., Masud, M., Ali, J., & Baz, M. (2021). Security, privacy, and reliability in digital healthcare systems using blockchain. *Electronics, 10*(16), 2034.

9. Wu, J. Y., Wang, Y., Ching, C. T. S., Wang, H. M. D., & Liao, L. D. (2023). IoT-based wearable health monitoring device and its validation for potential critical and emergency applications. *Frontiers in Public Health*, *11*, 1188304.

10. Rahaman, A., Islam, M. M., Islam, M. R., Sadi, M. S., & Nooruddin, S. (2019). Developing IoT based smart health monitoring systems: A review. *Revue d'Intelligence Artificielle*, *33*(6), 435–440.

11. Reddy, G. K., & Achari, K. L. (2015, January). A non invasive method for calculating calories burned during exercise using heartbeat. In *2015 IEEE 9th international conference on intelligent systems and control (ISCO)* (pp. 1–5). IEEE.

12. Malasinghe, L. P., Ramzan, N., & Dahal, K. (2019). Remote patient monitoring: A comprehensive study. *Journal of Ambient Intelligence and Humanized Computing*, *10*, 57–76.

13. Islam, M. M., Rahaman, A., & Islam, M. R. (2020). Development of smart healthcare monitoring system in IoT environment. *SN Computer Science*, *1*, 1–11.

14. Carnaz, G., & Nogueira, V. B. (2016). An overview of IoT and healthcare. https://core.ac.uk/download/pdf/75982102.pdf

15. Rohokale, V. M., Prasad, N. R., & Prasad, R. (2011, February). A cooperative Internet of Things (IoT) for rural healthcare monitoring and control. In *2011 2nd international conference on wireless communication, vehicular technology, information theory and aerospace & electronic systems technology (Wireless VITAE)* (pp. 1–6). IEEE.

16. Nikhita Reddy, G., & Ugander Reddy, G. J. (2014). Study of cloud computing in healthcare industry. *arXiv e-prints*, arXiv-1402.

17. Sultana, S. N., Ramu, G., & Reddy, B. E. (2014). Cloud-based development of smart and connected data in healthcare application. *International Journal of Distributed and Parallel Systems*, *5*(6), 1–11.

18. McGregor, C. (2011, June). A cloud computing framework for real-time rural and remote service of critical care. In *2011 24th international symposium on computer-based medical systems (CBMS)* (pp. 1–6). IEEE.

19. Youssef, A. E. (2014). A framework for secure healthcare systems based on big data analytics in mobile cloud computing environments. *International Journal of Ambient Systems and Applications*, *2*(2), 1–11.

20. Ashtari, S., Eydgahi, A., & Lee, H. (2015). Exploring cloud computing implementation issues in healthcare industry. https://scholarworks.wmich.edu/cgi/viewcontent.cgi?article=1051&context=ichita_transactions

21. AbuKhousa, E., Mohamed, N., & Al-Jaroodi, J. (2012). e-health cloud: Opportunities and challenges. *Future Internet*, *4*(3), 621–645.

22. Yew, H. T., Ng, M. F., Ping, S. Z., Chung, S. K., Chekima, A., & Dargham, J. A. (2020, February). IoT based real-time remote patient monitoring system. In *2020 16th IEEE international colloquium on signal processing & its applications (CSPA)* (pp. 176–179). IEEE.

23. Hu, Y., Duan, K., Zhang, Y., Hossain, M. S., Mizanur Rahman, S. M., & Alelaiwi, A. (2018). Simultaneously aided diagnosis model for outpatient departments via healthcare big data analytics. *Multimedia Tools and Applications*, *77*, 3729–3743.

24. Sandhu, R., Kaur, N., Sood, S. K., & Buyya, R. (2018). TDRM: Tensor-based data representation and mining for healthcare data in cloud computing environments. *The Journal of Supercomputing*, *74*, 592–614.

25. Kaur, K., Verma, S., & Bansal, A. (2021, October). IoT big data analytics in healthcare: Benefits and challenges. In *2021 6th international conference on signal processing, computing and control (ISPCC)* (pp. 176–181). IEEE.

26. Alhabi, F., Atkin, A., & Stanier, C. (2015, February). Strategic framework for cloud computing decision-making in healthcare sector in Saudi Arabia. In *The seventh international conference on ehealth, telemedicine, and social medicine* (Vol. 1, pp. 138–144).

27. Bai, L., Yang, D., Wang, X., Tong, L., Zhu, X., Zhong, N., & Tan, F. (2020). Chinese experts' consensus on the internet of things-aided diagnosis and treatment of coronavirus disease 2019 (COVID-19). *Clinical eHealth*, *3*, 7–15.
28. Rahman, M. S., Peeri, N. C., Shrestha, N., Zaki, R., Haque, U., & Ab Hamid, S. H. (2020). Defending against the novel coronavirus (COVID-19) outbreak: How can the internet of things (IoT) help to save the world? *Health Policy and Technology*, *9*(2), 136.
29. Liu, C. H., & Tu, J. F. (2020). Development of an IoT-based health promotion system for seniors. *Sustainability*, *12*(21), 8946.
30. Kumar, S., Tiwari, P., & Zymbler, M. (2019). Internet of things is a revolutionary approach for future technology enhancement: A review. *Journal of Big Data*, *6*(1), 1–21.

6 Cloud-Internet of Things in Smart Healthcare Systems

Ankit Garg, Anuj Kumar Singh, and Dhawan Singh

6.1 INTRODUCTION

The healthcare business is undergoing a major transition as a result of the combination of cloud computing and the Internet of Things (IoT). Cloud-IoT in healthcare refers to the confluence of these two technologies, which allows for the real-time gathering, analysis, and exchange of healthcare data. This combination has enormous potential to transform healthcare systems, improve patient outcomes, and improve overall care delivery [1]. This chapter elaborates on the concept of cloud-IoT in healthcare, attempts to clarify its key ideas, emphasizes its significance, and throws light on its disruptive potential. The chapter explores a wide range of applications where cloud-IoT has a significant impact, from remote patient monitoring and telemedicine to wearable devices and activity trackers, smart hospital infrastructure and asset management, ambient assisted living, medication adherence, and data-driven healthcare analytics and decision support [2]. Healthcare practitioners may remotely monitor patients using cloud-IoT technology, enabling continuous care outside of typical clinical settings. Individuals are empowered to actively participate in their own healthcare with IoT-enabled wearable gadgets and activity trackers, encouraging preventive care and better lives. Intelligent hospital infrastructure and asset management systems improve resource allocation, streamline operations, and improve patient safety [3]. Ambient assisted living technologies promote independent living for the elderly and those with impairments, improving their well-being while reducing the strain on healthcare services. Medication adherence technologies, such as smart pill bottles, increase patient compliance and treatment results [4]. Furthermore, cloud-IoT enables collecting and analyzing massive volumes of healthcare data, paving the way for data-driven analytics and decision support systems that can provide useful insights, improve diagnoses, personalize treatments, and overall healthcare delivery [5].

One of the most notable benefits is improved patient outcomes and personalized treatment, with real-time data collecting and analysis allowing healthcare practitioners to make better educated decisions about patient care. The seamless integration of cloud-IoT in healthcare systems improves efficiency and optimizes resource allocation, resulting in simplified processes, lower costs, and better patient experiences [6, 7]. Real-time monitoring and early intervention capabilities provide prompt reactions to changes in patient health status, allowing for proactive treatments and

DOI: 10.1201/9781032656694-6

reducing the risk of consequences. Individuals receive access to their own health data, allowing them to actively participate in their healthcare decisions and make educated choices, which increases patient empowerment and participation [8]. Furthermore, the cost savings and long-term benefits of cloud-IoT in healthcare are clear, with remote monitoring lowering the need for frequent hospital visits and optimized resource utilization leading to more effective healthcare delivery [9]. However, cloud-IoT adoption in healthcare is not without its obstacles and considerations. Concerns about security and privacy are prominent since collecting and transferring sensitive healthcare data need robust security measures and rigorous privacy regulations [10, 11]. Another problem is interoperability and data integration since various IoT devices and systems must smoothly connect and share data to allow successful healthcare delivery. The reliability and availability of cloud-IoT devices and networks are also critical factors that need to be addressed to ensure uninterrupted service and patient care.

Concerns emerge over the appropriate use of patient data, informed permission, and compliance with legislation and standards. The implementation landscape is further shaped by regulatory and compliance problems, such as conforming to medical device regulations and data protection legislation. The chapter presents case studies highlighting real-world applications and outcomes to provide a practical knowledge of cloud-IoT-enabled healthcare. Remote patient monitoring and telehealth platforms, cloud-IoT-based wearable devices for chronic disease management, smart hospital infrastructure and asset tracking systems, ambient assisted living and home healthcare solutions, and data analytics platforms for healthcare decision-making are all covered in these case studies. These examples show the various uses and benefits of cloud-IoT in healthcare, emphasizing its potential to alter care delivery, enhance patient outcomes, and address industry concerns. In addition, the chapter investigates the implications of cloud-IoT in healthcare, focusing on crucial issues such as data security and privacy protection, interoperability and data standardization, ethical concerns and patient permission, regulatory frameworks, and compliance needs [12]. These consequences highlight the significance of developing strong data security mechanisms, encouraging interoperability, assuring ethical data practices, and adhering to legislation in order to retain patient confidence and promote responsible and successful cloud-IoT adoption in healthcare.

Finally, the chapter presents emerging trends and future directions in cloud-IoT healthcare, including the incorporation of artificial intelligence and machine learning, the use of edge computing for real-time analytics, the use of blockchain technology for secure healthcare data sharing, and the potential for cloud-IoT-enabled personalized medicine and precision healthcare. These trends reflect the field's ongoing growth and innovation in cloud-IoT healthcare, opening the way for new potential and improvements. Finally, the purpose of this chapter is to offer a thorough review of cloud-IoT in healthcare, including its definition, applications, benefits, problems, case studies, consequences, and future directions. Understanding the vast potential and addressing the associated considerations is essential for harnessing the transformative power of cloud-IoT in revolutionizing smart healthcare systems, improving patient outcomes, and shaping the future of healthcare delivery.

6.2   IoT APPLICATIONS IN HEALTHCARE

By using networked devices and real-time data processing, healthcare IoT applications have revolutionized how patient care is delivered. Healthcare practitioners can remotely monitor patients' vital signs, medication adherence, and general health indicators via wearable sensors, medical equipment, and mobile health applications [13]. This permits early abnormality diagnosis, personalized treatment regimens, and proactive treatments. IoT also enables remote patient monitoring, allowing people with chronic illnesses to get ongoing treatment from the comfort of their own homes, minimizing hospital visits and enhancing disease management. Furthermore, smart hospitals powered by IoT optimize operations, automate inventory management, and improve patient flow. Overall, IoT applications in healthcare improve patient outcomes, increase efficiency, and empower people to take control of their health [14]. The most significant applications of IoT in the healthcare sector have been listed in Figure 6.1.

6.2.1   REMOTE PATIENT MONITORING AND TELEMEDICINE

When integrated with the IoT, remote patient monitoring and telemedicine transform healthcare by providing smooth and efficient healthcare delivery from a distance. Wearable sensors and smart medical devices, for example, may capture and send patient data in real-time, enabling healthcare practitioners to remotely monitor vital signs, medication adherence, and general health status [15]. Continuous monitoring improves early diagnosis of health conditions, allows for more personalized treatment regimens, and empowers patients to actively engage in their own care. Furthermore, telemedicine solutions use IoT connections to enable virtual consultations, allowing

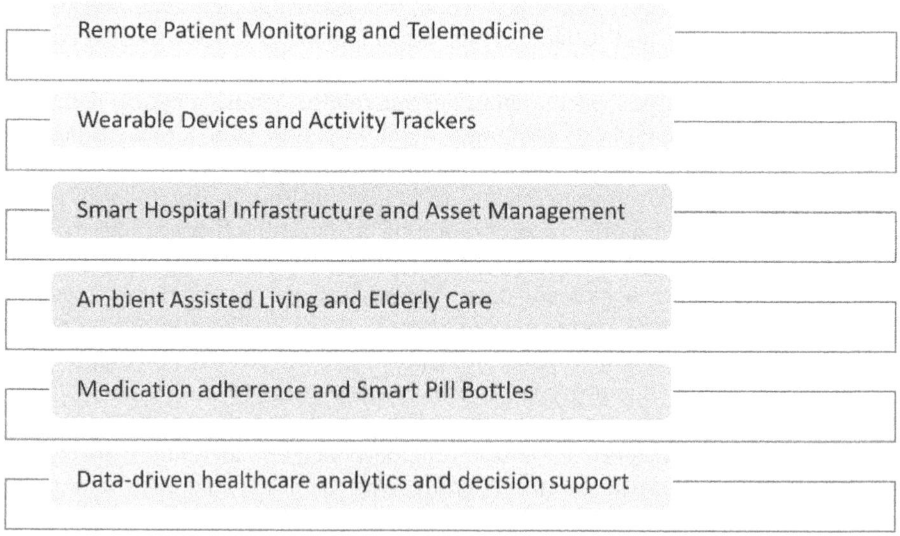

Remote Patient Monitoring and Telemedicine

Wearable Devices and Activity Trackers

Smart Hospital Infrastructure and Asset Management

Ambient Assisted Living and Elderly Care

Medication adherence and Smart Pill Bottles

Data-driven healthcare analytics and decision support

FIGURE 6.1   Applications of IoT in healthcare sector.

healthcare experts to diagnose, advise, and treat patients remotely. This removes geographical obstacles, increases access to healthcare treatments, and lessens the strain on traditional healthcare institutions. The use of IoT in remote patient monitoring and telemedicine increases healthcare efficiency, resource utilization, and patient outcomes, eventually altering the healthcare environment.

6.2.2 Wearable Devices and Activity Trackers

Wearable gadgets and activity trackers powered by the IoT have grown in popularity in recent years due to their promise to revolutionize personal health monitoring and encourage active lives. Smartwatches, fitness bands, and health trackers, for example, are outfitted with sensors that continually gather data on a variety of physiological characteristics, physical activities, and sleep patterns [16, 17]. These devices may communicate gathered data to smartphones or cloud platforms for real-time analysis and monitoring using an IoT connection. Individuals can use wearable devices to measure their daily steps, heart rate, calories burned, and sleep quality, allowing them to make more informed decisions regarding their health and well-being. Furthermore, IoT integration enables healthcare practitioners to remotely monitor and manage their patients' health problems, enabling personalized treatment and timely interventions. Furthermore, wearable gadgets make integrating gamification and social networking easier, boosting user engagement, motivation, and healthy competition.

6.2.3 Smart Hospital Infrastructure and Asset Management

Smart hospital infrastructure and asset management, utilizing the power of the IoT, has emerged as a disruptive method to increasing operational efficiency, resource utilization, and patient care in healthcare institutions [18]. Internet of Things-enabled technologies provide real-time monitoring and administration of important hospital assets such as medical equipment, supplies, and infrastructure components. These systems collect data on asset utilization, maintenance requirements, and performance indicators using sensors, actuators, and network connection. Hospitals may use IoT to track the location, condition, and availability of assets, resulting in decreased downtime, optimized processes, and enhanced patient safety. Furthermore, IoT-based asset management systems make predictive maintenance easier by continually monitoring asset health and generating alarms or automatic repair requests when abnormalities or failures are found. Using real-time asset data allows hospital managers to optimize inventory levels, enhance asset allocation, and expedite procurement procedures, resulting in cost savings and more effective resource utilization [19]. Furthermore, the incorporation of IoT in smart hospital infrastructure enables increased security via surveillance systems, access control mechanisms, and patient monitoring solutions, creating a safe environment for patients, employees, and visitors. To summarize, the use of IoT for smart hospital infrastructure and asset management has enormous promise for revolutionizing healthcare operations, increasing patient outcomes, and enabling cost-effective healthcare delivery.

6.2.4   Ambient Assisted Living and Elderly Care

The IoT has brought about a revolution in the realm of assisting and enhancing the quality of life for aging populations, achieved through ambient assisted living (AAL) and senior care. IoT technologies are critical in developing smart environments that give personalized, unobtrusive, and context-aware help to older people in their everyday activities [20]. AAL systems capture real-time data on the health, behavior, and living situations of the elderly through the use of sensors, wearables, and smart devices, allowing for remote monitoring and prompt interventions. These devices can identify possible dangers, including falls, aberrant vital signs, or crises, and send notifications to carers or healthcare providers. IoT-based solutions also make it easier to integrate assistive technology such as smart home automation, medication management, and cognitive support systems, which promotes independent living and reduces dependency on formal care services. Furthermore, the IoT offers social connectedness by enabling remote communication, social engagement, and telemedicine services, therefore alleviating the social isolation that the elderly frequently face. The application of IoT in AAL and senior care has great potential in increasing older individuals' well-being, safety, and autonomy, allowing them to age in place with dignity and an improved quality of life [21].

6.2.5   Medication Adherence and Smart Pill Bottles

Prescription adherence is a key feature of healthcare management, especially for patients with chronic diseases who rely on regular and timely prescription consumption. Smart pill bottles have emerged as a viable approach to improve medication adherence through real-time monitoring and reminders since the introduction of the IoT. These revolutionary gadgets connect with smartphones or other devices via IoT technology, allowing patients and healthcare practitioners to track drug intake trends and handle non-adherence concerns quickly. According to Mauro et al. [22], smart pill bottles with IoT capabilities considerably enhanced medication adherence rates among hypertensive patients. The study found that patients who used smart pill bottles had a 25% higher adherence rate than those who just used standard pill bottles. IoT-enabled features like automated reminders and real-time monitoring were critical in increasing adherence and avoiding prescription mistakes.

Furthermore, Park et al. [23] published a study that emphasized the potential of smart pill bottles to solve medication adherence issues in senior populations. The study found that smart pill bottles with IoT-based reminders and notifications enhanced prescription adherence by 30% among older persons. The use of IoT technology not only increased patient participation but also allowed carers and healthcare practitioners to remotely monitor medication consumption, allowing for prompt interventions as needed.

6.2.6   Data-driven Healthcare Analytics and Decision Support

The integration of IoT technology has considerably improved data-driven healthcare analytics and decision support. Wearable sensors, remote monitoring devices, and medical implants create massive volumes of real-time patient data, allowing healthcare practitioners to make more informed decisions and enhance patient outcomes. According to [24], IoT-based healthcare systems enable continuous monitoring of patients' vital signs, allowing for early diagnosis of anomalies and prompt treatments, lowering hospital readmissions and boosting patient satisfaction. Furthermore, modern data processing techniques are used in IoT-enabled healthcare analytics to derive significant insights from the acquired data. Nowadays machine learning algorithms can be used to analyze large-scale healthcare information, enabling predictive modeling for illness diagnosis, therapy optimization, and personalized healthcare delivery [25]. These analytics-driven techniques help healthcare professionals offer efficient and cost-effective treatment by facilitating evidence-based decision-making.

Remote patient monitoring is a significant use of IoT-driven healthcare analytics and decision assistance. IoT devices may be integrated with cloud-based analytics tools to allow healthcare practitioners to remotely monitor patients' health states, track medication adherence, and give timely treatments. Healthcare practitioners can minimize illness exacerbations, save healthcare costs, and enhance overall patient well-being by continually monitoring patients in their homes [26].

6.3 BENEFITS AND ADVANTAGES OF IoT IN HEALTHCARE

IoT in healthcare allows for real-time data gathering, remote monitoring, and personalized care, all of which contribute to better patient outcomes. Continuous vital sign monitoring and analysis allow for early identification and quick actions, decreasing problems. Individual patient data-driven personalized suggestions optimize treatment regimens and increase participation. Furthermore, giving patients access to their health data encourages informed decision-making and greater treatment adherence [27]. IoT has the potential to revolutionize healthcare delivery by combining continuous monitoring, personalized care, and patient empowerment. The advantages and benefits of IoT in healthcare have been shown in Figure 6.2 and are explained below.

6.3.1 IMPROVED PATIENT OUTCOMES AND PERSONALIZED CARE

The use of IoT technology in healthcare has a number of advantages, one of which is the possibility for improved patient outcomes and personalized treatment. IoT devices

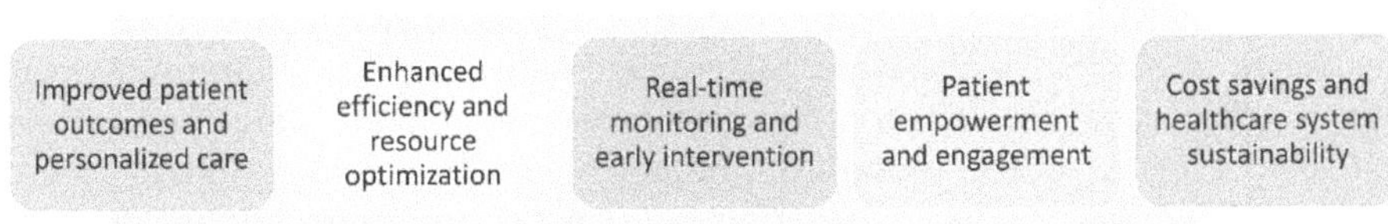

FIGURE 6.2 Advantages of IoT in healthcare.

may collect data from patients in real time, like as vital signs, medication adherence, and activity levels. Healthcare professionals may acquire significant insights into patients' health state and make better educated decisions regarding their care by continually monitoring and analyzing this data. Healthcare workers may remotely monitor patients and act as needed using IoT-enabled equipment. Wearable gadgets, for example, can monitor patients' heart rates, blood pressure, and other vital indicators, allowing for early identification and action in the event of an anomaly [28]. This proactive approach to healthcare can result in prompt treatments, decreasing complications and increasing patient outcomes. IoT also allows healthcare practitioners to adjust treatment programs to particular patients, facilitating personalized care. IoT devices may deliver personalized suggestions and therapies based on the individual requirements and circumstances of each patient through continuous data collecting and analysis. With this level of personalization, healthcare practitioners may administer focused treatments, optimize treatment programs, and promote patient involvement and satisfaction. Furthermore, IoT devices can allow patients to take an active part in their health management. Patients can obtain insights into their well-being and make educated lifestyle decisions if they have access to their own health data. This approach to self-monitoring and self-management can result in better adherence to treatment programs, enhanced patient empowerment, and, ultimately, better health outcomes.

6.3.2 Enhanced Efficiency and Resource Optimization

Another key benefit of IoT in healthcare is the possibility of increased efficiency and resource optimization. IoT devices have the potential to automate numerous processes, optimize workflows, and lessen the stress on healthcare personnel, resulting in increased operational efficiency. Healthcare organizations may optimize resource allocation by gathering real-time data from IoT devices. For example, IoT-enabled asset tracking systems can assist in swiftly locating medical equipment, saving time spent searching for supplies and increasing staff efficiency [29]. Furthermore, IoT devices may monitor inventory levels and automatically trigger orders when supplies run short, assuring prompt replenishment and minimizing stockouts. Remote patient monitoring is also possible with IoT, decreasing the need for frequent hospital visits and allowing healthcare staff to focus on high-priority situations. This remote monitoring capacity is especially useful for individuals with chronic diseases who need constant monitoring but do not need to be physically present at a healthcare institution. Healthcare practitioners can intervene when required by remotely tracking patients' health state, decreasing hospital readmissions and optimizing the use of healthcare resources. Furthermore, IoT-powered data analytics might reveal operational inefficiencies and opportunities for development. Healthcare organizations may detect bottlenecks in procedures, improve workflows, and optimize resource allocation by analyzing data acquired from IoT devices. This data-driven decision-making strategy can result in cost savings, improved patient flow, and better overall resource utilization [30].

Overall, the IoT in healthcare has the potential to improve patient outcomes, personalize care, increase efficiency, and optimize resource allocation. Healthcare organizations can reinvent healthcare delivery, achieve operational gains, and ultimately provide better care to patients by utilizing the potential of IoT devices and data analytics.

6.3.3 REAL-TIME MONITORING AND EARLY INTERVENTION

The capacity to offer real-time monitoring and early intervention is one of the key benefits of integrating IoT in healthcare. Wearable sensors and remote monitoring systems, for example, can continually gather and send patient data to healthcare practitioners in real-time. This real-time monitoring provides healthcare providers with a full and up-to-date picture of a patient's health condition, allowing them to notice any irregularities or changes as soon as they occur. Healthcare practitioners can receive alerts and messages when certain health metrics stray from typical ranges or when crucial events occur by employing IoT technology. This timely information enables healthcare providers to respond swiftly and deliver required interventions or treatments [31]. In the case of a chronic disease, for example, IoT-enabled devices may monitor vital signs and transmit notifications to healthcare practitioners if there is a rapid worsening, allowing for quick medical intervention and perhaps averting serious consequences or hospitalizations. Furthermore, real-time monitoring via IoT devices can aid in the early diagnosis and prevention of any health problems. Healthcare practitioners can discover patterns, trends, and subtle changes in patient data that may suggest the start of a health concern by continually monitoring patient data. Early identification allows for prompt treatments, such as medication changes or lifestyle changes, which can prevent disease development and improve patient outcomes [32].

6.3.4 PATIENT EMPOWERMENT AND ENGAGEMENT

IoT in healthcare is critical for empowering individuals and encouraging active participation in their own health management. IoT devices give patients access to personal health data, allowing them to actively engage in decision-making and take control of their health. Patient empowerment provides a number of advantages for both individuals and healthcare practitioners. Patients may monitor their health metrics, follow their progress, and obtain insights into their health state using IoT-enabled devices [33]. This real-time access to health information enables patients to make educated lifestyle, medication adherence, and overall self-care decisions. Patients who have a better grasp of their health issues can take proactive efforts to manage their health more efficiently, leading to better treatment outcomes. IoT devices also improve patient involvement by allowing patients and healthcare practitioners to communicate and collaborate in real time. Patients may engage with their healthcare team, ask questions, and receive prompt assistance using secure messaging platforms or telehealth programs. This increased communication not only improves the patient experience, but it also deepens the patient-provider connection, resulting in better adherence to treatment programs and better health outcomes. Furthermore, IoT devices may give instructional resources and personalized advice to patients directly. Healthcare practitioners may give personalized counseling, lifestyle changes, and self-care techniques by analyzing data acquired from IoT devices [34]. This personalized approach to healthcare provides patients with the knowledge and resources they need to make educated decisions, resulting in more involvement in their health management.

6.3.5 Cost Savings and Healthcare System Sustainability

Implementing IoT in healthcare can result in significant cost reductions and help to ensure the sustainability of healthcare systems. IoT devices provide several potential for cost reduction and resource utilization. Real-time monitoring and early intervention using IoT devices can assist to reduce illness development and the need for costly therapies or hospitalizations. By spotting health risks early on, healthcare practitioners may respond quickly, potentially lowering total healthcare costs and improving patient outcomes [35]. Early intervention can also help to avoid costly complications or emergency situations, saving both patients and healthcare systems money. Furthermore, IoT devices allow for remote patient monitoring, which reduces the need for frequent hospital visits and in-person consultations. Patients with chronic diseases that require continuing care and treatment may benefit from remote monitoring in particular. Healthcare professionals may discover possible difficulties in real time, give virtual consultations, and optimize the use of healthcare resources by remotely tracking patients' health state. This remote method saves money on hospital admissions, travel, and unneeded medical treatments. Furthermore, data analytics powered by IoT and predictive maintenance may optimize resource allocation and improve operational efficiency [36]. Healthcare organizations may find areas for improvement, optimize procedures, and decrease waste by analyzing data received from IoT devices. Medical equipment predictive maintenance can also help to avoid unexpected malfunctions, save downtime, and lower repair costs.

Finally, IoT in healthcare enables real-time monitoring and early intervention, allowing patients to actively control their health and participate in their treatment. Furthermore, IoT devices help to save costs by providing remote patient monitoring, early diagnosis of health risks, and resource optimization. Healthcare systems may improve patient outcomes, reduce costs, and assure the long-term viability of healthcare delivery by embracing IoT technologies.

6.4 CHALLENGES AND CONSIDERATIONS IN IoT-ENABLED HEALTHCARE

The problems and issues connected with IoT-enabled healthcare are numerous, but they are manageable with proper design and execution. Concerns about security and privacy are critical, needing strong methods like as encryption, authentication, and privacy rules to secure patient data. Interoperability and data integration difficulties necessitate standardization initiatives and standardized protocols for IoT device and system connectivity. IoT device reliability and availability necessitate redundancy mechanisms and frequent maintenance to assure uninterrupted service. Informed permission, responsible data usage, and compliance with legislation such as HIPAA and GDPR are all required by ethical and legal concerns. The following are the

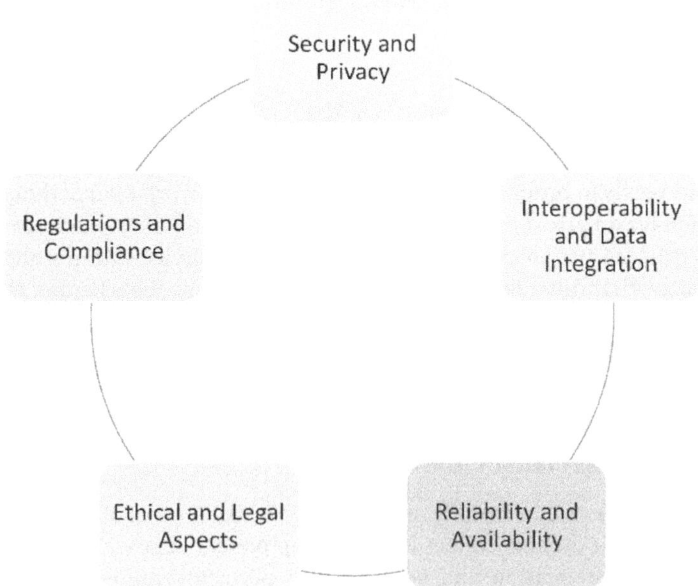

FIGURE 6.3   Challenges in IoT-enabled healthcare.

issues and considerations in IoT-enabled healthcare [37]. Figure 6.3 highlights the challenges in IoT-enabled healthcare system.

6.4.1 SECURITY AND PRIVACY CONCERNS

The security and privacy of patient data is one of the most difficult concerns in IoT-enabled healthcare. IoT devices capture and send critical medical data, rendering them vulnerable to hacks and data breaches. To secure the confidentiality, integrity, and availability of patient data, healthcare organizations must apply stringent security measures. Encryption methods, secure authentication procedures, and frequent vulnerability assessments to discover and resolve any security weaknesses are all part of this [38–40]. Strong privacy rules and permission methods should also be in place to guarantee that patient data is acquired, maintained, and shared in an ethical and compliant manner.

6.4.2   INTEROPERABILITY AND DATA INTEGRATION

IoT devices in healthcare frequently originate from multiple manufacturers and run on different platforms, posing interoperability issues. The seamless integration of various IoT devices and systems is critical for successful data exchange and communication among healthcare practitioners, devices, and patients. To solve these issues, standardization activities and the acceptance of interoperable frameworks are required [41]. Healthcare organizations should guarantee that IoT devices can communicate information easily by defining uniform data formats, communication protocols, and interfaces, resulting in enhanced care coordination and patient outcomes.

6.4.3 Reliability and Availability of IoT Devices

IoT devices rely on network connectivity and require a solid and dependable infrastructure to perform properly. Any interruptions in network connectivity or power supply can have an effect on the availability and dependability of IoT devices, possibly jeopardizing patient care. To reduce the risk of device failures and maintain continuous availability, healthcare organizations must incorporate redundancy mechanisms and backup systems [42]. Furthermore, frequent maintenance and monitoring of IoT devices are required to identify and fix any technical difficulties as soon as possible.

6.4.4 Ethical and Legal Considerations

The usage of IoT devices in healthcare generates ethical and legal concerns that must be addressed. For example, collecting and using patient data via IoT devices necessitates informed permission and adherence to privacy requirements like HIPAA and GDPR [37]. Patients must be informed about how their data will be collected, utilized, and safeguarded by healthcare organizations. Ethical issues include using patient data responsibly and transparently, ensuring confidentiality, and avoiding any biases or prejudice in the interpretation of IoT-generated data.

6.4.5 Regulatory and Compliance Issues

IoT device adoption in healthcare is subject to regulatory and compliance restrictions. To guarantee that their IoT activities comply with appropriate rules and regulations, healthcare organizations must negotiate complicated regulatory frameworks. Adherence to medical device regulations, data protection legislation, and industry-specific standards are all part of this. It is critical to be educated about new legislation and to proactively address compliance concerns to avoid legal risks and guarantee the ethical and responsible usage of IoT in healthcare [43]. Healthcare providers, technology suppliers, regulators, and legislators must work together to address these issues and considerations. To realize the full promise of IoT-enabled healthcare while protecting patient well-being and data privacy, comprehensive policies and frameworks that prioritize security, privacy, interoperability, dependability, ethics, and compliance are required.

6.5 IoT-ENABLED SMART HEALTHCARE SYSTEMS: CASE STUDIES

The case studies in this area demonstrate the actual application of IoT in smart healthcare systems. These technologies provide creative methods to remotely monitor patients, manage chronic illnesses, optimize hospital operations, enhance independent living, and use data analytics for decision-making by leveraging the potential of IoT devices. These case studies demonstrate the actual advantages and revolutionary possibilities of IoT in healthcare. Each case study in the following subsections delves deeper into examining the unique uses and consequences of IoT-enabled remote patient monitoring and telehealth systems.

6.5.1 Remote Patient Monitoring and Telehealth Platforms

Remote patient monitoring and telehealth systems use IoT technology to allow healthcare practitioners to remotely monitor and manage the health problems of their patients. These systems capture and send patient data in real-time using IoT-enabled devices such as wearable sensors or home monitoring kits. This data is accessible to healthcare practitioners via secure telemedicine systems, allowing them to monitor vital signs, medication adherence, and general health conditions. This method allows for quicker treatments, lowers the need for in-person visits, and enhances patient convenience and access to care, especially for people living in distant or disadvantaged locations [44].

6.5.2 IoT-based Wearable Devices for Chronic Disease Management

Wearable IoT devices are rapidly being used to manage chronic conditions such as diabetes, hypertension, and asthma. Smartwatches and glucose monitoring systems, for example, continually collect and communicate patient data, offering insights into health factors and prescription adherence. Remote monitoring of patient status, receiving notifications for bad data, and intervening proactively are all options for healthcare practitioners. By providing patients with real-time health information, these wearable devices promote self-management and enable personalized treatment programs, resulting in better disease control and a higher quality of life [45].

6.5.3 Smart Hospital Infrastructure and Asset Tracking Systems

IoT technology is used in smart hospital infrastructure to optimize hospital operations and improve patient care. Sensors and tracking systems enabled by the IoT can monitor different aspects of hospital surroundings, such as temperature, humidity, and air quality, assuring ideal conditions for patient comfort and safety. Asset tracking systems that use IoT devices can track the whereabouts of medical equipment in real time, lowering equipment search times, improving inventory management, and increasing staff efficiency. These intelligent solutions improve resource utilization, streamline workflows, and lead to more effective and patient-centered care delivery [46].

6.5.4 Ambient Assisted Living and Home Healthcare Solutions

AAL and home healthcare solutions powered by IoT are intended to help aging populations and people with disabilities live independently and safely in their own homes. Smart sensors, voice assistants, and IoT devices are used in these systems to monitor daily activities, identify crises, and give aid when needed. IoT-enabled fall detection sensors, for example, can warn carers or emergency services in the event of a fall, assuring rapid help. AAL systems also provide medication reminders, home security monitoring, and remote connection with healthcare professionals, fostering autonomy, safety, and a higher quality of life for those needing assistance [47].

6.5.5 DATA ANALYTICS PLATFORMS FOR HEALTHCARE DECISION-MAKING

Data analytics tools driven by IoT data provide useful insights and enhance evidence-based healthcare decision-making. These platforms collect and analyze data from a variety of sources, including as IoT devices, electronic health records, and healthcare systems. These systems can find patterns, trends, and correlations in massive datasets by utilizing advanced analytics approaches like machine learning and predictive modeling. This data allows healthcare clinicians to make educated decisions about treatment regimens, resource allocation, and population health management. Platforms for data analytics improve clinical outcomes, optimize resource utilization, and help to offer more efficient and proactive healthcare. These case studies demonstrate the numerous uses and advantages of IoT-enabled smart healthcare solutions. Healthcare organizations may reinvent care delivery, improve patient outcomes, and increase operational efficiency by utilizing IoT technologies [48]. These examples highlight the potential of IoT in revolutionizing healthcare and tackling industry concerns.

6.6 IMPLICATIONS OF IoT IN HEALTHCARE

The ramifications of IoT in healthcare are far-reaching and can potentially revolutionize the sector significantly. The following are the significant implications of IoT in healthcare.

6.6.1 DATA SECURITY AND PRIVACY PROTECTION

Concerns concerning data security and privacy have arisen as a result of the increased use of IoT in healthcare. Because IoT devices capture and send a growing quantity of sensitive patient data, healthcare organizations must employ comprehensive security measures to defend against unauthorized access, data breaches, and cyberattacks. To maintain the confidentiality, integrity, and availability of patient data, encryption techniques, secure authentication mechanisms, and frequent security audits are required [49]. Privacy rules and permission methods should also be in place to provide patients with a choice over how their data is gathered, used, and shared.

6.6.2 INTEROPERABILITY AND DATA STANDARDIZATION

Interoperability is a major issue in IoT-enabled healthcare systems. Standardization of data formats, communication protocols, and interfaces is required to integrate and easily transfer data across different IoT devices, healthcare providers, and systems. Data silos may impede effective care coordination, data exchange, and decision-making if they are not interoperable. Developing common standards and frameworks for data integration and interoperability is critical for realizing the full potential of IoT in healthcare and facilitating information transmission across platforms and devices [50].

6.6.3 ETHICAL CONSIDERATIONS AND PATIENT CONSENT

IoT devices in healthcare raises ethical issues with patient permission, autonomy, and data ownership. Healthcare organizations must ensure that patients are informed about the goal and extent of data gathering via IoT devices and that their informed consent is obtained. Transparent communication regarding data usage, sharing, and storage practices is critical for maintaining patient confidence and ethical norms [51]. Furthermore, ethical considerations should address potential biases in data collection and processing to avoid discrimination or unjust treatment based on IoT-generated data.

6.6.4 REGULATORY FRAMEWORKS AND COMPLIANCE REQUIREMENTS

The use of IoT in healthcare is constrained by legal frameworks and compliance constraints. To maintain compliance with privacy requirements (e.g., HIPAA, GDPR) and medical device restrictions, healthcare organizations must negotiate complicated regulatory environments. These rules require data protection, data breach reporting, informed consent, and the maintenance of data integrity and security. Keeping up with changing regulatory standards and proactively resolving compliance concerns are crucial for mitigating legal risks and ensuring ethical and responsible usage of IoT in healthcare. Considering the implications of IoT in healthcare necessitates a multifaceted strategy that includes technology solutions, legislative frameworks, and ethical issues [52]. Collaboration among healthcare providers, technology vendors, regulators, and legislators is essential in developing guidelines, standards, and best practices that prioritize data security, privacy protection, interoperability, ethical usage, and compliance. By properly addressing these consequences, healthcare organizations may fully realize the promise of IoT while guaranteeing patient safety, privacy, and confidence.

6.7 EMERGING TRENDS AND FUTURE DIRECTIONS

Emerging trends and future directions in cloud-IoT for healthcare show considerable potential for furthering the industry's transformation. Artificial intelligence (AI) and machine learning (ML) are projected to play critical roles in analyzing massive healthcare information, allowing accurate diagnoses, forecasting disease development, and enhancing personalized medication. Edge computing will minimize latency and enable real-time decision-making by moving data processing closer to the source, especially in time-sensitive healthcare applications. With its decentralized and tamper-resistant nature, blockchain technology will improve data security and privacy and enable secure exchange of healthcare records [53]. Furthermore, the future of cloud-IoT in healthcare will center on personalized medicine, precision healthcare, and utilizing technology for better patient outcomes and healthcare delivery. The emerging trends and future directions are illustrated in Figure 6.4 and explained below.

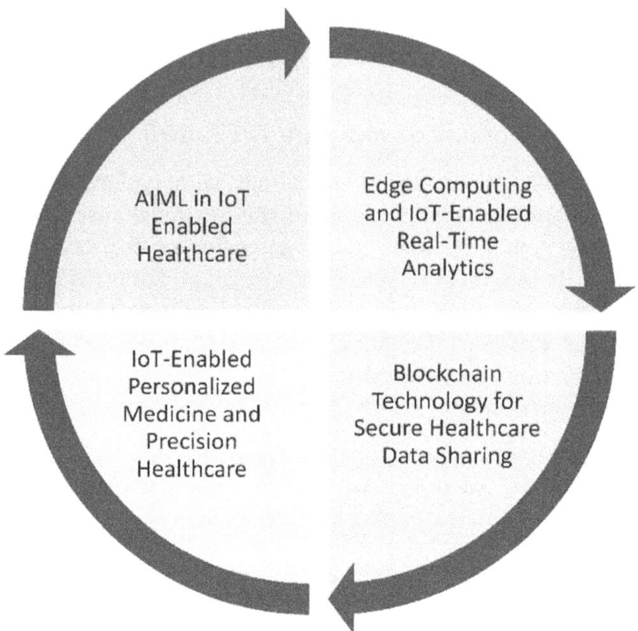

FIGURE 6.4   Emerging trends in IoT and cloud-enabled healthcare.

6.7.1 Artificial Intelligence and Machine Learning in IoT Healthcare

AI and ML have shown immense potential in revolutionizing the healthcare industry, particularly when combined with the IoT. The integration of AI and ML algorithms with IoT devices in healthcare has opened up new avenues for improving patient care, diagnostics, and overall operational efficiency [54]. The key aspects of AI and ML in IoT healthcare are summarized below in Table 6.1.

- Predictive Analytics: AI and ML algorithms can analyze massive volumes of patient data generated by IoT devices in order to uncover trends, detect abnormalities, and anticipate future health problems. This enables proactive and personalized healthcare actions, resulting in illness identification and prevention at an earlier stage.
- Remote Patient Monitoring: Sensor-equipped IoT devices may continually monitor patients' vital indicators such as heart rate, blood pressure, and glucose levels. AI and ML algorithms can analyze this real-time data, delivering useful insights to healthcare practitioners and enabling remote patient monitoring. This contributes to fewer hospital readmissions, better patient outcomes, and a better overall patient experience.
- Medical Imaging and Diagnostics: AI and ML algorithms can analyze medical pictures such as X-rays, MRIs, and CT scans to help radiologists spot anomalies and make correct diagnoses. This technique has the potential to improve diagnosis accuracy, reduce mistakes, and speed up treatment decisions.

TABLE 6.1

Key Aspects of AI and ML in IoT Healthcare

S. No.	Aspects of AI and ML in IoT Healthcare	Key Points
1.	Predictive Analytics	• AI and ML algorithms analyze IoT-generated patient data to identify trends, anomalies, and potential health issues. • AI and ML algorithms analyze IoT-generated patient data to identify trends, anomalies, and potential health issues.
2.	Remote Patient Monitoring	• IoT devices with sensors monitor patients' vital signs in real time. • IoT devices with sensors monitor patients' vital signs in real time. • Contributes to fewer hospital readmissions, improved patient outcomes, and enhanced patient experience.
3.	Medical Imaging and Diagnostics	• AI and ML algorithms analyze medical images (X-rays, MRIs, CT scans) to assist radiologists in detecting anomalies and making accurate diagnoses. • Enhances diagnostic accuracy, reduces errors, and accelerates treatment decisions.
4.	Drug Discovery and Personalized Therapy	• AI and ML analyze patient data, genetics, and clinical research to uncover personalized treatment options. • Precision medicine approach tailors therapies to individual patients based on their unique characteristics, leading to more effective outcomes with fewer side effects.

- Drug Discovery and Personalized therapy: To uncover personalized therapy alternatives, AI and ML algorithms may be used to analyze patient data, genetic information, and clinical studies. Precision medicine, in which therapies are personalized to individual patients based on their unique traits, offers more successful outcomes with fewer side effects [55].

6.7.2 EDGE COMPUTING AND IoT-ENABLED REAL-TIME ANALYTICS

The processing and analysis of data at the network's edge, closer to the data source, rather than transmitting it to a centralized cloud server, is referred to as edge computing. When integrated with IoT, edge computing offers real-time analytics and decision-making, providing various benefits in the healthcare domain, which are summarized below and summarized in Table 6.2.

- Reduced Latency: Edge computing reduces the latency associated with transferring data to a remote server for analysis by processing data closer to IoT devices. This is especially important in time-sensitive healthcare applications that need fast responses, such as emergency response systems or real-time patient monitoring [56].
- Capacity Optimization: Sending significant volumes of data from IoT devices to the cloud can burden network capacity. Edge computing enables data filtering and pre-processing at the edge, with only relevant insights

TABLE 6.2

Benefits of Edge Computing and IoT-Enabled Real-Time Analytics in Healthcare

S. No.	Benefits	Key Points
1.	Reduced Latency	• Edge computing processes data closer to IoT devices, reducing the delay associated with transferring data to a remote server for analysis. • Crucial for time-sensitive healthcare applications like emergency response and real-time patient monitoring, ensuring faster responses and decision-making.
2.	Capacity Optimization	• Edge computing filters and pre-processes data at the edge before sending relevant insights or summarized data to the cloud. • Prevents overwhelming network capacity with excessive data transmission, optimizing network bandwidth and reducing costs.
3.	Improved Privacy and Security	• Edge computing processes data locally, reducing the risk of data breaches during data transmission and cloud storage. • Particularly significant for healthcare data containing sensitive patient information, enhancing privacy and security in healthcare applications.
4.	Scalability and Reliability	• Edge computing can manage the high data volume generated by diverse IoT devices in healthcare. • Distributes processing power across multiple edge nodes, ensuring scalability, reliability, and reduced dependence on a single centralized infrastructure.

or summarized data sent to the cloud. This optimizes network bandwidth while lowering expenses.

• Improved Privacy and Security: Healthcare data frequently includes sensitive patient information. Edge computing decreases the danger of data breaches during transmission and storage in the cloud by processing data locally. It improves the privacy and security of healthcare applications.

• Scalability and dependability: Edge computing can handle the large data stream created by various IoT devices in healthcare. By dispersing processing power among several edge nodes, it assures scalability and stability while reducing reliance on a single centralized infrastructure [57].

6.7.3 Blockchain Technology for Secure Healthcare Data Sharing

Blockchain technology, originally developed for cryptocurrencies like Bitcoin, has found promising applications in healthcare data sharing [58]. To maintain security and privacy, the list of enhancements of blockchain in healthcare are listed below and summarized in Table 6.3.

• Data Integrity and Immutability: Blockchain provides a decentralized and distributed ledger in which healthcare data may be securely kept. This protects data integrity by preventing unauthorized changes or tampering.

TABLE 6.3

Blockchain Technology for Secure Healthcare Data Sharing

S. No.	Enhancements of Blockchain in Healthcare Data Sharing	Key Points
1.	Data Integrity and Immutability	• Blockchain provides a secure, decentralized ledger for healthcare data storage. • Ensures data integrity by preventing unauthorized changes or tampering. • Every transaction or data entry is recorded in a block, forming an immutable chain.
2.	Improved Privacy and Consent Management	• Patients gain more control over their medical data using blockchain. • Smart contracts and cryptographic techniques enable data access control and consent management. • Patients can share specific data with healthcare providers or researchers while remaining anonymous.
3.	Interoperability and Data Transmission	• Blockchain facilitates secure and seamless data transmission across healthcare institutions. • Creates a standardized and reliable platform for data exchange among hospitals, clinics, insurers, and more. • Boosts interoperability, reduces administrative overhead costs, and ensures trust in data exchange.
4.	Clinical Trials and Research	• Blockchain simplifies clinical trials by securely storing and exchanging trial data. • Enhances transparency, data quality, and auditability in clinical trial processes. • Improves reliability and efficiency of research results by creating a trusted data sharing environment.

Every transaction or data entry is recorded in a block connected to previous blocks to form an immutable chain.

- Improved Privacy and Consent Management: Blockchain gives patients more control over their medical data. Using smart contracts and cryptographic approaches, they may issue rights and regulate access to their data. This allows patients to share precise data with healthcare practitioners or researchers while remaining anonymous.
- Interoperability and Data Transmission: Blockchain allows for the secure and frictionless data transmission across various healthcare institutions such as hospitals, clinics, and insurers. It creates a standardized and trustworthy platform for data exchange across diverse systems, boosting interoperability and lowering administrative overhead [59].
- Clinical Trials and Research: By securely storing and exchanging trial data across researchers, sponsors, and regulatory agencies, blockchain technology can help simplify the clinical trial process. It improves openness, data quality, and auditability, resulting in more dependable and efficient research results.

6.7.4 IoT-Enabled Personalized Medicine and Precision Healthcare

IoT device integration in healthcare has cleared the path for personalized treatment and precision healthcare techniques. Table 6.4 presents various aspects of IoT-enabled personalized medicine and precision in healthcare.

- Continuous Monitoring and Data Collection: Sensor-enabled IoT devices capture real-time data on physiological metrics, medication adherence, activity levels, and environmental conditions. This continual data gathering allows a thorough assessment of a person's health state and behavioral tendencies [60].
- Personalized Treatment and Intervention: AI and ML algorithms may produce personalized insights and suggestions by analyzing data acquired from IoT devices. This might include anything from medication reminders and lifestyle changes to customized treatment regimens suited to an individual's specific requirements.
- Remote Patient Engagement and Care: IoT devices provide remote patient monitoring, allowing healthcare practitioners to monitor patients' health state and intervene as needed. This decreases the need for frequent hospital visits, increases patient convenience, and boosts overall patient participation in their own healthcare.

TABLE 6.4

IoT-Enabled Personalized Medicine and Precision Healthcare

S. No.	Aspects of IoT in Personalized Medicine	Key Points
1.	Continuous Monitoring and Data Collection	• IoT devices with sensors capture real-time physiological metrics, medication adherence, activity levels, and environmental conditions. • Provides continuous data to thoroughly assess individual health status and behavioral trends.
2.	Personalized Treatment and Intervention	• AI and ML algorithms analyze IoT-acquired data to generate personalized insights and recommendations. • Offers tailored suggestions, from medication reminders and lifestyle changes to individualized treatment plans.
3.	Remote Patient Engagement and Care	• IoT wearables and connected devices allow individuals to track health metrics like sleep patterns, activity, and stress. • Facilitates ongoing health state monitoring, reducing hospital visits, and increasing patient convenience and participation.
4.	Health and Wellness Tracking	• IoT wearables and connected devices allow individuals to track health metrics like sleep patterns, activity, and stress. • Promotes self-awareness, healthy behaviors, and preventive care.

- Health and Wellness Tracking: IoT wearables and linked gadgets enable people to measure their own health and wellness metrics, such as sleep patterns, activity levels, and stress levels. This increases self-awareness, promotes healthy behaviors, and aids in preventive care [61–63].

Overall, the confluence of IoT, AI, ML, edge computing, and blockchain technology has enormous promise for altering healthcare delivery, increasing patient outcomes, and driving medical research and innovation. These new trends and future directions have the potential to completely transform the healthcare business, making it more efficient, personalized, and secure.

6.8 CONCLUSION

Remote patient monitoring, telemedicine, wearable devices, smart hospital infrastructure, ambient supported living, medication adherence, and data-driven analytics have all been enabled by cloud-IoT. These breakthroughs empower individuals, optimize resource allocation, improve patient safety, encourage independent living, boost compliance, and provide important insights for enhanced diagnostics and personalized therapies. Cloud-IoT implementation in healthcare necessitates resolving important obstacles and concerns. Security and privacy are critical, necessitating stringent safeguards to safeguard sensitive data. Interoperability and data integration are difficult tasks that require standardized communication protocols and formats for easy data sharing. The dependability and availability of cloud-IoT devices and networks are crucial for ensuring continuous service. Informed permission, data ownership, and regulatory compliance (e.g., HIPAA) are all important ethical and legal considerations. Navigating these hurdles successfully will promote the safe, interoperable, and ethical use of cloud-IoT in healthcare. Looking ahead, the future of cloud-IoT in healthcare seems bright. The healthcare environment is projected to be further revolutionized by emerging trends and technology. AI and ML algorithms will be critical in extracting relevant insights from massive healthcare databases, enabling accurate diagnostics, forecasting disease development, and enhancing personalized medication. Edge computing, in which data processing and analytics take place closer to the data source, will minimize latency and enable real-time decision-making, especially in time-critical healthcare circumstances. Because of its decentralized and tamper-resistant nature, blockchain technology will improve data security and privacy while enabling the secure exchange of healthcare records among various providers and parties. Cloud-IoT will tremendously aid the notion of personalized medicine and precision healthcare. The combination of genetics, medical wearables, and AI-based analytics will enable personalized therapies, early illness identification, and proactive interventions, eventually leading to better patient outcomes. Finally, cloud-IoT has made considerable progress in altering healthcare systems, revolutionizing patient care, and optimizing healthcare delivery. However, resolving security, interoperability, dependability, and moral issues is critical for effective adoption. As time goes on, adopting emerging technologies and trends will help us realize the entire promise of cloud-IoT, paving the way for a smarter and more personalized future in healthcare.

REFERENCES

1. Botta, Alessio, Walter De Donato, Valerio Persico, and Antonio Pescapé. "Integration of cloud computing and internet of things: A survey." *Future Generation Computer Systems* 56 (2016): 684–700.
2. Shah, Junaid Latief, Heena Farooq Bhat, and Asif Iqbal Khan. "Integration of Cloud and IoT for smart e-healthcare." In *Healthcare paradigms in the internet of things ecosystem*, pp. 101–136. Academic Press, 2021.
3. Uslu, Banu Çalış, Ertuğ Okay, and Erkan Dursun. "Analysis of factors affecting IoT-based smart hospital design." *Journal of Cloud Computing* 9, no. 1 (2020): 1–23.
4. Aldeer, Murtadha, Mehdi Javanmard, and Richard P. Martin. "A review of medication adherence monitoring technologies." *Applied System Innovation* 1, no. 2 (2018): 14.
5. Selvaraj, Sureshkumar, and Suresh Sundaravaradhan. "Challenges and opportunities in IoT healthcare systems: A systematic review." *SN Applied Sciences* 2, no. 1 (2020): 139.
6. Kumar, Neelam Sanjeev, and P. Nirmalkumar. "A novel architecture of smart healthcare system on integration of cloud computing and IoT." In *2019 International Conference on Communication and Signal Processing (ICCSP)*, pp. 0940–0944. IEEE, 2019.
7. Garg, Ankit, Ashima Gambhir, and Prachi Goel. "IoT security, privacy, challenges, and solutions." *Trust-Based Communication Systems for Internet of Things Applications* (2022): 53–91.
8. Goyal, Prachi, Ankit Garg, and Prakhar Jindal. "Comparative analysis of indexing schemes used in cloud computing data management." *Trust-Based Communication Systems for Internet of Things Applications* (2022): 135–158.
9. Isravel, Deva Priya, and Salaja Silas. "A comprehensive review on the emerging IoT-cloud based technologies for smart healthcare." In *2020 6th International Conference on Advanced Computing and Communication Systems (ICACCS)*, pp. 606–611. IEEE, 2020.
10. Butpheng, Chanapha, Kuo-Hui Yeh, and Hu Xiong. "Security and privacy in IoT-cloud-based e-health systems—A comprehensive review." *Symmetry* 12, no. 7 (2020): 1191.
11. Sadek, Ibrahim, Shafiq Ul Rehman, Josué Codjo, and Bessam Abdulrazak. "Privacy and security of IoT based healthcare systems: concerns, solutions, and recommendations." In *How AI Impacts Urban Living and Public Health: 17th International Conference, ICOST 2019, New York City, NY, USA, October 14–16, 2019, Proceedings 17*, pp. 3–17. Springer International Publishing, 2019.
12. Yoo, Hyun, Roy C. Park, and Kyungyong Chung. "IoT-based health big-data process technologies: A survey." *KSII Transactions on Internet & Information Systems* 15, no. 3 (2021): 974–992.
13. Lakshmi, G. Jaya, Mangesh Ghonge, and Ahmed J. Obaid. "Cloud based IoT smart healthcare system for remote patient monitoring." *EAI Endorsed Transactions on Pervasive Health and Technology* 7, no. 28 (2021): e4–e4.
14. Kumar, J. Naveen Ananda, and Shivani Suresh. "A proposal of smart hospital management using hybrid cloud, IoT, ML, and AI." In *2019 International Conference on Communication and Electronics Systems (ICCES)*, pp. 1082–1085. IEEE, 2019.
15. Verma, Priyanka, and Rajan Mishra. "IoT based smart remote health monitoring system." In *2020 International Conference on Electrical and Electronics Engineering (ICE3)*, pp. 467–470. IEEE, 2020.
16. Singh, Anuj Kumar, Ankit Garg, and Anand Nayyar. "Blockchain for security and privacy in healthcare informatics." *Innovations in Healthcare Informatics: From Interoperability to Data Analysis* (2023): 157.
17. Garg, Ankit, Anuj Kumar Singh, and Mohit Garg. "Role of internet of things and artificial intelligence for healthcare informatics: An overview." *Innovations in Healthcare Informatics: From Interoperability to Data Analysis* (2023): 107.

18. Mahapatra, Bandana, Rajalakshmi Krishnamurthi, and Anand Nayyar. "Healthcare models and algorithms for privacy and security in healthcare records." *Security and Privacy of Electronic Healthcare Records: Concepts, Paradigms and Solutions* (2019): 183.
19. Fadi, Al-Turjman, Anand Nayyar, Ajantha Devi, and Piyush Kumar Shukla, eds. *Intelligence of things: AI-IoT based critical-applications and innovations.* New York, US: Springer, 2021.
20. Jara, Antonio J., Miguel A. Zamora, and Antonio FG Skarmeta. "An internet of things–based personal device for diabetes therapy management in ambient assisted living (AAL)." *Personal and Ubiquitous Computing* 15 (2011): 431–440.
21. Stojanova, Aleksandra, Saso Koceski, and Natasa Koceska. "Continuous blood pressure monitoring as a basis for ambient assisted living (AAL)–review of methodologies and devices." *Journal of Medical Systems* 43 (2019): 1–12.
22. Mauro, Joseph, Kelly B. Mathews, and Eric S. Sredzinski. "Effect of a smart pill bottle and pharmacist intervention on medication adherence in patients with multiple myeloma new to lenalidomide therapy." *Journal of Managed Care & Specialty Pharmacy* 25, no. 11 (2019): 1244–1254.
23. Park, Hyang Rang, Hee Sun Kang, Soo Hyun Kim, and Savitri Singh-Carlson. "Effect of a smart pill bottle reminder intervention on medication adherence, self-efficacy, and depression in breast cancer survivors." *Cancer Nursing* 45, no. 6 (2022): E874–E882.
24. Abdulmalek, Suliman, Abdul Nasir, Waheb A. Jabbar, Mukarram AM Almuhaya, Anupam Kumar Bairagi, Md Al-Masrur Khan, and Seong-Hoon Kee. "IoT-based healthcare-monitoring system towards improving quality of life: A review." *Healthcare* 10, no. 10 (1993): 2022.
25. Javaid, Mohd, Abid Haleem, Ravi Pratap Singh, Rajiv Suman, and Shanay Rab. "Significance of machine learning in healthcare: Features, pillars and applications." *International Journal of Intelligent Networks* 3 (2022): 58–73.
26. Pradhan, Bikash, Saugat Bhattacharyya, and Kunal Pal. "IoT-based applications in healthcare devices." *Journal of Healthcare Engineering* 2021 (2021): 1–18.
27. Bharathi, R., T. Abirami, S. Dhanasekaran, Deepak Gupta, Ashish Khanna, Mohamed Elhoseny, and K. Shankar. "Energy efficient clustering with disease diagnosis model for IoT based sustainable healthcare systems." *Sustainable Computing: Informatics and Systems* 28 (2020): 100453.
28. Mansour, Romany Fouad, Adnen El Amraoui, Issam Nouaouri, Vicente García Díaz, Deepak Gupta, and Sachin Kumar. "Artificial intelligence and internet of things enabled disease diagnosis model for smart healthcare systems." *IEEE Access* 9 (2021): 45137–45146.
29. Khanna, Ashish, Pandiaraj Selvaraj, Deepak Gupta, Tariq Hussain Sheikh, Piyush Kumar Pareek, and Vishnu Shankar. "Internet of things and deep learning enabled healthcare disease diagnosis using biomedical electrocardiogram signals." *Expert Systems* 40, no. 4 (2023): e12864.
30. Alzubi, Omar A., Jafar A. Alzubi, K. Shankar, and Deepak Gupta. "Blockchain and artificial intelligence enabled privacy-preserving medical data transmission in internet of things." *Transactions on Emerging Telecommunications Technologies* 32, no. 12 (2021): e4360.
31. Abirami, S., and P. Chitra. "Energy-efficient edge based real-time healthcare support system." In *Advances in computers*, vol. 117, no. 1, pp. 339–368. Elsevier, 2020.
32. Nguyen, Hoa Hong, Farhaan Mirza, M. Asif Naeem, and Minh Nguyen. "A review on IoT healthcare monitoring applications and a vision for transforming sensor data into real-time clinical feedback." In *2017 IEEE 21st International Conference on Computer Supported Cooperative Work in Design (CSCWD)*, pp. 257–262. IEEE, 2017.

33. Khuntia, Jiban, Dobin Yim, Mohan Tanniru, and Sanghee Lim. "Patient empowerment and engagement with a health infomediary." *Health Policy and Technology* 6, no. 1 (2017): 40–50.

34. Graffigna, Guendalina, Serena Barello, and Stefano Triberti. *Patient engagement: A consumer-centered model to innovate healthcare.* Walter de Gruyter GmbH & Co KG, 2016.

35. Qi, Jun, Po Yang, Geyong Min, Oliver Amft, Feng Dong, and Lida Xu. "Advanced internet of things for personalised healthcare systems: A survey." *Pervasive and mobile Computing* 41 (2017): 132–149.

36. Demirkan, Haluk. "A smart healthcare systems framework." *It Professional* 15, no. 5 (2013): 38–45.

37. Shuaib, Mohammed, Shadab Alam, Mohammad Shabbir Alam, and Mohammad Shahnawaz Nasir. "Compliance with HIPAA and GDPR in blockchain-based electronic health record." *Materials Today: Proceedings*, 2021.

38. Singh, Anuj Kumar, Anand Nayyar, and Ankit Garg. "A secure elliptic curve based anonymous authentication and key establishment mechanism for IoT and cloud." *Multimedia Tools and Applications* (2022): 1–52.

39. Singh, Anuj Kumar, and B. D. K. Patro. "Security of low computing power devices: A survey of requirements, challenges & possible solutions." *Cybernetics and Information Technologies* 19, no. 1 (2019): 133–164.

40. Alasmari, Sultan, and Mohd Anwar. "Security & privacy challenges in IoT-based health cloud." In *2016 International Conference on Computational Science and Computational Intelligence (CSCI)*, pp. 198–201. IEEE, 2016.

41. Balakrishna, Sivadi, M. Thirumaran, and Vijender Kumar Solanki. "IoT sensor data integration in healthcare using semantics and machine learning approaches." *A Handbook of Internet of Things in Biomedical and Cyber Physical System* (2020): 275–300.

42. Balakrishna, Sivadi, and M. Thirumaran. "Semantic interoperability in IoT and big data for health care: A collaborative approach." In *Handbook of data science approaches for biomedical engineering*, pp. 185–220. Academic Press, 2020.

43. Iyengar, Arun, Ashish Kundu, and George Pallis. "Healthcare informatics and privacy." *IEEE Internet Computing* 22, no. 2 (2018): 29–31.

44. Abdellatif, Mohammad M., and Walaa Mohamed. "Telemedicine: An IoT based remote healthcare system." *International Journal of Online & Biomedical Engineering* 16, no. 6 (2020): 72–81.

45. Xie, Yi, Lin Lu, Fei Gao, Shuang-jiang He, Hui-juan Zhao, Ying Fang, Jia-ming Yang, Ying An, Zhe-Wei Ye, and Zhe Dong. "Integration of artificial intelligence, blockchain, and wearable technology for chronic disease management: A new paradigm in smart healthcare." *Current Medical Science* 41 (2021): 1123–1133.

46. Thangaraj, Muthuraman, Pichaiah Punitha Ponmalar, and Subramanian Anuradha. "Internet Of Things (IOT) enabled smart autonomous hospital management system-A real world health care use case with the technology drivers." In *2015 IEEE International Conference on Computational Intelligence and Computing Research (ICCIC)*, pp. 1–8. IEEE, 2015.

47. Memon, Mukhtiar, Stefan Rahr Wagner, Christian Fischer Pedersen, Femina Hassan Aysha Beevi, and Finn Overgaard Hansen. "Ambient assisted living healthcare frameworks, platforms, standards, and quality attributes." *Sensors* 14, no. 3 (2014): 4312–4341.

48. López-Martínez, Fernando, Edward Rolando Núñez-Valdez, Vicente García-Díaz, and Zoran Bursac. "A case study for a big data and machine learning platform to improve medical decision support in population health management." *Algorithms* 13, no. 4 (2020): 102.

49. Gupta, Ayan Das, Shaik Mohammad Rafi, Balaji Ramkumar Rajagopal, T. Milton, and S. G. Hymlin. "Comparative analysis of internet of things (IoT) in supporting the health care professionals towards smart health research using correlation analysis." *Bulletin of Environment, Pharmacology and Life Sciences* 1 (2022): 701–708.

50. Kaur, Pavleen, Ravinder Kumar, and Munish Kumar. "A healthcare monitoring system using random forest and internet of things (IoT)." *Multimedia Tools and Applications* 78 (2019): 19905–19916.

51. Mishra, Sushruta, Brojo Kishore Mishra, Hrudaya Kumar Tripathy, and Arijit Dutta. "Analysis of the role and scope of big data analytics with IoT in health care domain." In *Handbook of data science approaches for biomedical engineering*, pp. 1–23. Academic Press, 2020.

52. Darwish, Ashraf, Aboul Ella Hassanien, Mohamed Elhoseny, Arun Kumar Sangaiah, and Khan Muhammad. "The impact of the hybrid platform of internet of things and cloud computing on healthcare systems: Opportunities, challenges, and open problems." *Journal of Ambient Intelligence and Humanized Computing* 10 (2019): 4151–4166.

53. Tuli, Shreshth, Shikhar Tuli, Gurleen Wander, Praneet Wander, Sukhpal Singh Gill, Schahram Dustdar, Rizos Sakellariou, and Omer Rana. "Next generation technologies for smart healthcare: Challenges, vision, model, trends and future directions." *Internet Technology Letters* 3, no. 2 (2020): e145.

54. Sharma, Vahiny, Ankur Gupta, Najam Ul Hasan, Mohammad Shabaz, and Isaac Ofori. "Blockchain in secure healthcare systems: State of the art, limitations, and future directions." *Security and Communication Networks* 2022 (2022): 1–15.

55. Sharma, Ravi, and Nir Kshetri. "Digital healthcare: Historical development, applications, and future research directions." *International Journal of Information Management* 53 (2020): 102105.

56. Sarangi, Anish Kumar, Ambarish Gajendra Mohapatra, Tarini Charan Mishra, and Bright Keswani. "Healthcare 4.0: A voyage of fog computing with IoT, cloud computing, big data, and machine learning." *Fog Computing for Healthcare 4.0 Environments: Technical, Societal, and Future Implications* (2021): 177–210.

57. Lakshmi, L., A. Naga Kalyani, G. Naga Satish, D. Swapna, and M. Purushotham Reddy. "The preeminence of Fog Computing and IoT enabled Cloud Systems in Health care." In *2021 Third International Conference on Intelligent Communication Technologies and Virtual Mobile Networks (ICICV)*, pp. 368–375. IEEE, 2021.

58. Siyal, Asad Ali, Aisha Zahid Junejo, Muhammad Zawish, Kainat Ahmed, Aiman Khalil, and Georgia Soursou. "Applications of blockchain technology in medicine and healthcare: Challenges and future perspectives." *Cryptography* 3, no. 1 (2019): 3.

59. Veeramakali, T., R. Siva, B. Sivakumar, P.C. Senthil Mahesh, and N. Krishnaraj. "An intelligent internet of things-based secure healthcare framework using blockchain technology with an optimal deep learning model." *The Journal of Supercomputing* 77 (2021): 1–21.

60. Ahmed, Imran, Gwanggil Jeon, and Abdellah Chehri. "An IoT-enabled smart health care system for screening of COVID-19 with multi layers features fusion and selection." *Computing* 105 (2022): 1–18.

61. Jagatheesaperumal, Senthil Kumar, Preeti Mishra, Nour Moustafa, and Rahul Chauhan. "A holistic survey on the use of emerging technologies to provision secure healthcare solutions." *Computers and Electrical Engineering* 99 (2022): 107691.

62. Garg, Ankit, and Anuj Kumar Singh. "Internet of things (IoT): Security, cybercrimes, and digital forensics." In *Internet of things and cyber physical systems*, pp. 23–50. CRC Press, 2022.

63. Garg, Ankit, and Anuj Kumar Singh. "Applications of internet of things (IoT) in green computing." *Intelligence of Things: AI-IoT Based Critical-Applications and Innovations* (2021): 1–34.

7 Bridging the Gap
Harnessing Cloud Computing and IoT for Wildlife Conservation

Yogita Yashveer Raghav and Vaibhav Vyas

7.1 INTRODUCTION

Wildlife conservation refers to the practice of protecting and preserving the natural habitats, populations, and diversity of wild animals and plants. The primary goal is to ensure the sustainable existence of various species and maintain the health and balance of ecosystems. This involves the management of habitats, prevention of overexploitation, and the implementation of measures to address threats such as habitat loss, pollution, climate change, and illegal activities like poaching [1].

7.1.1 SIGNIFICANCE OF WILDLIFE CONSERVATION

7.1.1.1 Biodiversity Preservation

Wildlife conservation is crucial for maintaining biodiversity, which refers to the variety of life on Earth. Biodiversity ensures ecosystem stability, resilience, and adaptability to environmental changes.

7.1.1.2 Ecological Balance

Wild animals and plants play specific roles within ecosystems, contributing to ecological balance. Conservation efforts help prevent disruptions to these systems, preserving the relationships between species and their environments [2].

7.1.1.3 Ecosystem Services

Wildlife provides essential ecosystem services, including pollination, seed dispersal, and regulation of insect populations. Conservation ensures the continuation of these services, benefiting both natural and human-made environments.

7.1.1.4 Scientific and Medical Discoveries

Many species of plants and animals have unique biochemical compounds that contribute to scientific and medical advancements. Conservation efforts can lead to discovering new species or valuable genetic resources for pharmaceutical and medical research [3].

DOI: 10.1201/9781032656694-7

7.1.1.5 Cultural and Aesthetic Value

Wildlife has cultural and aesthetic significance. Many species are integral to indigenous cultures, folklore, and traditions. Additionally, people derive recreational and aesthetic value from observing and interacting with diverse wildlife.

7.1.1.6 Economic Benefits

Sustainable wildlife management contributes to the economy through activities such as ecotourism, which relies on the presence of diverse and healthy ecosystems. Conservation can enhance the long-term economic viability of such activities.

7.1.1.7 Climate Change Mitigation

Healthy ecosystems, supported by diverse wildlife, contribute to climate change mitigation. Forests, for example, act as carbon sinks, helping to regulate the Earth's climate and mitigate the impacts of global warming [4].

7.1.1.8 Education and Research

Wildlife conservation provides opportunities for education and research. Studying wildlife helps scientists understand ecosystems, species interactions, and the broader natural world. This knowledge is vital for making informed conservation decisions.

7.1.1.9 Human Well-Being

The well-being of human populations is closely linked to the health of ecosystems and wildlife. Conservation efforts contribute to sustainable resource management, clean water, and overall environmental quality, benefiting both present and future generations.

7.1.1.10 Ethical Responsibility

Many consider the conservation of wildlife and ecosystems an ethical responsibility. It reflects a commitment to preserving the intrinsic value of all living organisms and recognizing the interconnectedness of life on Earth.

In summary, wildlife conservation is essential for maintaining the planet's health, supporting human well-being, and preserving the rich tapestry of life for future generations. It involves combining scientific research, habitat protection, sustainable management practices, and global cooperation to address the complex challenges facing ecosystems and species worldwide [5].

7.1.2 Overview of Cloud Computing and IoT in Wildlife Preservation

Wildlife preservation has entered a new era by integrating cutting-edge technologies, including cloud computing and the Internet of Things (IoT). These technologies offer innovative solutions to address the challenges faced by conservationists and researchers in monitoring, protecting, and understanding wildlife. Here's an overview of how cloud computing and IoT are transforming wildlife preservation:

7.1.2.1   Cloud Computing in Wildlife Preservation

7.1.2.1.1   Data Storage and Management

Cloud platforms provide scalable and secure storage solutions for the massive amounts of data generated in wildlife preservation efforts. This includes data from satellite imagery, sensor networks, and tracking devices.

7.1.2.1.2   Data Processing and Analytics

Advanced analytics on the cloud enable researchers to process and analyze large datasets rapidly. This is crucial for gaining insights into animal behavior, habitat changes, and population dynamics [6].

7.1.2.1.3   Collaboration and Information Sharing

Cloud-based collaboration platforms facilitate information sharing among researchers, conservation organizations, and governmental agencies. This promotes a more integrated and cooperative approach to wildlife preservation on a global scale.

7.1.2.1.4   Real-time Monitoring

Cloud computing supports real-time monitoring applications, allowing conservationists to track and respond to events such as illegal activities, natural disasters, or changes in animal behavior promptly.

7.1.2.1.5   Resource Optimization

Cloud-based solutions help optimize resource management by providing on demand access to computational resources. This is particularly beneficial for running complex simulations, predictive modeling, and other resource-intensive tasks [7].

7.1.2.2   IoT in Wildlife Preservation

7.1.2.2.1   GPS Tracking and Telemetry

IoT devices, such as global positioning system (GPS) collars and satellite tags, are used to track the movements of individual animals. These devices provide real-time location data, contributing to the study of migration patterns and habitat use.

7.1.2.2.2   Sensor Networks

Environmental sensors, including temperature, humidity, and sound sensors, form networks in wildlife habitats. These sensors collect data on the ecosystem's health and dynamics, providing valuable information for conservation planning.

7.1.2.2.3   Camera Traps and Smart Cameras

IoT-connected cameras, commonly known as camera traps, capture images and videos of wildlife. These devices, equipped with motion sensors, contribute to population monitoring and anti-poaching efforts.

7.1.2.2.4   Acoustic Monitoring

IoT-enabled audio sensors are used to monitor sounds in wildlife habitats. This technology aids in identifying species, detecting distress calls, and monitoring animal behavior.

7.1.2.2.5 Citizen Science and Mobile Apps

IoT is leveraged in citizen science initiatives through mobile apps. Individuals can use their smartphones to contribute data on wildlife sightings, helping researchers gather information on species distribution and behavior [8].

7.1.2.2.6 Wildlife Health Monitoring

Implantable or wearable biometric sensors on animals provide real-time health data, allowing researchers to monitor the well-being of individuals and populations.

7.1.2.2.7 Anti-Poaching Technologies

Smart camera networks, drones, and other IoT devices are employed to detect and deter illegal activities, providing an additional layer of security for protected areas.

7.1.2.2.8 Connectivity and Communication

IoT devices enhance communication among different components of a conservation system. This connectivity ensures that data from various sensors and devices can be seamlessly integrated to provide a comprehensive understanding of ecosystems.

Benefits of the Integration

Comprehensive Data Insights: The combination of cloud computing and IoT enables the collection, storage, and analysis of diverse data types, providing conservationists with a more comprehensive understanding of wildlife and ecosystems.

Real-time Decision-Making: Cloud-based analytics and IoT sensors contribute to real-time decision-making, allowing for swift responses to emerging threats or changing environmental conditions.

Global Collaboration: Cloud platforms facilitate global collaboration by providing a centralized space for data sharing and collaboration among researchers, conservation organizations, and policymakers.

Efficient Resource Allocation: Cloud computing allows for the efficient allocation of computational resources, optimizing the processing of large datasets and enabling sophisticated modeling for conservation planning [9–11].

Enhanced Monitoring and Protection: IoT devices contribute to enhanced monitoring, protection, and management of wildlife and ecosystems, offering a more proactive and responsive approach to conservation challenges.

The integration of cloud computing and IoT in wildlife preservation is a dynamic and evolving field, offering new possibilities for conservationists to leverage technology for the sustainable management of biodiversity and the protection of our planet's ecosystems.

7.2 LITERATURE SURVEY

The study [1] explores the deployment and performance of GPS collars on nonhuman primates. Over the past two decades, GPS collars have become invaluable tools for studying the movement ecology of nonhuman primates. The review outlines the wide range of primate taxa on which GPS collars have been deployed, offering insights into the diverse applications and effectiveness of this technology. It provides an overview of GPS collar manufacturers and collar designs, contributing to the understanding of the technical aspects involved in primate tracking. The authors highlight the significance of GPS collars in advancing research on nonhuman primates, emphasizing their role in studying movement patterns, behavior, and ecological interactions. With 30 citations, this review consolidates knowledge in the field, making it a valuable resource for researchers and practitioners interested in primate ecology and conservation.

The study [2] addresses the challenges faced by early researchers using GPS technology for terrestrial mammal tracking in Australia. Collated data from 24 studies cover 280 GPS collar deployments on 13 species, including dingoes and other wild dogs. The paper discusses issues related to movement patterns and behavior of wild mammals tracked with GPS wildlife telemetry collars. The authors highlight the increasing use of GPS collars to understand the movement patterns of wild mammals globally. Despite the challenges faced, the success of these deployments demonstrates the efficacy of GPS technology in studying wildlife. The comprehensive survey provides valuable insights for researchers and wildlife practitioners involved in mammal ecology and conservation.

The authors [3] explore the synergies between cloud computing and the IoT two domains, presenting a comprehensive analysis of their main properties, features, underlying technologies, and open issues. This study contributes to the emerging field by introducing the CloudIoT paradigm, emphasizing the transformative potential of merging cloud and IoT technologies. The authors discuss the disruptive nature of this integration, foreseeing its role as an enabler for a myriad of application scenarios.

The study [4] explores the transformative roles of cloud computing, the IoT, and artificial intelligence (CIA) in sustainable aquaculture development. The authors highlight the revolutionary impact of these technologies on aquaculture, leveraging tools such as drones to enhance various aspects of the industry. The review delves into motivations and challenges related to food production in aquaculture, shedding light on the integration of advanced technologies. By examining the applications and implications of cloud computing, IoT, and artificial intelligence, the study provides valuable insights into their collective contributions to sustainability in aquaculture.

Research authors [5] conducted an insightful literature survey in Big Data and Cognitive Computing on IoT technologies for livestock management. The review comprehensively addresses the current state of IoT applications in this domain, shedding light on present opportunities and paving the way for future trends. The authors explore the utilization of Machine Learning (ML) and highlight the synergies with Deep Learning (DL) for predictive abilities in livestock management. The study provides a valuable resource for understanding the advancements in IoT, emphasizing its potential impact on enhancing efficiency and sustainability in livestock practices.

Authors [6] conducted a comprehensive literature survey on the intersection of Industry 4.0 and health, specifically focusing on the IoT, big data, and cloud computing for Healthcare 4.0. The study systematically explores the adoption and integration of these technologies, providing insights into their roles and applications in the healthcare domain. By selectively analyzing the existing literature, the authors shed light on the advancements and potential benefits of leveraging Industry 4.0 technologies to enhance healthcare practices. The paper contributes to understanding how integrating IoT, big data, and cloud computing can transform healthcare systems, paving the way for what is referred to as Healthcare 4.0.

A study [7] explored the potential of the IoT in agriculture, focusing on its contributions to sustainable rural development. The research investigates the role of IoT technologies in poverty reduction in rural areas. By examining the applications and impacts of IoT in the agricultural sector, the study provides insights into how these technologies can enhance productivity, efficiency, and overall rural development.

Another research [8] delved into the relationship between big data analytics and smart cities, exploring whether this coupling is loose or tight. This scrutinizes the dynamics and interplay between big data analytics and the evolving concept of smart cities. Investigating the challenges and opportunities posed by integrating big data analytics in urban development, the authors contribute valuable insights to the discourse on the synergy between data analytics and smart city initiatives.

One study [9] provides a comprehensive review of the challenges and opportunities posed by the massive growth of big data in the context of cloud computing. The authors highlight the significant scale of data generated through cloud computing and emphasize the complexity and time-demanding nature of addressing big data. They delve into open research issues, shedding light on the implications of big data on cloud computing. This literature serves as a valuable resource for understanding the evolving landscape of big data and cloud computing, offering insights into potential avenues for further exploration and research in this dynamic field.

7.3 CASE STUDIES ON THE SUCCESSFUL DEPLOYMENT OF GPS COLLARS AND ENVIRONMENTAL SENSORS

- **Yellowstone National Park Wolf Tracking:**
  Objective: Monitoring and studying the behavior and movement patterns of gray wolves in Yellowstone National Park.
  Technology Used: GPS collars were deployed on gray wolves to track their movements and analyze their behavior.
  Outcome: Researchers gained valuable insights into wolf territories, migration patterns, and interactions with other wildlife. The data collected helped in understanding the ecological impact of wolf reintroduction to Yellowstone [11].
- **Cheetah Conservation in Namibia:**
  Objective: Conservation of cheetah populations in Namibia, focusing on understanding their behavior, habitat use, and threats.
  Technology Used: GPS collars were attached to cheetahs to track their movements, and environmental sensors were deployed to monitor habitat conditions.
  Outcome: The data collected contributed to the development of effective conservation strategies. Researchers were able to identify critical habitats, assess human-wildlife conflict areas, and implement measures to reduce conflicts and promote coexistence [1, 12].
- **Rainforest Environmental Monitoring in the Amazon:**
  Objective: Monitoring environmental changes in the Amazon rainforest to understand deforestation patterns and their impact on biodiversity.

Technology Used: Environmental sensors were deployed to collect data on temperature, humidity, and other environmental factors. GPS collars were used on certain species to track their movements in response to habitat changes.

Outcome: Integrating GPS tracking and environmental sensors provided a comprehensive understanding of how different species respond to environmental changes. The data collected aided in conservation planning, including identifying areas at high risk of deforestation [2, 13].

- **Elephant Conservation in Africa:**

Objective: Protecting elephant populations and minimizing human-elephant conflict in certain regions of Africa.

Technology Used: GPS collars were fitted on elephants to track their migration patterns and identify potential conflict zones. Environmental sensors were used to monitor vegetation changes.

Outcome: The data obtained from GPS collars and environmental sensors helped conservationists map out elephant corridors, predict migration routes, and implement strategies to reduce conflicts with local communities. This contributed to the development of more effective conservation policies.

These case studies highlight the versatility of GPS collars and environmental sensors in wildlife conservation. By combining location tracking with environmental data, researchers and conservationists can gain a deeper understanding of species behavior, habitat conditions, and the impact of human activities on wildlife. Ongoing advancements in technology continue to enhance the effectiveness of these monitoring tools in conservation efforts [11–14].

7.4 ETHICAL CONSIDERATIONS IN THE USE OF TECHNOLOGY FOR CONSERVATION

The use of technology in conservation presents various ethical considerations that require careful attention to ensure responsible and sustainable practices. Here are some key ethical considerations in the use of technology for conservation:

Privacy and Data Security

Issue: Tracking technologies, such as GPS collars and camera traps, may collect data on individual animals. There's a risk of infringing on the privacy of wildlife, particularly in species with small populations.

Ethical Approach: Conservationists must prioritize the privacy and welfare of individual animals. Data should be handled securely, and efforts should be made to anonymize or aggregate data to minimize the risk of harm.

Informed Consent for Tracking

Issue: Tracking technologies may involve attaching devices to animals, potentially causing stress or discomfort. In some cases, researchers might not obtain explicit consent from the animals.

Ethical Approach: Researchers should prioritize the welfare of animals, ensuring that tracking methods are minimally invasive and do not compromise the health or behavior of the tracked individuals. Protocols should be in place to evaluate and minimize potential negative impacts [15, 16].

Accuracy and Reliability of Technology

Issue: The use of technology, such as ML algorithms, may be prone to errors or biases. Incorrect data could lead to flawed conservation decisions.

Ethical Approach: Conservation practitioners should critically evaluate the accuracy and reliability of technological tools. Transparency in algorithmic decision-making is essential, and efforts should be made to address biases and improve accuracy.

Open Data and Accessibility

Issue: The sharing of conservation data, including sensitive information on species locations or vulnerable ecosystems, raises concerns about potential exploitation or harm.

Ethical Approach: While sharing data is essential for collaborative conservation efforts, the potential risks should be carefully considered. Data sharing should follow ethical guidelines, ensuring that sensitive information is appropriately protected and shared only with trusted partners.

Community Engagement and Consent

Issue: Implementing technology in conservation projects may impact local communities. There might be ethical concerns if communities are not adequately informed or involved in decision-making.

Ethical Approach: Conservation projects should prioritize community engagement, respecting local knowledge, and obtaining informed consent. Communities should be involved in designing and implementing technology-based initiatives, ensuring that their perspectives and concerns are considered.

Equitable Access to Technology

Issue: Unequal access to technology may result in disparities in conservation efforts. Some regions or communities may benefit more from technological solutions than others.

Ethical Approach: Efforts should be made to ensure equitable access to technology, addressing issues of accessibility, affordability, and digital literacy. Conservation initiatives should strive to bridge technological gaps and avoid exacerbating existing inequalities.

Long-Term Environmental Impact of Technology

Issue: The production, use, and disposal of technological devices may contribute to environmental degradation and electronic waste.

Ethical Approach: Conservationists should consider the life cycle of technology, from manufacturing to disposal. Sustainable practices, such as using eco-friendly materials and recycling, should be prioritized [17, 18].

Consent for Citizen Science and Crowdsourcing

Issue: Citizen science initiatives and crowdsourcing projects involve the public in data collection. Ethical concerns may arise if participants are not fully informed or their contributions are misused.

Ethical Approach: Clear guidelines and informed consent processes should be established for citizen science projects. Participants should understand the project's purpose, how their data will be used, and the potential impacts.

Ethical Use of AI in Conservation

Issue: The use of artificial intelligence in conservation, such as in predictive modeling or automated monitoring, raises ethical concerns related to accountability, transparency, and unintended consequences.

Ethical Approach: Transparency in developing and deploying artificial intelligence (AI) systems is crucial. Conservation practitioners should consider the potential biases in AI algorithms and actively work to mitigate them. Regular ethical assessments should be conducted to ensure responsible AI use.

Cultural Sensitivity

Issue: Conservation technologies may be deployed in areas with diverse cultural contexts. Failure to consider local beliefs and practices may lead to cultural insensitivity.

Ethical Approach: Conservation projects should respect and integrate local knowledge and cultural values. Collaborative approaches that involve local communities in decision-making can help ensure cultural sensitivity [18–21].

Ethical considerations in using technology for conservation involve a balance between the benefits of technological advancements and the potential risks and ethical concerns. Conservation practitioners should adopt a holistic and inclusive approach, incorporating ethical frameworks that prioritize the well-being of wildlife, ecosystems, and human communities. Regular ethical assessments and stakeholder engagement are essential components of responsible conservation technology implementation.

7.4.1 Addressing Challenges in Implementing Cloud-Based IoT Solutions

Implementing cloud-based IoT solutions come with its set of challenges, ranging from technical complexities to security concerns. Addressing these challenges is crucial to ensure the successful deployment and operation of IoT systems in the cloud. Here are common challenges and strategies to overcome them:

Security and Privacy Concerns

Challenge: IoT devices generate and transmit sensitive data, making them potential targets for cyberattacks. Data integrity, confidentiality, and user privacy are critical concerns.

Strategy: Implement robust security measures, including encryption, secure access controls, and regular security audits. Employ secure communication protocols and use trusted cloud service providers with a focus on data protection.

Scalability

Challenge: As the number of IoT devices increases, ensuring that the cloud infrastructure can scale to handle the growing volume of data becomes a challenge [19].

Strategy: Choose cloud platforms that offer scalable services and infrastructure. Use auto-scaling features to dynamically adjust resources based on demand. Implement efficient data processing and storage strategies to handle scalability effectively.

Interoperability

Challenge: IoT solutions often involve devices from different manufacturers and use diverse communication protocols, leading to interoperability challenges.

Strategy: Adopt standardized communication protocols such as MQTT or CoAP. Ensure compatibility with industry standards and use middleware solutions that support interoperability. Choose IoT platforms that provide flexibility in integrating devices with different protocols.

Reliability and Availability

Challenge: IoT applications often require high levels of reliability and availability. Downtime or disruptions can have significant consequences.

Strategy: Select cloud providers with a proven track record of reliability and high availability. Implement redundancy and failover mechanisms for critical components. Use geographically distributed data centers to enhance availability [20].

Latency and Real-time Processing

Challenge: Some IoT applications, particularly those requiring real-time processing, are sensitive to latency. Delays in data processing can impact the effectiveness of the solution.

Strategy: Utilize edge computing solutions to process data closer to the source, reducing latency. Implement real-time analytics and processing capabilities within the cloud infrastructure. Optimize network configurations to minimize communication delays.

Data Management and Storage

Challenge: Handling and storing large volumes of IoT-generated data efficiently is a significant challenge. Traditional databases may struggle with the velocity and variety of IoT data.

Strategy: Leverage cloud-based databases and storage solutions designed for handling high volumes of time-series data. Implement data compression and aggregation techniques. Utilize data lifecycle management strategies to archive or purge older, less relevant data.

Cost Management

Challenge: Cloud services can incur substantial costs, especially as IoT deployments scale. Managing costs effectively while ensuring optimal performance is a challenge.

Strategy: Monitor resource usage closely and use cloud provider tools to analyze and optimize costs. Utilize serverless architectures and pay-as-you-go models to align costs with actual usage. Consider reserved instances for predictable workloads [22–24].

Regulatory Compliance

Challenge: IoT solutions often involve the collection and processing of sensitive data, requiring compliance with various regulations and standards.

Strategy: Stay informed about data protection regulations relevant to your industry and geography. Choose cloud providers with compliance certifications. Implement data encryption, access controls, and auditing features to meet regulatory requirements.

Integration with Existing Systems

Challenge: Integrating IoT solutions with existing enterprise systems, such as enterprise resource planning (ERP) or customer relationship management (CRM), can be complex and may require significant modifications.

Strategy: Choose IoT platforms that offer integration capabilities through APIs and support common integration patterns. Implement a well-defined integration strategy and conduct thorough testing to ensure seamless interoperability.

Human Factors and Training

Challenge: The successful deployment of IoT solutions requires skilled personnel for design, implementation, and maintenance.

Strategy: Invest in training programs for staff to acquire IoT-specific skills. Leverage external expertise through partnerships or consulting services. Foster a culture of continuous learning and adaptation within the organization [24–27].

Addressing these challenges requires a comprehensive approach that involves technology, processes, and organizational considerations. Regular assessment and adaptation to evolving technologies and security threats are essential for the long-term success of cloud-based IoT implementations.

7.5 FUTURE TRENDS AND OPPORTUNITIES FOR FURTHER ADVANCEMENTS

Here are some anticipated future trends and opportunities for further advancements:

- Edge Computing Evolution
  Trend: The continued growth of edge computing, bringing computation closer to the data source to reduce latency and enhance real-time processing.
  Opportunities: Increased efficiency in IoT applications, improved response times, and the development of edge-native applications.

- 5G Technology Integration:
  Trend: Widespread deployment of 5G networks, providing faster and more
    reliable connectivity.
  Opportunities: Enhanced IoT capabilities, support for more connected
    devices, and the development of new applications in areas such as aug-
    mented reality (AR) and virtual reality (VR).
- AI Advancements:
  Trend: Continued advancements in AI and ML, including more sophis-
    ticated algorithms, explainable AI, and AI applications in diverse
    fields.
  Opportunities: Improved data analysis, personalized user experiences, and
    AI-driven automation in various industries [2, 28].
- Blockchain Beyond Cryptocurrencies:
  Trend: Expanding use cases for blockchain technology beyond crypto-
    currencies, including supply chain management, smart contracts, and
    decentralized finance (DeFi).
  Opportunities: Increased transparency, security, and efficiency in various
    business processes and financial transactions.
- Quantum Computing Development:
  Trend: Ongoing research and development in quantum computing, focusing
    on achieving practical quantum advantage.
  Opportunities: Solving complex problems, optimizing processes, and
    advancements in cryptography and cybersecurity.
- Extended Reality (XR):
  Trend: Growth in extended reality technologies, including AR, VR, and
    mixed reality (MR).
  Opportunities: Enhanced user experiences in gaming, education, training,
    and virtual collaboration [29–32].
- Biotechnology and Genomics:
  Trend: Advancements in biotechnology, gene editing, and genomics, with
    applications in healthcare, agriculture, and environmental conservation.
  Opportunities: Precision medicine, gene therapies, and improved crop
    resilience.
- Sustainable Technology Solutions:
  Trend: Increasing focus on sustainable and eco-friendly technology solu-
    tions driven by environmental concerns.
  Opportunities: Development of energy-efficient technologies, green com-
    puting, and sustainable practices in various industries.
- Human Augmentation:
  Trend: Growing interest in technologies that enhance human capabilities,
    including brain-computer interfaces and exoskeletons.
  Opportunities: Improved healthcare solutions, assistance for individuals with
    disabilities, and advancements in human performance augmentation.
- Resilient and Adaptive Cybersecurity:
  Trend: Evolving cybersecurity strategies to address new threats, including
    the integration of AI and ML for threat detection.

> Opportunities: Enhanced protection against cyber threats, improved incident response, and the development of proactive cybersecurity measures.

- Digital Health and Telemedicine:
  Trend: Accelerated adoption of digital health solutions and telemedicine, driven by the need for remote healthcare services.
  Opportunities: Increased accessibility to healthcare, remote patient monitoring, and the integration of health technologies for personalized care [32].
- Autonomous Systems and Robotics:
  Trend: Advancements in autonomous systems, robotics, and unmanned vehicles across various industries.
  Opportunities: Automation of repetitive tasks, increased efficiency in logistics, and advancements in autonomous transportation.

Staying informed about these trends and actively exploring how they can be leveraged in specific domains will be crucial for organizations and professionals looking to embrace future opportunities and advancements. It's important to note that the technology landscape is continually evolving, and new trends may emerge as research and development progress [33–35].

7.6 CONCLUSION

This chapter concludes by summarizing the key insights into the transformative role of cloud computing and IoT in wildlife conservation. It emphasizes the need for continued research, collaboration, and ethical considerations to ensure the responsible and effective use of technology in preserving our planet's diverse ecosystems.

REFERENCES

1. Dore, K. M., Hansen, M. F., Klegarth, A. R., Fichtel, C., Koch, F., Springer, A., & Fuentes, A. (2020). Review of GPS collar deployments and performance on nonhuman primates. *Primates, 61*, 373–387.
2. Matthews, A., Ruykys, L., Ellis, B., FitzGibbon, S., Lunney, D., Crowther, M. S., & Wiggins, N. (2013). The success of GPS collar deployments on mammals in Australia. *Australian Mammalogy, 35*(1), 65–83.
3. Botta, A., De Donato, W., Persico, V., & Pescapé, A. (2016). Integration of cloud computing and internet of things: A survey. *Future Generation Computer Systems, 56*, 684–700.
4. Mustapha, U. F., Alhassan, A. W., Jiang, D. N., & Li, G. L. (2021). Sustainable aquaculture development: A review on the roles of cloud computing, internet of things and artificial intelligence (CIA). *Reviews in Aquaculture, 13*(4), 2076–2091.
5. Akhigbe, B. I., Munir, K., Akinade, O., Akanbi, L., & Oyedele, L. O. (2021). IoT technologies for livestock management: A review of present status, opportunities, and future trends. *Big Data and Cognitive Computing, 5*(1), 10.
6. Aceto, G., Persico, V., & Pescapé, A. (2020). Industry 4.0 and health: Internet of things, big data, and cloud computing for healthcare 4.0. *Journal of Industrial Information Integration, 18*, 100129.

7. Dlodlo, N., & Kalezhi, J. (2015, May). The internet of things in agriculture for sustainable rural development. In *2015 international conference on emerging trends in networks and computer communications (ETNCC)* (pp. 13–18). IEEE.

8. Osman, A. M. S., Elragal, A., & Bergvall-Kåreborn, B. (2017). Big data analytics and smart cities: A loose or tight couple? In *10th International Conference on Connected Smart Cities 2017 (CSC 2017), Lisbon, 20–22 July 2017* (pp. 157–168). IADIS.2 w2w2.

9. Banotra, A., Ghose, S., Mishra, D., & Modem, S. (2023). Energy harvesting in self-sustainable IoT devices and applications based on cross-layer architecture design: A survey. *Computer Networks, 236,* 110011.

10. Hashem, I. A. T., Yaqoob, I., Anuar, N. B., Mokhtar, S., Gani, A., & Khan, S. U. (2015). The rise of "big data" on cloud computing: Review and open research issues. *Information Systems, 47,* 98–115.

11. Shivaprakash, K. N., Swami, N., Mysorekar, S., Arora, R., Gangadharan, A., Vohra, K., & Kiesecker, J. M. (2022). Potential for artificial intelligence (AI) and machine learning (ML) applications in biodiversity conservation, managing forests, and related services in India. *Sustainability, 14*(12), 7154.

12. Galle, N. J., Nitoslawski, S. A., & Pilla, F. (2019). The internet of nature: How taking nature online can shape urban ecosystems. *The Anthropocene Review, 6*(3), 279–287.

13. Ahmad, T., Ahmad, S., & Jamshed, M. (2015, October). A knowledge based Indian agriculture: With cloud ERP arrangement. In *2015 International Conference on Green Computing and Internet of Things (ICGCIoT)* (pp. 333–340). IEEE.

14. Nundloll, V., Porter, B., Blair, G. S., Emmett, B., Cosby, J., Jones, D. L., & Ullah, I. (2019). The design and deployment of an end-to-end IoT infrastructure for the natural environment. *Future Internet, 11*(6), 129.

15. Kaur, U., Malacco, V. M., Bai, H., Price, T. P., Datta, A., Xin, L., & Voyles, R. M. (2023). Invited review: Integration of technologies and systems for precision animal agriculture—A case study on precision dairy farming. *Journal of Animal Science, 101,* skad206.

16. Ashir, D. M. N. A., Ahad, M. T., Talukder, M., & Rahman, T. (2022). Internet of Things (IoT) based smart agriculture aiming to achieve sustainable goals. *arXiv preprint arXiv:2206.06300.*

17. Kalatzis, N., Routis, G., Marinellis, Y., Avgeris, M., Roussaki, I., Papavassiliou, S., & Anagnostou, M. (2019). Semantic interoperability for IoT platforms in support of decision making: An experiment on early wildfire detection. *Sensors, 19*(3), 528.

18. Raghav, Y. Y., Vyas, V., & Rani, H. (2022). Load balancing using dynamic algorithms for cloud environment: A survey. *Materials Today: Proceedings, 69,* 349–353.

19. Dewan, M., Mudgal, A., Pandey, P., Raghav, Y. Y., & Gupta, T. (2023). Predicting Pregnancy Complications Using Machine Learning. In D. Satishkumar & P. Maniiarasan (Eds.), *Technological Tools for Predicting Pregnancy Complications* (pp. 141–160). IGI Global. https://doi.org/10.4018/979-8-3693-1718-1.ch008

20. Gupta, T., Pandey, P., & Raghav, Y. Y. (2023). Impact of Social Media Platforms on the Consumer Decision-Making Process in the Food and Grocery Industry. In T. Tarnanidis, M. Vlachopoulou, & J. Papathanasiou (Eds.), *Influences of Social Media on Consumer Decision-Making Processes in the Food and Grocery Industry* (pp. 119–139). IGI Global. https://doi.org/10.4018/978-1-6684-8868-3.ch006

21. Raghav, Y. Y., & Vyas, V. (2023). ACBSO: A hybrid solution for load balancing using ant colony and bird swarm optimization algorithms. *International Journal of Information Technology,* 1–11.

22. Raghav, Y. Y., Tipu, R. K., Bhakhar, R., Gupta, T., & Sharma, K. (2024). The Future of Digital Marketing: Leveraging Artificial Intelligence for Competitive Strategies and Tactics. In *The Use of Artificial Intelligence in Digital Marketing: Competitive Strategies and Tactics* (pp. 249–274). IGI Global.

23. Raghav, Y. Y., & Vyas, V. (2019). A comparative analysis of different load balancing algorithms on different parameters in cloud computing. In *2019 3rd International Conference on Recent Developments in Control, Automation & Power Engineering (RDCAPE)* (pp. 628–634), Noida, India. https://doi.org/10.1109/RDCAPE47089.2019.8979122
24. Dauvergne, P. (2020). *AI in the Wild: Sustainability in the Age of Artificial Intelligence.* MIT Press.
25. Verma, S., Sharma, V., Kukreja, V., Shankar, A., & Elngar, A. A. (2024). Research Potentials and Future Trends of Digital Sustainability. In *Sustainable Digital Technologies* (pp. 187–204). CRC Press.
26. Gebresenbet, G., Bosona, T., Patterson, D., Persson, H., Fischer, B., Mandaluniz, N., & Nasirahmadi, A. (2023). A concept for application of integrated digital technologies to enhance future smart agricultural systems. *Smart Agricultural Technology, 5,* 100255.
27. Ray, P. P., Dash, D., & De, D. (2019). Edge computing for internet of things: A survey, e-healthcare case study and future direction. *Journal of Network and Computer Applications, 140,* 1–22.
28. Maraveas, C., Piromalis, D., Arvanitis, K. G., Bartzanas, T., & Loukatos, D. (2022). Applications of IoT for optimized greenhouse environment and resources management. *Computers and Electronics in Agriculture, 198,* 106993.
29. Almalki, F. A., Alsamhi, S. H., Sahal, R., Hassan, J., Hawbani, A., Rajput, N. S., & Breslin, J. (2023). Green IoT for eco-friendly and sustainable smart cities: Future directions and opportunities. *Mobile Networks and Applications, 28*(1), 178–202.
30. Lăzăroiu, G., & Harrison, A. (2021). Internet of things sensing infrastructures and data-driven planning technologies in smart sustainable city governance and management. *Geopolitics, History & International Relations, 13*(2), 23–36.
31. Kour, V. P., & Arora, S. (2020). Recent developments of the internet of things in agriculture: A survey. *IEEE Access, 8,* 129924–129957.
32. George, A. S., George, A. H., & Baskar, T. (2023). Edge computing and the future of cloud computing: A survey of industry perspectives and predictions. *Partners Universal International Research Journal, 2*(2), 19–44.
33. Omoniwa, B., Hussain, R., Javed, M. A., Bouk, S. H., & Malik, S. A. (2018). Fog/edge computing-based IoT (FECIoT): Architecture, applications, and research issues. *IEEE Internet of Things Journal, 6*(3), 4118–4149.
34. Hussain, F., Hassan, S. A., Hussain, R., & Hossain, E. (2020). Machine learning for resource management in cellular and IoT networks: Potentials, current solutions, and open challenges. *IEEE Communications Surveys & Tutorials, 22*(2), 1251–1275.
35. Bakker, K., & Ritts, M. (2018). Smart earth: A meta-review and implications for environmental governance. *Global Environmental Change, 52,* 201–211.

8 Smart Farming Solutions
Unveiling the Power of Cloud-IoT Integration

Sneha, Prabh Deep Singh, and Vikas Tripathi

8.1 INTRODUCTION

Cloud computing and the Internet of Things (IoT) are two novel technologies that have transformed numerous industries, and their integration has been particularly vital in modern agriculture. Consider what these technologies imply and how their cooperation transformed the agricultural setting [1].

8.1.1 CLOUD COMPUTING

Cloud computing offers computing services and resources via the internet. It empowers individuals, businesses, and organizations to gain access to and utilize diverse computing capabilities, including storage, processing power, networking, and software applications, rather than the requirement for on-site infrastructure or hardware.

Cloud computing is a virtualized environment where users can store, manage, and interact with data and applications remotely [2]. This remote accessibility offers scalability, flexibility, and cost-efficiency, as users can adjust their resource usage according to their needs without the burden of managing physical equipment.

Cloud computing promotes collaboration and remote access, facilitating seamless interaction and data sharing across locations. Users can work together on projects, access information, and use applications from virtually anywhere with an internet connection. This approach reduces the limitations of geographical boundaries and enables efficient workflows.

8.1.1.1 Fundamental Models in Cloud Computing

Infrastructure as a Service, Platform as a Service, and Software as a Service is essential to cloud computing models that offer users various services and resources, each adapting to varying purposes and demands. Figure 8.1 shows the fundamental models of cloud computing.

8.1.1.2 Infrastructure as a Service (IaaS)

IaaS conveys virtualized computing resources to consumers across the internet. Users can lease these resources from IaaS providers against owning and managing physical servers, storage, and networking components. This architecture provides a

DOI: 10.1201/9781032656694-8

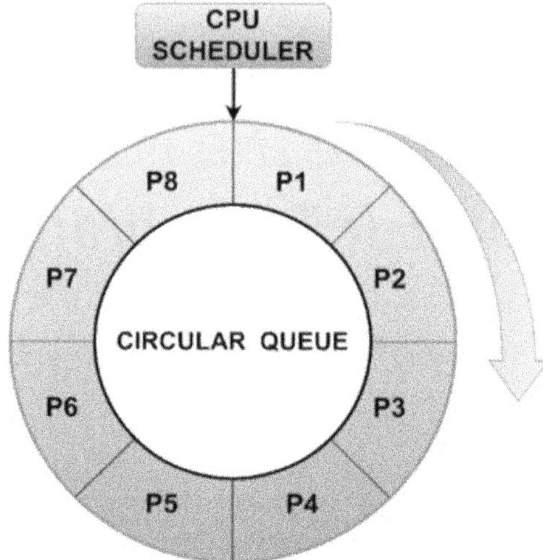

FIGURE 8.1   Fundamental models in cloud computing [3].

highly flexible and scalable solution that allows users to deliver and manage virtual machines, storage, and network resources on demand. Users choose the operating systems, apps, and configurations to run on the infrastructure offered. IaaS is particularly valuable for companies that depend on the underlying hardware and networking infrastructure to perform unique apps or approach specific workloads [4].

8.1.1.3   Platform as a Service (PaaS)

PaaS extends the cloud computing paradigm by providing a platform containing the underlying infrastructure, tools, and services necessary for application creation, deployment, and management. PaaS providers provide development frameworks, databases, middleware, and other tools, enabling developers to focus solely on designing applications without worrying about the complexities of infrastructure maintenance [5]. This methodology accelerates the development process, encourages collaboration through development teams, and automates application deployment. PaaS is ideal for developers who prefer to concentrate on writing, testing, and supplying applications rather than managing and upholding the infrastructure that underlies them.

8.1.1.4   Software as a Service (SaaS)

The most user-centric model is SaaS, which immediately supplies fully established software programs to end users throughout the internet. SaaS permits customers to use apps without paying for installation, setup, or maintenance. These apps are maintained by SaaS providers, who additionally deal with upgrades, security, and infrastructure [6]. This tactic provides convenience and accessibility, facilitating users to easily implement software for various functions such as email, customer relationship management (CRM), project management, etc. SaaS benefits both companies and individuals who want ready-to-use solutions without the difficulties of software maintenance.

8.1.2 IoT

In the context of smart agriculture, the IoT refers to an interconnected network of physical goods, sensors, and devices beneath the agricultural environment suited to collecting, transferring, and exchanging data via the internet. This technological integration facilitates real-time monitoring, data analysis, and automation, making agricultural processes effective, productive, and sustainable.

IoT has been utilized in smart agriculture for integrating sensors and devices in various parts of the farming ecosystem, such as soil, crops, livestock, and equipment. Temperature, humidity, soil moisture, nutrient levels, crop growth rates, weather conditions, and animal behavior are a few examples of data these devices can collect. This information is then delivered to cloud platforms or local systems for assessment and processing.

Essential objectives of integrating IoT into smart agriculture: Farmers can utilize IoT devices to make data-driven decisions, improve precision farming, optimize resource management, and streamline tasks [7]. Monitoring historical and real-time data permits remote monitoring, predictive analytics, and sustainability. IoT-enabled substances also support traceability by tracking and recording agricultural goods' origin, growth, and handling, assuring transparency and minimizing resource waste. These advancements help enhance production in agriculture, resource management, and traceability.

8.2 LITERATURE REVIEW

Yuyanto et al. [8] proposed an early warning system, the significance of diverse environmental factors in agricultural success. Proposes developing an early warning system that leverages these factors to detect critical values that might lead to crop failure. Recognizing the rapid evolution of information technology, the study highlights its potential as a tool to address agricultural challenges. The research aims to provide solutions that enhance agricultural outcomes by integrating technology into data collection and analysis.

George Suciu et al. [9] evaluated the integration of M2M communication and Cloud IoT, addressing energy efficiency and resilience. Measurement results from the 2014 winegrowing season were obtained through M2M telemetry and processed via a Big Data platform. The study shows that Cloud IoT and Big Data are capable of helping anticipate disease in viticulture. It increases the system's applicability to include additional agricultural uses. This study highlights the potential of M2M and Cloud IoT integration for enhanced precision agriculture and its applicability beyond grape vineyards.

Meenaxi M Raikar et al. [10] investigated the convergence of cloud computing with IoT in agriculture, stressing the significance of Cloud IoT in smart applications. The paper illustrates the relevance of disruptive technologies such as cloud computing, IoT, machine learning, and data analytics across various industries. It reviews a case study of a smart irrigation system that employs the MQTT lightweight protocol, showcasing its energy efficiency and speed amenities.

Vandana Satam et al. [11] investigated the possible application of microstrip antennas as temperature sensors in Internet of Things deployments. The study highlights the significance of temperature sensing across diverse sectors. It presents a novel approach wherein the resonant frequency change of a microstrip antenna correlates with temperature variation, offering a self-contained temperature-sensing mechanism. This method eliminates the need for human intervention. The research demonstrates the feasibility of transmitting sensed data through an IoT platform, employing Arduino and a GSM module for cloud data logging. ThingSpeak, a Cloud IoT analytics platform, is used for real-time data visualization and analysis. The study showcases the microstrip antenna's potential for temperature sensing within IoT contexts.

Fang Wu et al. [12] evaluated implementing an intelligent remote monitoring system for slaughter production lines using IoT and cloud technology. The research addresses the challenges of low information levels and costly downtimes in slaughter lines. The study proposes an IoT-based system where real-time equipment status data is collected by PLCs on-site. This data is encapsulated into MQTT messages through a gateway and sent to the Alibaba Cloud IoT platform. Cloud servers analyze and store the data, enabling remote monitoring through a web server. Test results demonstrate successful system functionality, stability, and clear real-time data monitoring, meeting the requirements for industrial remote monitoring.

Praveen Kumar Reddy Maddikunta et al. [13] evaluated the vast potential of unmanned aerial vehicles (UAVs) in smart agriculture. While the prospects are promising, cost-effectiveness and ease of control remain pivotal factors driving UAV adoption. Various technologies, including Wi-Fi, ZigBee, and Smart Bluetooth (Bluetooth Low Energy), are explored for UAV control. The cost-efficiency and ubiquity of Smart Bluetooth make it an attractive option for UAV operation via smartphones. The study outlines sensor types suitable for smart farming, addresses operational requirements and challenges, and identifies future UAV applications in agriculture. As smart farming gains global momentum, this research underscores the importance of addressing key considerations for successful UAV integration in the agricultural sector.

Zeng Hu et al. [14] evaluated the integration of NOMA in Wireless Sensor Networks (WSN) for optimizing smart agriculture. WSNs play a role in collecting and passing valuable data from farm fields to servers. Relay-aided NOMA is used in the uplink transmission stage to relieve spectrum constraints and improve efficiency. Using power domain splitting, this kind of technology enables the simultaneous transmission of several symbols on the same resource element. The work gives a theoretical analysis of the average cumulative data rate and outage probability for relay-aided NOMA. The simulation observations show that it outperforms typical orthogonal multiple access (OMA) networks used for uplink transmission within agricultural WSNs, enhancing the data throughput and efficiency of the network.

Yuanhao Sun et al. [15] evaluated that MCS could revolutionize data collection for smart agriculture. The report underscores the three key benefits of MCS: cost-effectiveness, scalability, and mobility. As cell phones become ubiquitous, the synergy between MCS characteristics and plug-and-play infrastructure opens up opportunities for MCS-enabled smart agriculture. The study thoroughly analyzes Agriculture Mobile Crowd Sensing (AMCS), favorably comparing it to existing data collection strategies regarding flexibility, implicit data capture, and cost-effectiveness.

However, the analysis notes possible drawbacks in data integrity and quality. The study highlights seven potential AMCS-enabled applications for agriculture, emphasizing the importance of MCS for deciding the precision of the agricultural sector's future.

S. Rajeswari et al. [16] evaluated the integration of IoT, mobile, and cloud-based Big Data analytics in a smart agricultural model. The study highlights the insufficiency of traditional databases in storing IoT-generated data, necessitating cloud storage. It synergizes for effective smart agriculture feasibility studies. This convergence enhances agriculture by aiding crop monitoring, fertilizer selection, and irrigation decision-making. Data mining predicts agricultural information accessible via a mobile app. The model aims to optimize crop production and cost management through data-driven insights, advancing precision agriculture practices.

Mohamed Rawidean et al. [17] evaluated the integration of WSN, cloud computing, and Big Data technologies for agricultural environment applications. The study acknowledges the growing demand for precision agriculture and automation, emphasizing the importance of real-time control and communication with the physical world in this context. The integration of WSN and cloud computing enables comprehensive data collection and processing across the agricultural process. The paper outlines the proposed integrated architecture, highlighting its benefits and applications in the agricultural sector. The research underscores the potential of this integrated approach to address challenges and enhance efficiency in precision agriculture and automation.

8.3 AGRICULTURE AND THE USE OF CLOUD COMPUTING

Agriculture has long been the foundation of human civilization, supporting populations and fueling economies. Agriculture has entered a new era due to technological advancements, with cloud computing becoming indispensable.

8.3.1 INTEGRATING CLOUD IoT IN AGRICULTURAL

Integrating Cloud IoT with agriculture is a significant step in transforming traditional farming into a technologically permitted and data-driven effort. The melding of cloud computing with agricultural techniques throws up numerous potential ways of managing the sector's struggles and optimizing operations.

Cloud computing lends a centralized platform for collecting, storing, and analyzing massive amounts of agricultural data generated through IoT devices and sensors. This permits farmers to make informed decisions by monitoring crops, soil conditions, atmospheric trends, and machinery effectiveness in real-time. The scalability of cloud resources allows farmers to tailor their computational infrastructure to meet fluctuating demands, promoting cost-efficiency and accessibility.

Incorporating cloud computing into agriculture underscores a paradigm shift toward smarter, more sustainable, and resilient farming practices [18]. By harnessing the potential of the cloud, agriculture stands poised to embrace innovation, optimize resource utilization, and contribute significantly to global food security in an increasingly dynamic and interconnected world.

8.3.2 Advantages of Integrating Cloud IoT in Agriculture

Agriculture has been a cornerstone of human civilization for millennia, ensuring sustenance and forming the basis of our economies. Over the centuries, this crucial sector has evolved, adapting to varying climatic conditions, population growth, and technological advancements. Today, in the digital age, agriculture is poised for another transformative leap through the integration of cloud computing. As farms become more digitized and data-driven, cloud computing is a powerful ally, offering many advantages.

8.3.2.1 Cost-Effectiveness

Historically, agricultural ventures, especially small to medium-scale ones, have operated on tight budgets. Investing in traditional IT infrastructure—servers, storage units, and software—demands substantial capital expenditure. Here, cloud computing shines by introducing a pay-as-you-use model. This approach eliminates hefty upfront costs and makes high-tech solutions accessible to even small-scale farmers. Furthermore, with cloud services, maintenance, upgrades, and IT personnel expenses are shifted to the service providers. This redistribution allows agricultural entities to focus their resources on core farming activities rather than IT management.

8.3.2.2 Scalability and Flexibility

Agriculture is inherently cyclical, with different seasons demanding varied computational resources. For instance, planting season might require intense data analytics for soil health, weather predictions, and seed optimization. Conversely, off-seasons might see a reduced demand. Traditional IT setups, with their static infrastructure, find it challenging to adapt to such fluctuations efficiently [19]. Cloud computing offers unparalleled scalability. Resources can be ramped up or down based on real-time requirements, ensuring that farmers only pay for what they use. This elastic nature of cloud systems ensures that they are future-proof, ready to adapt as the agricultural enterprise grows or diversifies.

8.3.2.3 Data Storage, Management, and Accessibility

The modern agricultural landscape is awash with data. Soil sensors, drones, satellite imagery, and IoT devices churn out vast information daily. Storing this data securely and ensuring it's readily accessible is paramount. Cloud platforms excel in this domain. They offer virtually limitless storage, ensuring every data byte is securely housed. More importantly, the cloud ensures that data is accessible anytime, anywhere. A farmer in the field can access real-time data on a tablet, make informed decisions, or collaborate with experts globally, breaking geographical constraints.

8.3.2.4 Real-time Data Analysis and Decision-Making

Data is of limited use in its raw form. Its true value is unlocked when processed and analyzed. Cloud platforms, with their robust computational power, are tailor-made for this. They can analyze vast datasets in real-time, offering actionable insights on the go. Consider a scenario where sensors detect a drop in soil moisture levels. Cloud systems can instantly analyze this data, cross-reference it with weather forecasts, and offer irrigation recommendations. Such real-time decision-making capabilities can be the difference between a fruitful harvest and crop failure.

8.3.2.5 Collaborative Platforms

Agriculture isn't an isolated endeavor. It involves many stakeholders - agronomists, farm workers, and supply chain partners. Effective collaboration is key. Cloud platforms facilitate this with centralized data repositories and shared workspaces [20]. A farm manager could share crop yield data with a supplier in real-time, ensuring timely deliveries of storage units. Alternatively, an unexpected pest infestation's data could be shared with experts worldwide, pooling knowledge to find an effective countermeasure.

8.3.2.6 Security and Data Backup

One of the primary concerns with digitization is security. Farms, with their sprawling landscapes, seem vulnerable to data breaches. However, leading cloud service providers prioritize security. They employ advanced encryption techniques, ensuring data remains confidential and shielded from malicious threats.

8.3.3 APPLICATIONS OF CLOUD IoT IN AGRICULTURE

8.3.3.1 Precision Agriculture

Precision agriculture involves the precise management of resources based on real-time data. Cloud IoT facilitates collecting and analyzing data from various sources, such as soil sensors, weather stations, drones, and satellite imagery. Farmers can monitor soil moisture, nutrient levels, and crop health remotely. This data-driven approach enables targeted irrigation, optimized fertilizer application, and timely pest management, resulting in increased yields, reduced costs, and minimized environmental impact.

8.3.3.2 Smart Irrigation

Cloud IoT-enabled smart irrigation systems monitor soil moisture levels and weather forecasts. These systems can determine when and how much to irrigate by analyzing this data, ensuring optimal water usage. This conserves water resources and prevents over-irrigation, which can lead to soil degradation and water wastage.

8.3.3.3 Livestock Monitoring and Management

Cloud IoT technologies empower livestock farmers to track the health and behavior of their animals closely. Wearable sensors can monitor animals' vital signs, activity patterns, and feeding habits. Farmers will get real-time notifications in instances of anomalies, allowing for quick responses to mitigate disease outbreaks. Furthermore, GPS-enabled trackers aid in managing animal movements and supply patterns.

8.3.3.4 Supply Chain Management

The agricultural supply chain reaps the advantages of enhanced traceability and transparency thanks to Cloud IoT. Data on crop origins, production processes, shipping, and storage conditions can be saved and retrieved from farm to fork on the cloud. This supplies food safety, quality control, and regulatory compliance, developing confidence among shoppers.

8.3.3.5 Crop Monitoring and Predictive Analytics

Cloud IoT platforms convey continuous crop monitoring using remote sensors and cameras. By examining historical and real-time data, predictive analytics programs can forecast yield estimations, disease outbreaks, and pest infestations. This enables farmers to make proactive selections, effectively allocate resources, and avoid future losses.

8.3.3.6 Autonomous Machinery and Robotics

Cloud IoT connectivity facilitates the integration of autonomous machinery and robotic systems in agriculture. Drones can survey big fields and acquire crop health data, while robotic harvesting tools can harvest crops precisely and effectively. These technologies improve labor efficiency while minimizing the need for human involvement.

8.3.3.7 Environmental Monitoring and Sustainability

To evaluate the effect of farming activities on the ecosystem, Cloud IoT systems might track environmental indicators such as air quality, temperature, and humidity. Farmers can employ environmental metrics to adopt sustainable practices, lessen their carbon footprint, and safeguard biodiversity.

8.3.3.8 Data-Driven Decision-Making

Cloud IoT lends a plethora of data that can be demonstrated and analyzed to yield valuable information. Farmers can make informed decisions about growing, harvesting, and distributing resources based on earlier patterns and real-time observations.

8.4 PROPOSED ARCHITECTURE

The Round Robin (RR) scheduling algorithm [21] is a time-sharing technique widely employed in operating systems to manage and allocate CPU time among multiple processes. This algorithm has noteworthy implications in traditional operating systems and when integrated into contemporary paradigms like Cloud IoT in agriculture.

In an RR scheduling approach, each process is assigned a fixed time quantum during which it executes on the CPU. When the time quantum elapses, the scheduler preempts the current process and switches to the next one in line, ensuring equitable distribution of CPU resources. This cycle continues until all processes have received their fair share of execution time.

When examining the relationship between RR scheduling and Cloud IoT in agriculture, parallels can be drawn regarding resource allocation and fairness. Just as RR ensures each process receives its allotted time on the CPU, Cloud IoT integration in agriculture enables the precise distribution of resources such as water and nutrients and monitoring capabilities for crops. This alignment of principles underscores the potential of Cloud IoT to act as a dynamic, real-world application of the RR concept.

In agriculture, Cloud IoT is the orchestrator, allocating resources based on real-time data collected from sensors, drones, and other IoT devices. For instance, consider a scenario where different farm sections require varying amounts of irrigation. By integrating Cloud IoT, a Round Robin-inspired approach could be adopted to ensure each section receives an equitable amount of water, optimizing water usage while promoting crop health.

8.4.1 Working on RR Scheduling

In time-sharing systems, the RR scheduling strategy is a commonly used methodology. Like FCFS scheduling, preemption is a technique employed to switch between processes. A precise quantum or time slice has been designated, frequently prolonged between 10 and 100 milliseconds. The CPU scheduler successively sweeps the ready queue, apportioning the CPU to each job for a maximum of one-time quantum (TQ). The ready queue is presumed to be a circular queue. The system will investigate whether process P1 has finished execution regardless of whether it passes through the end of the queue.

If P1 has accomplished its execution, it is unnecessary to request the CPU again. However, if P1 does not finish its execution within the initial TQ, the CPU will be returned to P1, granting it another TQ to execute before moving on to the subsequent process. Figure 8.2 shows the working of RR scheduling.

8.5 PERFORMANCE EVALUATION OF PROPOSED FRAMEWORK

Interpreting the influence of multiple cloudlet counts on the system's average waiting time constitutes a part of the performance assessor's review of the recommended RR framework.

The number of cloudlets and the typical wait time are two gauges of performance for cloud computing that have been demonstrated in the table. To improve the performance and experience for users of cloud computing, decision-making processes

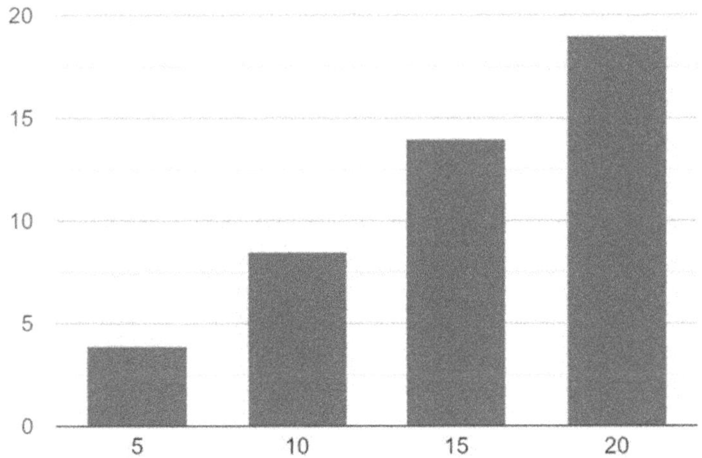

FIGURE 8.2 Working of RR scheduling [21].

TABLE 8.1

Average Waiting Time of Cloudlets

	Average Waiting Time (WT)	
S. No.	No. Of Cloudlets	Avg. WT on Total
1.	5	3.90
2.	10	8.46
3.	15	14.00
4.	20	19.00

and optimization tactics have the potential to be guided by the scrutiny of this data. Table 8.1 and Figure 8.3 shows the average waiting time of cloudlets. Table 8.2 and Figure 8.4 shows the average TAT of cloudlets.

8.5.1 Burst Time of Cloudlets

Burst time (BT) is an evaluation of the true duration of time a process must spend finished running, including the time required for computation and resource access. Table 8.3 and Figure 8.5 show the BT cloudlets.

8.6 RESULT

The chapter "Cloud IoT in Agriculture" delves into the dynamic convergence of cloud computing and the IoT within the agricultural domain. Through a comprehensive exploration of various facets, the chapter underscores the transformative impact of this integration on modern farming practices. The chapter demonstrates

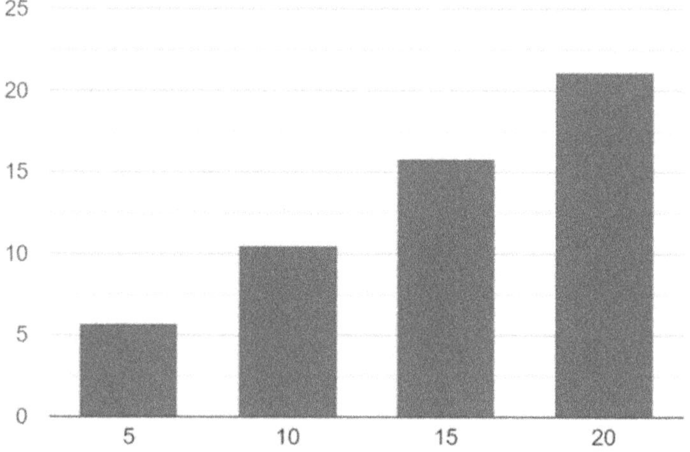

FIGURE 8.3 Average waiting time of cloudlets.

TABLE 8.2
Average Turnaround Time (TAT) of Cloudlets

	Average TAT	
S. No.	No. Of Cloudlets	Avg. TAT on Total
1.	5	5.72
2.	10	10.47
3.	15	15.84
4.	20	21.10

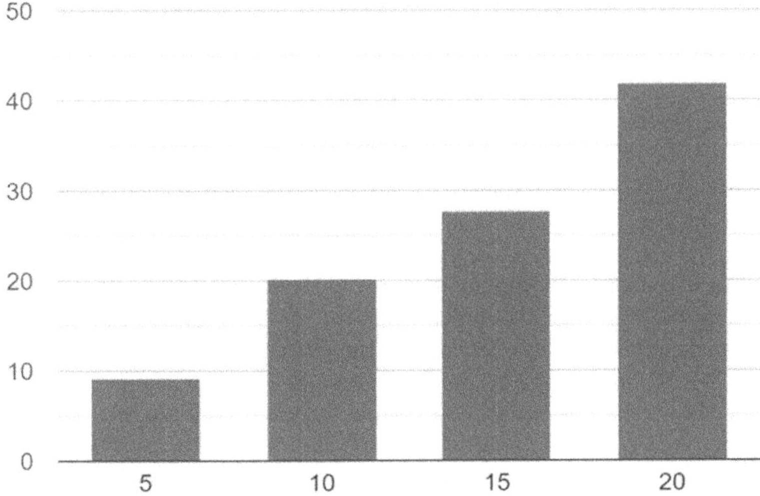

FIGURE 8.4 Average turnaround time of cloudlets.

TABLE 8.3
Burst Time of Cloudlets

	Burst Time (BT)	
S. No.	No. Of Cloudlets	Avg. BT on Total
1.	5	9.09
2.	10	20.15
3.	15	27.71
4.	20	41.91

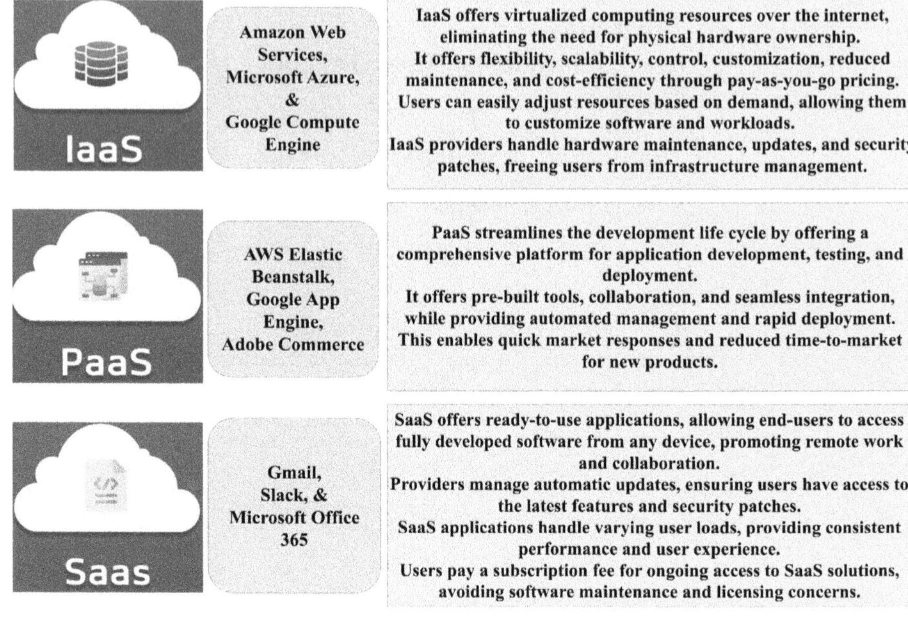

FIGURE 8.5   Burst time of cloudlets.

how Cloud IoT coordination revolutionizes agriculture by dissecting fundamental concepts, examining practical implementations, and evaluating performance.

Transitioning seamlessly, the chapter integrates the concept of cloud computing with IoT, illuminating the potent teamwork between these two paradigms. With a focus on the agricultural sector, the chapter explores the symbiotic relationship between Cloud Computing and IoT, showcasing their collaborative prowess in enabling data-driven decision-making, remote monitoring, and resource optimization.

The chapter introduces a RR scheduling algorithm and blends it with the Cloud IoT paradigms, resulting in an amazing architectural framework. The architecture exhibits the seamless orchestration of resources, data, and intelligence to ensure equitable allocation and effective agricultural operations.

The chapter concludes with a performance examination of the presented framework. The chapter explores the system's responsiveness, efficiency, and resource consumption to demonstrate Cloud IoT integration's practicality and economic importance in agriculture. These analyses' findings show the likelihood of enhanced yields, reduced wasteful use of resources, and a higher level of sustainability.

REFERENCES

1. A. Goldstein, L. Fink, and G. Ravid, "A Cloud-Based Framework for Agricultural Data Integration: A Top-Down-Bottom-Up Approach," IEEE Access, vol. 10, pp. 88527–88537, 2022, https://doi.org/10.1109/access.2022.3198099

2. H. A. Alharbi, and M. Aldossary, "Energy-Efficient Edge-Fog-Cloud Architecture for IoT-Based Smart Agriculture Environment," IEEE Access, vol. 9, pp. 110480–110492, 2021, https://doi.org/10.1109/access.2021.3101397

3. S. Liu, L. Guo, H. Webb, X. Ya, and X. Chang, "Internet of Things Monitoring System of Modern Eco-Agriculture Based on Cloud Computing," IEEE Access, vol. 7, pp. 37050–37058, 2019, https://doi.org/10.1109/access.2019.2903720

4. W. Zhao, X. Wang, B. Qi, and T. Runge, "Ground-Level Mapping and Navigating for Agriculture Based on IoT and Computer Vision," IEEE Access, vol. 8, pp. 221975–221985, 2020, https://doi.org/10.1109/access.2020.3043662

5. N. K. Rajpoot, P. Singh, and B. Pant, "Nature-Inspired Load Balancing Algorithms for Resource Allocation in Cloud Computing," Apr. 2023, https://doi.org/10.1109/cises58720.2023.10183630

6. N. K. Rajpoot, P. Singh, and B. Pant, "Load Balancing in Cloud Computing: A Simulation-Based Evaluation," Apr. 2023, https://doi.org/10.1109/cises58720.2023.10183622

7. M. Mudholkar, P. Mudholkar, H. Reddy, K. Sreenivasulu, K. Joshi, and B. Pant, "A Novel Approach to IoT-based Plant Health Monitoring System in Smart Agriculture," Dec. 2022, https://doi.org/10.1109/ic3i56241.2022.10072603

8. Y. Yuyanto, and S. Liawatimena, "Implementation of Data Collecting Platform Over Distributed Sensors for Global Open Data for Agriculture and Nutrition," Aug. 2018, https://doi.org/10.1109/citsm.2018.8674057

9. G. Suciu, A. Vulpe, O. Fratu, and V. Suciu, "M2M Remote Telemetry and Cloud IoT Big Data Processing in Viticulture," International Conference on Wireless Communications and Mobile Computing, Oct. 2015, https://doi.org/10.1109/iwcmc.2015.7289239

10. M. M. Raikar, P. Desai, N. Kanthi, and S. Bawoor, "Blend of Cloud and Internet of Things (IoT) in Agriculture Sector using the Lightweight Protocol," 2018 International Conference on Advances in Computing, Communications, and Informatics (ICACCI), Sep. 2018, https://doi.org/10.1109/icacci.2018.8554406

11. V. Satam, C. V. Kulkarni, and A. Kholapure, "Microstrip Antenna as a Temperature Sensor for IoT Applications," Dec. 2022, https://doi.org/10.1109/mapcon56011.2022.10047111

12. F. Wu, Z. Miao, and C. He, "Remote Monitoring System for Intelligent Slaughter Production Line Based on Internet of Things and Cloud Platform," 2020 11th International Conference on Prognostics and System Health Management (PHM-2020 Jinan), Oct. 2020, https://doi.org/10.1109/phm-jinan48558.2020.00104.

13. P. K. R. Maddikunta et al., "Unmanned Aerial Vehicles in Smart Agriculture: Applications, Requirements, and Challenges," IEEE Sensors Journal, pp. 1–1, 2021, https://doi.org/10.1109/jsen.2021.3049471

14. Z. Hu et al., "Application of Non-Orthogonal Multiple Access in Wireless Sensor Networks for Smart Agriculture," IEEE Access, vol. 7, pp. 87582–87592, 2019, https://doi.org/10.1109/access.2019.2924917

15. Y. Sun et al., "On Enabling Mobile Crowd Sensing for Data Collection in Smart Agriculture: A Vision," IEEE Systems Journal, vol. 16, no. 1, pp. 132–143, Mar. 2022, https://doi.org/10.1109/jsyst.2021.3104107

16. S. Rajeswari, K. Suthendran, and K. Rajakumar, "A Smart Agricultural Model by Integrating IoT, Mobile and Cloud-Based Big Data Analytics," IEEE Xplore, Jun. 01, 2017, https://ieeexplore.ieee.org/abstract/document/8321902

17. M. Rawidean, and A. Nahar Harun, "Wireless Sensor Networks and Cloud Computing Integrated Architecture for Agricultural Environment Applications," Dec. 2017, https://doi.org/10.1109/icsenst.2017.8304445

18. P. Rawat, P. Singh, and V. Tripathi, "Load Balancing in Cloud Computing Leading Us Towards Green Cloud Computing," SSRN Electronic Journal, 2022, https://doi.org/10.2139/ssrn.4031990

19. P. Madan, V. Singh, D. P. Singh, M. Diwakar, B. Pant, and A. Kishor, "A Hybrid Deep Learning Approach for ECG-Based Arrhythmia Classification," Bioengineering, vol. 9, no. 4, p. 152, Apr. 2022, https://doi.org/10.3390/bioengineering9040152
20. S. Ghosh, and C. Banerjee, "Dynamic Time Quantum Priority Based Round Robin for Load Balancing in Cloud Environment," Nov. 2018, https://doi.org/10.1109/icrcicn.2018.8718694
21. K. Surendhar, L. S. Chamarthy, D. M. Priyadharshini, G. Keerthana, G. Sinha, and A. Vijay Kumar, "Enhancement of Round Robin Algorithm with Dynamic Quantum in Cloud Computing," Apr. 2023, https://doi.org/10.1109/icict57646.2023.10133995

9 Cloud-IoT Applications in Agriculture

Mohit Angurala, Mohit Lalit, and Shivani Aggarwal

9.1 INTRODUCTION

The convergence of Cloud Computing and Internet of Things (IoT) technologies has catapulted agriculture into a new era of enhanced efficiency, sustainability, and accuracy. This introduction serves as the starting point for further investigation of Cloud-IoT applications in agriculture in order to comprehensively investigate the transformational potential of these technologies. The book Evolution of Agriculture in the Cloud-IoT Era begins by examining agriculture's historical trajectory and highlighting the paradigm change brought about by the introduction of Cloud-IoT technology. The story emphasizes evolutionary milestones, stressing crucial times that transformed traditional farming techniques. This section lays the groundwork for the next examination of the essential role that Cloud-IoT plays in challenging traditional agricultural paradigms. The chapter also delves into the motives driving the use of Cloud and IoT technologies in agriculture. The talk will center on the compelling reasons that have prompted the agriculture business to embrace these technological breakthroughs. This section seeks to provide a full understanding of why Cloud-IoT integration has become critical for agricultural modernization by clarifying the possible advantages and resolving difficulties.

As the chapter progresses, we clarify the broader aspirations that motivate this technological convergence by unraveling the layers of Cloud-IoT applications in agriculture. The major goals emerge as improved resource management and enhanced agricultural productivity. The introduction elaborates on how Cloud-IoT contributes to accomplishing these goals, providing readers with a road map to appreciate the broad goals of adopting these technologies in an agricultural environment. Each application dissects the benefits and developments of Cloud-IoT technology, from precision agriculture to smart irrigation systems and animal monitoring.

This section guides readers through the historical backdrop, rationale, aims, and available literature around Cloud-IoT in agriculture. It establishes the tone for the following parts, which move the focus to specific applications and their revolutionary influence on the agricultural environment.

9.2 LITERATURE SURVEY

The literature on Cloud-IoT applications in agriculture has seen a boom in interest and study, reflecting the industry's understanding of these technologies' revolutionary potential. This study synthesizes data from a wide range of scholarly publications,

DOI: 10.1201/9781032656694-9

including research articles, conference papers, and journal articles, to offer a thorough knowledge of the present level of Cloud-IoT integration in agriculture.

Goldstein et al. [1] propose a top-down-bottom-up cloud-based paradigm for agricultural data integration. This paradigm provides a comprehensive view of data management in agriculture, employing cloud computing to combine disparate datasets. The study underlines the necessity of a complete strategy consistent with the larger aims of improved resource management and increased agricultural output.

Liu et al. [2] investigate the development of a Cloud-based IoT monitoring system for modern eco-agriculture. The study emphasizes the need for real-time monitoring and data analytics in agricultural process optimization. Cloud-based architecture emerges as a significant enabler, allowing for the seamless integration of IoT devices for improved agricultural decision-making.

Rajpoot et al. [3] present nature-inspired load balancing methods for cloud computing resource allocation, covering essential areas of resource optimization. While not directly related to agriculture, load balancing techniques are critical in guaranteeing the efficient operation of Cloud-IoT systems in agriculture. This research adds to our grasp of cloud computing fundamentals.

Rajpoot et al. [4] undertake a simulation-based study of cloud computing load balancing. The research investigates the performance of several load balancing algorithms, offering insights on their usefulness and applicability. Understanding load balancing methods is critical for improving resource allocation in agricultural Cloud-IoT systems.

Yuyanto and Liawatimena [5] concentrate on developing a data collection infrastructure based on networked sensors for global open data in agriculture. The report emphasizes the need for Cloud-IoT in building a strong data infrastructure for the agriculture industry. The use of Cloud technology allows for the smooth collection and analysis of data from dispersed sensors, promoting data-driven decision-making in agriculture.

Suciu et al. [6] propose a case study in viticulture on M2M remote telemetry and Cloud-IoT big data processing. The study demonstrates how Cloud-IoT technologies may be utilized in the context of viticulture, highlighting the importance of real-time data processing and analytics in enhancing agricultural methods.

Raikar et al. [7] use a lightweight protocol to investigate the use of Cloud and IoT in agriculture. The research focuses on the practical deployment of Cloud-IoT technologies, demonstrating their promise in tackling the agriculture sector's particular issues. The lightweight protocol allows efficient data transmission and communication, which is crucial in resource-constrained agricultural situations.

Satam et al. [8] present a unique application of a microstrip antenna as a temperature sensor for IoT applications. While the study is not directly related to agriculture, it contributes to a larger knowledge of sensor technologies that may be integrated into Cloud-IoT systems for agricultural monitoring, underscoring the diversity of IoT applications.

Maddikunta et al. [9] conduct a thorough investigation of Unmanned Aerial Vehicles (UAVs) in smart agriculture. The research explores the uses, needs, and constraints of UAVs in agricultural contexts, offering insight into how Cloud IoT technologies might exploit aerial data for precision agriculture.

Hu et al. [10] explore how non-orthogonal multiple access may be used in wireless sensor networks for smart agriculture. The research looks into communication protocols that are important for IoT devices in agriculture, focusing on the role of cloud computing in managing and processing data from wireless sensor networks.

Sun et al. [11] provide a concept for mobile crowd sensing in smart agricultural data collecting. The study uses an innovative method to gather data, emphasizing the importance of public participation in agricultural monitoring. Cloud IoT technologies are crucial in gathering and analyzing data from many sources.

Rajeswari et al. [12] develop a smart agriculture model that combines IoT, mobile, and cloud-based big data analytics. The research provides a comprehensive look at the synergies between various technologies, demonstrating their overall influence on agricultural efficiency and decision-making.

Rawidean and Harun [13] describe an integrated wireless sensor network and cloud computing architecture for agricultural environment applications. The research gives insights into the design and deployment of Cloud-IoT systems for agricultural monitoring and management.

Rawat et al. [14] investigate cloud load balancing, stressing its importance in advancing green cloud computing. While not directly related to agriculture, the study contributes to a greater understanding of sustainable cloud computing methods, which has implications for Cloud-IoT applications in agriculture.

Ghosh and Banerjee [15] present a dynamic time quantum priority-based Round Robin technique for cloud load balancing. The work offers a novel load balancing technique, emphasizing its potential application in Cloud-IoT systems with dynamic resource allocation.

In conclusion, the evaluation of literature derives insights from a broad range of research, demonstrating the complex nature of Cloud-IoT applications in agriculture. These studies, which range from data integration frameworks and monitoring systems to load balancing algorithms and creative sensor applications, all contribute to a better understanding of how Cloud-IoT technologies are altering and optimizing agricultural operations. The following parts will expand on these foundations by delving into particular applications and consequences for Cloud-IoT in agriculture.

9.3 APPLICATIONS OF CLOUD IoT IN AGRICULTURE

Cloud computing and IoT technologies have ushered in a new age in agriculture, providing novel answers to long-standing problems. The confluence of these technologies has resulted in a multitude of applications that improve agricultural efficiency, sustainability, and accuracy. The following section delves into significant Cloud-IoT applications in agriculture, demonstrating how these synergies are transforming the landscape of modern farming.

1. **Precision Agriculture:** Precision agriculture uses Cloud-IoT technology to allow data-driven farming decisions. Sensors and IoT devices capture data on soil conditions, weather patterns, and crop health in real-time. This data is sent to cloud platforms, which are processed by powerful analytics and machine learning algorithms to deliver actionable insights. Farmers

may then make educated decisions about irrigation, fertilization, and pest management, maximizing resource use and crop output.

2. **Smart Irrigation Systems:** Cloud-IoT technologies enable smart irrigation systems, which change irrigation techniques. IoT devices capture soil moisture sensors and meteorological data, which are then transferred to cloud platforms. This data is analyzed by the cloud-based system to develop ideal irrigation schedules, ensuring that crops receive the proper quantity of water at the right time. This not only saves water, but also improves crop health and reduces farmers' water-related expenditures.

3. **Livestock Monitoring and Management:** Cloud-IoT apps for animal husbandry provide solutions for livestock monitoring and management. Sensor-equipped wearable devices are fitted to animals, gathering data on their health, location, and activity. This data is sent to the cloud, where farmers may remotely monitor the health of their cattle. Cloud analytics enables the early diagnosis of health concerns, the optimization of breeding programs, and the general improvement of herd management.

4. **Crop Monitoring and Disease Detection:** Cloud-IoT apps enable farmers to more efficiently monitor crops and identify illnesses. Drones with sensors take high-resolution photos of fields, while ground-based IoT devices collect plant health data. This data is sent to cloud platforms, where machine learning algorithms evaluate it to detect illness or pest indications. Early identification enables farmers to take targeted interventions, lowering pesticide use and crop losses.

5. **Farm Management Systems:** Cloud-based farm management systems enable farmers to monitor and control numerous parts of their operations from a centralized platform. These systems incorporate data from IoT devices such as equipment sensors, weather stations, and crop monitors. Farmers may access this information via the cloud from anywhere, allowing for remote administration of chores such as equipment maintenance, field operations, and inventory tracking.

6. **Agricultural Supply Chain Optimization:** Cloud-IoT technologies help to optimize the agricultural supply chain. Sensors installed in storage facilities and transportation vehicles track crop status along the supply chain. Cloud technologies provide real-time tracking and traceability, ensuring crops are kept and transported in the best circumstances possible.

7. **Monitoring using UAVs:** UAVs, often known as drones, outfitted with IoT sensors has become essential in agricultural monitoring. Drones take high-resolution photos as well as multispectral data, allowing for a complete perspective of broad agricultural fields. This data is processed and analyzed by cloud systems to generate insights about crop health, growth trends, and prospective problems. UAVs provide a cost-effective and efficient way of monitoring large areas of agricultural.

8. **Environmental Monitoring and Sustainability:** Cloud-IoT apps help to monitor the environment in agriculture and promote sustainable practices. Environmental characteristics such as soil quality, air quality, and water use are measured via sensors. The collected data is sent to cloud platforms and

evaluated to determine the environmental impact of agricultural activities. This information assists farms in adopting sustainable methods, reducing environmental impacts, and meeting regulatory requirements.

9. **Mobile and Cloud-Based Agricultural Advisory Services:** Cloud-IoT apps enable farmers to get agricultural advisory services via mobile and online platforms. These services provide farmers with real-time information, weather predictions, market pricing, and expert suggestions. Cloud-based databases store and process massive volumes of agricultural data, enabling farmers to make educated decisions.

10. **Integration with Autonomous Machinery:** Cloud-IoT technologies are critical in integrating autonomous machinery in agriculture. Smart tractors, harvesters, and other self-driving vehicles outfitted with IoT sensors get real-time data from the cloud, allowing them to make autonomous decisions depending on environmental circumstances. This connection improves farm operations' efficiency and decreases the need for human involvement.

9.4 CONCLUSION

The chapter on Cloud-IoT Applications in Agriculture examined the revolutionary influence of cloud computing and IoT technology on current farming techniques. The applications discussed in this chapter range from precision agriculture to smart irrigation, animal monitoring, and beyond, and they highlight the dramatic changes occurring in the agricultural sector. Cloud-IoT integration in agriculture marks a paradigm change, propelling the industry toward higher efficiency, sustainability, and accuracy. This chapter has covered the history of agriculture, the reasoning for Cloud-IoT integration, and the overarching aim of improved resource management and increased agricultural output. The literature study gave a broad overview of current research and trends, laying the groundwork for a thorough examination of applications. The application exploration emphasized how Cloud-IoT transforms precision agriculture, makes irrigation smarter, effectively monitors livestock, and enhances crop management through data-driven decision-making. These applications increase production and help ensure sustainability by reducing resource waste and environmental effects.

As we close this chapter, it is critical to remember that the Cloud-IoT journey in agriculture is continuous. Future research paths may include more sophisticated analytics, artificial intelligence integration, and the creation of effective cybersecurity methods to secure sensitive agricultural data. Addressing concerns like connection and data interoperability, as well as assuring the inclusion of small-scale farmers in the use of these technologies, would be critical for general success.

REFERENCES

1. A. Goldstein, L. Fink, and G. Ravid, "A Cloud-Based Framework for Agricultural Data Integration: A Top-Down-Bottom-Up Approach," IEEE Access, vol. 10, pp. 88527–88537, 2022, doi: https://doi.org/10.1109/access.2022.3198099
2. S. Liu, L. Guo, H. Webb, X. Ya, and X. Chang, "Internet of Things Monitoring System of Modern Eco-Agriculture Based on Cloud Computing," IEEE Access, vol. 7, pp. 37050–37058, 2019, doi: https://doi.org/10.1109/access.2019.2903720

3. N. K. Rajpoot, P. Singh, and B. Pant, "Nature-Inspired Load Balancing Algorithms for Resource Allocation in Cloud Computing," Apr. 2023, doi: https://doi.org/10.1109/cises58720.2023.10183630

4. N. K. Rajpoot, P. Singh, and B. Pant, "Load Balancing in Cloud Computing: A Simulation-Based Evaluation," Apr. 2023, doi: https://doi.org/10.1109/cises58720.2023.10183622

5. Y. Yuyanto, and S. Liawatimena, "Implementation of Data Collecting Platform Over Distributed Sensors for Global Open Data for Agriculture and Nutrition," Aug. 2018, doi: https://doi.org/10.1109/citsm.2018.8674057

6. G. Suciu, A. Vulpe, O. Fratu, and V. Suciu, "M2M Remote Telemetry and Cloud IoT Big Data Processing in Viticulture," International Conference on Wireless Communications and Mobile Computing, Oct. 2015, doi: https://doi.org/10.1109/iwcmc.2015.7289239

7. M. M. Raikar, P. Desai, N. Kanthi, and S. Bawoor, "Blend of Cloud and Internet of Things (IoT) in Agriculture Sector using the Lightweight Protocol," 2018 International Conference on Advances in Computing, Communications, and Informatics (ICACCI), Sep. 2018, doi: https://doi.org/10.1109/icacci.2018.8554406

8. V. Satam, C. V. Kulkarni, and A. Kholapure, "Microstrip Antenna as a Temperature Sensor for IoT Applications," Dec. 2022, doi: https://doi.org/10.1109/mapcon56011.2022.10047111

9. P. K. R. Maddikunta et al., "Unmanned Aerial Vehicles in Smart Agriculture: Applications, Requirements, and Challenges," IEEE Sensors Journal, pp. 1–1, 2021, doi: https://doi.org/10.1109/jsen.2021.3049471

10. Z. Hu et al., "Application of Non-Orthogonal Multiple Access in Wireless Sensor Networks for Smart Agriculture," IEEE Access, vol. 7, pp. 87582–87592, 2019, doi https://doi.org/10.1109/access.2019.2924917

11. Y. Sun et al., "On Enabling Mobile Crowd Sensing for Data Collection in Smart Agriculture: A Vision," IEEE Systems Journal, vol. 16, no. 1, pp. 132–143, Mar. 2022, doi: https://doi.org/10.1109/jsyst.2021.3104107

12. S. Rajeswari, K. Suthendran, and K. Rajakumar, "A Smart Agricultural Model by Integrating IoT, Mobile and Cloud-Based Big Data Analytics," IEEE Xplore, Jun. 01, 2017. https://ieeexplore.ieee.org/abstract/document/8321902

13. M. Rawidean, and A. Nahar Harun, "Wireless Sensor Networks and Cloud Computing Integrated Architecture for Agricultural Environment Applications," Dec. 2017, doi: https://doi.org/10.1109/icsenst.2017.8304445

14. P. Rawat, P. Singh, and V. Tripathi, "Load Balancing in Cloud Computing Leading Us Towards Green Cloud Computing," SSRN Electronic Journal, 2022, doi: https://doi.org/10.2139/ssrn.4031990

15. S. Ghosh, and C. Banerjee, "Dynamic Time Quantum Priority Based Round Robin for Load Balancing in Cloud Environment," Nov. 2018, doi: https://doi.org/10.1109/icrcicn.2018.8718694

10 Cloud IoT in Railways

Mohit Angurala and Harmeet Singh

10.1 INTRODUCTION

The railway sector is critical to contemporary transportation, connecting people and products across long distances. Integration of Cloud Computing with the Internet of Things (IoT) has emerged as a transformational solution to satisfy the rising demands of efficiency, safety, and sustainability. In this chapter, we discuss the significance of cloud IoT integration in railroads and outline the goals that will guide our investigation.

10.2 IMPORTANCE OF CLOUD IoT INTEGRATION IN RAILWAYS

The importance of cloud IoT integration in the railway industry cannot be emphasized. Railway systems have traditionally depended on centralized control and manual monitoring, posing issues in real-time data processing, maintenance, and resource efficiency. Railways may change their operations by utilizing Cloud Computing and IoT technology to create a smart and linked rail network. Cloud computing has unrivaled scalability, cost-effectiveness, and computational power, making it a perfect platform for processing and storing massive volumes of railway data. IoT devices, on the other hand, integrated with sensors and actuators, allow for real-time data collection and smooth communication across various railway components. The use of cloud and IoT in railroads enables rail authorities to utilize data-driven insights, improve decision-making, forecast maintenance requirements, and optimize train timetables and resource allocation [1]. Furthermore, cloud IoT integration allows train firms to deliver improved passenger services such as real-time trip information, customized experiences, and increased safety measures. Railways may update their infrastructure, increase operational efficiency, minimize downtime, and improve overall safety by embracing cloud IoT integration, fulfilling the demands of 21st-century travelers and enterprises.

10.3 OBJECTIVES OF THE CHAPTER

This chapter's goals are to offer a thorough grasp of the possibilities and advantages of cloud IoT integration in railroads. We intend to introduce the essential ideas of Cloud Computing and IoT, building the framework for their use in the railway area. A focus will be on the examination of cloud IoT architecture created expressly for railroads, where we will address design considerations for IoT devices and their smooth interaction with cloud-based systems. We will explore the numerous cloud IoT applications in railroads, emphasizing the transformative influence on predictive maintenance, asset tracking, train scheduling, and passenger services. Furthermore,

DOI: 10.1201/9781032656694-10

we will investigate cloud-based data analytics and insights, with a focus on the role of machine learning in optimizing railway operations. To give practical insights, we will provide case studies demonstrating successful cloud IoT deployments in real-world railway applications. While cloud IoT integration has many advantages, it is not without obstacles. As a result, we will talk about potential roadblocks and future possibilities in cloud IoT integration for railroads, addressing challenges including connection, data security, scalability, and legacy system integration. We want to present a complete roadmap by the conclusion of this chapter for tapping the full potential of cloud IoT integration in the railway sector, driving its digital transformation, and allowing the construction of smarter, more efficient, and passenger-centric train networks.

10.4   FUNDAMENTALS OF CLOUD COMPUTING AND IoT

The IoT is a paradigm that links physical devices and items to the internet, allowing them to gather and exchange data as well as interact with one another and their surroundings [2]. Understanding IoT architecture and components is critical for understanding its applications in railroads.

10.4.1   IoT ARCHITECTURE AND COMPONENTS

Internet of Things Architecture: IoT architecture may be divided into three layers: perception layer, network layer, and application layer. The perception layer is made up of IoT devices outfitted with various sensors and actuators that collect data from the physical world. Examples of these technologies include simple temperature sensors to complicated video cameras, and vibration sensors used in trains. The network layer supports data connectivity between IoT devices and gateways, guaranteeing safe and efficient data transmission. Gateways serve as middlemen, gathering data from different devices and sending it to the cloud for processing. The application layer is the highest layer, where data is evaluated and used to gain insights, make choices, and initiate actions.

IoT Components: Sensors, actuators, gateways, cloud platforms, and user interfaces are examples of IoT system components. Sensors are basic components that detect and transform physical changes like temperature, humidity, and vibration into electrical impulses. Actuators, on the other hand, respond to cloud or other IoT device commands by starting physical activities depending on the data received. As previously said, gateways are critical in gathering data from various sensors, preprocessing it, and transferring it to the cloud. Cloud platforms serve as the backbone of IoT systems, storing, processing, and analyzing data. These systems provide the scalability, processing power, and data storage needed to manage the massive volumes of data created by IoT devices. User interfaces allow users to interact with the IoT system by monitoring and controlling connected devices and accessing real-time insights.

10.4.2 IoT Applications in Railways

The integration of IoT in railroads has created new opportunities for revolutionizing railway operations, improving passenger experience, and maintaining the transportation network's safety and efficiency.

Predictive analytics and smart maintenance: IoT-enabled sensors deployed on railway rails and train stock can continually monitor asset status. Railways can apply predictive maintenance plans by collecting data on track wear, vibrations, and temperature [3]. This method enables the early diagnosis of probable defects, reducing downtime and improving maintenance efficiency.

Asset Tracking and Management in Real-Time: IoT-based asset tracking systems allow real-time monitoring of railway assets such as locomotives, freight wagons, and maintenance equipment. This tracking system promotes asset use, maintenance schedule optimization, and overall resource management.

Passenger Safety and Security: IoT equipment, such as surveillance cameras and crowd sensors, can be installed at railway stations and trains to improve passenger safety and security. The ability to monitor these gadgets in real-time allows authorities to respond promptly to situations and safeguard the safety of passengers.

Energy Efficiency and Resource Allocation: IoT technologies may be used to optimize energy usage and resource allocation in trains [4]. Railways may find possibilities for energy savings and apply sustainable practices by monitoring energy consumption at stations, depots, and trains.

IoT-based solutions can facilitate automated train control and signaling, allowing trains to connect with one another and with railway infrastructure. This technology improves train scheduling, reduces delays, and ensures safe and efficient railway operations.

Finally, knowing IoT architecture and components is critical for realizing its applications in railroads. IoT integration in railroads has the potential to change the sector by improving safety, efficiency, and passenger experience. Railways may begin on a path toward a smart and linked transportation network by embracing IoT technology, revolutionizing the way people and commodities are moved, and satisfying modern-day needs.

10.5 CLOUD IoT ARCHITECTURE FOR RAILWAY SYSTEMS

The cloud IoT architecture for railway systems adds revolutionary capabilities, allowing railways to streamline operations, improve passenger experience, and assure safety and efficiency. The basic architecture of cloud IoT in railways is as shown in Figure 10.1.

The architecture shows three different layers at which the cloud, IoT, and Train sensors operate. They are interconnected to exchange data with the cloud, and at the

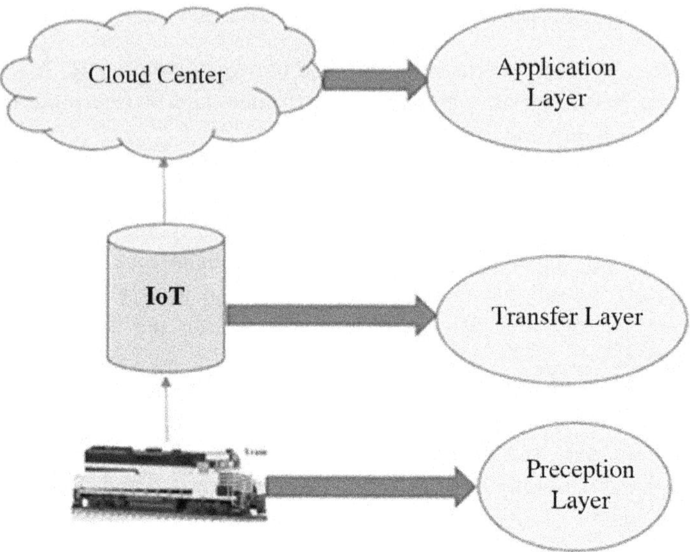

FIGURE 10.1 Architecture of cloud IoT in railways.

cloud, analysis of data is performed. Robustness, stability, power efficiency, scalability, and interoperability are key design factors for IoT devices, facilitating the easy implementation of IoT infrastructure in railway contexts. Railways can manage massive volumes of data, analyze it in real time, and draw useful insights to improve decision-making by storing and processing it on the cloud. Integrating real-time data from railway sensors into the cloud enables continuous monitoring and management, leading to proactive safety and operational efficiency measures. Robust security mechanisms, data encryption, secure communication routes, and compliance with privacy rules are required to handle security and privacy concerns.

10.5.1 Design Considerations for IoT Devices in Railways

The proper design and deployment of IoT devices are critical to effectively integrating cloud and IoT in railway systems. To ensure the efficacy and efficiency of IoT devices in the railway environment, several major factors must be addressed.

Robustness and dependability: Railway settings are characterized by difficult circumstances such as vibrations, high temperatures, and dust and moisture exposure. To enable ongoing data collection and transmission, IoT devices must be built to withstand these difficulties and retain dependable functioning.

Scalability and Modularity: Railways cover vast geographical areas, and the number of IoT devices required can be extensive. Therefore, IoT devices should be designed with scalability and modularity in mind, allowing for easy deployment, expansion, and maintenance of the overall IoT infrastructure.

Interoperability and Standardization: To ensure seamless communication and data exchange, IoT devices in railways should adhere to standardized communication protocols and data formats [5]. This promotes interoperability among various devices and systems, enabling easy integration with cloud platforms and other railway infrastructure.

10.5.2   Cloud-based Data Storage and Processing for Railway Applications

The cloud IoT architecture in railroads is built on cloud-based data storage and processing. Cloud computing provides scalable and cost-effective methods for dealing with massive volumes of data produced by IoT devices.

Data Storage: Cloud platforms offer safe and dependable data storage solutions, allowing railway authorities to store and manage a wide range of data types, including sensor readings, maintenance records, and operating logs. Cloud systems with scalable storage can handle the growing volume of data from an expanding network of IoT devices.

Data Processing and analyses: Cloud-based data processing provides real-time and batch analyses of acquired data, providing railway operators with useful insights. Machine learning algorithms may be used to assess data patterns, discover abnormalities, and forecast maintenance requirements, assisting in proactive decision-making and resource efficiency [6].

Integration with Railway Information Systems: Cloud platforms can be connected with current railway information systems such as Enterprise Resource Planning (ERP) systems and Customer Relationship Management (CRM) systems. This interface allows for easy data transmission and gives a comprehensive perspective of railway operations, allowing for better decision-making at all levels.

10.5.3   Integrating Real-time Data from Railway Sensors into the Cloud

The connection of real-time data from railway sensors to the cloud is a critical component of the cloud IoT architecture [7]. Data transmission that is timely and accurate supports swift decision-making and rapid responsiveness to changing situations.

Data Transmission Protocols: IoT devices communicate data to the cloud using several data transmission protocols, such as MQTT (Message Queuing Telemetry Transport) and CoAP (Constrained Application Protocol). These lightweight protocols enable effective communication while reducing data transmission overhead.

Continuous Monitoring and Control: Real-time data integration enables railway operators to monitor crucial metrics in real-time, such as track conditions, train positions, and passenger counts. This enables proactive safety and operational efficiency initiatives.

10.5.4 Ensuring Security and Privacy in Cloud IoT Railway Systems

Because of the huge volumes of sensitive data created and transferred, the integration of cloud and IoT in railway systems creates security and privacy problems. Strong security measures are required to secure railway operations and passenger data [8].

Data Encryption and Authentication: IoT devices should use strong encryption techniques to safeguard data during transmission and storage. Secure keys and digital certificates, for example, ensure that only authorized devices may access the cloud platform.

Secure Communication Channels: Between IoT devices and the cloud, Secure Virtual Private Networks (VPNs) or Secure Socket Layer (SSL) connections may be formed, protecting data integrity and preventing unwanted access during data transfer.

Identity Access Management (IAM): IAM solutions govern user access to cloud resources, ensuring that only authorized people may access and alter data and system configurations. IAM policies assist to reduce the risk of data breaches and unauthorized system modification.

Data Anonymization and Privacy legislation: To preserve passenger privacy, railway authorities should follow privacy legislation and use data anonymization procedures. Users' personal information should be maintained securely and processed with their permission.

Overall, the cloud IoT architecture offers railways an unprecedented chance to update their infrastructure and move toward smart and efficient transportation networks.

10.6 ENHANCING RAILWAY OPERATIONS WITH CLOUD IoT INTEGRATION

The use of Cloud Computing and the IoT in railway systems opens up a plethora of potential to improve operations, efficiency, and the overall passenger experience. In this section, we will look at how cloud IoT integration helps modernize railway operations.

10.6.1 Predictive Maintenance and Fault Detection

Predictive maintenance is a key use of cloud IoT integration in the railway industry [9]. Real-time data on the state of railway assets including, rails, rolling stock, and signaling systems, may be continually gathered and communicated to the cloud by placing IoT sensors on them. This data is processed by cloud-based data analytics and machine learning algorithms to estimate maintenance needs and identify possible flaws before they become serious concerns.

Data-driven Maintenance Strategies: Fixed-schedule maintenance techniques are generally inefficient and expensive. Railway operators may use predictive maintenance to implement data-driven maintenance programs, completing repair activities only when necessary, saving downtime, and optimizing maintenance resources.

Preventing unforeseen breakdowns: The capacity to identify early indicators of wear and tear allows for prompt intervention and proactive maintenance, hence preventing unforeseen breakdowns that might interrupt railway operations and pose safety risks.

10.6.2 Real-time Asset Tracking and Management

Cloud IoT connection enables real-time asset tracking and management, giving railway operators complete visibility into asset inventory and movement.

Asset Location and Utilization: IoT sensors installed on locomotives, freight wagons, and maintenance equipment allow real-time tracking of their positions and statuses [10]. This data assists in optimizing asset use and streamlining asset deployment for optimal efficiency.

Railway operators may optimize the allocation of resources such as locomotives and wagons by monitoring asset movement and usage trends, assuring smooth operations and cost savings.

10.6.3 Optimizing Train Scheduling and Operations

Train scheduling efficiency is crucial for ensuring on-time operations and reducing delays. Cloud IoT integration helps to optimize train timetables and improve overall railway operations.

Real-time Data for Decision-Making: By using real-time data on train positions, track conditions, and weather, railway operators can make educated decisions about train timetables and rerouting, reducing delays and disruptions.

Congestion Reduction: Cloud-based analytics assist in identifying bottlenecks and regions of congestion, allowing for more effective traffic management and lowering the chance of congestion-related delays.

10.6.4 Improving Passenger Experience and Safety

The integration of cloud IoT has a significant influence on passenger experience and safety, resulting in increased consumer satisfaction and trust in rail travel.

Real-time Travel Information: IoT-enabled displays and smartphone applications give passengers with real-time travel information such as train timetables, platform assignments, and service updates, allowing them to better plan their travels.

Safety and emergency response: IoT sensors and surveillance cameras installed in railway stations and trains allow for continuous monitoring of possible safety issues and crises. Real-time data transfer to the cloud enables speedy response and coordination in crucial situations.

Tailored Services: Cloud-based analytics enable train operators to analyze passenger data and preferences, allowing them to provide tailored services, including targeted offers, travel recommendations, and increased accessibility for customers with special requirements.

10.7　CLOUD IoT APPLICATIONS IN RAILWAY INFRASTRUCTURE

The use of Cloud Computing and the IoT in railway infrastructure opens up a slew of possibilities for developing smart and efficient systems. In this part, we will look at numerous cloud IoT applications that have the potential to change the railway business.

10.7.1　SMART RAILWAY STATIONS AND TICKETING SYSTEMS

The integration of cloud IoT delivers innovation to railway stations and ticketing systems, boosting passenger comfort and operational efficiency [11].

Automated Ticketing and Fare Collection: IoT-enabled ticketing devices and contactless payment systems make it simple for passengers to purchase tickets and collect fares, eliminating lineups and wait times at ticket counters.

Smart Station Management: In railway stations, cloud-based sensors and surveillance cameras offer real-time monitoring of passenger movement, allowing station managers to maximize crowd management, improve security, and expedite station operations.

Intelligent Passenger Assistance: IoT-powered kiosks and mobile applications deliver real-time travel information, platform upgrades, and navigation aids to travelers, improving the entire passenger experience.

10.7.2　INTELLIGENT SIGNALING AND TRAIN CONTROL SYSTEMS

Cloud IoT integration transforms railroad signaling and control systems, making trains safer and more efficient.

Automated Train Control: IoT devices outfitted with train location and monitoring sensors provide real-time data to the cloud, enabling automated train control and signaling systems [12]. This technology lowers human mistakes while also improving railway safety and dependability. Cloud-based analytics evaluate data on train positions, track conditions, and traffic, enabling dynamic route optimization to avoid crowded regions and save travel time.

Enhanced Safety Measures: IoT sensors identify possible safety concerns on the railway, such as track irregularities.

10.7.3　RAILWAY TRACK CONDITION MONITORING AND MANAGEMENT

The integration of cloud IoT provides proactive solutions for monitoring and regulating railway track conditions, guaranteeing safe and efficient train movements.

Continuous Track Monitoring: IoT sensors put along railway tracks analyze track characteristics such as track alignment, wear, and faults in real-time. Data sent to the cloud allows for real-time analysis and early diagnosis of track problems. Cloud-based analytics evaluate track condition data to forecast maintenance needs and prioritize repairs, lowering the risk of track failures and minimizing delays to train services.

Optimized Track Maintenance Planning: Cloud-based data storage and processing make historical track maintenance data easier to organize, allowing railway operators to plan and schedule maintenance tasks more efficiently.

10.7.4 Energy Management and Efficiency in Railways

The integration of cloud IoT plays a critical role in optimizing energy use and increasing energy efficiency in railway operations.

Smart Energy Monitoring: IoT sensors at railway stations, depots, and trains monitor energy use and broadcast real-time data to the cloud. Cloud analytics give insights into energy consumption trends, assisting in identifying opportunities for energy reduction.

Energy-efficient Train Operations: The combination of the cloud and IoT allows for real-time monitoring of train performance data such as energy usage and speed profiles. This data supports railroad operators in implementing energy-efficient driving methods and improving train operations.

Integration of Renewable Energy Sources: Cloud-based energy management systems enable the integration of renewable energy sources such as solar and wind into railway operations. The cloud promotes sustainability by facilitating efficient energy delivery and storage.

Cloud IoT applications in railway infrastructure provide enormous potential for industrial transformation. Intelligent signaling and train control technologies improve safety and efficiency, while smart railway stations and ticketing systems boost passenger convenience. Railway track condition monitoring and management allow for preventive maintenance, decreasing downtime, and maintaining train safety. Solutions for energy management and efficiency offer sustainability and cost savings. The railway sector may transform into a smart and efficient transportation network by embracing cloud IoT integration, improving passenger experience, and fulfilling the needs of the digital era.

10.8 DATA ANALYTICS AND INSIGHTS IN RAILWAYS WITH CLOUD IoT

The use of Cloud Computing and the IoT by the railway sector generates vast amounts of data. Railway personnel may get important insights and make smart decisions if this data is evaluated using current analytics. In this section, we'll look at how cloud IoT may help the railway sector with data analytics and insights.

10.8.1   Data Aggregation and Analytics in the Cloud

Aggregation of Data: Cloud platforms serve as a central repository for data received from numerous IoT devices installed across the railway network. In the cloud, real-time data from sensors, ticketing systems, trains, and infrastructure components is pooled to create a comprehensive dataset.

Big Data Analytics: The computational power and scalability of the cloud enable big data analytics on aggregated datasets [13]. Advanced analytics software processes and analyzes massive amounts of data, finding patterns, trends, and anomalies.

10.8.2   Leveraging Machine Learning for Railway Analytics

To estimate maintenance needs, machine learning systems analyze sensor data from trains and infrastructure. Early detection of potential faults allows for proactive maintenance, which reduces operational disruptions and expenditures. Machine learning algorithms constantly analyze data for deviations from expected patterns, allowing the detection of faults and irregularities that may indicate safety concerns or potential breakdowns.

Machine learning algorithms analyze real-time data on train locations, passenger demand, and track conditions in order to optimize train scheduling for reduced delays and increased efficiency.

10.8.3   Real-time Decision Support for Railway Authorities

Real-time Data Insights: Cloud IoT integration enables railway authorities to monitor key performance indicators and respond quickly to changing situations.

Emergency: In an emergency or interruption, real-time data analytics provides railway officials with the information they need to make swift decisions and coordinate emergency response activities.

Real-time analytics: It aids in the more effective allocation of resources such as trains, employees, and maintenance teams, maximizing resource usage and enhancing overall system efficiency.

Improved Passenger Experience: With real-time data on train timetables, platform assignments, and service updates, railway authorities can provide passengers with timely and accurate information, boosting their experience and satisfaction.

Therefore, data analytics and insights enabled by cloud IoT integration are revolutionizing railway operations. Data aggregation and big data analytics in the cloud give a comprehensive perspective of railway operations, allowing for operational insights and data-driven decision-making. Railway authorities may use machine learning to forecast maintenance requirements, detect defects, optimize train timetables, and increase overall system performance. Real-time data insights help in emergency response and resource allocation. By fully using data analytics and insights, railway authorities may optimize operations, improve safety and efficiency, and enhance passenger experience, ushering in a new era of smart and data-driven transportation networks.

10.9 CASE STUDIES OF SUCCESSFUL CLOUD IoT IMPLEMENTATIONS IN RAILWAYS

The revolutionary potential of Cloud Computing and the IoT has resulted in a paradigm change in the railway industry's approach to operations and passenger experience in recent years. Railway operators have reimagined several aspects of their services by leveraging the possibilities of cloud-based platforms and IoT devices, ranging from predictive maintenance and real-time asset tracking to intelligent ticketing systems and better passenger experiences. These case studies show how the seamless integration of Cloud Computing and Internet of Things technology has catapulted railways into a new era of efficiency, dependability, and passenger-centric services. We reveal the real advantages of cloud IoT installations in railroads by fully examining each case study.

10.9.1 CASE STUDY 1: PREDICTIVE MAINTENANCE AND ASSET TRACKING IN A RAILWAY COMPANY

10.9.1.1 Introduction

The railway firm in this case study, a prominent industry participant, encountered difficulties with traditional maintenance techniques, which frequently resulted in unexpected breakdowns, increased downtime, and rising maintenance expenses. To address these concerns and assure the best possible performance and safety of their railway assets, the firm opted to use cloud IoT integration for predictive maintenance and asset tracking.

10.9.1.2 Predictive Maintenance Implementation

Throughout the railway network, IoT sensors were strategically deployed on trains, tracks, and important infrastructure components. These sensors gathered real-time data on a variety of characteristics such as temperature, vibration, humidity, and wear. The data from these sensors was transferred in real-time to the cloud, resulting in a consolidated store of crucial information.

10.9.1.3 Advanced Analytics and Machine Learning

To analyze the massive volume of sensor data, the railway business used advanced analytics and machine learning algorithms in the cloud. Machine learning models were developed to discover patterns and abnormalities in data that could suggest possible problems or impending breakdowns.

10.9.1.4 Early Detection and Proactive Maintenance

When aberrant patterns or deviations were observed, the predictive maintenance system was meant to warn maintenance workers in real time. Because of the early identification, the organization was able to proactively plan maintenance operations, detecting and fixing possible concerns before they became severe failures.

10.9.1.5   Advantages of Predictive Maintenance

The railway firm reaped many important benefits from the introduction of predictive maintenance:

1. **Reduced Downtime:** By proactively addressing maintenance needs, the train business significantly reduced unexpected breakdowns and related downtime. Trains were kept in excellent condition, resulting in increased dependability and operating efficiency.
2. **Cost Savings:** The organization was able to optimize its maintenance operations thanks to predictive maintenance. Unnecessary maintenance tasks were reduced, resulting in cost savings and better resource allocation.
3. **Better Safety:** Early defect identification reduced safety hazards, contributing to better passenger safety and trust in the train operation.

10.9.1.6   Asset Tracking and Management

In addition to predictive maintenance, the train business implemented cloud-based asset tracking technologies. IoT devices were put on train stock, wagons, and other vital assets to track their position and use in real-time.

10.9.1.7   Optimizing Resource Allocation

The asset monitoring data was connected with the maintenance management system, allowing the organization to allocate resources better. The train firm could deploy resources more efficiently by correctly tracking asset locations and consumption patterns, ensuring the right assets were accessible when and where they were required.

1. **Improved Resource use:** The real-time tracking of assets enabled the organization to improve its deployment and use. Idle assets were identified, and improved resource management was achieved, resulting in cost savings.
2. **Effective Asset Management:** The railway firm got important insights into asset utilization trends, which resulted in improved asset management techniques. Maintenance and replacements were scheduled in advance, increasing the life of important equipment.

10.9.1.8   Conclusion

The case study focuses on how cloud IoT integration enables predictive maintenance and asset tracking, revolutionizing the train company's operations and increasing overall efficiency, safety, and dependability. The use of predictive maintenance resulted in less downtime, lower costs, and increased safety. Cloud-based asset tracking enabled better resource allocation and asset management. The railway firm demonstrated how technology-driven solutions can move the railway sector toward a more intelligent and data-driven future by embracing cloud IoT integration for predictive maintenance and asset tracking.

10.9.2 Case Study 2: Cloud IoT-Enabled Train Scheduling and Operations Optimization

10.9.2.1 Introduction

In this case study, we look at how cloud IoT integration has transformed train scheduling and operations for a major railway firm. Faced with manual scheduling issues and poor operations, the railway business wanted to use cloud IoT technology to improve train scheduling procedures and maximize operational efficiency.

10.9.2.2 Cloud IoT-Enabled Train Scheduling

Real-time Data Collection: To gather real-time data on train movements, track conditions, and passenger loads, the railway firm installed IoT sensors on trains, tracks, and important infrastructure. This data was communicated in real-time to the cloud, resulting in a consolidated and up-to-date store of important information.

Data Analysis and Predictive Modeling: Powerful data analytics and machine learning techniques were used to handle real-time data in the cloud. These algorithms used historical and current data to create predicted models for train delays, track repair needs, and passenger demand patterns.

Train Scheduling Optimization: The predictive models were utilized to optimize train scheduling. Based on real-time data and forecast insights, the railway firm can now dynamically alter train frequencies, routes, and stops.

10.9.2.3 The Advantages of Cloud IoT-Enabled Train Scheduling

1. **Enhanced Punctuality:** The railway business could better anticipate and reduce potential delays by leveraging real-time data and predictive models, resulting in enhanced train punctuality and dependability.
2. **Improved Resource Allocation:** Because of the enhanced train schedule, the organization was able to distribute resources more efficiently, such as trains and staff. Idle resources were reduced, resulting in cost savings and better resource use.
3. **Improved Passenger Experience:** With more precise train timetables and fewer delays, travelers had a more predictable and comfortable ride, which improved satisfaction and loyalty.

10.9.2.4 Operations Optimization with IoT

Intelligent Asset Monitoring: IoT sensors were deployed on trains and critical infrastructure to monitor the health and performance of assets, such as locomotives and signaling systems. The data from these sensors was transmitted to the cloud, enabling real-time monitoring and analysis.

Proactive Maintenance and Asset Management: With real-time asset monitoring, the railway company could detect potential faults or performance issues early on. Proactive maintenance measures were implemented to reduce the risk of unexpected breakdowns and ensure optimal asset performance.

Dynamic Resource Management: The railway company used cloud-based platforms to manage resources, such as fuel, crew, and maintenance teams, dynamically. Real-time data allowed for better resource planning and allocation, minimizing waste and inefficiencies.

10.9.2.5 The Advantages of IoT-Based Operations Optimization

1. **Less unexpected Downtime:** Proactive maintenance and asset monitoring reduced unexpected downtime, resulting in better asset dependability and operational efficiency.
2. **Cost reductions:** Dynamic resource management and streamlined operations resulted in cost reductions in a variety of railway activities.
3. **Improved Safety:** Real-time monitoring and preventive maintenance contributed to higher safety standards, lowering the danger of accidents and maintaining passenger safety.

10.9.2.6 Conclusion

The case study demonstrates the substantial impact of cloud IoT integration on train scheduling and operational optimization. The railway firm improved train punctuality, resource allocation, and passenger experiences by leveraging the power of real-time data analytics and predictive modeling. Furthermore, IoT-enabled intelligent asset monitoring and preventive maintenance boosted asset dependability, decreased downtime, and improved safety. The train firm demonstrated how data-driven solutions can change traditional railway operations into efficient, flexible, and customer-centric systems by embracing cloud IoT technology.

10.10 CHALLENGES AND FUTURE DIRECTIONS

The integration of Cloud Computing and the IoT in railroads offers the sector several obstacles as well as interesting prospects. One of the major issues is dealing with connection and network dependability. Railway networks traverse huge geographical areas, including isolated and rugged terrains where regular communication can be difficult to maintain. To ensure uninterrupted data transfer, railway firms must invest in robust communication infrastructure and investigate technologies such as 5G and satellite connectivity. Furthermore, redundant communication channels and failover procedures can improve network resilience by reducing interruptions and assuring real-time data availability.

Another significant issue is data security and privacy. As railways become more networked through cloud IoT integration, the security and integrity of sensitive data become increasingly important. Passenger data, operational data, and vital infrastructure details must all be protected from cyber-attacks and unlawful access. To preserve passengers' privacy and develop confidence among stakeholders, railway firms must implement strong security measures such as encryption, authentication, and access controls while also complying with data privacy legislation.

Another problem is scalability and interaction with old railway infrastructure. As railway businesses install IoT devices and extend their cloud-based platforms, maintaining seamless coexistence with older systems becomes increasingly important.

Because legacy systems may employ various protocols, data formats, and communication standards, careful integration planning is required. Railway firms should invest in scalable cloud solutions that can handle the growing volume of data produced by IoT devices. Adopting open standards and interoperable technologies would make integration easier, allowing for more efficient operations and reducing interruptions.

Looking ahead, railway firms will have an intriguing opportunity to investigate possible synergies with upcoming technologies such as 5G and Edge Computing. The introduction of 5G networks has the potential to revolutionize real-time data transfer by enabling quicker and more reliable connections between IoT devices, paving the door for new applications and services. Edge computing may bring data processing closer to the source, lowering latency and improving responsiveness, which is especially important for crucial railway operations. Railways can unleash the full potential of cloud IoT integration by closely monitoring technical changes and promoting cooperation with industry partners and technology suppliers. This will drive the sector toward a more efficient, safe, and connected future. Accepting these problems and looking ahead will put railways on a revolutionary path.

10.11 CONCLUSION

The integration of cloud IoT provides a transformational potential for the railway sector, reinventing how rail systems are managed, operated, and experienced. Railways may accomplish predictive maintenance, optimize operations, and improve passenger safety and experience by utilizing cloud-based data storage, real-time analytics, and IoT devices. The case studies that were successful demonstrated the practical benefits of cloud IoT integration, ranging from greater efficiency to cost reductions. However, addressing connection, security, scalability, and interaction with legacy systems difficulties is crucial for widespread use. In the future, ongoing collaboration between railway authorities, technology suppliers, and researchers will be critical in pushing the advancement of cloud IoT in railroads. Cloud IoT integration has enormous potential to transform rail transport, ushering in a new era of smart and efficient railway systems.

REFERENCES

1. S. Arivazhagan, R. N. Shebiah, J. S. Magdalene, and G. Sushmitha, "Railway Track Derailment Inspection System Using Segmentation Based Fractal Texture Analysis," ICTACT Journal on Image and Video Processing, vol. 6, no. 1, pp. 1060–1065, 2015.
2. S. M. Ujjan, I. H. Kalwar, B. S. Chowdhry, T. D. Memon, and D. K Soother, "Adhesion Level Identification in Wheel-Rail Contact Using Deep Neural Networks," 3C Technologia, 2020. https://3ciencias.com/wp-content/uploads/2020/04/12_ee_3c-tecno_abril-2020-1.pdf
3. M. Hussain et al., "Stator Winding Fault Detection and Classification in Three-Phase Induction motor," Intelligent Automation, vol. 29, no. 3, pp. 869–883, 2021.
4. D. K. Soother et al., "The Importance of Feature Processing in Deep-Learning-Based Condition Monitoring of Motors," Mathematical Problems in Engineering, vol. 2021, pp. 1–23, 2021.
5. D. K. Soother, I. H. Kalwar, T. Hussain, B. S. Chowdhry, S. M. Ujjan, and T. D. Memon, "A Novel Method Based on UNET for Bearing Fault Diagnosis," Computer Materials and Continua, vol. 69, no. 1, pp. 393–408, 2021.

6. K. Mal, I. Hussain, B. S. Chowdhry, and T. D. Memon, "Extended Kalman Filter for Estimation of Contact Forces at Wheel-Rail Interface," 3C Tecnología, pp. 280–301, 2020.
7. S. Hu, and T. Dai, "Online Map Application Development Using Google Maps API SQL Database and ASP. NET," vol. 3, no. 3, pp. 102–110, 2013.
8. M. A. Dar, and J. Parvez, "A Live-Tracking Framework for Smartphones," 2015 International Conference on Innovations in Information Embedded and Communication Systems (ICIIECS), pp. 1–4, 2015.
9. J.-E. Lee, and H.-J. Mun, "Design of Deep Learning-Based Automatic Drone Landing Technique Using Google Maps API," JIC, vol. 18, no. 1, pp. 79–85, 2020.
10. A. M. Qadir, and P. Cooper, "GPS-Based Mobile Cross-Platform Cargo Tracking System with Web-Based Application," 2020 8th International Symposium on Digital Forensics and Security (ISDFS), pp. 1–7, 2020.
11. A. Chandra, S. Jain, and M. A. Qadeer, "Implementation of Location Awareness and Sharing System Based on GPS and GPRS using J2ME PHP and MYSQL," 2011 3rd International Conference on Computer Research and Development, vol. 1, pp. 216–220, 2011.
12. T. E. Kaplinger, J. L. Lentz, C. C. Mitchell, and A. K. Shook, "Automatically Replacing Localhost as Hostname in URL With Fully Qualified Domain Name or IP Address," Google Patents, 2017.
13. K. Chen, G. Tan, J. Cao, M. Lu, and X. Fan, "Modeling and Improving the Energy Performance of GPS Receivers for Location Services," IEEE Sensors Journal, vol. 20, no. 8, pp. 4512–4523, 2019.

11 Cloud IoT in Railways
Advancements, Challenges, and Future Prospects

Tanusha Mittal, Jyoti Agarwal, and Upma Jain

11.1 INTRODUCTION

In today's era, the world is changed beyond imagination because of the Internet of Things (IoT). The internet has a great impact not only on a daily basis of human beings but also on almost every domain, including railways, smart cities, healthcare, government, education, business, infrastructure, transportation, etc. [1]. In recent decades, the internet has evolved so much that it has helped to improve the living standards of human beings and provide various types of safety measures with the integration of different types of technologies [2].

The IoT, as its name suggests, is a network of physical items or "things" outfitted with sensors, software, and other technologies to gather and exchange data with other systems and devices across a network [3]. These items, which are all connected via the common medium of the internet, might range from commonplace goods like home appliances to sophisticated industrial machines. The potential of IoT to connect the physical and digital worlds is one of its defining characteristics. IoT gives commonplace items and processes unprecedented interconnectedness and intelligence by giving them the ability to communicate and work together autonomously [4].

The IoT, cloud computing, and logistics have steadily taken on greater significance in fostering social and economic development as a result of the rapid development of big data, and they are continuously moving toward the creation of a new intelligent information visualization brand. Yi has created an intelligent logistics information management model based on the IoT and cloud computing in response to the current economic and social development trend. This model can grant intelligent users with various privileges and query logistics data following the requirements of system users [5].

Based on their intended applications, IoT-based technologies can be divided into two main categories: (1) enabling building block technologies that directly contribute to the development of the IoT and (2) synergistic technologies that enhance the IoT. Electronic communication protocols, machine-to-machine interfaces, microcontrollers, RFID technologies, wireless communication, sensors, location technologies, actuators, and software are examples of enabling building block technologies. Geotagging/ geocaching, machine vision, biometrics, robotics, mirror worlds, augmented reality, adjustable autonomy, life recorders, tangible user interfaces, private black boxes, and clean technologies are examples of synergistic technologies on the other side [6].

DOI: 10.1201/9781032656694-11

The idea of utilizing computers and networks to monitor and manage objects is not new, but the name "IoT" is. This has been the case for decades. For instance, systems for remotely monitoring electricity grid meters via telephone lines were already in the business world by the late 1970s. The "machine-to-machine" (M2M) corporate and industrial solutions for equipment monitoring and operation became widely used in the 1990s thanks to advancements in wireless technology. In contrast to Internet Protocol (IP)-based networks and internet standards, many of these early M2M solutions were based on closed purpose-built networks and proprietary or industry-specific standards [7].

IoT-based digital rail asset inspection, monitoring, and control offer enormous potential to boost crew and passenger safety, ensure rail service reliability, increase maintenance efficiency, and prevent asset breakdown. A solid understanding of the opportunities and risks associated with IoT implementation is necessary before implementing it for the maintenance of rail assets, even though it has the potential to have enormous benefits for rail asset management. Additionally, designing and developing IoT systems for maximum efficiency could be difficult, particularly for a risky industry like rail. The capacity of IoT technology to collect and interpret data in real-time is key to any chances for its use in the rail industry [8].

The IoT opens the door for more easily integrating the physical world into PC-based frameworks, and, when expanded with sensors and actuators, leads to improved system, accuracy, and economic advantage. The IoT enables objects to be detected and controlled remotely over existing networks [9].

IoT's applications span various sectors. Some of them are discussed below:

- **Smart Homes:** IoT-enabled home appliances, lighting, security systems, and other features can all be controlled remotely.
- **Healthcare:** Wearable IoT devices track health indicators and send information to medical specialists.
- **Manufacturing:** By tracking equipment health and anticipating maintenance requirements, IoT enhances manufacturing operations.
- **Agriculture:** Precision irrigation, crop health monitoring, and resource use optimization are all made possible by smart agriculture.
- **Transportation:** Traffic management systems and connected cars improve safety and effectiveness.
- **Smart Cities:** Through energy efficiency, trash management, and smart infrastructure, IoT improves urban living.
- **Smart Railway Sleepers:** Railway sleepers are crucial elements of ballasted railways that, along with the fastening system, keep the rails at a specific track gauge and alignment while attenuating dynamic train loads and transmitting and distributing loads to the sub-structure [10].

IoT systems consist of several key components. Some of them are elaborated on below points:

- **Sensors and Actuators:** These tools collect environmental data and launch activities in response to that data. The three pillars of information technology—sensor detection technology, computer technology, and communication technology—comprise mostly sensor and signal processing technology [11].

- **Connectivity:** IoT devices have connectivity, allowing easy data interchange over the internet or other networks.
- **Data processing:** It consists of organizing, analyzing, and drawing conclusions from the collected data.
- **User Interface:** Through interfaces like mobile apps or web portals, users communicate with IoT systems.
- **Cloud computing:** Cloud platforms manage and store the enormous volumes of data that IoT devices produce. Cloud computing services are required to store the massive amounts of data produced by IoT devices. It would be helpful at this point to have access to a cloud service. Problems like electrical failures and errors are more likely to be found after the data has been analyzed and learned from [12].

Today, the technologies are evolving so rapidly that the convergence of IoT and cloud computing has given rise to cloud-based Internet of Things, i.e., CloudIoT. This innovative concept can revolutionize industries. It also can transform the way we interact with technology [13]. CloudIoT is the integration of cloud computing services with IoT devices and systems. It represents the merging of the computational power of the cloud, its storage capabilities, and accessibility with the interconnectedness and data-generating potential of IoT devices.

11.1.1 What Is CloudIoT in Railway?

The integration of IoT devices with cloud computing services is referred to as CloudIoT. Connecting numerous physical items, sensors, and gadgets to the internet allows them to gather and share data. This process is known as the "IoT." By using cloud computing platforms to process, analyze, store, and manage the data produced by these IoT devices, CloudIoT takes this idea a step further.

In a CloudIoT system:

- IoT devices are physical things or sensors with connectivity capabilities (such as Wi-Fi, cellular, Bluetooth, or other communication protocols) to send data over the internet. These gadgets might be anything from cameras and temperature sensors to cars and industrial equipment.
- A cloud platform is a type of remote computer infrastructure that offers a range of services, such as analytics tools, processing power, and data storage. IoT devices can transfer data to central servers housed in data centers using cloud platforms, which are reachable from any location with an internet connection.

IoT devices in a CloudIoT system frequently have sensors or actuators that collect data from or interact with the physical world. These devices send information to cloud servers, which process, store, and make the information available to users and apps. Platforms for the IoT in the cloud provide a range of services, including data storage, real-time analytics, machine learning, and remote device administration [14]. This makes it possible for businesses and individuals to leverage the power of sophisticated analytics, automation, and large-scale data processing, which can result in more effective operations, better decision-making, and creative applications.

A cloud platform, in its most basic sense, is a group of distributed computing and storage servers that are networked together and offer virtual environments for various operating systems, application containers (like Google's app engine), and computing services. By putting a software system's instances on the cloud platform's virtual machines, a software system can run on that platform. However, the benefit of doing so is limited to doing away with the need to buy and maintain physical hardware [15].

Thermal and vision imaging applications of artificial intelligence (AI) have been observed in airports, and this safety measure offers vital data input for public health in identifying the potentially affected individuals in crowds and identifying travelers who are not wearing masks. AI-based computer vision has also been used to identify suspicious people and for social distance metrics. The detection of fever and a temperature, as well as the diagnosis and monitoring of geographical induction spread and public health and safety monitoring, are all made possible by technologies like AI applications and sensors with biometrics. This might also be used in other public places, including railroad station platforms and entrance sections [16].

The distribution of cloud computing is extremely effective, storage is constantly evolving, and several organizations are now transferring their data from internal databases to cloud computing vendor hubs. While utilizing cloud computing resources, intensive IoT applications for workloads and data are prone to difficulties [17].

CloudIoT in the railway sector refers to the combination of two emerging technologies, i.e., cloud computing and IoT that creates a smart and interconnected railway ecosystem. It involves the deployment of sensors, devices, and data analytics platforms to gather, process, and manage the huge amount of information generated from various railway assets and systems [18]. There are some key components and benefits of CloudIoT in the domain of railway:

i. **Connected Devices and Sensors:** Intelligent sensors can be mounted on trains, tracks, signaling systems, and other essential infrastructure components thanks to CloudIoT. Real-time data from these sensors is gathered on things like train speed, position, temperature, vibration, and equipment health.

ii. **Data Analytics and Insights:** Advanced data analytics algorithms and machine learning models in the cloud process the data collected from linked devices. The performance of the trains, the status of the tracks, the need to predict maintenance, and potential safety hazards are all useful data provided by these analytics.

iii. **Predictive Maintenance:** Predictive maintenance is one of the most significant benefits of CloudIoT for railroads. The system may anticipate equipment failures through data pattern analysis, enabling proactive maintenance, decreasing downtime, and optimizing maintenance costs.

iv. **Enhanced Safety and Security:** Safety in railroad operations can be significantly improved with CloudIoT. In order to protect the safety of passengers and crew, real-time monitoring and analytics can help uncover potential dangers, unauthorized access, or unexpected situations.

v. **Improved Operational Efficiency:** Improved resource management and optimal train schedules are made possible by CloudIoT, which increases operational effectiveness. Data-driven choices improve overall service for passengers and cargo transport by minimizing delays and bottlenecks.

Many different industries, including manufacturing, agriculture, healthcare, smart cities, transportation, and more, find use for CloudIoT. For instance, CloudIoT might be used in agriculture to track crop health, weather information, and soil conditions, enabling farmers to make educated decisions regarding irrigation, fertilizing, and harvesting.

The internet-based interconnection of machines, sensors, and devices is a key component of CloudIoT. These parts might be anything from wearable technology and cars to home and office furnishings and machinery. It is seamlessly transported to cloud systems to handle, store, and analyze the data produced by these devices. There are various characteristics of CloudIoT. Some of them are shown in Figure 11.1.

a. **Scalability:** The systems of CloudIoT can scale to accommodate a large number of connected devices. Also, it can handle huge amounts of data.
b. **Real-time Processing:** The data can be processed in a real-time system using cloud computing, enabling decision-making and responses.
c. **Data Analytics:** On merging the CloudIoT with data analytics and machine learning, we can derive valuable outputs from the collected data.
d. **Remote Management:** With the internet, CloudIoT provides the facility to users that from anywhere they can remotely monitor, control, and manage connected devices and systems.

The impact of CloudIoT is in different sectors and industries. Some of them are listed in Figure 11.2.

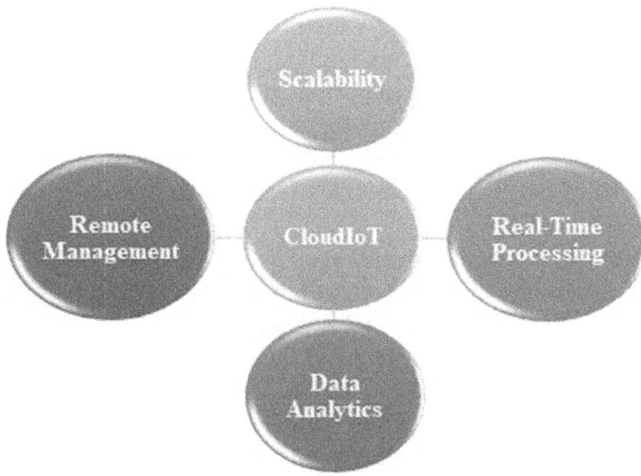

FIGURE 11.1   Characteristics of CloudIoT.

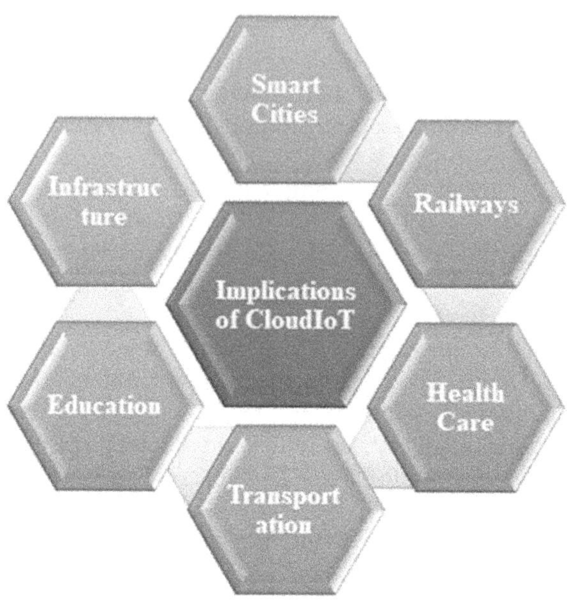

FIGURE 11.2 Implications of CloudIoT.

While there are many benefits to CloudIoT as shown in Figure 11.2, there are also some challenges, such as concerns about data security and privacy, the compatibility of devices from various manufacturers, and potential data overload. To ensure the success of CloudIoT deployments, reliable security mechanisms, and standardized communication protocols are required.

Overall, CloudIoT enables data-driven insights, automation, and optimization of numerous processes and operations by empowering enterprises to harness the value of data produced by IoT devices.

There are various applications of CloudIoT in different domains. This chapter focuses on one of the applications of CloudIoT, i.e., CloudIoT in Railways.

The railway industry plays an important role in the era of transportation, and upon merging it with CloudIoT technologies, it becomes a smarter and more connected system. In this chapter, we explore the various applications of CloudIoT from the perspective of railways. The chapter also focuses on the future scope of the railway sector using CloudIoT.

The structure of this chapter is arranged as follows: Section 11.2 briefly introduces the application of CloudIoT technology in railways. Section 11.3 is focused on the advancement in CloudIoT technologies, including edge computing and real-time analytics, machine learning and predictive maintenance, fusion of sensors, and digital twin technology. Section 11.4 introduces the challenges and limitations of CloudIoT in the railways sector. Section 11.5 contributes to the impact of CloudIoT on Railway Operations. Concluding remarks are presented in Section 11.6. Lastly, Section 11.7 contributes to future prospects and research directions.

11.2   CLOUDIoT APPLICATIONS IN RAILWAY

As shown in Figure 11.3, Applications for the CloudIoT have been a major driver of innovation in the field of contemporary transportation, particularly in the railway industry. This section explores the numerous and significant ways that CloudIoT technologies are changing the face of the railway industry, providing fresh approaches to venerable problems and ushering in a new era of effectiveness, safety, and passenger experience.

a. **Remote Monitoring and Diagnostics:** Real-time data such as temperature, speed etc can be collected with the help of IoT sensors installed on railway tracks and trains. This data is very useful for the prediction. This data will be stored in the cloud from where it can be accessed by railway operators anywhere, which can be used to prevent breakdowns.
b. **Passenger Information:** Real-time information about seat availability, passenger details, the temperature inside the coach, etc, can be recorded using IoT sensors in a more efficient manner. This information can be used by passengers to get exact information about train schedules, seat availability, and all other important updates, which make a better journey experience for the passenger.
c. **Traffic Management:** Traffic can be managed by getting updated and real-time information using IoT sensors installed at railway tracks, reducing train delays and providing a better journey experience for the passengers.

FIGURE 11.3   CloudIoT application areas in railway.

    d. **Energy Management:** Energy usage can be optimized by monitoring energy usage in different trains. Recorded energy consumption data can be used to identify areas of improvement and cost-saving methods.

    e. **Data Analytics and Reporting:** Data captured through IoT sensors are stored in the cloud. This stored data can be processed and analyzed by railway operators over time to generate reports, which help in decision-making for future plans and improvements.

    f. **Inventory and Supply Chain Management:** Train inventories and equipment can be tracked by placing IoT devices on trains so that we can track their location. This will help with inventory management and reduce the chances of misplaced or lost items.

    g. **Smart Infrastructure maintenance:** The benefits of CloudIoT extend to maintaining train infrastructure. Structure, vibration, and stress are all monitored by sensors installed on railroads and bridges. This data is processed using cloud analytics to anticipate maintenance requirements, extending the life of infrastructure and boosting operational safety.

    h. **Data-Driven Decision-Making:** Railway operators are given data-driven insights using CloudIoT. Operators may make wise judgments about scheduling, maintenance, resource allocation, and route optimization with the use of historical and real-time data analysis. This improves both general effectiveness and service level.

    i. **Regulatory Compliance:** Because of real-time data provided by IoT sensors, railway companies are able to maintain compliance with all the safety and environment regulations.

The industry is changing as a result of the adoption of CloudIoT applications in the railway sector, which is raising passenger satisfaction, efficiency, and safety to previously unheard-of heights. Railways are positioned to develop into more intelligent, connected, and responsive systems as CloudIoT technologies advance, providing a glimpse into the future of transportation, where data-driven decision-making governs every facet of operations.

All these applications show the importance of cloud-based IoT in the railway sector, which requires careful planning by considering all the security and environmental safety aspects.

11.3   ADVANCEMENT IN CLOUDIoT TECHNOLOGIES

The use of CloudIoT technologies in the railway sector has been a major turning point for the railway industry, transforming railroad operations and posing new problems. This section examines the extraordinary developments in CloudIoT technologies in the context of railways, emphasizing their profound impacts on passenger experience, safety, efficiency, and maintenance.

    a. **Edge Computing and Real-Time Analytics:** A key development in CloudIoT, edge computing, tackles the latency problems related to sending data to far-off cloud servers. Edge computing is used in the railway industry to process data closer to the point of origin, such as on trains, trackside sensors, and signaling systems. This guarantees in-the-moment analysis and prompt reactions to urgent situations. Edge devices, for instance, can assess train health parameters quickly, alerting maintenance teams to fix arising issues swiftly, thereby reducing downtime.

b. **Machine learning and predictive maintenance:** Predictive maintenance has reached new heights due to machine learning algorithms. Sensors on trains, rails, and other infrastructure are used by CloudIoT systems to collect enormous volumes of data. After being submitted to machine learning models that identify patterns of equipment deterioration, these data are subsequently analyzed to provide insights into probable breakdowns. Railways can reduce disruptions, maximize uptime, and optimize maintenance schedules by spotting anomalies and anticipating maintenance needs.

c. **Fusion of Sensors for Detailed Insights:** A comprehensive understanding of railroad operations is made possible by the interaction of several sensors. To build thorough operating profiles, sensor fusion combines data from accelerometers, temperature sensors, GPS devices, and other sensors. For instance, this fusion can identify how different weather conditions affect track stability, improve risk assessment, and allow for more focused maintenance methods.

d. **Digital Twin Technology:** Through the use of digital twin technology, the operational effectiveness of railway assets is increased. These cloud-hosted digital twins replicate how trains, rails, and other components behave in the actual world. Before making changes to the actual railway network, engineers and operators can model scenarios, test adjustments, and forecast performance outcomes. This reduces the need for trial-and-error methods and guarantees affordable improvements.

e. **Augmented Reality (AR) for Maintenance and Training:** Augmented reality applications are also improved by CloudIoT. Equipment schematics may be seen, real-time diagnostic data can be overlaid, and step-by-step repair instructions can be accessed by maintenance teams using AR devices. In addition, AR technology facilitates training for new employees by providing immersive, practical experiences without interfering with actual operations.

f. **Improved Traveler Experience:** The advantages of CloudIoT technologies extend to the passenger experience. Passengers may access services for business, entertainment, and real-time trip information on trains due to Wi-Fi connectivity driven by CloudIoT technologies. Dynamic pricing methods are made possible by cloud-based ticketing and reservation systems, which provide seamless booking experiences.

11.4 CHALLENGES AND LIMITATIONS

In recent years, the IoThas proven to be a powerful technology for analyzing the world from a different perspective. Though there are vast applications of IoT in real life, such as monitoring the manufacturing process [19], identifying problems, and detecting transportation, which provide the safety and quality materials to the consumers from producers. In the railway sector, we have seen many advantages of using CloudIoT. But they are also implemented with a number of drawbacks. Some of the limitations are discussed below:

i. **Data Security and Privacy Issues:** A huge volume of data is generated and transmitted in the CloudIoT system, which leads to security and privacy issues as the railway system stores sensitive information, including the details of passengers. Therefore, a secure data storage mechanism is required to prevent unauthorized access to information.

ii. **Latency and Real-time Processing:** Real-time decision-making is necessary for rail operations, but it can be hampered by the latency created while transmitting data to and from cloud servers. While edge computing helps to reduce this problem, some applications may still need extremely low latency, which makes it difficult to maintain ideal responsiveness.

iii. **Interoperability and Standardization:** The railway sector has many different systems, technology, and producers. It might be challenging to guarantee flawless compatibility and integration across various IoT devices, legacy systems, and cloud platforms. Smooth data exchange may be hampered by the absence of established communication protocols across distinct devices.

iv. **Connectivity and Network Reliability:** For CloudIoT systems, consistent connectivity is essential. Continuous data transmission may be difficult on remote or rural railway sections where there is a lack of reliable internet connectivity. Data flow might be disrupted, and network failures can affect real-time decision-making processes.

v. **Scalability and System Complexity:** Railroad networks can span large geographic areas with many interconnected devices. It takes careful design and resource allocation to make sure that CloudIoT systems scale successfully to handle increasing data loads and support expanding numbers of devices.

vi. **Integration of Legacy Systems:** Legacy infrastructure that wasn't intended to work with CloudIoT technology is present in many railway systems. It can be expensive and time-consuming to retrofit these systems with contemporary IoT sensors and cloud connectivity.

vii. **Cost Factors:** Investments in sensors, hardware, software development, and cloud infrastructure are required to implement CloudIoT systems. The upfront expenditures can be high, and it's also important to consider continuing fees for cloud service subscriptions, maintenance, and updates.

viii. **Data Overload and Analysis Complexity:** IoT devices produce an enormous amount of data, which can overwhelm systems and cause data overload. These data need to be managed, stored, and analyzed effectively, which calls for advanced data analytics tools and algorithms that can cope with massive datasets.

ix. **Regulatory Compliance:** Standards for data management, privacy, and safety are mandated by regulatory frameworks that apply to railroad operations. Implementing CloudIoT solutions while adhering to these rules necessitates paying close attention to compliance and regulatory issues.

11.5 IMPACT ON RAILWAY OPERATIONS

The adoption of CloudIoT technologies in railway operations has a significant impact, transforming the sector's landscape and improving several aspects of performance, including safety, effectiveness, maintenance, and overall quality. The impact of CloudIoT on railroad operations is examined in this section.

a. **Improved Safety and Risk Management:** CloudIoT is essential to raising safety standards across railway networks. Detecting potential safety threats, such as track flaws, signaling issues, or track obstructions, is made possible through real-time monitoring and data processing. Operators can prevent mishaps and ensure the safety of the crew and passengers thanks to immediate alerts.

b. **Predictive Maintenance for Less Downtime:** CloudIoT-powered predictive maintenance is transforming the management of railway assets. Maintenance staff can anticipate equipment failures by examining data from onboard sensors, tracks, and infrastructure. As a result, unplanned downtime is minimized, service interruptions are decreased, and maintenance schedules are optimized, ultimately increasing service reliability.

c. **Operational effectiveness and resource management:** By providing real-time insights into train movements, passenger flow, and equipment utilization, CloudIoT solutions maximize resource allocation. With these data, operators may modify schedules, distribute resources more effectively, and simplify operations, reducing traffic, shortening travel times, and making better use of the infrastructure.

d. **Real-time traffic management:** It is made possible by CloudIoT, allowing for the in-the-moment tracking of train positions, speeds, and routes [20]. This data enables dynamic adjustments to train routes and schedules based on unforeseen circumstances, reducing delays and ensuring a more seamless flow of train traffic.

e. **Better Maintenance Planning:** The CloudIoT revolutionizes maintenance planning by making it very data-driven and effective. Sensor data that is both historical and current can be used to understand the health and wear patterns of an object. This knowledge contributes to maintenance schedule optimization, maintenance cost reduction, and asset longevity.

f. **Enhancing the passenger experience:** CloudIoT apps provide better information access and improved services for travelers. Through mobile apps and digital displays, real-time updates on train schedules, delays, and service modifications are easily accessible. Onboard entertainment options and cloud-connected Wi-Fi make traveling more enjoyable.

g. **Environmental Sustainability:** By maximizing energy consumption and lowering emissions, CloudIoT helps make railway operations more environmentally sustainable. By streamlining train timetables, eliminating idle time, and running trains more effectively, operators can lower the carbon footprint of rail transportation.

h. **Data-Driven Decision-Making:** CloudIoT systems provide a ton of data that can be used to make decisions. Operators can examine patterns, pinpoint problem areas, and implement strategies for ongoing operation improvement thanks to historical and real-time data insights.

i. **System Integration:** By smoothly connecting diverse train systems and parts, CloudIoT fosters interoperability. The integration of contemporary IoT devices, cloud platforms, and legacy systems results in fewer silos and a more complete picture of railway operations.

With previously unprecedented levels of safety, effectiveness, and passenger happiness, the adoption of CloudIoT technologies into railway operations signifies a transformative phase. CloudIoT is reshaping future railways by providing real-time data analysis, predictive insights, and dynamic decision-making, resulting in a more connected, intelligent, and effective transportation network.

11.6　CONCLUSION

Wide-ranging developments in the realms of sensors, radio access, networks, and hardware/software platforms have been made possible by the IoT rapidly expanding demand. Despite recent developments, IoT device limitations in terms of coverage and battery life for persistent connections continue to be a major obstacle to developing useful service applications [21].

To avoid a significant negative influence on its recommended application, it is critical to carefully examine railway tracks, their consequent flaws, and their causes. This chapter presents a thorough analysis of the various non-destructive testing techniques for railways. The benefits, drawbacks, and potential applications of these methodologies have all been examined, and a comparison study has been conducted [22].

The incredible rate of technological innovation, particularly in the sphere of ICT, has paved the way for new railway transportation applications, and these cutting-edge technologies provide more advantages than older ones. These advantages are quantified not only at the application level but also during the integration of various systems and modes of transportation that rely on data interchange. Linking railroad trucks to infrastructure and figuring out how to connect infrastructures and vehicles from other forms of transportation are unique approaches [23]. The term "digital railway" refers to a fresh idea and paradigm that can be used to alter the design and development of existing railway systems as well as change the architecture of existing ones.

In recent years, almost every sector has drawn some interest in IoT. It is now an important component of our daily lives. It has the potential to link everything with one another in this world. The design of IoT systems is very complex, and it also has limited capacity for storage and retrieval. Therefore, merging cloud computing with IoT offers various advantages to IoT applications. As a result of the digital transformation, emerging technologies are present throughout the public transportation and railroad sectors. Although these sub-sectors have historically been resistant to change, they are now inextricably linked to these trends [24].

Trains go faster, carry more passengers, or have bigger axle loads than ever before on the world's increasingly crowded railway networks. In order to keep up with its progress, the railway sector needs modern information technology (ITs). Railway systems already rely on ITs almost as much as they do on physical assets. As these systems must meet increasing demands for robustness, stability, and capacity, their reliance on ITs is only increasing [25].

The most widely utilized and one of the least expensive ways of transportation in India and other nations is the railway system [26]. As this exploration of CloudIoT in railways comes to a close, it is clear that the combination of these two technologies has enormous promise for the future of rail transportation. The influence of CloudIoT on railways goes beyond simple technological development and entails a complete overhaul. The once-traditional railway ecosystem is developing into a dynamic, linked network where data serves as the lifeblood, and insights drawn from this data help guide wise decision-making. The benefits of CloudIoT are evident in every aspect of railroad operations, from proactive maintenance that prevents disruptions to real-time traffic control that reduces delays.

Future possibilities for CloudIoT in the railway industry are very bright. The real-time capabilities and data processing prowess of CloudIoT will only increase as technologies like 5G and edge computing evolve [27], enabling even more complex applications and finer-grained control over operations. This path of development prepares the way for a rail network that is genuinely integrated and runs smoothly and intelligently.

By continuously monitoring train parts, tracks, and infrastructure, CloudIoT can help railways deploy predictive maintenance techniques. Sensor data can be examined to forecast equipment failures, minimize downtime, enhance dependability, and improve maintenance plans [28]. Real-time tracking of train movements, track conditions, and passenger loads is possible thanks to CloudIoT. Analytics of this data can increase resource allocation, route design, and train scheduling. CloudIoT can aid in the real-time detection of anomalies or dangerous conditions, such as track obstacles or equipment failures. This makes it possible to take prompt action to stop accidents and raise general safety. Energy use in trains and stations can be optimized via CloudIoT. Data analytics can spot areas where energy can be saved, resulting in lower operating costs and a smaller carbon footprint. By providing real-time updates on train schedules, delays, and platform information, CloudIoT can improve the passenger experience. This information can be provided to passengers via mobile apps or digital displays to help them plan their trips better. The tracking and controlling freight trains and cargo can be streamlined via CloudIoT. Logistics and supply chain efficiency can be increased by using real-time data about cargo circumstances, position, and transit times [29]. Remote diagnosis of train systems and parts is made possible by CloudIoT. By evaluating problems remotely and advising on-site staff, maintenance teams may cut downtime and minimize service interruptions. Through real-time monitoring of train stations and infrastructure, CloudIoT can improve security. Sensors and cameras for surveillance can be combined to follow intruders, identify unauthorized entry, and react quickly to security risks. The monitoring and tracking of railway assets, such as rolling stock and equipment, is made easier by CloudIoT [30]. This may involve keeping an eye on asset locations, use rates, and maintenance records. Rail systems can be integrated with larger smart city projects using CloudIoT. A more integrated urban environment can result from data exchange and collaboration across various forms of transportation, urban planning, and infrastructure. Informed decisions about operations, maintenance, and resource allocation can be made using the plethora of data generated by CloudIoT.

In conclusion, the adoption of CloudIoT by the railway industry represents an evolution rather than only a transition. CloudIoT has transformed railways into a more resilient, efficient, and responsive system thanks to its capacity to increase safety, optimize operations, advance maintenance procedures, and elevate passenger experiences. The journey is ongoing, and the destination is an intelligent railway system that harnesses data and technology to provide a seamless and superior travel experience for all.

11.7 FUTURE PROSPECTS AND RESEARCH DIRECTION

Innovative research projects and CloudIoT technologies will shape the future of railways. This section explores the fascinating possibilities that lie ahead and highlights important research directions that will define the evolution of this paradigm in rail transportation as the railway sector continues to embrace the transformational power of CloudIoT.

The future of railways is promising in terms of CloudIoT, which defines the future of traveling. Continuous research in this area will provide significant advancements in terms of safety, operation, management, and experience. Still, there is a large scope to incorporate the use of CloudIoT to transform the railway sector in the following ways:

- In case of breakage track, tracking status can be linked to the signal network so that alerts can be sent to the concerned person and miss happenings can be avoided.
- Future studies will investigate how the combination of 5G and CloudIoT might provide real-time data flow between infrastructure, cloud platforms, and trains, enabling applications with greater accuracy and responsiveness.
- More work can be done on cloud security to protect data from unauthorized access.
- A developing area in CloudIoT is edge computing's potential to process data closer to the source. It will focus on enhancing edge computing capabilities for railways in order to speed up decision-making and reduce data latency, especially for safety-critical applications.
- New low-cost and small-size hardware devices can be developed or assessed that can work with new generations of IoT technologies, and work can be done to investigate low-power consumption.
- AI-based algorithms should be used for self-optimization of the IoT network, which will be helpful for reducing time and effort, and efficient results will be generated.
- In the future, work should be explored for the integration of drones with IoT technologies to improve the monitoring system.

- The development of linked and autonomous trains will be significantly influenced by the growth of CloudIoT. The goal of research is to develop systems that enable real-time communication between trains and infrastructure, as well as to improve collision avoidance, increase energy efficiency, and optimize train timetables.
- The spread of airborne diseases is always a concern in public railway systems. In the future, work should be done to use IoT technology to prevent the spread of viruses by tracking and monitoring the information.

By utilizing data-driven insights and automation, CloudIoT has the ability to optimize railway operations, increase safety, and improve the passenger experience. As with any technology, however, its adoption necessitates thorough planning, addressing cybersecurity issues, and cooperation amongst railway industry parties.

REFERENCES

1. Chellaswamy, C., T. S. Geetha, A. Vanathi, and K. Venkatachalam. "An IoT Based Rail Track Condition Monitoring and Derailment Prevention System." International Journal of RF Technologies 11, no. 2 (2020): 81–107.
2. Vaidya, Gauri, Prabhleen Bindra, Meghana Kshirsagar, and Sharvari Chandrashekhar Tamane. "Privacy and Security Technologies for Smart City Development." Security and Privacy Applications for Smart City Development (2021): 3–23.
3. Villamil, Sebastian, Cesar Hernández, and Giovanny Tarazona. "An Overview of Internet of Things." Telkomnika (Telecommunication Computing Electronics and Control) 18, no. 5 (2020): 2320–2327.
4. Yuehong, Y. I. N., Yan Zeng, Xing Chen, and Yuanjie Fan. "The Internet of Things in Healthcare: An Overview." Journal of Industrial Information Integration 1 (2016): 3–13.
5. Lv, Xiaojing, and Minghai Li. "Application and Research of the Intelligent Management System Based on Internet of Things Technology in the Era of Big Data." Mobile Information Systems 2021 (2021): 1–6.
6. Singh, Prashant, Zeinab Elmi, Vamshi Krishna Meriga, Junayed Pasha, and Maxim A. Dulebenets. "Internet of Things for Sustainable Railway Transportation: Past, Present, and Future." Cleaner Logistics and Supply Chain 4 (2022): 100065.
7. Rose, Karen, Scott Eldridge, and Lyman Chapin. "The Internet of Things: An Overview." The Internet Society (ISOC) 80 (2015): 1–50.
8. Gbadamosi, Abdul-Quayyum, Lukumon O. Oyedele, Juan Manuel Davila Delgado, Habeeb Kusimo, Lukman Akanbi, Oladimeji Olawale, and Naimah Muhammed-Yakubu. "IoT for Predictive Assets Monitoring and Maintenance: An Implementation Strategy for the UK Rail Industry." Automation in Construction 122 (2021): 103486.
9. Sagar, K., A. Kumar, G. Ankush, Thota Harika, Madireddy Saranya, and Dasaraju Hemanth. "Implementation of IoT Based Railway Calamity Avoidance System Using Cloud Computing Technology." Indian Journal of Science and Technology 9, no. 17 (2016): 1–5.
10. Jing, Guoqing, Mohammad Siahkouhi, J. Riley Edwards, Marcus S. Dersch, and N. A. Hoult. "Smart Railway Sleepers-a Review of Recent Developments, Challenges, and Future Prospects." Construction and Building Materials 271 (2021): 121533.
11. Zhong, Guidong, Ke Xiong, Zhangdui Zhong, and Bo Ai. "Internet of Things for High-Speed Railways." Intelligent and Converged Networks 2, no. 2 (2021): 115–132.
12. Ghosh, Pinaki, Kirti Jain, Mr Chandan Kumar, and Ms Kanchan Jha. Introduction to IOT and Its Applications. Academic Guru Publishing House, 2023.

13. Memon, Tarique Rafique, Tayab Din Memon, Bhawani Shankar Chowdhry, Imtiaz H. Kalwar, and Khakoo Mal. "Development of Specialized IoT Cloud Platform for Railway Track Condition Monitoring." In 2021 International Conference on Robotics and Automation in Industry (ICRAI), pp. 1–4. IEEE, 2021.

14. Saini, Komal, and Sandeep Sharma. "Smart City: Road Traffic Monitoring System Based on the Integration of IoT and ML." In International Conference on Communication and Intelligent Systems, pp. 137–148. Singapore: Springer Nature Singapore, 2022.

15. Roehl, Nathan. "Cloud Based IoT Architecture." (2019).

16. Alawad, Hamad, and Sakdirat Kaewunruen. "5G Intelligence Underpinning Railway Safety in the COVID-19 Era." Frontiers in Built Environment 7 (2021): 639753.

17. Sadeeq, Mohammed Mohammed, Nasiba M. Abdulkareem, Subhi RM Zeebaree, Dindar Mikaeel Ahmed, Ahmed Saifullah Sami, and Rizgar R. Zebari. "IoT and Cloud Computing Issues, Challenges and Opportunities: A Review." Qubahan Academic Journal 1, no. 2 (2021): 1–7.

18. Le, Dac-Nhuong, Chintan Bhatt, and Mani Madhukar, eds. Security Designs for the Cloud, IoT, and Social Networking. John Wiley & Sons, 2019.

19. Kamboj, Pradeep, T. Ratha Jeyalakshmi, P. Thillai Arasu, S. Balamurali, and A. Murugan. "Smart Applications of IoT." The Smart Cyber Ecosystem for Sustainable Development (2021): 131–151.

20. Minoli, Daniel, and Benedict Occhiogrosso. "Internet of Things (IoT)-Based Apparatus and Method for Rail Crossing Alerting of Static or Dynamic rail Track Intrusions." In ASME/IEEE Joint Rail Conference, vol. 50718, p. V001T06A016. American Society of Mechanical Engineers, 2017.

21. Jo, Ohyun, Yong-Kyu Kim, and Juyeop Kim. "Internet of Things for Smart Railway: Feasibility and Applications." IEEE Internet of Things Journal 5, no. 2 (2017): 482–490.

22. Salvi, Sanket, and Shashank Shetty. "AI Based Solar Powered Railway Track Crack Detection and Notification System with Chatbot Support." In 2019 Third International conference on I-SMAC (IoT in Social, Mobile, Analytics and Cloud) (I-SMAC), pp. 565–571. IEEE, 2019.

23. Nemtanu, Florin Codrut, and Marin Marinov. "Digital Railway: Trends and Innovative Approaches." In Sustainable Rail Transport: Proceedings of RailNewcastle 2017, pp. 257–268. Springer International Publishing, 2019.

24. Möller, Dietmar, Lukas Iffländer, Michael Nord, Patrik Krause, Bernd Leppla, Kristin Mühl, Nikolai Lenski, and Peter Czerkewski. "Emerging technologies in the era of digital transformation: state of the art in the railway sector." In Proceedings of the 19th International Conference on Informatics in Control, Automation and Robotics-ICINCO, pp. 721–728. 2022.

25. Li, Qingyong Y., Zhangdui D. Zhong, Ming Liu, and Weiwei W Fang. "Smart Railway Based on the Internet of Things." In Big Data Analytics for Sensor-Network Collected Intelligence, pp. 280–297. Academic Press, 2017.

26. Sunitha, Mrs P., M. Eswar Krishna, D. Deepika, D. Divya, and K. Bhargav. "Automated Railway Level Crossing System Using IoT." International Research Journal of Engineering and Technology (IRJET) 7, no. 3 (2020): 1255–1258.

27. Liberg, Olof, Marten Sundberg, Eric Wang, Johan Bergman, and Joachim Sachs. Cellular Internet of things: Technologies, Standards, and Performance. Academic Press, 2017.

28. Sarkar, Debasis, Harsh Patel, and Bhargav Dave. "Development of Integrated Cloud-Based Internet of Things (IoT) Platform for Asset Management of Elevated Metro Rail Projects." International Journal of Construction Management 22, no. 10 (2022): 1993–2002.

29. Eiza, Mahmoud Hashem, Martin Randles, Princy Johnson, Nathan Shone, Jennifer Pang, and Amhmed Bhih. "Rail Internet of Things: An architectural platform and assured requirements model." In 2015 IEEE International Conference on Computer and Information Technology; Ubiquitous Computing and Communications; Dependable, Autonomic and Secure Computing; Pervasive Intelligence and Computing, pp. 364–370. IEEE, 2015.
30. Bhuyan, Muhibul Haque, Sheikh Md Mamunur Rahman, and Md Tofayel Tarek. "Design and Simulation of a PLC and IoT-Based Railway Level Crossing Gate Control and Track Monitoring System Using LOGO." IOSR Journal of Electrical and Electronics Engineering IOSR-JEEE 17 (2022): 13–23.

12 Cloud IoT Integration for Enhanced Underwater Communication

Harmeet Singh, Mohit Angurala, and Nipun Chhabra

12.1 INTRODUCTION

Long problems, constraints, and ambiguities cloaked the communication environment under the ocean's surface. As our oceans contain a treasure mine of vital data, the necessity to reinvent undersea communication becomes clearer. Fortunately, the rise of cloud computing and the Internet of Things (IoT) has created new opportunities for transformation in this industry. This book chapter delves into the possibility of Cloud-IoT integration in lifting underwater communication systems to new levels of efficiency and dependability. Underwater communication has a unique set of challenges, ranging from low bandwidth and latency to unstable connectivity. These roadblocks have hampered our capacity to access the abundance of information stored in our seas, which includes crucial environmental data, navigation intelligence, and surveillance insights [1]. However, by integrating Cloud-IoT, the undersea communication paradigm may be significantly transformed, allowing for real-time data insights, predictive analytics, and seamless connectivity.

We begin with a review of underwater communication, illuminating the obstacles and restrictions that have long stymied development in this sector [2]. This lays the groundwork for understanding how cloud computing and IoT may help, ushering in a new era of improvements in undersea data transmission and analysis. To lay a solid basis, we go into the principles of cloud computing and IoT, explaining their significance and potential applications in underwater communication. We set the framework for smooth integration into underwater situations by investigating several cloud services models such as Software-as-a-Service (SaaS), Platform-as-a-Service (PaaS), and Infrastructure-as-a-Service (IaaS), as well as cloud deployment patterns such as Public, Private, and Hybrid. Simultaneously, we introduce the Internet of Things (IoT) architecture and components, as well as underwater communication protocols tailored to this specific environment. The essence of this chapter is found in the collaboration of cloud computing, IoT, and underwater communication. It is vital to ensure the security and privacy of data in Cloud-IoT underwater systems, and we underline the methods implemented to protect this sensitive information. The following part, cloud-enabled data management and analytics, delves into how data aggregation and fusion on the cloud enable enhanced underwater monitoring and prediction. We show how predicted insights from the cloud contribute to enhanced

DOI: 10.1201/9781032656694-12

underwater communication methods by leveraging machine learning (ML) and artificial intelligence (AI) approaches.

Various use cases for Cloud-IoT integration in underwater communication are emerging, including underwater environmental monitoring, cloud-assisted navigation, surveillance, and exploration. These examples demonstrate the actual benefits of this seamless connectivity. We prove the usefulness of Cloud-IoT underwater systems by comparing them to traditional ways and displaying successful real-world deployments as we go into performance evaluation and case studies [3]. Along with our successes, we discuss obstacles and prospects, including latency, dependability, power management, and possible synergies with upcoming technologies such as AI and blockchain.

12.1.1 OVERVIEW OF UNDERWATER COMMUNICATION

Our seas' vastness creates a complicated network of communication issues that have long attracted experts and developers alike. Underwater communication, an important sector for ocean exploration, environmental monitoring, defense, and marine research, has restrictions and difficulties that distinguish it from terrestrial communication systems.

12.1.2 CHALLENGES AND LIMITATIONS

Underwater communication has various challenges, making it a difficult arena for data and information sharing. The attenuation of electromagnetic waves in water, which drastically limits the range and bandwidth of typical radio-frequency communication, is one of the key problems. Furthermore, the acoustic medium frequently utilized for underwater data transmission is susceptible to interference from environmental variables such as ocean currents, background noise, and changing water conditions.

The ocean's breadth and depth cause signal propagation and latency challenges, resulting in lengthier communication delays. Furthermore, deploying and maintaining communication equipment in isolated underwater sites becomes logistically difficult and costly. These constraints impede real-time data transmission and interpretation, making it difficult to quickly gather critical information from underwater environments [4]. As a result, underwater communication systems encounter challenges in delivering continuous monitoring, navigation support, and dependable connectivity, all of which are required for a wide range of underwater applications.

12.1.3 EMERGENCE OF CLOUD COMPUTING AND IoT IN UNDERWATER
ENVIRONMENTS

As the globe embraces the era of cloud computing and the IoT, novel solutions to the long-standing issues of underwater communication are emerging. The rise of cloud computing and IoT technology provides a glimmer of optimism, promising to transform the way we interact and communicate under the ocean's surface. With its immense computational capacity and storage capacities, cloud computing provides a perfect platform for processing and storing massive amounts of data created in underwater environments. Underwater communication systems can overcome limits associated with on-site data storage and processing by employing cloud services [5].

The IoT provides intelligent and networked gadgets, sensors, and actuators that can interact with one another and cloud-based systems in real time. Underwater IoT devices with enhanced sensing and communication capabilities have the ability to gather real-time data from the ocean's depths and effectively transport it to the cloud. The integration of cloud computing and IoT in underwater environments ushers in a new era of improved efficiency, dependability, and data insights. The integration of cloud and IoT provides the way for real-time data streaming, predictive analytics, and adaptive communication techniques, revealing the tremendous potential hidden beneath the ocean's surface.

In this book chapter, we dig into the complexities of Cloud-IoT integration for underwater communication, looking at how these cutting-edge technologies work together to address the issues that this area faces. We look at the principles of cloud computing and IoT, how they apply to underwater communication, and how they may alter underwater applications. Furthermore, we go into the architectural issues, security concerns, and practical applications of Cloud-IoT integration, imagining a future in which seamless underwater communication is a reality [6].

12.2 FUNDAMENTALS OF CLOUD COMPUTING, IoT, AND THEIR APPLICATION IN UNDERWATER COMMUNICATION

In this section, we will look at the fundamental principles of cloud computing and the IoT and how they may be applied to underwater communication. Because underwater settings present unique problems, we investigate how cloud computing and IoT technologies might work together to overcome these constraints and improve data transfer, analysis, and networking under the ocean's surface.

12.2.1 CLOUD COMPUTING: A POWERFUL ENABLER

Cloud computing has emerged as a game changer in a variety of businesses by providing scalable computer resources, storage, and services over the internet. Understanding the fundamentals of cloud computing is critical to understanding its applicability in underwater communication.

12.2.2 CLOUD SERVICE MODELS (SaaS, PaaS, IaaS)

We start by deconstructing the various cloud service models, which include SaaS, PaaS, and IaaS. Each model meets certain needs by providing customers access to software applications, development platforms, and virtualized infrastructure. We investigate how these service models aid in effectively processing, analyzing, and transferring data for undersea applications.

12.2.3 CLOUD DEPLOYMENT MODELS (PUBLIC, PRIVATE, HYBRID)

Following that, we will look at the various cloud deployment methods, including Public, Private, and Hybrid Clouds [7]. The deployment strategy chosen is critical for data security, accessibility, and control. Selecting the most appropriate cloud deployment type for underwater communication systems has important consequences for guaranteeing smooth and dependable data connection.

12.3 IoT IN UNDERWATER COMMUNICATION

As we go toward the IoT, we see how it has the potential to transform undersea communication. The IoT presents a network of networked devices, sensors, and actuators capable of automatically gathering and transmitting data.

12.3.1 IoT Architecture and Components

We look at the fundamental elements of IoT architecture, such as sensors, communication protocols, gateways, and cloud platforms. Understanding the architecture sets the groundwork for integrating IoT devices into underwater communication networks, allowing for real-time data capture and transfer [8].

12.3.2 IoT Communication Protocols Suitable for Underwater Scenarios

Specialized communication protocols that can efficiently penetrate the aquatic environment are required for underwater communication. We investigate IoT communication protocols that are well-suited for underwater settings, such as acoustic and optical approaches. These protocols allow for smooth data transmission while also addressing the issues associated with underwater signal propagation.

12.4 SYNERGIES BETWEEN CLOUD COMPUTING, IoT, AND UNDERWATER COMMUNICATION

This section delves into the potent synergy that exists between cloud computing, IoT, and underwater communication. The combination of these technologies overcomes the particular constraints of underwater data transmission while also creating a platform for real-time data insights and predictive analytics.

12.4.1 Enhanced Efficiency and Reliability

Underwater communication systems offer improved efficiency and reliability by leveraging cloud computing's computing capacity, data storage, and processing capabilities in conjunction with IoT's intelligent sensing devices. Data may be seamlessly captured, processed, and transferred, allowing for continuous monitoring and rapid reaction to changing undersea circumstances.

12.4.2 Real-time Data Insights

The convergence of cloud computing and IoT opens up the possibility of real-time data insights from underwater environments [9]. Cloud-based advanced data analytics and ML algorithms deliver actionable information for navigation, environmental monitoring, and scientific research, empowering decision-makers with fast and accurate information.

12.5 CLOUD IoT ARCHITECTURE FOR UNDERWATER COMMUNICATION

Underwater communication has issues that necessitate a well-designed Cloud-IoT system. This section delves into the fundamental components and design factors required to construct an efficient and dependable underwater communication system. The basic architecture consists of sensor nodes installed underwater connected to the base station, which stores data in the cloud at the cloud layer (refer Figure 12.1).

12.5.1 DESIGN CONSIDERATIONS FOR UNDERWATER IoT DEVICES

Underwater IoT devices are essential for data collecting and transfer. Designing these devices necessitates careful consideration of the hostile underwater environment, which includes elements such as pressure, temperature, and salt. Miniaturization and power efficiency are essential for extending battery life and reducing device footprint. Furthermore, appropriate communication methods, such as acoustic or optical, must be used for effective data transfer underwater. Furthermore, underwater IoT devices should have cognitive data pretreatment capabilities to decrease data quantities before sending them to the cloud. Underwater IoT devices can become durable data collectors at the ocean's depths if certain design issues are addressed.

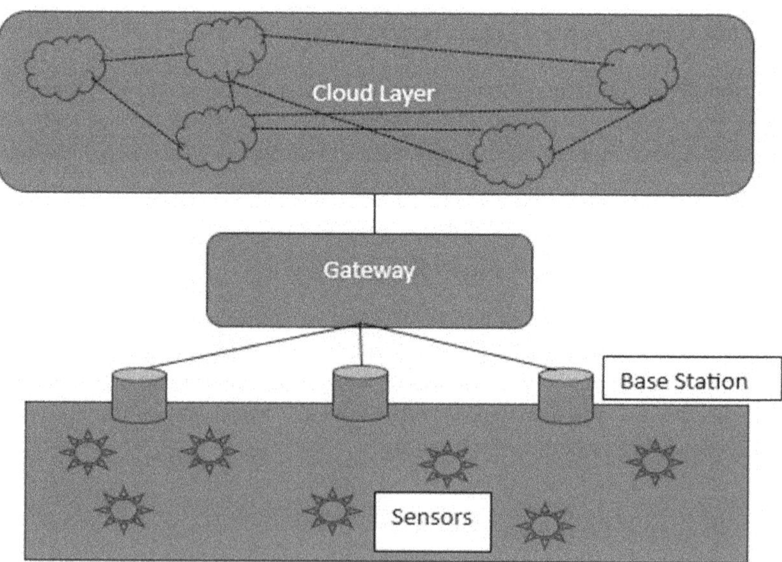

FIGURE 12.1 Architecture of cloud IoT underwater communication.

12.5.2 CLOUD-BASED DATA STORAGE AND PROCESSING FOR UNDERWATER APPLICATIONS

Cloud-based data storage and processing are required to deal with the massive volumes of data generated by undersea IoT devices. Cloud storage services have nearly infinite capacity, allowing for efficient data preservation and retrieval. Cloud Processing as a Service (CPaaS) solutions process data in the cloud, conserving resources on IoT devices and boosting response times [10, 11]. Sophisticated data analytics and ML techniques may be used to derive important insights from acquired data by exploiting cloud computing's computing capabilities. This provides real-time and predictive analytics, allowing researchers and maritime operators to make educated decisions based on current data.

12.5.3 INTEGRATING REAL-TIME DATA STREAMING FROM UNDERWATER SENSORS INTO THE CLOUD

Real-time data from underwater sensors is used by underwater communication systems. By integrating real-time data streams into the cloud, researchers and marine operators will be able to access live data feeds and respond to dynamic changes in underwater conditions more swiftly. This integration involves deploying sophisticated communication protocols capable of transferring data in near real-time from sensors to the cloud. Acoustic communication is commonly utilized for this purpose because of its ability to travel long distances underwater. The continuous flow of real-time data enables continuous monitoring, navigation support, and rapid actions, improving the overall efficacy of underwater communication systems.

12.5.4 ENSURING SECURITY AND PRIVACY IN CLOUD IoT UNDERWATER SYSTEMS

Underwater communication systems are no exception when it comes to security and privacy. The combination of cloud computing with IoT introduces new issues for protecting data security, integrity, and availability. Robust encryption and authentication procedures are used to protect data during transmission and storage. Furthermore, access control and user authentication methods are in place to restrict data access to only authorized people. In the event of an unanticipated occurrence, the cloud service provider's security procedures, including data backup and disaster recovery plans, are crucial. Ensuring the privacy of acquired data, particularly in sensitive underwater situations, necessitates strict adherence to data protection standards.

Underwater communication requires careful design considerations for underwater IoT devices, efficient cloud-based data storage and processing, smooth real-time data streaming, and strong security and privacy safeguards. The seamless integration of cloud computing and IoT technologies allows for fast data transfer, real-time insights, and data-driven decision-making, propelling underwater communication systems to new heights of efficiency and dependability. Researchers and marine operators may unleash the full potential of undersea habitats by using the possibilities of this design, paving the path for revolutionary discoveries, environmental monitoring, and sustainable marine exploration [12].

12.6 CLOUD-ENABLED DATA MANAGEMENT AND ANALYTICS IN UNDERWATER COMMUNICATION

Underwater communication creates massive volumes of data that must be managed and analyzed efficiently. Cloud-based data management and analytics provide a strong option for dealing with the intricacies of undersea data, delivering real-time insights and predictive capabilities. This section delves into the many aspects of cloud-based data management and analytics as they apply to undersea communication.

12.6.1 DATA AGGREGATION AND FUSION IN THE CLOUD

Underwater habitats are home to a plethora of sensors, each of which generates constant streams of data. Data aggregation and fusion on the cloud become critical to make sense of this onslaught of data. Cloud-based data aggregation entails gathering information from many sources, such as underwater IoT devices and satellite feeds, and storing it in a centralized repository. This aggregation streamlines data access, allowing academics and maritime operators to easily access large datasets.

Data fusion is the practice of combining data from several sources in order to gain more meaningful insights. This requires merging data from many sensors, such as temperature, salinity, and current speed, to generate a more comprehensive image of the underwater environment in underwater communication. The combination of multimodal data improves comprehension of underwater conditions, allowing for more accurate environmental monitoring, navigation assistance, and scientific study [13].

12.6.2 CLOUD-BASED DATA ANALYTICS FOR UNDERWATER MONITORING AND PREDICTION

Cloud-based data analytics is critical in converting raw data into actionable insight. Data analytics algorithms may handle and analyze big datasets fast by exploiting cloud computing's tremendous computational capabilities. This enables researchers to discover patterns, trends, and anomalies in real-time, allowing them to respond to changes in underwater conditions quickly.

Data analytics enables the detection of possible dangers and risks, such as hazardous algal blooms or pollution breaches, for underwater monitoring, providing prompt interventions to protect marine ecosystems and human activities. Furthermore, cloud-based data analytics enables predictive analytics, which uses previous data to estimate future underwater conditions. Marine operators may use predictive insights to forecast environmental changes, enhance underwater operations, and reduce hazards.

12.6.3 ML AND AI TECHNIQUES FOR UNDERWATER DATA ANALYSIS

ML and AI are essential techniques for undersea data processing. ML algorithms may learn from previous data patterns to generate accurate predictions and judgments. In underwater communication, ML algorithms may be taught to distinguish certain underwater occurrences, such as the presence of marine organisms, geological characteristics, or underwater noise signatures [14].

Deep Learning and other AI algorithms provide even more complex analytical capabilities. Deep Learning models can automatically extract detailed patterns and characteristics from undersea data, allowing for sophisticated detection and classification tasks. AI-powered underwater data analysis aids ecological research and conservation efforts by providing greater knowledge of underwater habitats.

12.7 ENHANCING UNDERWATER COMMUNICATION WITH PREDICTIVE INSIGHTS FROM THE CLOUD

By offering predictive insights, the combination of cloud-based data analytics and ML/AI approaches revolutionizes underwater communication. Predictive analytics allows researchers and marine operators to anticipate future problems, such as changes in ocean currents or severe weather conditions. With this knowledge, they can optimize undersea operations, change navigation paths, and efficiently deploy resources.

Furthermore, predictive insights improve communication dependability by identifying probable communication outages and enabling proactive efforts to preserve continuous connectivity. Underwater communication systems that use predictive analytics may assure continuous data transfer, making them more efficient and robust.

Underwater communication has enormous possibilities with cloud-enabled data management and analytics. Researchers and marine operators may acquire full insights into underwater ecosystems through data aggregation, fusion, and cloud-based analytics. ML and AI approaches augment data processing capabilities, offering predictive insights that improve the efficiency and reliability of underwater communication systems. The seamless integration of cloud-based data management and analytics increases undersea research and monitoring, informs decision-making, and promotes sustainable marine practices.

12.8 CLOUD-IoT APPLICATIONS IN UNDERWATER COMMUNICATION

The combination of cloud and IoT enables a multitude of novel applications in the field of underwater communication. Underwater habitats may be better monitored, navigated, explored, and communicated with the help of the cloud and intelligent IoT devices. This section delves into the various uses of Cloud-IoT integration in underwater communication.

12.8.1 UNDERWATER ENVIRONMENTAL MONITORING USING CLOUD-IoT SYSTEMS

Environmental monitoring is an important part of underwater communication because it helps marine researchers and conservationists understand and protect vulnerable aquatic ecosystems. Cloud-IoT technologies improve environmental monitoring capabilities by allowing for real-time data collecting and processing. Cloud-IoT systems, which are outfitted with a network of intelligent sensors and underwater IoT devices, collect critical data such as water temperature, salinity, dissolved oxygen levels, and contaminants.

The acquired data is sent in real-time to the cloud, where advanced analytics and ML algorithms process it. Researchers learn about the health of undersea ecosystems, identify possible environmental concerns, and track changes over time. Researchers may access historical data using cloud-based data storage, allowing for thorough trend analysis and long-term environmental investigations [15]. The seamless integration of Cloud-IoT systems guarantees continuous and dependable environmental monitoring, allowing for more informed decision-making for marine conservation and sustainable resource management.

12.8.2 CLOUD-ASSISTED UNDERWATER NAVIGATION AND LOCALIZATION

Underwater operations, marine exploration, and navigation support require precise navigation and localization. Cloud-assisted underwater navigation improves existing approaches by providing real-time positional information to underwater vehicles and divers. Underwater devices can have access to high-precision global positioning data and underwater mapping databases by utilizing cloud-based navigation systems.

Underwater IoT devices with acoustic and optical communication capabilities provide navigation data to the cloud, where sophisticated algorithms determine accurate locations and trajectories. Furthermore, the cloud gives dynamic navigation updates to underwater vehicles depending on changing environmental conditions and ocean currents. This integration improves the efficiency of underwater exploration and navigation, lowering mission risks and assuring effective underwater operations.

12.8.3 UNDERWATER SURVEILLANCE AND EXPLORATION THROUGH CLOUD-IoT INTEGRATION

The integration of Cloud-IoT enables intelligent underwater surveillance systems, which improves security and monitoring capabilities in maritime settings. Underwater IoT devices with high-resolution cameras, acoustic sensors, and environmental monitoring tools provide real-time data to the cloud, which is examined for abnormalities and possible dangers.

Unauthorized underwater activity can be detected, underwater infrastructure monitored, and underwater items or structures of interest identified using cloud-based monitoring tools. Real-time notifications are provided to enable quick intervention in urgent circumstances. This integration improves underwater security by supporting law enforcement, military, and maritime operators in safeguarding critical coastal regions, offshore assets, and undersea infrastructure.

Furthermore, Cloud-IoT connection enables improved underwater investigation. Uncharted ocean depths can be explored by autonomous underwater vehicles (AUVs) outfitted with advanced sensors and cameras. The data is uploaded to the cloud, where AI-powered analysis assists in identifying new species, geological features, and underwater occurrences. Cloud-based exploration contributes to scientific discoveries and fosters a better knowledge of our waters.

12.9 IMPROVING EFFICIENCY AND RELIABILITY IN UNDERWATER COMMUNICATION

Underwater communication systems must be efficient and reliable, especially for critical applications such as ocean monitoring, offshore enterprises, and disaster response. By streamlining data transmission and processing, Cloud-IoT integration improves communication efficiency. Data may be preprocessed locally before being transmitted to the cloud by underwater IoT devices, lowering data quantities and enhancing overall system efficiency.

The scalability of cloud computing means that underwater communication systems can successfully handle variable workloads and data needs. Furthermore, the cloud's redundancy and backup procedures enable resilience, ensuring continued data connection even if a device fails. Predictive insights produced from cloud-based analytics can assist in predicting probable communication interruptions and the preventive steps required to sustain connectivity. The combination of cloud and IoT technologies improves communication dependability, allowing for continuous and uninterrupted data transfer.

Cloud-IoT applications in underwater communication provide a wide range of practical applications. Cloud-IoT integration alters underwater communication systems, providing real-time insights and informed decision-making across the board, from environmental monitoring and navigation support to surveillance and exploration. As cloud and IoT technologies advance, their potential to transform underwater communication grows, ushering in a new era of efficiency, dependability, and creativity under the ocean's surface.

12.10 METRICS FOR ASSESSING THE PERFORMANCE OF CLOUD IoT UNDERWATER SYSTEMS

12.10.1 EVALUATING THE PERFORMANCE OF CLOUD-IoT UNDERWATER SYSTEMS

The performance of Cloud-IoT underwater systems must be evaluated in order to understand their usefulness and suggest areas for development. This section highlights critical performance indicators used to evaluate the performance of underwater communication systems.

12.10.2 DATA TRANSMISSION LATENCY

The time it takes for data to go from underwater IoT devices to the cloud and vice versa is measured as data transmission latency. Low latency is critical in underwater communication for real-time applications such as navigation support and environmental monitoring. Due to signal propagation delays and the necessity for strong communication methods, achieving low latency in underwater environments can be difficult. Data transmission latency is used to assess the responsiveness of the Cloud-IoT system, guaranteeing timely data insights and decision-making.

12.10.3 COMMUNICATION RELIABILITY

The ability of the Cloud-IoT system to sustain the continuous and steady connection between underwater devices and the cloud is referred to as communication reliability. Because underwater communication is vulnerable to signal attenuation and interference, dependability is crucial in underwater operations. Measuring communication reliability entails evaluating data packet loss, network stability, and the system's adaptability to changing undersea circumstances. A dependable Cloud-IoT solution guarantees continuous data transfer, lowering the risk of data loss and mission failure.

12.10.4 DATA PROCESSING AND STORAGE EFFICIENCY

Data processing and storage efficiency are critical in Cloud-IoT underwater systems. Data collected by undersea sensors may be massive, and the computing capabilities of the cloud are critical in processing and evaluating this data. The time it takes to analyze data and extract insights is one of the performance criteria for data processing efficiency. Furthermore, assessing the time it takes to save and retrieve data from the cloud is part of evaluating data storage efficiency. Data processing and storage optimization improves system responsiveness and scalability.

12.10.5 ENERGY CONSUMPTION

Because underwater IoT devices are often powered by batteries with limited capacity, energy efficiency is a crucial factor. Evaluating the energy consumption of IoT devices and cloud-based processing aids in identifying energy-intensive jobs and optimizing power management measures. Cloud-IoT systems that use less energy prolong the operating lifespan of underwater equipment, decreasing the need for regular battery changes and maintenance.

12.10.6 COMPARATIVE ANALYSIS OF TRADITIONAL VS. CLOUD-IoT APPROACHES

A comparison of traditional underwater communication technologies and Cloud-IoT integration gives useful insights into the benefits and drawbacks of each strategy.

12.10.6.1 Traditional Underwater Communication

Point-to-point acoustic or optical communication between underwater equipment is common in traditional underwater communication systems. While these technologies may be useful for short-range and low-data-rate applications, scaling up for large-scale underwater networks and long-distance data transmission may be difficult. Traditional methods rely on onboard data processing and storage, which limits data volume and real-time analytics capabilities.

12.10.6.2 Cloud-IoT Integration

Many of the constraints of traditional underwater communication are addressed by integrating Cloud-IoT. Cloud-IoT systems can manage massive volumes of data and provide real-time analytics by shifting data processing and storage to the cloud.

The scalability and processing power of the cloud allow for the smooth extension of underwater networks and adaptive communication mechanisms. Furthermore, Cloud-IoT solutions provide remote monitoring and control, reducing the requirement for underwater human presence.

12.10.6.3   Trade-offs and Considerations

While Cloud-IoT integration has various advantages, there are trade-offs and factors to consider. Certain real-time applications may be concerned about the latency created by data transfer to the cloud. Edge computing and hybrid Cloud-IoT systems, on the other hand, can alleviate latency concerns by conducting certain data processing operations closer to the underwater devices.

In underwater situations, energy efficiency is crucial, and Cloud-IoT systems must optimize communication protocols and data transfer to reduce energy usage. Furthermore, protecting the security and privacy of data transported to the cloud is critical, especially in sensitive undersea applications.

12.11   CASE STUDIES SHOWCASING SUCCESSFUL CLOUD IoT IMPLEMENTATIONS IN UNDERWATER COMMUNICATION

Real-world case studies demonstrate the practical uses and advantages of integrating Cloud-IoT in underwater communication.

12.11.1   CASE STUDY 1: ENVIRONMENTAL MONITORING IN MARINE RESERVES

Marine reserves are regions of the ocean where human activity is prohibited in order to preserve marine biodiversity and safeguard fragile ecosystems. In these reserves, effective environmental monitoring is critical for understanding the health of marine ecosystems, detecting possible threats, and making informed conservation choices. In this case study, we look at how a Cloud-IoT system for environmental monitoring in a marine reserve was successfully implemented.

12.11.1.1   Objective

The major goal of this case study was to use Cloud-IoT integration to create a comprehensive environmental monitoring system in a marine reserve. The system was designed to gather and analyze data from multiple underwater sensors to acquire insights about water quality, temperature, and marine life. The real-time data would assist marine researchers and conservationists in making evidence-based choices to maintain and manage the biodiversity of the reserve.

12.11.1.2   System Setup

A network of clever underwater IoT sensors strategically distributed around the marine reserve comprises the Cloud-IoT environmental monitoring system. Each gadget came with a variety of sensors, including:

1. Water Quality Sensors: To determine the health of the marine ecosystem by measuring factors such as pH, dissolved oxygen, turbidity, and salinity.

2. Temperature Sensors: To monitor changes in water temperature, which affect marine life distribution and breeding habits.
3. Underwater Cameras: Capture high-resolution photos and movies of marine ecosystems to allow for visual evaluation of the reserve's biodiversity.

12.11.1.3   Data Transmission and Cloud Integration

The underwater IoT devices were linked to a buoy-based communication hub, which served as a data gateway. The hub employed sound communication to transport data from the underwater gadgets to the surface buoy. The data was then transferred to the cloud via satellite, enabling continuous connectivity even in distant marine reserve regions.

12.11.1.4   Cloud-Based Data Analytics

The cloud-based data analytics platform received the sent data in real time and processed and analyzed it immediately. ML algorithms were used by the platform to discover patterns, trends, and anomalies in environmental data. The system might, for example, detect abrupt changes in water quality or temperature, suggesting probable pollution events or dangerous algal blooms.

12.11.1.5   User Interface and Decision Support

Marine researchers and conservationists might utilize a user-friendly online interface to access the cloud-based technology. The interface presented real-time data visualizations such as graphs, charts, and interactive maps, allowing users to monitor the environmental conditions of the marine reserve at a glance.

In addition, the platform created automatic alerts and notifications to notify users of crucial changes in environmental factors. For example, if water quality data exceeded acceptable levels or if the temperature suddenly rose, the system would promptly inform the proper authorities to take required conservation measures.

12.11.1.6   Outcomes and Impact

The adoption of the Cloud-IoT environmental monitoring system has a substantial beneficial influence on the management and conservation activities of the marine reserve:

1. **Real-Time Insights:** Marine scientists and conservationists now have access to real-time environmental data, allowing them to respond to changes in the marine ecosystem quickly. Data availability in real time permitted swift responses to alleviate possible hazards to marine life and environments.
2. **Evidence-Based Conservation:** The system's data collection and analysis gave evidence-based insights into the marine reserve's health. This data served as the foundation for conservation efforts such as habitat restoration, pollution management, and so on.
3. **Long-Term Monitoring:** Researchers could study long-term trends and changes in the environmental conditions of the marine reserve, laying the groundwork for continued monitoring and adaptive conservation management.

12.11.1.7   Conclusion

The successful deployment of a Cloud-IoT environmental monitoring system in the marine reserve revealed the revolutionary power of Cloud-IoT integration in marine conservation. The system performed a critical role in maintaining marine biodiversity, ensuring sustainable marine resource management, and raising public awareness about the value of marine reserves by delivering real-time insights and evidence-based decision support. The case study demonstrates the substantial influence that Cloud-IoT integration may have on environmental monitoring and conservation efforts in fragile underwater environments.

12.11.2   CASE STUDY 2: AUTONOMOUS UNDERWATER EXPLORATION

12.11.2.1   Introduction

Autonomous Underwater Exploration is a discipline that uses sophisticated AUVs to improve our understanding of the undersea world. These AUVs are outfitted with sophisticated sensors and cameras, allowing them to explore new ocean depths without the need for human interaction. We look at a successful Cloud-IoT deployment in the context of autonomous undersea exploration in this case study.

12.11.2.2   Objective

The major goal of this case study was to demonstrate how Cloud-IoT integration might improve autonomous underwater exploring capabilities. The Cloud-IoT system was designed to provide real-time data transfer, processing, and analysis, allowing quick decision-making during underwater operations. The mission was to explore previously unknown ocean depths, collect vital data, and make significant scientific discoveries.

12.11.2.3   System Setup

The autonomous underwater exploration system was made from a cutting-edge AUV outfitted with modern sensors and high-resolution cameras. The AUV was built to move independently, using powerful navigation algorithms to plot its path while avoiding obstructions.

12.11.2.4   Data Transmission and Cloud Integration

The AUV acquired massive amounts of data during the underwater exploration missions, including sonar data, images, and environmental information. The AUV was outfitted with an underwater acoustic communication module to ensure continuous data transfer. This module enabled the AUV to send data to a surface buoy, which served as a communication gateway.

The data from the surface buoy was relayed to the cloud through satellite communication. The data sent was received in real time by the cloud-based platform, allowing for continuous monitoring and data processing during the exploratory missions.

12.11.2.5   Cloud-Based Data Processing and Analysis

To evaluate the data acquired by the AUV, the cloud-based platform used strong data processing capabilities. Advanced data analytics methods were used to analyze sonar data, allowing the detection of underwater elements such as seabed topography and geological formations.

The AUV's cameras gathered high-resolution footage, which was then analyzed in the cloud using computer vision algorithms. Researchers were able to identify and categorize marine species, examine their behavior, and witness interactions between diverse marine animals thanks to these analyses.

12.11.2.6   Real-Time Insights and Decision-Making

During the exploratory flights, real-time data processing and analysis provided researchers with quick findings. The cloud platform created real-time visualizations and reports while the AUV explored undiscovered regions, allowing researchers to make educated decisions about changing the AUV's trajectory or focusing on areas of particular interest.

Because of the quick availability of insights, the AUV was able to adjust its exploration approach on the fly, boosting mission efficiency and increasing the odds of finding significant discoveries [16].

12.11.2.7   Outcomes and Impact

The effective integration of Cloud-IoT in autonomous undersea exploration resulted in a number of notable outputs and impacts:

1. **Scientific breakthroughs:** The Cloud-IoT system aided in the identification of previously undiscovered seabed structures and marine animals. The data gathered and evaluated during exploratory flights resulted in scientific advances, which improved our understanding of underwater ecosystems and geological formations.
2. **Efficient Resource Utilization:** Researchers were able to optimize the AUV's exploration paths thanks to real-time data insights, ensuring that vital data was captured efficiently. This resulted in greater resource usage and reduced exploration time, resulting in cost and resource savings.
3. **Remote Monitoring and Collaboration:** The integration of Cloud-IoT allows researchers to remotely monitor exploratory missions from any location with an internet connection.

12.11.2.8   Conclusion

The case study of Cloud-IoT integration in autonomous underwater exploration demonstrates how this technology can revolutionize marine research and exploration. By providing real-time data transmission, processing, and analysis capabilities, Cloud-IoT integration enhances the efficiency, productivity, and scientific impact of underwater missions. The ability to make informed decisions in real-time during exploration missions opens up new possibilities for discovering and understanding the mysteries of the underwater world, making Cloud-IoT integration a transformative force in autonomous underwater exploration.

12.11.3   CASE STUDY 3: DISASTER RESPONSE AND UNDERWATER SURVEILLANCE

12.11.3.1   Introduction

Underwater monitoring and disaster response are key components of protecting coastal regions, maritime infrastructure, and offshore assets from possible threats. The capacity to monitor underwater infrastructure in real time and discover abnormalities is critical for timely action and catastrophe mitigation. We look at a successful Cloud-IoT solution for disaster response and underwater monitoring in this case study.

12.11.3.2   Objective

The major goal of this case study was to create an effective disaster response and underwater monitoring system through Cloud-IoT integration. The system's goal was to continually monitor underwater infrastructure, including pipelines, offshore platforms, and coastal installations, for possible threats or abnormalities. Furthermore, the Cloud-IoT system aims to deliver real-time alerts and notifications to appropriate authorities for faster response times and reduce the effect of future calamities.

12.11.3.3   System Setup

The disaster response and underwater surveillance system was built with a network of intelligent underwater IoT devices and surveillance equipment strategically positioned around submerged structures in mind. The following were the system's primary components:

1. **Underwater IoT Devices:** The underwater IoT devices, which were outfitted with acoustic sensors, cameras, and environmental monitoring tools, were deployed near vital underwater infrastructure to monitor structural integrity and environmental conditions.
2. **Surface Buoy Communication Hub:** A surface buoy served as a communication hub, sending data between underwater IoT devices and the cloud. The buoy relayed real-time data and alarms to the cloud-based platform through a satellite connection.

12.11.3.4   Data Transmission and Cloud Integration

The underwater IoT devices collected data from the undersea environment continuously, including auditory signals, visual data, and environmental indicators. This information was sent to the surface buoy through acoustic transmission.

The data from the surface buoy was effortlessly sent to the cloud through a satellite connection. The data was received in real-time by the cloud-based platform, allowing for continuous monitoring and analysis during disaster response and surveillance operations.

12.11.3.5   Cloud-Based Data Analytics and Anomaly Detection

To handle the data acquired from the underwater IoT devices, the cloud-based platform used powerful data analytics and ML algorithms. The platform was designed to identify abnormalities or possible threats such as structural deformations, underwater leaks, and unusual environmental conditions.

When predetermined criteria were surpassed, the platform could send automatic warnings and messages in real-time. These signals were instantly conveyed to the appropriate authorities, allowing for a quick response and mitigating activities.

12.11.3.6   Real-Time Alerts and Decision Support

The Cloud-IoT system alerted and notified selected stakeholders and reaction teams in real-time. When an anomaly was discovered, the platform issued alerts that included the location, severity level, and recommended reaction steps.

Authorities were able to make educated judgments and establish appropriate disaster response methods due to the prompt availability of real-time data insights. In the event of a pipeline leak or structural deformation, for example, reaction teams might be dispatched immediately to examine and remedy the issue.

12.11.3.7   Outcomes and Impact

The successful deployment of the Cloud-IoT disaster response and underwater monitoring system resulted in a number of major outcomes and impacts:

1. **Early Detection and Mitigation:** Because of the real-time anomaly detection capabilities, possible risks or structural flaws were detected early. This early detection enabled quick mitigation efforts, lowering the severity of disasters and averting more damage.
2. **Increased Safety and Security:** Continuous observation of undersea structures increased safety and security by reducing the chance of accidents and protecting maritime infrastructure and coastal zones.
3. **Resource Optimization:** The Cloud-IoT system facilitated the optimal utilization of resources in disaster response activities. Response teams might focus their efforts on locations with the most danger by delivering real-time data insights and maximizing resource allocation.

12.11.3.8   Conclusion

The case study of Cloud-IoT integration in disaster response and underwater surveillance demonstrates how important this technology is in protecting underwater structures and coastal regions. Cloud-IoT integration improves the efficiency, reactivity, and efficacy of disaster response activities by offering real-time monitoring, data analytics, and automated notifications. The capacity to detect abnormalities and possible dangers in real-time allows for increasing prompt actions, reducing catastrophic effects, and assuring the safety and security of marine infrastructure and coastal populations. The convergence of cloud and IoT is a game-changing tool for disaster response and underwater monitoring, providing authorities and reaction teams with actionable data insights in crucial situations.

12.12   CHALLENGES AND FUTURE DIRECTIONS

The use of Cloud Computing and the IoT in underwater communication creates great prospects, but it also offers some obstacles that must be solved in order for the area to continue to grow. In this part, we will look at important problems and future approaches to overcoming them to ensure the smooth and effective deployment of Cloud-IoT systems in underwater environments.

12.12.1 Addressing Latency and Reliability Concerns in Cloud-IoT Underwater Setups

Latency, or data transmission latency, is a major problem in real-time underwater applications. The time it takes for data to travel from underwater IoT devices to the cloud and back may have an impact on the system's responsiveness. Latency is critical in applications like autonomous navigation and real-time environmental monitoring. Future research should concentrate on optimizing communication protocols and edge computing approaches in order to reduce data transmission delays and provide reliable connectivity between underwater devices and the cloud.

12.12.2 Power Management and Energy Efficiency in Resource-Constrained Underwater IoT Devices

Underwater IoT devices are frequently fueled by energy-constrained sources such as batteries or energy harvesting equipment. Power management is a critical issue in long-term underwater operations. Researchers should investigate new power-saving approaches, such as duty cycling and low-power components, to increase the operating lifespan of underwater gadgets. Furthermore, advances in energy-efficient sensors and communication modules might help to dramatically reduce power usage in resource-constrained underwater IoT devices.

12.12.3 Scaling Cloud-IoT Systems for Large-Scale Underwater Deployments

As the size of underwater deployments grows, Cloud-IoT solutions must become more scalable to support the expanding number of IoT devices and the vast volume of data collected. Scaling the cloud infrastructure to meet the data processing and storage requirements of large-scale undersea networks is a difficult task. Future research should concentrate on developing cloud architectures that can dynamically scale up and down in response to the operational needs of undersea operations, guaranteeing smooth data management and real-time insights.

12.12.4 Potential Synergies with Other Emerging Technologies (e.g., AI, Blockchain) in Underwater Communication

Synergies with other new technologies can assist the field of underwater communication. AI can improve data analytics by allowing for more accurate pattern identification and predictive modeling for underwater environmental monitoring and exploration. Integrating AI with Cloud-IoT systems can result in better decision-making and autonomous operations in underwater environments.

Additionally, blockchain technology enables decentralized and secure data storage and transmission. The use of blockchain in undersea communication can improve data integrity and security by providing tamper-proof recordings of environmental data and mission logs. Future studies should look at how blockchain technology might be used to secure data integrity and trustworthiness in Cloud-IoT underwater systems.

The obstacles and future prospects in Cloud-IoT integration for underwater communication are critical concerns for the field's advancement. Researchers and industry personnel may realize the full potential of Cloud-IoT systems in underwater environments by resolving concerns about latency, dependability, power management, and scalability. Exploring synergies with other developing technologies, such as AI and blockchain, might further enhance underwater communication capabilities, leading to safer, more efficient, and ecologically responsible underwater exploration and surveillance. Cloud-IoT integration will play a critical role in creating the future of underwater communication, transforming our understanding and exploitation of the vast undersea environment via sustained study and cooperation.

12.13 CONCLUSION

Unauthorized underwater activity can be detected, underwater infrastructure monitored, and underwater items or structures of interest identified using cloud-based monitoring tools. Real-time notifications are provided to enable quick intervention in urgent circumstances. This integration improves underwater security by supporting law enforcement, military, and maritime operators in safeguarding critical coastal regions, offshore assets, and undersea infrastructure. Cloud-IoT integration is a game changer in underwater communication, providing significant advantages in environmental monitoring, navigation, surveillance, and communication dependability. As technology advances, the prospects for Cloud-IoT integration in underwater environments look promising, potentially opening up new vistas for marine exploration, environmental protection, and sustainable practices. By embracing these improvements and encouraging collaborative research, we can realize the full promise of Cloud-IoT integration and build a future where undersea communication is more efficient, dependable, and impactful.

REFERENCES

1. Z. Zhou, C. Gao, C. Xu, Y. Zhang, S. Mumtaz, and J. Rodriguez, "Social Big-Data-Based Content Dissemination in Internet of Vehicles," IEEE Trans. Ind. Informat., vol. 14, no. 2, pp. 768–777, Feb. 2018.
2. T. Qiu, X. Wang, C. Chen, M. Atiquzzaman, and L. Liu, "TMED: A Spider Web-Like Transmission Mechanism for Emergency Data in Vehicular Ad Hoc Networks," IEEE Trans. Veh. Technol., vol. 67, no. 9, pp. 8682–8694, Sep. 2018.
3. D. Saxena, and V. Raychoudhury, "Design and Verification of an NDN-Based Safety-Critical Application: A Case Study With Smart Healthcare," IEEE Trans. Syst. Man Cybern. Syst., vol. 49, no. 5, pp. 991–1005, May 2019.
4. H. Nguyen, F. Mirza, M. Naeem, and M. Nguyen, "A Review on IoT Healthcare Monitoring Applications and a Vision for Transforming Sensor Data into Real-Time Clinical Feedback," Proc. IEEE 21st Int. Conf. Comput. Supported Cooperative Work Des., pp. 257–262, Apr. 2017.

5. M. Wollschlaeger, T. Sauter, and J. Jasperneite, "The Future of Industrial Communication: Automation Networks in the Era of the Internet of Things and Industry 4.0," IEEE Ind. Electron. Mag., vol. 11, no. 1, pp. 17–27, Mar. 2017.

6. T. Qiu, H. Wang, K. Li, H. Ning, A. Sangaiah, and B. Chen, "SIGMM: A Novel Machine Learning Algorithm for Spammer Identification in Industrial Mobile Cloud Computing," IEEE Trans. Ind. Informat., vol. 15, no. 4, pp. 2349–2359, Apr. 2019.

7. L. Yang, S. Yang, and L. Plotnick, "How the Internet of Things Technology Enhances Emergency Response Operations," Technological Forecasting Soc. Change, vol. 80, no. 9, pp. 1854–1867, 2013.

8. T. Qiu, R. Qiao, and D. Wu, "EABS: An Event-Aware Backpressure Scheduling Scheme for Emergency Internet of Things," IEEE Trans. Mobile Comput., vol. 17, no. 1, pp. 72–84, Jan. 2018.

9. R. Naylor, and M. Burke, "Aquaculture and Ocean Resources: Raising Tigers of the Sea," Annu. Rev. Environ. Resour, vol. 30, pp. 185–218, 2005.

10. G. Besio, L. Mentaschi, and A. Mazzino, "Wave Energy Resource Assessment in the Mediterranean Sea on the Basis of a 35-Year Hindcast," Energy, vol. 94, pp. 50–63, 2016.

11. R. Roslynna, and D. Brunei, "A Review of Tidal Current Energy Resource Assessment: Current Status and Trend," Proc. 5th Int. Conf. Renewable Energy Gener. Appl., pp. 34–40, Feb. 2018.

12. M. Xu, and L. Liu, "Sender-Receiver Role-Based Energy-Aware Scheduling for Internet of Underwater Things," IEEE Trans. Emerg. Topics Comput., vol. 7, no. 2, pp. 324–336, Apr./Jun. 2019.

13. M. Heesemann, T. Insua, M. Scherwath, K. Juniper, and K. Moran, "Ocean Networks Canada: From Geohazards Research Laboratories to Smart Ocean Systems," Oceanography, vol. 27, no. 2, pp. 151–153, 2014.

14. L. Liu, R. Wang, and F. Xiao, "Topology Control Algorithm for Underwater Wireless Sensor Networks Using GPS-Free Mobile Sensor Nodes," J. Netw. Comput. Appl., vol. 35, no. 6, pp. 1953–1963, 2012.

15. W. Kim, H. Moon, and Y. Yoon, "Adaptive Triangular Deployment of Underwater Wireless Acoustic Sensor Network Considering the Underwater Environment," J. Sensors, vol. 2019, pp. 1–11, 2019.

16. F. Senel, K. Akkaya, M. Erol-Kantarci, and T. Yilmaz, "Self-Deployment of Mobile Underwater Acoustic Sensor Networks for Maximized Coverage and Guaranteed Connectivity," Ad Hoc Netw, vol. 34, pp. 170–183, 2015.

13 Integration and Applications of Cloud IoT in the Education Sector

Yojna Arora

13.1 INTRODUCTION

The delivery and experience of education will change dramatically as a result of the Internet of Things (IoT) and cloud computing integration in the education industry. The education sector utilizes cloud and IoT technologies as technology develops to provide cutting-edge learning environments, increase productivity, and improve educational outcomes.

Cloud IoT can potentially transform various industries, including the education sector. IoT devices, combined with cloud computing capabilities, can offer innovative solutions to enhance teaching, learning, campus management, and overall educational experiences. Here are some ways in which Cloud IoT can be applied in the education sector:

13.2 LITERATURE REVIEW

13.2.1 EVOLUTION OF EDUCATIONAL TECHNOLOGY

To understand the significance of IoT and cloud computing in education, it is crucial to trace the evolution of educational technology [1, 2]. The history of educational technology reveals a series of key milestones as shown in Figure 13.1.

13.2.2 TRADITIONAL CLASSROOM MODELS

Historically, education primarily relied on traditional classroom models, with physical textbooks, chalkboards, and face-to-face interactions. This section explores the limitations of this model and the factors that drove the transition to digital education.

- Traditional education often suffered from limited access to up-to-date learning resources, especially in remote or underserved areas.
- One-Size-Fits-All Approach:
- Traditional classrooms typically followed a one-size-fits-all approach, making it challenging to cater to individual learning needs.

DOI: 10.1201/9781032656694-13

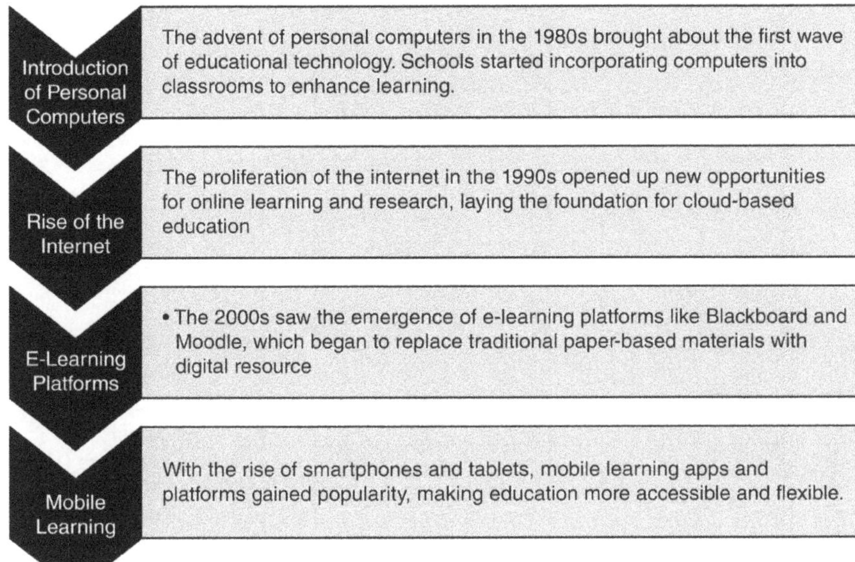

FIGURE 13.1   Evolution of educational technology.

13.2.3   EMERGENCE OF DIGITAL LEARNING

The emergence of digital learning in education has been a transformative and ongoing process that has significantly impacted the way students learn and educators teach. This shift toward digital learning has been driven by various technological advancements and changes in educational paradigms.

13.3   INTRODUCTION OF CLOUD COMPUTING

Cloud computing is a technology that allows users to access and utilize computing resources and services over the internet instead of relying on local or on-premises hardware and software. This model of computing has become increasingly popular due to its scalability, flexibility, cost-effectiveness, and convenience. Here are some key aspects of cloud computing [3, 4].

13.3.1   CLOUD SERVICE MODELS

Cloud computing is built on various service models, each offering unique advantages, as shown in Figure 13.2.

- **Infrastructure as a Service (IaaS):** With IaaS, educational institutions can rent virtualized computing resources like servers and storage. For example, Amazon Web Services (AWS) provides IaaS solutions universities can use to host their applications and data.
- **Platform as a Service (PaaS):** PaaS platforms, such as Microsoft Azure, simplify application development and deployment. Educational institutions

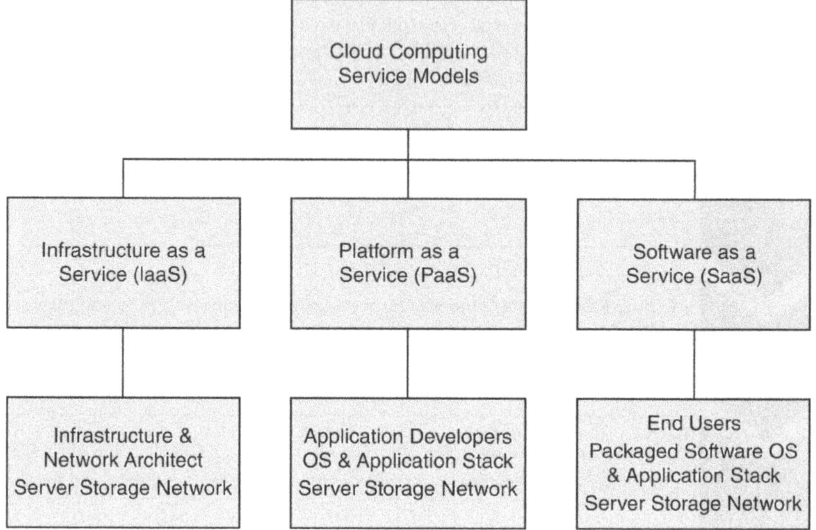

FIGURE 13.2   Cloud computing service models.

can build custom educational software or apps without managing the underlying infrastructure.

- **Software as a Service (SaaS):** SaaS solutions like Google Workspace for
  Education provide cloud-based applications like Gmail and Google Docs.
  These apps enable collaboration and document sharing among students and
  faculty.

13.3.2   Cloud Computing in Education

Cloud computing has profoundly impacted education, transforming how students
learn, how educators teach, and how educational institutions operate [5, 6]. A few
applications of cloud computing in education are mentioned in Figure 13.3.

13.3.2.1   The Rise of Cloud-Based Learning Management Systems (LMS)

Cloud-based Learning Management Systems (LMS) have become the backbone
of online education. Prominent examples include Canvas, Blackboard Learn, and
Google Classroom. A recent survey conducted by Educause Review found that 84%
of higher education institutions in the United States use a cloud-based LMS. This
indicates the widespread adoption of cloud technologies in education [7–10].

13.4   APPLICATIONS OF CLOUD COMPUTING IN THE EDUCATION SECTOR

Cloud computing offers numerous applications and benefits in the field of education.
It has transformed the way educational institutions operate, collaborate, and deliver
content to students.

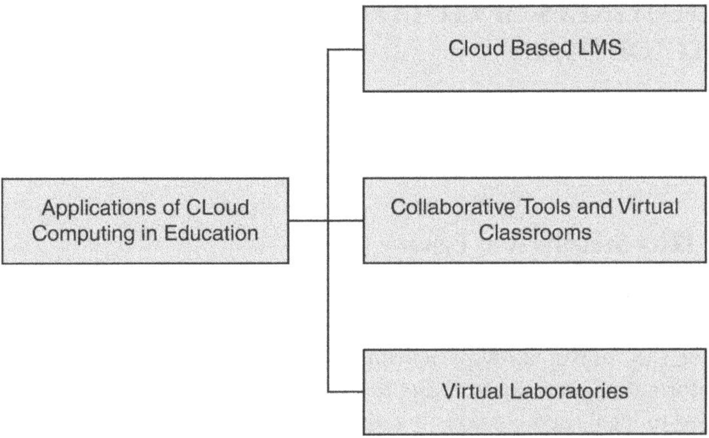

FIGURE 13.3   Applications of cloud computing in education.

13.4.1   CLOUD-BASED LMS

Instructure's Canvas Learning Management System is widely used in higher education institutions [11, 12]. Here's how Canvas has transformed the learning experience:

- **Anywhere, Anytime Access:** Students can access course materials, assignments, and discussions from any device with internet access, promoting flexible learning.
- **Integration with Third-Party Tools:** Canvas integrates with a wide range of educational tools and services, enhancing the teaching and learning experience.

13.4.2   COLLABORATIVE TOOLS AND VIRTUAL CLASSROOMS

Google Classroom has gained popularity in K-12 education for its user-friendly interface and integration with Google Workspace [13]. Key benefits include:

- **Seamless Collaboration:** Google Classroom enables teachers and students to collaborate on assignments and projects, fostering engagement and teamwork.
- **Paperless Workflow:** Assignments and feedback are exchanged digitally, reducing paper usage and simplifying administrative tasks for educators.

13.4.3   VIRTUAL LABORATORIES [14]

Cloud computing allows educators to create virtual laboratories, eliminating the need for physical equipment. For instance, universities can offer virtual chemistry labs, enabling students to conduct experiments online safely.

13.5 CHALLENGES OF CLOUD COMPUTING IN EDUCATION SECTOR [15, 16]

While cloud computing offers numerous benefits to the education sector, it also presents certain challenges and considerations that educational institutions need to address, as depicted in Figure 13.4.

13.5.1 DATA SECURITY AND PRIVACY [17]

Data security is a paramount concern in the cloud computing era, particularly in educational institutions. Several high-profile data breaches serve as cautionary tales.

A major U.S. university experienced a data breach where sensitive student and faculty information was exposed due to inadequate security measures. This breach not only led to legal repercussions but also damaged the institution's reputation.

13.5.2 REGULATORY COMPLIANCE

Education institutions must comply with various data protection laws, such as the Family Educational Rights and Privacy Act (FERPA) in the United States or the General Data Protection Regulation (GDPR) in Europe. Non-compliance can result in severe penalties.

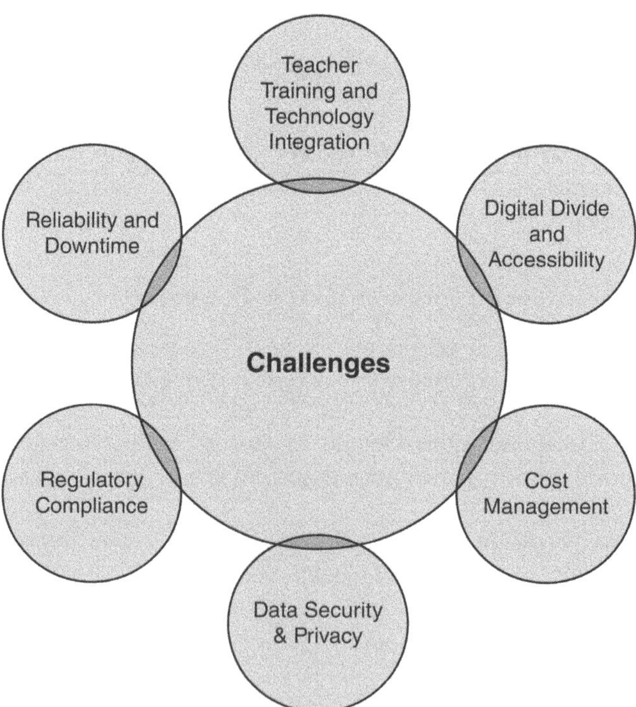

FIGURE 13.4 Challenges of cloud computing.

Case studies showcasing institutions that successfully navigated regulatory challenges and implemented robust data protection policies can offer valuable insights.

13.5.3   DIGITAL DIVIDE AND ACCESSIBILITY [18]

Bridging the Digital Divide: The digital divide is a significant challenge in education, with disparities in access to devices and high-speed internet. However, case studies reveal successful initiatives to bridge this divide:

Providing Devices to Underserved Students: A school district in a low-income area partnered with local businesses to provide laptops and internet access to students in need, enabling them to participate in online learning during the pandemic.

Ensuring Inclusivity: Accessibility for students with disabilities is another aspect of the digital divide. Educational institutions need to ensure that cloud-based resources are accessible to all. Case studies of universities implementing accessible e-learning platforms and tools can demonstrate how to accommodate diverse student needs.

13.5.4   TEACHER TRAINING AND TECHNOLOGY INTEGRATION

The successful integration of cloud technologies in education relies heavily on teacher readiness and competence. Case studies of teacher professional development programs can highlight best practices: A school district implemented a comprehensive training program that empowered teachers with the skills needed to effectively use cloud-based tools. This program improved teacher confidence and student outcomes.

13.5.5   COST MANAGEMENT

While cloud computing offers scalability and cost-efficiency, managing cloud expenses can be challenging. A case study of a university's cost optimization efforts can provide insights:

Cloud Cost Analysis: A university thoroughly analyzed its cloud spending and identified areas where cost savings could be realized. By optimizing their cloud resources, they reduced overall expenditure.

13.5.6   RELIABILITY AND DOWNTIME

Cloud service outages can disrupt online learning and administrative operations. Case studies on how institutions have prepared for and mitigated the impact of cloud service downtime can illustrate strategies for maintaining continuity:

Business Continuity Planning: A college developed a robust business continuity plan with failover mechanisms and backup solutions. During a cloud service outage, they seamlessly switched to backup resources, minimizing disruption.

13.6   INTRODUCTION TO IoT AND IoT DEVICES

The IoT is a concept that refers to the interconnection of everyday physical objects to the internet. These objects, often called "smart" or "IoT devices," are equipped with sensors, software, and connectivity capabilities, allowing them to collect and exchange data with other devices and systems over the internet [19].

Sensors and Actuators: Sensors gather data from the environment (e.g., temperature, humidity, motion), while actuators enable devices to perform actions (e.g., turning on lights, adjusting thermostats).

Connectivity: IoT devices use various communication protocols (e.g., Wi-Fi, Bluetooth, cellular) to transmit data to the cloud or other devices.

Cloud Computing: Data generated by IoT devices is often sent to cloud platforms for storage, analysis, and processing.

Data Analytics: Data collected from IoT devices can be analyzed to extract valuable insights and inform decision-making.

13.7   IoT IN EDUCATION

The IoT can potentially transform the education sector by enhancing teaching, learning, and administrative processes. IoT refers to the network of physical objects or "things" embedded with sensors, software, and connectivity that enables them to collect and exchange data [20–22]. IoT is being applied in education in the following ways:

Smart Classrooms: Smart projectors, sensors, and interactive whiteboards all contribute to the immersive learning environment of a smart classroom. For instance, teachers might encourage in-the-moment collaboration by using IoT-enabled whiteboards to instantaneously communicate notes and diagrams with students' devices [23–26].

Smart Campus Management: IoT sensors can be deployed across the campus to monitor various aspects such as energy usage, temperature, lighting, air quality, and occupancy levels. This data can be analyzed through cloud-based platforms to optimize resource allocation, reduce energy consumption, and create a more sustainable and efficient campus environment [23, 27].

Enhancing Safety and Security: IoT devices like smart cameras and sensors can be utilized to enhance security on campus. These devices can detect and alert authorities in unusual activities or emergencies, providing a safer learning environment for students and staff.

Data-Driven Decision-Making: Educational institutions can use IoT-generated data to make informed decisions about curriculum, resource allocation, and student support services.

Personalized Learning: IoT devices can be integrated into the classroom to collect data on students' learning patterns, preferences, and progress. This information can then be analyzed through cloud-based platforms to deliver personalized learning experiences, adaptive content, and targeted interventions for each student.

Remote Learning and Virtual Classrooms: IoT-enabled devices can support remote learning by providing interactive and immersive virtual classrooms. Cloud-based IoT platforms can facilitate seamless communication and collaboration between teachers and students, allowing them to interact in real time regardless of their physical locations.

Smart Libraries: Libraries can benefit from IoT solutions by incorporating smart bookshelves, tracking systems, and automated check-in/check-out processes. Cloud-based IoT platforms can manage the inventory, analyze user preferences, and recommend relevant reading materials to students and faculty.

Monitoring and Improving Equipment: IoT sensors can be installed on laboratory equipment, machinery, and other assets, enabling real-time monitoring of their status and performance. This proactive approach allows educational institutions to schedule maintenance and minimize downtime, ensuring efficient use of resources.

Environmental Education: IoT devices can be used to monitor and study environmental changes on campus or in nearby areas. This data can be stored, analyzed, and used to educate students about environmental sustainability and the impact of human activities on the ecosystem.

Smart Wearables and Health Monitoring: In physical education or sports programs, students can use smart wearables equipped with IoT sensors to track their physical activities and health metrics. The data collected can be synced to cloud-based platforms, providing insights into individual and group performance over time.

Campus Security: Security systems at educational facilities enabled by the IoT can spot irregularities and improve security. For instance, in the event of an active shooter scenario, gunshot detection systems can inform authorities, potentially saving lives.

Research and Development: Educational institutions can leverage IoT devices and cloud platforms for scientific research and experiments. IoT sensors can gather data from various experiments, and cloud computing can process and store this data efficiently for analysis and sharing.

However, while Cloud IoT offers numerous benefits, it is crucial to address security and privacy concerns. Educational institutions must ensure proper data protection measures and protocols to safeguard sensitive information collected through IoT devices.

13.7.1 IoT for Special Education

IoT has shown immense promise in special education. Inclusive classrooms benefit from IoT devices that can assist students with disabilities. The examples are shown in Figure 13.5 and explained further in detail.

Overall, Cloud IoT has the potential to revolutionize the education sector by offering innovative solutions to improve learning outcomes, enhance campus management, and foster a more engaging and interconnected educational experience for students and educators alike.

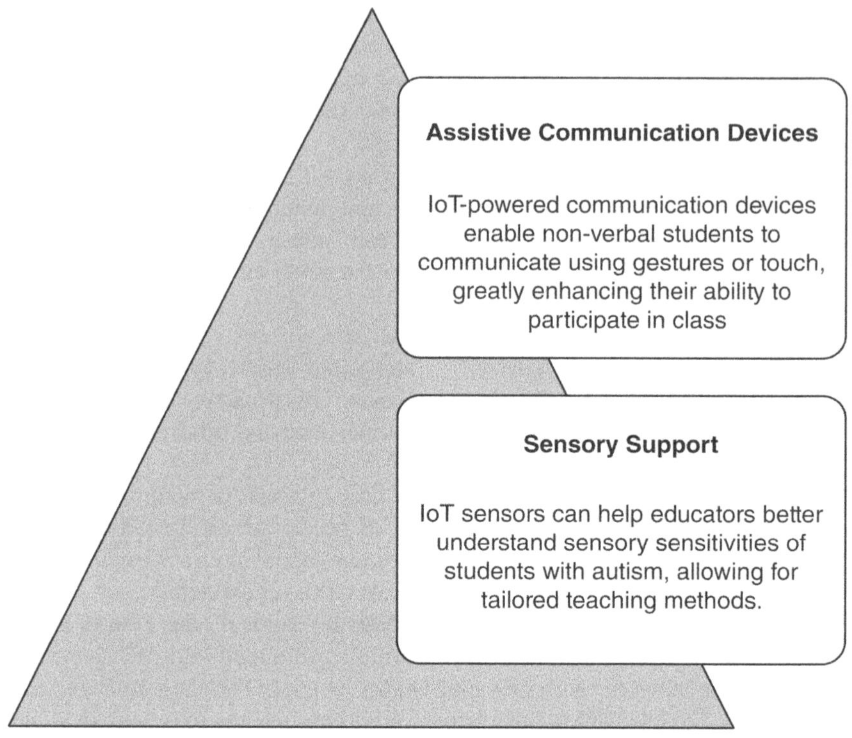

FIGURE 13.5   IoT for special education.

13.7.2 IoT for Inclusive Learning

The Seattle Inclusive Education Initiative is a remarkable case study that showcases the potential of IoT in creating inclusive learning environments. The initiative introduced IoT devices into classrooms to support students with a range of disabilities:

- **Smart Desks:** Equipped with IoT sensors, these desks adjusted their height and settings to accommodate students with physical disabilities, ensuring a comfortable learning experience.
- **IoT Wearables:** Some students wore IoT-enabled wearable devices that monitored stress levels. Educators used this data to provide timely interventions and support.

13.8 APPLICATIONS OF IoT IN THE EDUCATION SECTOR

The IoT has the potential to revolutionize the education sector by introducing innovative applications and enhancing various aspects of teaching, learning, and administration [19].

- **Campus Safety:** IoT technology can significantly enhance campus safety.

- **Real-time Surveillance:** Smart cameras and sensors across the campus can monitor suspicious activities and trigger alerts to security personnel in real time.
- **Emergency Response:** IoT-connected panic buttons and smart alarms can be placed strategically, allowing immediate emergency response in case of threats or incidents.
- **Equipment Tracking:** Schools and universities can use IoT to track the location and usage of valuable equipment like projectors, computers, and laboratory instruments.
- **Resource Optimization:** Smart sensors can monitor resource usage, such as water and electricity, helping institutions reduce costs and minimize waste.

13.9 CHALLENGES OF IoT IN EDUCATION INDUSTRIES

While the adoption of IoT in the education sector brings numerous benefits, it also presents several challenges that educational institutions need to address.

13.9.1 DATA PRIVACY AND CONSENT

IoT devices collect vast amounts of data, raising concerns about privacy:

Informed Consent: Obtaining consent from students and ensuring they understand how their data will be used is essential but can be challenging to implement effectively.

Data Ownership: Determining who owns the data generated by IoT devices in educational settings can be legally and ethically complex.

13.9.2 SECURITY RISKS

The interconnected nature of IoT can create security vulnerabilities:

DDoS Attacks: IoT devices are susceptible to being used in Distributed Denial of Service (DDoS) attacks, disrupting network operations. A university experienced a data breach when hackers exploited vulnerabilities in their IoT-connected access control system. This breach compromised student and faculty data, highlighting the need for robust security measures.

13.9.3 INFRASTRUCTURE AND COST CHALLENGES

Network Scalability: Ensuring a robust network infrastructure capable of handling a large number of IoT devices can be costly.

Ongoing Maintenance: Regular maintenance and updates for IoT devices and networks add to the operational costs.

Digital Divide: The digital divide became more apparent during the pandemic, with students lacking access to devices or reliable internet and struggling to keep up with their peers.

Privacy Concerns: The increased use of online platforms raised concerns about data privacy and security. Case studies of data breaches in educational institutions highlighted the need for robust cybersecurity measures.

Teacher Training: Teachers needed to adapt to new technologies and teaching methods quickly. Comprehensive training programs and ongoing support became crucial.

13.10 INTEGRATION OF IoT AND CLOUD COMPUTING IN THE EDUCATION SECTOR

Integrating IoT and cloud computing offers a powerful combination for educational institutions. The contribution of Cloud Computing & IOT in Education [28, 29] based on various aspects is shown in Table 13.1.

- **Scalable Storage:** IoT devices generate vast amounts of data. Cloud platforms provide scalable storage solutions to manage and store this data efficiently.
- **Real-time Analytics:** Cloud-based analytics tools can process IoT-generated data in real time, providing valuable insights into student behavior, resource usage, and more.

TABLE 13.1

Aspects Supporting Integration of Cloud Computing and IoT in Education

Aspect	Integration of Cloud Computing and IoT in Education
Enhanced Learning Environments	Combines cloud resources and IoT devices to create dynamic and interactive learning environments.
Real-time Data and Analytics	IoT sensors collect data on student engagement, attendance, and behavior, which is processed in the cloud for analysis.
Personalized Learning	Data from IoT devices and cloud analytics enable personalized learning plans and content recommendations.
Cost-Effective Infrastructure	Cloud provides scalable and cost-efficient infrastructure for storing and analyzing IoT-generated data.
Efficient Resource Management	IoT sensors optimize resource usage, such as energy and facility maintenance, with data analyzed in the cloud.
Remote and Hybrid Learning Support	IoT-enabled remote labs and cloud-based virtual classrooms support remote and hybrid learning scenarios.
Student Safety and Security	IoT-based campus security systems and cloud-based monitoring enhance student safety and campus security.
Asset Tracking and Management	IoT devices track assets like laptops and projectors, with data managed in the cloud to reduce loss and theft.
Library Automation	Cloud-based library systems can integrate with IoT solutions for improved book sorting and inventory management.
Accessibility and Equity	Careful planning ensures equitable access to IoT devices and cloud resources, addressing the digital divide.
Privacy and Data Security	Robust security measures are crucial to protect sensitive student data collected by IoT devices and stored in the cloud.

A university integrated IoT sensors into its campus facilities to monitor classroom occupancy and energy usage. By sending this data to the cloud for analysis, the institution identified patterns in classroom utilization and optimized heating, ventilation, and air conditioning (HVAC) systems, resulting in significant cost savings.

13.10.1 ENHANCING REMOTE LEARNING

IoT can enhance the quality of remote learning:

- **Smart Cameras and Microphones:** IoT-enabled cameras and microphones can provide a more immersive and interactive remote learning experience.
- **Internet Connectivity:** IoT can improve internet connectivity for remote learners in underserved areas, ensuring a more equitable online education experience [30].

During the COVID-19 pandemic, a high school used IoT-enabled lab equipment in conjunction with cloud-based virtual lab platforms. Students could conduct experiments remotely by controlling lab equipment via the internet. This allowed for continuity in science education during lockdowns.

13.10.2 DATA-DRIVEN DECISION-MAKING

IoT and cloud computing can enable predictive analytics in education.

Early Warning Systems: By analyzing IoT-generated data, educational institutions can identify students at risk of falling behind and intervene early to provide support.

Resource Allocation: Data-driven insights can inform decisions about resource allocation, helping schools optimize budgets and allocate resources where they are needed most.

13.10.2.1 Predictive Analytics for Student Success

A university implemented a predictive analytics system that combined IoT data from smart classrooms with cloud-based analytics tools. The system identified factors contributing to student success and allowed instructors to intervene when students showed signs of struggling. This resulted in higher retention rates and improved graduation rates.

13.10.3 SECURITY CONSIDERATIONS

Integrating IoT and cloud computing requires robust security measures:

Encryption and Access Controls: Data transmitted between IoT devices and the cloud must be encrypted, and strict access controls should be in place to prevent unauthorized access.

Vulnerability Assessment: Regular vulnerability assessments and penetration testing are essential to identify and mitigate security risks.

13.11 CASE STUDIES

IOT and cloud computing both have their tremendous contribution to education files. Real case study examples where these have been used are explained further.

13.11.1 University Cloud Migration

The University's cloud migration is an illustrative case study of how cloud computing can streamline operations:

- **Cost Saving:** By moving to the cloud, the university reduced infrastructure costs by 30%. It no longer needed to maintain on-premises servers and data centers.
- **Scalability:** During peak enrollment periods, the university could easily scale its cloud resources to accommodate increased demand for online services.

13.11.2 Google Workspace for Education

Google Workspace for Education is an excellent example of cloud computing's impact on education. This suite of cloud-based tools includes Google Docs, Sheets, and Classroom, facilitating collaboration and document sharing among students and teachers:

- Collaborative Learning: Google Workspace promotes collaborative learning, allowing students to work together on assignments and projects in real time.
- Paperless Classrooms: The cloud-based platform reduces the need for physical paper and textbooks, contributing to sustainability efforts.

13.11.3 Smart Classrooms

Imagine a scenario where smart classrooms are equipped with IoT sensors and cloud-connected devices:

- Smartboards: Interactive whiteboards with IoT capabilities allow teachers to share real-time notes, diagrams, and quizzes with students' devices, promoting active engagement.
- IoT Attendance Tracking: Sensors automatically record student attendance when they enter the classroom, reducing administrative workload.

13.11.4 Adaptive Learning

In this scenario, IoT and cloud computing support adaptive learning:

- IoT Wearables: Students wear IoT-enabled fitness trackers that monitor their stress levels and attention spans. Data from these wearables is sent to the cloud for analysis.

- Personalized Recommendations: The cloud uses this data to recommend study breaks or relaxation exercises to help students stay focused and manage stress during their learning sessions.

13.11.5 SMART CAMPUS

Imagine a smart campus scenario where IoT and cloud technologies are used for efficient campus management:

- Energy Efficiency: IoT sensors monitor energy usage in campus buildings. Data is sent to the cloud, where machine learning algorithms analyze it to optimize heating, cooling, and lighting systems for energy savings.
- Maintenance Predictions: IoT sensors on equipment and infrastructure detect anomalies and predict maintenance needs. Maintenance teams receive alerts through the cloud, reducing downtime and improving overall campus efficiency.

13.11.6 KHAN ACADEMY

Khan Academy, founded by Salman Khan, is a prime example of how technology has transformed education. Khan Academy offers a vast library of free educational videos and exercises on a wide range of subjects [31]. The platform has been instrumental in democratizing education, particularly in underserved communities. Some key takeaways from the Khan Academy case are shown in Figure 13.6.

13.11.7 THE SINGAPORE EDUCATION SYSTEM

Singapore's education system offers an insightful case study on digital transformation. The government's commitment to integrating technology into education has yielded several outcomes, as mentioned in Figure 13.7.

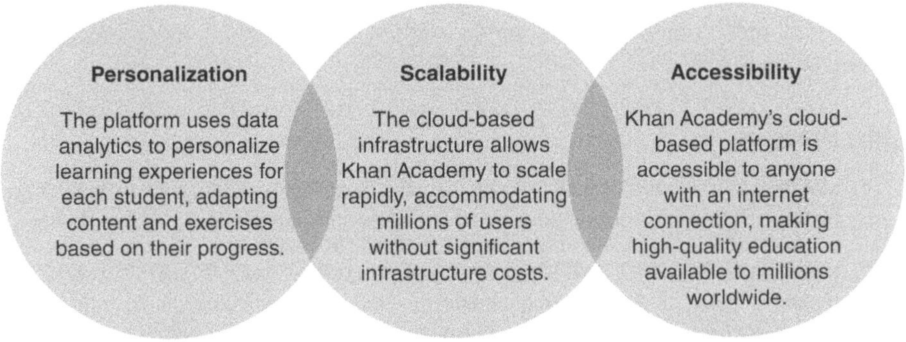

Personalization

The platform uses data analytics to personalize learning experiences for each student, adapting content and exercises based on their progress.

Scalability

The cloud-based infrastructure allows Khan Academy to scale rapidly, accommodating millions of users without significant infrastructure costs.

Accessibility

Khan Academy's cloud-based platform is accessible to anyone with an internet connection, making high-quality education available to millions worldwide.

FIGURE 13.6 Outcome from Khan Academy case.

E-Government Initiatives	**Blended Learning**
• Singapore introduced the "Parents Gateway" app, allowing parents to monitor their child's attendance and academic progress in real-time	The Ministry of Education in Singapore actively promotes blended learning, where students combine digital and face-to-face learning. This approach has proven to be effective in adapting to different learning styles.

FIGURE 13.7 Outcome from Singapore education system case.

13.12 CONCLUSION AND FUTURE SCOPE

The integration of IoT and Cloud Computing in the Education sector has the potential to revolutionize teaching and learning. Numerous applications, ranging from smart classrooms to predictive analytics for student success, have been explored. However, acknowledging the challenges, including data privacy and security concerns, is crucial.

The future of education lies in even more personalized learning experiences. IoT and cloud computing will continue to play a pivotal role in tailoring education to individual student needs. Augmented reality (AR) and virtual reality (VR) technologies, integrated with IoT and the cloud, will enable immersive and experiential learning, transporting students to virtual labs, historical events, and distant locations for interactive learning. The future will see a greater emphasis on ethical considerations, including responsible data use and the ethical design of IoT devices, ensuring that the benefits of these technologies are realized without compromising privacy and security. Integrating IoT and cloud computing will support lifelong learning, enabling individuals to acquire new skills and knowledge throughout their lives, fostering continuous personal and professional growth.

In conclusion, the integration of IoT and cloud computing is poised to reshape education, making it more accessible, engaging, and effective. To ensure its success, educational institutions must navigate the challenges and adapt to the evolving technological landscape.

REFERENCES

1. McGarr, O., and Johnston K. "Exploring the Evolution of Educational Technology Policy in Ireland: From Catching-Up to Pedagogical Maturity." Educational Policy 35, no. 6: (2021): 841–865. https://doi.org/10.1177/0895904819843597
2. Haji, Lailan, Omar Ahmad, Subhi R. M. Zeebaree, Hivi Dino, Rizgar R. Zebari, and Hanan Shukur. "Impact of cloud computing and internet of things on the future internet." Technology Reports of Kansai University 62 (2020): 2179–2190.
3. Rashid, Aaqib, and Amit Chaturvedi. "Cloud computing characteristics and services: a brief review." International Journal of Computer Sciences and Engineering 7, no. 2 (2019): 421–426.

4. Mohammed, Chnar Mustafa, and Subhi R.M Zeebaree. "Sufficient comparison among cloud computing services: IaaS, PaaS, and SaaS: A review." International Journal of Science and Business 5, no. 2 (2021): 17–30.

5. Sharma, Mahak, Ruchita Gupta, and Padmanav Acharya. "Analysing the adoption of cloud computing service: A systematic literature review." Global Knowledge, Memory and Communication 70, no. 1/2 (2021): 114–153.

6. Thavi, Riddhi, Rujuta Jhaveri, Vaibhav Narwane, Bhaskar Gardas, and Nima Jafari Navimipour. "Role of cloud computing technology in the education sector." Journal of Engineering, Design and Technology (2021): 1–33.

7. Malhi, Muhammad Saqib, Usman Iqbal, Muhammad Mustafa Nabi, and Muhammad Aaqib-Ishtiaq Malhi. "E-learning based on cloud computing for educational institution: Security issues and solutions." International Journal of Electronics and Information Engineering 12, no. 4 (2020): 162–169.

8. Qasem, Yousef AM, Rusli Abdullah, Yusmadi Yah Jusoh, Rodziah Atan, and Shahla Asadi. "Cloud computing adoption in higher education institutions: A systematic review." IEEE Access 7 (2019): 63722–63744.

9. Naved, Mohd, Domenic T. Sanchez, Adeline P. Dela Cruz, Larry B. Peconcillo Jr, and Emerson D. Peteros. "Identifying the role of cloud computing technology in management of educational institutions." Materials Today: Proceedings 51 (2022): 2309–2312.

10. Al-Malah, Duha Khalid Abdul-Rahman, Ibtisam A. Aljazaery, Haider Th Salim Alrikabi, and Hussain Ali Mutar. "Cloud computing and its impact on online education." In *IOP Conference Series: Materials Science and Engineering*, vol. 1094, no. 1, p. 012024. IOP Publishing, 2021.

11. Surameery, Nigar M. Shafiq, and Mohammed Y. Shakor. "CBES: Cloud Based Learning management System for Educational Institutions." In *2021 3rd East Indonesia Conference on Computer and Information Technology (EIConCIT)*, pp. 270–275. IEEE, 2021.

12. Chatterjee, Paramita, Rajesh Bose, and Sandip Roy. "A review on architecture of secured cloud based learning management system." Journal of Xidian University 14, no. 7 (2020): 365–376.

13. Raes, Annelies, Pieter Vanneste, Marieke Pieters, and Ine Windey. "Wim Van Den Noortgate, and Fien Depaepe. "Learning and instruction in the hybrid virtual classroom: An investigation of students' engagement and the effect of quizzes." Computers & Education 143 (2020): 103682.

14. Glassey, Jarka, and Fernão D. Magalhães. "Virtual labs–love them or hate them, they are likely to be used more in the future." Education for Chemical Engineers 33 (2020): 76.

15. M. Sadeeq, Mohammed A., Nasiba M. Abdulkareem, Subhi RM Zeebaree, Dindar Ahmed, Ahmed Saifullah Sami, and Rizgar R. Zebari. "IoT and cloud computing issues, challenges and opportunities: A review." Qubahan Academic Journal 1, no. 2 (2021): 1–7.

16. Kumari, Chandini, Gagandeep Singh, Gursharan Singh, and Ranbir Singh Batth. "Security issues and challenges in cloud computing: A mirror review." In *2019 International Conference on Computational Intelligence and Knowledge Economy (ICCIKE)*, pp. 701–706. IEEE, 2019.

17. Rizwan, Syed, and Muhammad Zubair. "Basic security challenges in cloud computing." In *2019 4th International Conference on Emerging Trends in Engineering, Sciences and Technology (ICEEST)*, pp. 1–4. IEEE, 2019.

18. Verma, Deepak Kumar, and Tanya Sharma. "Issues and challenges in cloud computing." International Journal of Advanced Research in Computer and Communication Engineering 8 (2019): 188–195.

19. Perwej, Yusuf, Kashiful Haq, Firoj Parwej, M. Mumdouh, and Mohamed Hassan. "The internet of things (IoT) and Its application domains." International Journal of Computer Applications 975, no. 8887 (2019): 182.

20. Arora, Jyoti Batra, and Suruchi Kaushik. "IoT in education: A future of sustainable learning." In *Handbook of Research on the Internet of Things Applications in Robotics and Automation*, pp. 300–317. IGI Global, 2020.

21. Ramlowat, Dosheela Devi, and Binod Kumar Pattanayak. "Exploring the internet of things (IoT) in education: A review." In *Information Systems Design and Intelligent Applications: Proceedings of Fifth International Conference INDIA 2018 Volume 2*, pp. 245–255. Springer Singapore, 2019.

22. Kassab, Mohamad, Joanna DeFranco, and Phillip Laplante. "A systematic literature review on internet of things in education: Benefits and challenges." Journal of Computer Assisted Learning 36, no. 2 (2020): 115–127.

23. Revathi, R., M. Suganya, and Gladiss Merlin NR. "IoT based cloud integrated smart classroom for smart and a sustainable campus." Procedia Computer Science 172 (2020): 77–81.

24. Saini, Mukesh Kumar, and Neeraj Goel. "How smart are smart classrooms? A review of smart classroom technologies." ACM Computing Surveys (CSUR) 52, no. 6 (2019): 1–28.

25. Alfoudari, Aisha M., Christopher M. Durugbo, and Fairouz M. Aldhmour. "Understanding socio-technological challenges of smart classrooms using a systematic review." Computers & Education 173 (2021): 104282.

26. Kwet, Michael, and Paul Prinsloo. "The 'smart' classroom: A new frontier in the age of the smart university." Teaching in Higher Education 25, no. 4 (2020): 510–526.

27. Polin, Ken, Tan Yigitcanlar, Mark Limb, and Tracy Washington. "The making of smart campus: A review and conceptual framework." Buildings 13, no. 4 (2023): 891.

28. Fahmideh, Mahdi, Farhad Daneshgar, Fethi Rabhi, and Ghassan Beydoun. "A generic cloud migration process model." European Journal of Information Systems 28, no. 3 (2019): 233–255.

29. Stergiou, Christos, Kostas E. Psannis, Byung-Gyu Kim, and Brij Gupta. "Secure integration of IoT and cloud computing." Future Generation Computer Systems 78 (2018): 964–975.

30. Arnavut, Ahmet, Hüseyin Bicen, and Cahit Nuri. "Students' approaches to massive open online courses: The case of Khan Academy." BRAIN. Broad Research in Artificial Intelligence and Neuroscience 10, no. 1 (2019): 82–90.

31. Santos, Juan Carlos Luna, and Angel Onzari Luna Santos. "The use of the Khan Academy virtual platform and the learning of mathematics in a private university of Peru." Inicc-Perú (2021): 84–89.

14 Cloud-IoT in Underwater Communication

Swathi Tejah Yall and Wuriti Kavya

14.1 INTRODUCTION

The convergence of cloud computing and Internet of Things (IoT) technologies has ushered in a transformative era in underwater communication systems. As the depths of our oceans remain largely unexplored and inaccessible, harnessing these technological advancements has become essential to unlock the mysteries of the underwater world. This chapter delves into the dynamic landscape of Cloud-IoT integration in underwater communication, exploring its applications, challenges, and the remarkable impact it brings to various domains.

14.1.1 THE UNSEEN FRONTIER

The ocean covers over two-thirds of Earth's surface, yet our understanding of its vast and complex ecosystems is limited. Beneath the waves lies a realm of uncharted territories, intricate marine life, and geological wonders. Traditionally, communicating and gathering data from underwater environments presented immense challenges due to the harsh conditions, lack of infrastructure, and limited data transmission capabilities. Cloud-IoT integration changes this narrative by bridging the gap between the underwater world and human comprehension.

14.1.2 THE POWER OF CLOUD-IoT INTEGRATION

At its core, Cloud-IoT integration empowers us to collect, transmit, analyze, and respond to real-time data from underwater environments. The cloud acts as a central hub where data from underwater sensors, vehicles, and platforms is aggregated and processed. IoT technologies enable these devices to communicate with each other and the cloud seamlessly. This integration is not only enhancing our capacity to explore the ocean's depths but is also catalyzing advancements in scientific research, environmental monitoring, disaster management, and industrial operations.

14.1.3 A MULTIDIMENSIONAL APPROACH

The scope of Cloud-IoT integration in underwater communication is multidimensional. It encompasses deploying advanced underwater sensors, transmitting data through acoustic and optical communication methods, applying real-time analytics for insights, and remote control of underwater platforms. These components

DOI: 10.1201/9781032656694-14

converge to create a synergy that transforms how we perceive, interact with, and benefit from the underwater environment.

14.1.4 CHAPTER OVERVIEW

This chapter delves into the multifaceted aspects of Cloud-IoT integration in underwater communication. It begins by establishing the context with a background on underwater communication challenges and the motivation for adopting Cloud-IoT solutions. The subsequent sections explore the implications of this integration in environmental monitoring, scientific research, disaster management, and industrial applications. Moreover, it delves into the advancements in underwater communication technologies, the role of cloud computing and IoT, and the opportunities that lie on the horizon.

In conclusion, the introduction sets the stage for exploring the exciting world of Cloud-IoT integration in underwater communication. By harnessing the power of technology, we are poised to unlock unprecedented insights into the oceans, drive sustainable practices, and address the challenges that have eluded us for so long. The subsequent sections will delve into the intricacies of this integration, illuminating the path toward a new era of understanding and harnessing the potential of the underwater world.

14.2 SCOPE AND OBJECTIVES

14.2.1 SCOPE

The scope of this chapter encompasses a comprehensive exploration of the integration of cloud computing and IoT technologies within underwater communication systems. It delves into the intricate ways these technologies intersect, their impact on various domains, and the transformative potential they offer for understanding, utilizing, and safeguarding underwater environments.

This will explore the applications, challenges, advancements, and implications of Cloud-IoT integration in underwater communication. It will cover topics such as underwater sensor networks, data transmission methods, real-time analytics, remote control of underwater platforms, and the global collaboration facilitated by these technologies. While the primary focus is on the integration itself, the chapter will also touch on broader implications such as scientific research, environmental conservation, disaster management, and industrial applications.

14.2.2 OBJECTIVES

The objectives are multifold:

1. Understanding Cloud-IoT Integration: The chapter aims to provide a clear understanding of what Cloud-IoT integration entails in underwater communication. It will elucidate this integration's core concepts, mechanisms, and benefits, helping readers grasp the technical and conceptual aspects.

2. Exploring Applications: By exploring a range of applications, the chapter seeks to demonstrate the diverse ways in which Cloud-IoT integration is transforming underwater communication. It will highlight how these technologies are applied in scientific research, environmental monitoring, disaster response, and industrial operations.

3. Examining Technological Advancements: The chapter will delve into the advancements in underwater communication technologies that enable Cloud-IoT integration. It will discuss the evolution of acoustic and optical communication methods, the development of advanced sensors and underwater platforms, and the role of edge computing in enhancing real-time data processing.

4. Analyzing Implications: A critical objective is to analyze the implications of Cloud-IoT integration in underwater communication. This includes understanding its impact on scientific discoveries, environmental conservation, sustainable resource management, disaster preparedness, and global collaboration.

5. Envisioning the Future: The chapter will conclude by envisioning the future directions of Cloud-IoT integration in underwater communication. By considering the ongoing advancements, potential challenges, and emerging trends, it aims to provide readers with insights into the transformative possibilities on the horizon.

14.3 UNDERWATER COMMUNICATION TECHNOLOGIES

14.3.1 CHALLENGES OF UNDERWATER COMMUNICATION

The underwater realm presents a host of unique challenges that have long hindered effective communication in this environment. This section delves into the intricate web of challenges that underwater communication systems face, shedding light on the complexities that have prompted the exploration of Cloud-IoT integration as a solution.

14.3.2 HOSTILE AND VAST ENVIRONMENT

Underwater communication encounters a hostile and vast environment that poses significant barriers to conventional methods. Water's high density, salinity, and pressure interfere with signal propagation, leading to signal attenuation, distortion, and reduced communication range. Additionally, underwater communication must contend with obstacles such as underwater terrain features, marine life, and human-made structures that impede signal transmission.

14.3.3 LIMITED BANDWIDTH AND DATA RATES

The limited bandwidth for underwater communication restricts the data rates that can be achieved. Traditional communication methods, such as acoustic waves, offer lower data rates compared to their terrestrial counterparts. This limitation inhibits the real-time transmission of large datasets, high-definition video, and complex sensor readings, compromising the depth and accuracy of information that can be relayed from the ocean's depths.

14.3.4 Latency and Delays

The propagation speed of signals underwater, especially in acoustic communication, leads to significant latency and delays in data transmission. This delay can hinder time-sensitive applications, such as remote control of underwater vehicles and real-time monitoring of dynamic events. The challenge lies in developing solutions that minimize latency while ensuring data accuracy and integrity.

14.3.5 Noise and Interference

Underwater communication systems contend with various sources of noise and interference. Ambient noise from natural phenomena like ocean currents, waves, and marine life, as well as anthropogenic activities such as ship traffic and underwater construction, can overwhelm weak signals. Filtering out noise and distinguishing desired signals from interference becomes a complex task in underwater communication.

14.3.6 Energy Constraints

Energy availability is a critical constraint in underwater communication systems. Battery-operated devices, such as underwater sensors and vehicles, have limited energy reserves and face challenges in recharging or replacing batteries in remote locations. Balancing the energy requirements of communication with other operational needs, such as data collection and propulsion, presents a significant design challenge.

14.3.7 Lack of Infrastructure

The absence of established communication infrastructure underwater necessitates creative solutions for establishing connectivity. Unlike terrestrial environments with existing cellular networks, underwater communication often relies on deploying dedicated sensors, buoys, and vehicles equipped with communication equipment. Establishing and maintaining these infrastructures in the harsh underwater environment requires careful planning and resource allocation.

14.3.8 Harsh Environmental Conditions

Underwater communication systems must withstand the harsh conditions of the deep ocean, including extreme pressures, temperature fluctuations, and corrosive saltwater. These conditions can impact the reliability and longevity of communication equipment. Developing robust, waterproof, and corrosion-resistant devices that can endure these conditions is essential for ensuring consistent communication.

14.4 ACOUSTIC COMMUNICATION

14.4.1 PRINCIPLES OF ACOUSTIC COMMUNICATION

Acoustic communication relies on the propagation of sound waves through water as the medium of transmission. Sound waves travel as pressure variations, or acoustic signals, that can carry data over considerable distances underwater. Devices designed for acoustic communication, such as underwater modems, emit sound signals that encode information. These signals are received by other devices that decode the information from the variations in pressure.

14.4.2 CHALLENGES IN ACOUSTIC COMMUNICATION

While acoustic communication offers a means of transmitting data in underwater environments, it is not without challenges. Sound waves can be subject to attenuation, where they lose energy and weaken over distance due to water's density and viscosity. Additionally, sound waves are susceptible to scattering and reflection off underwater obstacles, such as the seabed or marine life. This can lead to signal loss, distortion, and reduced communication range.

14.4.3 INTEGRATION WITH CLOUD-IoT

Cloud-IoT integration brings a transformative dimension to acoustic communication. Underwater modems equipped with IoT capabilities can transmit acoustic signals to cloud platforms, where the data is aggregated and analyzed. This integration enables real-time monitoring of acoustic communication networks, allowing operators to assess signal quality, detect anomalies, and optimize data transmission protocols.

14.4.4 APPLICATIONS OF ACOUSTIC COMMUNICATION

Acoustic communication finds diverse applications in underwater environments. It facilitates underwater navigation for autonomous vehicles, enables remote control of underwater platforms, and supports scientific data collection from remote sensors. Acoustic communication also serves to relay real-time information from underwater observatories to cloud platforms, offering insights into seismic activities, marine life behavior, and oceanographic phenomena.

14.5 OPTICAL COMMUNICATION

14.5.1 PRINCIPLES OF OPTICAL COMMUNICATION

Optical communication employs light as the medium to transmit data underwater. Unlike acoustic communication, which uses sound waves, optical communication utilizes light waves to encode and transmit information. Devices such as underwater modems equipped with LEDs (light-emitting diodes) emit modulated light signals that carry data. These signals are detected by light-sensitive receivers on the receiving end.

14.5.2   ADVANTAGES OF OPTICAL COMMUNICATION

Optical communication offers several advantages over traditional acoustic methods. Light travels faster through water than sound, reducing latency and enabling higher data rates. Optical signals are less prone to interference and scattering than acoustic signals, resulting in improved signal integrity and longer communication ranges. Optical communication also allows bi-directional data transmission, enabling more complex and interactive underwater applications.

14.5.3   CHALLENGES IN OPTICAL COMMUNICATION

While optical communication presents numerous advantages, it also faces challenges. Water's optical properties, such as absorption and scattering, can limit the transmission distance of light. The presence of suspended particles and water turbidity can attenuate optical signals, affecting signal strength and clarity. These challenges necessitate careful consideration of the communication environment and the development of adaptive communication protocols.

14.5.4   INTEGRATION WITH CLOUD-IoT

The integration of optical communication with cloud computing and IoT technologies marks a significant advancement in underwater connectivity. Optical modems equipped with IoT capabilities can transmit light signals to cloud platforms, where data is aggregated, analyzed, and interpreted. This integration enables real-time monitoring of optical communication networks, allowing operators to assess signal quality, optimize data transmission, and detect potential obstacles.

14.5.5   APPLICATIONS OF OPTICAL COMMUNICATION

Optical communication finds applications in a variety of underwater scenarios. It supports high-definition video streaming from remotely operated vehicles (ROVs) and autonomous underwater vehicles (AUVs), enabling real-time exploration and surveillance of underwater environments. Optical communication is also utilized for underwater sensor networks, which relay complex environmental data and images from sensors to the cloud for analysis.

14.6   ELECTROMAGNETIC COMMUNICATION

14.6.1   PRINCIPLES OF ELECTROMAGNETIC COMMUNICATION

Electromagnetic communication employs electromagnetic waves to transmit data underwater. Unlike acoustic and optical methods, electromagnetic waves include radio waves and certain microwave frequencies. Devices equipped with antennas emit electromagnetic signals that carry data encoded in variations of the electromagnetic field. These signals are detected by receivers tuned to the specific frequency, allowing data extraction.

14.6.2 Challenges in Electromagnetic Communication

While electromagnetic communication offers advantages, it also faces challenges specific to underwater environments. Electromagnetic waves, particularly higher frequencies like radio waves, are absorbed by water and have limited penetration depth. This limits the communication range and effectiveness of electromagnetic communication in deeper and more turbid waters. Additionally, the presence of metal structures, such as ships and underwater infrastructure, can interfere with electromagnetic signals.

14.6.3 Integration with Cloud-IoT

Cloud-IoT integration extends the capabilities of electromagnetic communication in underwater environments. Devices with electromagnetic transmitters and IoT functionalities can transmit signals to cloud platforms, where data is aggregated, analyzed, and interpreted. This integration allows real-time monitoring of electromagnetic communication networks, enabling operators to assess signal strength, optimize data transmission, and identify potential sources of interference.

14.6.4 Applications of Electromagnetic Communication

Electromagnetic communication finds applications in various underwater scenarios. It is suitable for short-range communication, making it useful for underwater vehicles, underwater sensor networks, and remotely operated platforms. Electromagnetic communication can also serve as a complementary method alongside acoustic and optical communication, enhancing communication redundancy and reliability.

14.7 IoT AND CLOUD COMPUTING

14.7.1 IoT Overview and Architecture

At its core, IoT refers to the network of interconnected devices, sensors, and objects that communicate and exchange data over the internet. These devices, often equipped with sensors, actuators, and embedded technology, collect and transmit data to central systems for analysis and decision-making. IoT extends the capabilities of traditional computing by allowing real-world objects to interact with the digital realm, opening up a multitude of applications across industries.

14.7.2 IoT Architecture

IoT architecture is a conceptual framework that defines the components and relationships within an IoT system. It is composed of several layers, each serving a specific purpose in the communication and data flow process. The typical layers in an IoT architecture include:

1. Perception Layer: This layer consists of sensors and actuators interacting with the physical world. Sensors collect data from the environment, while actuators enable devices to act based on received instructions.

2. Network Layer: The network layer transmits data between devices and gateways. It encompasses various communication protocols, such as Wi-Fi, Bluetooth, Zigbee, and cellular networks, ensuring seamless data transmission.
3. Middleware Layer: This layer facilitates communication between devices and the cloud. It includes components like data brokers, message queues, and application enablement platforms that manage data flow, filtering, and routing.
4. Application Layer: The application layer encompasses software and services that process and analyze the collected data. It can include data analytics, machine learning algorithms, and visualization tools that provide insights and actionable information.
5. Business Layer: This layer involves decision-making based on the insights derived from the data. It influences the actions of devices or users based on the analyzed information.

14.7.3 INTEGRATION WITH CLOUD-IoT

Cloud-IoT integration builds upon the foundation of IoT architecture, extending its capabilities by incorporating cloud computing technologies. Cloud platforms serve as repositories for the massive amounts of data generated by IoT devices, enabling storage, processing, and analysis. This integration empowers real-time data analysis, remote control, and global accessibility, which is particularly valuable in underwater communication where physical presence is challenging.

14.8 CLOUD COMPUTING IN IoT

14.8.1 UNDERSTANDING CLOUD COMPUTING IN IoT

Cloud computing involves delivering computing resources such as processing power, storage, and software applications over the internet. In the context of IoT, cloud computing offers a centralized platform where the data collected from IoT devices can be processed, analyzed, and stored. This offloads the computational burden from individual devices and enables data to be processed on powerful Cloud servers.

14.8.2 BENEFITS OF CLOUD COMPUTING IN IoT

The integration of Cloud computing with IoT brings several benefits that are especially relevant to underwater communication:

1. Scalability: Cloud platforms provide the ability to scale resources up or down based on demand. This is essential in IoT deployments where the number of connected devices can vary widely, ensuring that the infrastructure can accommodate fluctuations in data traffic.
2. Data Storage: IoT generates massive amounts of data, and cloud computing offers virtually limitless storage capacity. This is particularly valuable for

underwater communication, where continuous data collection from sensors and platforms generates significant data volumes.

3. Processing Power: Cloud platforms boast high computational power, capable of processing complex analytics and machine learning algorithms. This enables real-time data analysis from underwater sensors, transforming raw data into actionable insights.

4. Remote Accessibility: Cloud computing allows authorized users to access and manage IoT data from anywhere with an internet connection. This is crucial for underwater communication, where remote access is necessary due to the challenging and remote nature of underwater environments.

5. Data Analytics: Cloud platforms offer a range of data analytics tools and frameworks that can uncover patterns, anomalies, and trends within IoT data. This capability enhances decision-making by providing valuable insights from the vast amount of collected data.

14.8.3 INTEGRATION WITH UNDERWATER COMMUNICATION

In the context of underwater communication, cloud computing enhances the efficiency, accuracy, and accessibility of data collected from underwater sensors, vehicles, and platforms. Data transmitted from underwater environments can be securely stored in the cloud, enabling real-time analytics, remote control, and global collaboration. Cloud-IoT integration empowers researchers, scientists, and stakeholders to access and interpret underwater data, driving advancements in scientific research, environmental conservation, disaster management, and industrial applications.

14.8.4 BENEFITS AND CHALLENGES OF CLOUD-IoT INTEGRATION

The integration of cloud computing and IoT technologies offers a host of benefits while also posing certain challenges. This section explores both the advantages and obstacles that come with Cloud-IoT integration in the context of underwater communication systems.

14.8.5 BENEFITS OF CLOUD-IoT INTEGRATION

1. Real-Time Data Analysis: Cloud-IoT integration empowers real-time data analysis by offloading computational tasks to powerful cloud servers. This enables quick insights, actionable information, and timely decision-making based on continuous data collection from underwater sensors and platforms.

2. Scalability: Cloud platforms can easily scale resources up or down based on demand. This is vital in underwater communication systems where the number of connected devices, such as sensors and vehicles, can vary widely over time.

3. Global Accessibility: Cloud-IoT integration facilitates remote access to underwater data and control of devices. Scientists, researchers, and stakeholders can collaborate across geographical boundaries, sharing real-time information and insights for scientific exploration and disaster management.

4. Enhanced Storage: The cloud offers virtually limitless storage capacity, accommodating the vast amounts of data generated by underwater sensors and platforms. This storage capacity is crucial for maintaining historical data and facilitating long-term analysis.
5. Data Security: Cloud platforms provide robust security measures, including encryption, authentication, and access controls. This ensures that sensitive underwater data remains confidential and protected from unauthorized access.
6. Cost Efficiency: Cloud-IoT integration eliminates the need for extensive on-site infrastructure and hardware. This can lead to cost savings in terms of hardware procurement, maintenance, and system upgrades.

14.8.6 CHALLENGES OF CLOUD-IoT INTEGRATION

1. Latency: The transmission of data from underwater environments to the cloud can introduce latency due to the time taken for data to travel through communication channels. Minimizing latency is crucial in time-sensitive applications, such as remote control of underwater vehicles.
2. Data Transmission Reliability: Underwater communication faces challenges such as signal attenuation, interference, and underwater obstacles. Ensuring reliable data transmission from underwater devices to the cloud requires robust communication protocols and error correction mechanisms.
3. Power Constraints: Underwater devices like sensors and vehicles, operate on limited power sources. Sending data to the cloud for processing can consume additional energy, potentially affecting the operational lifespan of these devices.
4. Privacy Concerns: Cloud-IoT integration raises privacy concerns as sensitive underwater data is transmitted and stored offsite. Ensuring data privacy and compliance with data protection regulations is paramount, especially when dealing with scientific research data or sensitive industrial information.
5. Dependence on Connectivity: Cloud-IoT integration relies heavily on internet connectivity. Maintaining continuous communication with the Cloud can be a significant challenge in remote underwater locations where reliable internet access is challenging.

14.9 CLOUD-IoT INTEGRATION FOR UNDERWATER COMMUNICATION

The integration of cloud computing and IoT technologies brings a revolutionary approach to underwater communication systems. This section delves into the various aspects of Cloud-IoT integration within the context of underwater communication, highlighting its implementation stages, data processing pipeline, and the transformative impact it has on data collection, transmission, and analysis in underwater environments.

14.9.1 Sensor Deployment and Data Collection

The process begins with the strategic deployment of advanced sensors equipped with IoT capabilities. These sensors are carefully placed on various underwater platforms such as buoys, AUVs, and underwater gliders. Depending on the specific application, sensors can be designed to measure parameters such as temperature, pressure, salinity, dissolved oxygen levels, and pH, and even detect the presence of marine life. Sensors may also be equipped with cameras, hydrophones, and other specialized equipment to capture visual and acoustic data from the underwater environment.

14.9.2 Role of IoT Capabilities

The integration of IoT capabilities equips these sensors with the ability to communicate, transmit, and receive data over the internet. Each sensor becomes an intelligent node in the larger IoT ecosystem, capable of collecting data and processing, encoding, and sending it to remote cloud platforms. This connectivity transforms the once-isolated sensors into valuable contributors to a global network of underwater data collection and analysis.

14.9.3 Continuous Data Collection

Equipped with IoT capabilities, sensors continuously collect data from their immediate underwater surroundings. This data collection operates autonomously, eliminating the need for frequent manual intervention. This is particularly crucial in underwater environments where access is challenging, and conditions rapidly change. Continuous data collection enables the capture of real-time events, patterns, and trends in the underwater ecosystem.

14.9.4 Data Types and Diversity

The sensors deployed in underwater communication systems can capture a wide range of data types. This includes environmental parameters such as water temperature variations, pressure fluctuations, and salinity changes. Additionally, sensors can capture visual data through cameras, audio data through hydrophones, and even capture data related to marine life behavior and activity. The diverse range of data collected provides a holistic understanding of underwater ecosystems and phenomena.

14.9.5 Data Transmission to the Cloud

After collecting data, the integration of IoT capabilities enables sensors to transmit the collected information to cloud platforms. This transmission can be achieved through various communication methods, including acoustic waves, optical signals, or electromagnetic waves. Once the data reaches the cloud, it undergoes further processing, analysis, and interpretation, offering valuable insights into the underwater world.

14.9.6   DATA AGGREGATION AND PRE-PROCESSING

Data collected from multiple sensors distributed across underwater platforms is often fragmented and diverse. Data aggregation involves gathering and organizing this dispersed data into a centralized repository within the cloud. Aggregating data makes it accessible in one location, streamlining the analysis and decision-making process. This centralized approach allows researchers, scientists, and stakeholders to access a comprehensive view of underwater conditions, trends, and events.

14.9.7   DATA CLEANING AND FILTERING

During data aggregation, data cleaning and filtering take place to ensure the accuracy and integrity of the collected information. This involves identifying and rectifying inconsistencies, outliers, and erroneous data points that may arise from sensor malfunctions, interference, or other factors. By eliminating noisy or unreliable data, the quality of the dataset is improved, ensuring that subsequent analysis is based on accurate and valid information.

14.9.8   DATA TRANSFORMATION AND COMPRESSION

Once the data is cleaned and organized, it may undergo transformation and compression. Data transformation involves converting raw sensor readings into standardized formats, units, or scales for meaningful comparison and analysis. Compression reduces the size of the data while retaining essential information, optimizing storage and transmission efficiency. These processes prepare the data for further analysis and interpretation.

14.9.9   CONTEXTUALIZATION

Data collected from underwater sensors often lacks contextual information necessary for proper interpretation. Pre-processing may involve enriching the dataset with contextual metadata, such as timestamps, geographic coordinates, and sensor identifiers. Contextualization ensures that data is properly understood and can be correlated with other relevant data sources, enhancing the accuracy and relevance of subsequent analysis.

14.9.10   ENHANCING DATA QUALITY

The pre-processing phase significantly enhances the quality of the data by reducing noise, addressing inconsistencies, and standardizing formats. High-quality data is essential for accurate and meaningful analysis, allowing researchers to draw reliable conclusions and make informed decisions based on the collected information.

14.9.11   PREPARING FOR ANALYSIS

Data pre-processing sets the stage for data analysis in the cloud. By aggregating, cleaning, and transforming the data, the cloud platforms are equipped with a refined dataset ready for various analytical techniques. This enhances the efficiency and accuracy of subsequent analysis and interpretation, facilitating the extraction of valuable insights from the underwater data.

14.9.12 DATA TRANSMISSION TO THE CLOUD

After sensors collect and preprocess data, the next challenge is to transmit this data to cloud platforms for analysis and interpretation. This transmission bridges the gap between the underwater world and the digital realm, enabling scientists, researchers, and stakeholders to access real-time insights, make informed decisions, and respond to dynamic underwater conditions.

14.9.13 COMMUNICATION METHODS

Data transmission from underwater environments to the Cloud can be achieved through various communication methods:

1. Acoustic Communication: Utilizing sound waves, underwater modems transmit data acoustically. Acoustic communication is effective for relatively short distances and can navigate through challenging underwater environments, although it might be limited by factors such as attenuation and interference.
2. Optical Communication: Optical modems use light signals to transmit data through water. Optical communication offers higher data rates and lower latency compared to acoustic methods, but it is sensitive to water turbidity and absorption, limiting its effective range.
3. Electromagnetic Communication: Electromagnetic waves, including radio waves or microwaves, can be used to transmit data underwater. This method offers potential advantages in terms of data rates, but it may have limitations due to the absorption of electromagnetic waves in water.

14.9.14 CHALLENGES IN DATA TRANSMISSION

Data transmission in underwater environments presents several challenges:

1. Attenuation: Both acoustic and optical signals can experience attenuation, where the signals weaken over distance due to water's density and characteristics. This can limit the range of effective communication.
2. Interference and Reflection: Underwater obstacles, seabed structures, and marine life can cause interference and signal reflection, leading to signal distortion or loss.
3. Latency: The time taken for data to travel from sensors to cloud platforms introduces latency. Minimizing latency is crucial for applications that require real-time response.

14.9.15 REAL-TIME VS. BATCH TRANSMISSION

Data transmission can occur in real-time or through batch processing. Real-time transmission enables immediate access to data, allowing for quick responses to dynamic underwater events. Batch transmission involves storing and transmitting data locally in periodic batches, optimizing bandwidth usage and reducing the strain on communication channels.

14.10　CLOUD-IoT APPLICATIONS IN UNDERWATER ENVIRONMENTS

14.10.1　ENVIRONMENTAL MONITORING AND CONSERVATION

Environmental monitoring and conservation represent a cornerstone application of Cloud-IoT integration in underwater communication systems. This section delves into how Cloud-IoT integration empowers the monitoring and preservation of underwater ecosystems, enabling real-time insights, informed decision-making, and sustainable practices to protect the delicate balance of marine environments.

14.10.2　IMPORTANCE OF ENVIRONMENTAL MONITORING

Underwater ecosystems play a vital role in maintaining the planet's health. Monitoring these ecosystems is crucial for understanding changes in water quality, temperature, marine life behavior, and the impact of human activities. Environmental monitoring not only provides insights into the state of marine environments but also offers early warnings of potential disruptions, such as pollution, climate change, and habitat degradation.

14.10.3　REAL-TIME DATA COLLECTION AND ANALYSIS

Cloud-IoT integration revolutionizes environmental monitoring by enabling real-time data collection and analysis. Equipped with sensors and IoT capabilities, underwater platforms collect continuous data streams related to water chemistry, temperature, salinity, currents, and marine life presence. Cloud platforms process this data in real time, providing immediate insights into the conditions and changes occurring in underwater ecosystems.

14.10.4　EARLY DETECTION OF ANOMALIES

Cloud-IoT integration enhances the ability to detect anomalies and deviations from normal environmental conditions. Algorithms and machine learning models analyze incoming data, flagging sudden changes that might indicate pollution events, harmful algal blooms, or abnormal temperature variations. This early detection allows authorities to respond swiftly and take preventive measures to mitigate potential damage.

14.10.5　PRESERVING BIODIVERSITY AND HABITATS

Environmental monitoring via Cloud-IoT integration contributes to preserving marine biodiversity and habitats. Continuous data collection allows scientists to track the migration patterns of marine species, monitor breeding behaviors, and understand the interactions between different species. This information is vital for designing conservation strategies that protect vulnerable species and their habitats.

14.10.6 SUSTAINABLE RESOURCE MANAGEMENT

Cloud-IoT integration aids in sustainable resource management by monitoring the health of fisheries and marine resources. Data on fish populations, migration routes, and habitat conditions help fisheries managers implement sustainable fishing practices that prevent overfishing and maintain the balance of marine ecosystems.

14.10.7 EDUCATIONAL AND OUTREACH OPPORTUNITIES

The insights from environmental monitoring data can be shared with the public, educational institutions, and policymakers. Cloud platforms offer visualizations and interactive tools that simplify complex data, allowing stakeholders to understand the underwater world's intricate dynamics. This awareness drives advocacy for marine conservation and fosters a deeper appreciation for the importance of preserving underwater ecosystems.

14.11 UNDERWATER SURVEILLANCE AND SECURITY

14.11.1 IMPORTANCE OF UNDERWATER SURVEILLANCE AND SECURITY

The vast and remote nature of underwater environments poses challenges to monitoring and ensuring security. Threats can arise from illegal fishing, unauthorized access to sensitive areas, maritime piracy, and potential security breaches near critical infrastructure. Underwater surveillance and security are pivotal in preventing and mitigating these risks.

14.11.2 CONTINUOUS MONITORING AND DATA FUSION

Cloud-IoT integration enables continuous monitoring of underwater areas using an array of sensors, cameras, and detectors placed on underwater vehicles, buoys, and fixed platforms. These sensors collect data on vessel movement, diver activity, water quality, and potential anomalies. The cloud serves as a central hub where data from multiple sources is fused, providing a comprehensive picture of the underwater environment.

14.11.3 REAL-TIME THREAT DETECTION

By integrating data from various sensors and platforms, cloud platforms facilitate real-time threat detection. Algorithms analyze incoming data, identifying unusual behaviors, suspicious vessels, or unauthorized access. This enables prompt responses to potential security breaches, deterring illicit activities and safeguarding sensitive marine areas.

14.11.4 CRITICAL INFRASTRUCTURE PROTECTION

Underwater surveillance extends to the protection of critical infrastructure such as underwater cables, pipelines, and energy installations. Cloud-IoT integration ensures that data from sensors placed near these assets is transmitted in real time, enabling early detection of potential threats such as anchor damage, corrosion, or tampering.

14.11.5 EMERGENCY RESPONSE AND CRISIS MANAGEMENT

Cloud-IoT integration enhances emergency response and crisis management in underwater environments. In the event of an accident, oil spill, or security breach, real-time data collection and transmission enable rapid deployment of response teams. Cloud platforms facilitate coordination, data sharing, and communication among stakeholders, streamlining crisis management efforts.

14.11.6 NATIONAL SECURITY AND DEFENSE

For military and defense purposes, underwater surveillance and security play a vital role in protecting national interests. Cloud-IoT integration enhances the Navy's ability to monitor underwater activities, identify potential threats, and safeguard maritime boundaries. This integration improves naval intelligence, anti-submarine warfare capabilities, and strategic awareness.

14.12 CASE STUDY

14.12.1 REAL-TIME TSUNAMI DETECTION AND ALERT SYSTEM

14.12.1.1 Implementation of the System

Deployment of Sensor Network: To establish an effective real-time tsunami detection and alert system, a network of sensors is strategically positioned in oceanic zones prone to seismic activity. These sensors include seismometers to detect earthquakes, tide gauges to measure water level changes, and underwater pressure sensors to monitor seafloor movements. Each sensor is equipped with IoT capabilities for data transmission.

Integration with Cloud Platforms: The collected sensor data is transmitted to cloud platforms through wireless communication. These cloud platforms serve as centralized hubs for data aggregation, analysis, and alert generation. The integration of cloud computing technologies ensures seamless data processing and real-time insights.

14.12.1.2 Early Warning and Alert Generation

Data Analysis and Event Identification: Cloud-based algorithms process incoming data streams to identify seismic events that might trigger tsunamis. By analyzing seismic wave patterns, water level changes, and pressure fluctuations, the system distinguishes normal activities from potential tsunami-inducing events.

Threshold Detection and Alert Triggering: The system establishes predefined thresholds for different parameters. An alert is triggered if any threshold is crossed due to unusual data patterns. This initiates a series of actions for further analysis and dissemination of warnings.

14.12.1.3 Real-time Analysis and Impact Prediction

Modeling and Prediction: Upon detecting a potential tsunami-triggering event, the cloud platform employs oceanographic models and historical data to predict the tsunami's behavior. Parameters like wave height, propagation speed, and potential inundation areas are estimated, providing crucial information for assessing the threat's severity.

14.12.1.4 Communication and Response

Alert Dissemination: The system rapidly communicates alerts to coastal communities and relevant authorities through various channels such as mobile apps, text messages, sirens, and emergency broadcasts. This ensures that residents receive timely warnings and can take immediate action to evacuate.

Emergency Response Coordination: Emergency management agencies receive real-time data from the cloud platform. This data informs their decision-making, allowing them to allocate resources efficiently, organize evacuations, and coordinate first responders based on the predicted impact of the tsunami.

14.12.1.5 Community Engagement and Preparedness

Educational Initiatives: The cloud platform also hosts educational resources, maps, and interactive tools to raise public awareness about tsunamis and disaster preparedness. Regular drills and public outreach campaigns promote a culture of readiness, ensuring that communities are well-informed and equipped to respond effectively.

14.12.1.6 Impact and Significance

The implementation of a real-time tsunami detection and alert system powered by Cloud-IoT integration holds profound implications. By providing early warnings, accurate impact predictions, and efficient emergency response coordination, this system can save countless lives, prevent damage to infrastructure, and minimize chaos during a disaster. It exemplifies how technology, when integrated intelligently, can bridge the gap between natural hazards and human safety, demonstrating the immense potential of Cloud-IoT integration in disaster management scenarios.

14.13 FUTURE TRENDS AND CHALLENGES

14.13.1 Advancements in Underwater Communication Technologies

Advancements in underwater communication technologies have revolutionized our ability to gather, transmit, and exchange information beneath the waves. This section explores the remarkable strides made in underwater communication, from overcoming inherent challenges to enabling real-time data transfer and collaborative exploration of the underwater world.

14.13.2 TRADITIONAL CHALLENGES IN UNDERWATER COMMUNICATION

Underwater communication has historically grappled with challenges such as signal attenuation, limited bandwidth, and interference caused by water's characteristics. The high density and conductivity of water can significantly weaken signals, making long-distance communication difficult. Additionally, the underwater environment introduces acoustic noise and scattering that further degrade signal quality.

14.13.3 ACOUSTIC COMMUNICATION ADVANCEMENTS

1. Multi-Path Communication: Advancements in acoustic communication involve mitigating the effects of signal attenuation and scattering. Techniques like multi-path communication leverage reflections from the seafloor and surface to create multiple signal paths, enhancing reliability and coverage.
2. Beamforming: Beamforming technology focuses acoustic signals in specific directions, increasing signal strength and reducing interference. This enables targeted communication with underwater platforms and vehicles.
3. Underwater Modems: Modern underwater modems employ advanced signal processing algorithms and error correction mechanisms to overcome the challenges of noise and signal degradation. These modems offer improved data rates, allowing faster data transmission between underwater devices and surface stations.

14.13.4 OPTICAL COMMUNICATION ADVANCEMENTS

1. High-Speed Data Transfer: Optical communication has witnessed significant advancements in terms of data rates. Laser-based optical communication offers much higher bandwidth compared to acoustic methods, enabling real-time streaming of high-definition video and large datasets.
2. Turbidity Compensation: To address the sensitivity of optical signals to water turbidity, researchers have developed techniques that adaptively adjust the power and wavelength of the transmitted light. This compensates for variations in water clarity, enhancing the reliability of optical communication.
3. Hybrid Systems: Hybrid communication systems integrate both acoustic and optical technologies to harness their complementary strengths. Acoustic communication offers robustness over longer distances, while optical communication excels in high-bandwidth, short-range scenarios.

14.13.5 ELECTROMAGNETIC COMMUNICATION ADVANCEMENTS

1. Submarine Communication Cables: Electromagnetic communication has evolved significantly with the deployment of submarine communication cables. These cables enable high-speed data transfer between continents, supporting global data networks and enabling real-time data transmission from remote underwater sensors.

2. AUVs: AUVs equipped with electromagnetic communication capabilities can relay data to surface vessels or satellites, allowing researchers to collect data remotely from underwater environments without physical connections.

14.13.6 Integration with Cloud and IoT Technologies

Advancements in cloud computing and IoT technologies have seamlessly integrated with underwater communication systems. Cloud-IoT integration enables real-time data transmission, remote control of underwater platforms, and centralized data storage and analysis. This integration empowers researchers, scientists, and stakeholders to monitor, analyze, and respond to underwater data in unprecedented ways.

14.14 EDGE COMPUTING IN UNDERWATER ENVIRONMENTS

Edge computing has emerged as a transformative paradigm in underwater communication systems, enabling data processing, analysis, and decision-making at the source of data collection. This section delves into the application of edge computing in underwater environments, highlighting its benefits, challenges, and role in overcoming the limitations of traditional centralized processing.

14.14.1 Understanding Edge Computing

Edge computing shifts the traditional data processing paradigm from centralized cloud servers to the "edge" of the network, closer to where data is generated. In underwater environments, this means processing data on underwater platforms, vehicles, or buoys before transmitting it to the cloud. This approach reduces latency, conserves bandwidth, and enhances real-time decision-making.

14.14.2 Benefits of Edge Computing in Underwater Environments

1. Latency Reduction: Edge computing drastically reduces the time it takes to process and analyze data. In applications such as underwater surveillance or real-time environmental monitoring, reducing latency is crucial for quick response and timely decision-making.
2. Bandwidth Efficiency: Transmitting large volumes of raw sensor data to the cloud can strain communication channels. With edge computing, only relevant, pre-processed data or summarized insights are sent to the cloud, optimizing bandwidth usage.
3. Real-Time Analytics: Edge devices can analyze data on the spot, providing real-time insights without waiting for data to reach the cloud. This is particularly valuable in applications requiring immediate action, such as detecting anomalies or triggering emergency alerts.
4. Resilience: In underwater environments with intermittent connectivity, edge devices can continue processing data even when communication links are temporarily disrupted. This ensures that data is not lost and critical operations continue uninterrupted.

14.14.3 CHALLENGES OF EDGE COMPUTING IN UNDERWATER ENVIRONMENTS

1. Limited Resources: Underwater platforms often have limited processing power, memory, and energy resources. Implementing advanced analytics on resource-constrained devices can be challenging.
2. Data Variability: Underwater data can be highly variable and complex, requiring adaptive algorithms and models that can handle changing conditions and sensor inputs.
3. Scalability: Deploying edge devices across a network of underwater platforms requires careful management and scalability planning to ensure consistent performance and synchronization.
4. Data Security: Data processed at the edge could potentially be exposed to physical risks, necessitating robust security measures to protect against tampering and unauthorized access.

14.14.4 APPLICATIONS OF EDGE COMPUTING IN UNDERWATER ENVIRONMENTS

1. Real-Time Decision Support: Edge computing enables underwater vehicles to analyze environmental data on-site and make decisions based on predefined criteria, such as adjusting mission parameters or redirecting to specific areas of interest.
2. Autonomous Navigation: Underwater robots equipped with edge devices can process sensor data to navigate safely through dynamic underwater environments, avoiding obstacles and adapting to changing conditions.
3. Environmental Monitoring: Edge devices can detect anomalies in underwater ecosystems, trigger local alerts, and initiate data transmission to the cloud for further analysis.
4. Disaster Response: In the event of a sudden underwater event like an oil spill, edge devices on buoys can analyze water quality data and send immediate alerts to response teams, accelerating the deployment of cleanup efforts.

14.15 SECURITY AND PRIVACY CONCERNS

1. Data Breaches: One of the most prominent threats is data breaches occur when unauthorized parties gain access to sensitive information. This can result in stolen data, financial loss, reputational damage, and legal ramifications.
2. Unauthorized Access: Weak authentication mechanisms or compromised credentials can lead to unauthorized access to cloud resources and IoT devices. Malicious actors could exploit this to manipulate data, launch attacks, or steal valuable information.
3. Malware and Cyberattacks: Cloud platforms and IoT devices are susceptible to various cyberattacks, including malware, ransomware, Distributed Denial of Service (DDoS) attacks, and phishing attempts. These attacks can disrupt operations, compromise data, and expose vulnerabilities.

4. Privacy Breaches: The collection and processing of vast amounts of data in Cloud-IoT systems raise concerns about user privacy. Inadequate data anonymization or unintentional data sharing could expose sensitive personal or corporate information.
5. Vendor Lock-in: Depending heavily on a specific cloud provider can result in vendor lock-in, limiting flexibility and making it challenging to migrate to other platforms without disruptions.

14.15.1 PRIVACY CONCERNS IN CLOUD-IoT INTEGRATION

1. Data Collection and Profiling: The continuous data collection by IoT devices can lead to extensive profiling of individuals or entities. This raises concerns about personal data protection and potential misuse or unauthorized surveillance.
2. Data Sharing and Consent: Sharing data among cloud services or IoT devices without explicit user consent can lead to the unintended exposure of personal information. Proper consent mechanisms should be in place to ensure compliance with privacy regulations.
3. IoT Device Insecurity: Inadequately secured IoT devices could serve as entry points for attackers to access sensitive data or gain control over connected systems. Unsecured devices might be vulnerable to remote manipulation, endangering user privacy.

14.16 CONCLUSION

Summary of Key Points

1. Introduction:
   Cloud-IoT integration revolutionizes underwater communication by enabling real-time data collection, analysis, and decision-making in challenging aquatic environments.
2. Challenges of Underwater Communication:
   - Underwater communication faces obstacles such as signal attenuation, interference, and limited bandwidth due to the unique properties of water.
3. Cloud Computing in IoT:
   - Cloud platforms provide the computational power and storage needed for processing and analyzing vast volumes of underwater data.
   - IoT devices collect, transmit, and receive data, creating a network that interacts with cloud platforms for seamless integration.
4. Cloud-IoT Integration for Underwater Communication:
   - Cloud-IoT integration bridges the gap between data collection in underwater environments and data analysis in the cloud.
   - It offers real-time insights, enhanced decision-making, and remote control of underwater platforms.

5. Applications:
   - Environmental Monitoring and Conservation: Cloud-IoT integration enables real-time monitoring of marine ecosystems, aiding in conservation efforts and early anomaly detection.
   - Underwater Surveillance and Security: Cloud-IoT integration enhances situational awareness, threat detection, and emergency response in underwater security scenarios.
   - Real-time Tsunami Detection and Alert System: A case study demonstrates how Cloud-IoT integration can save lives by providing early warnings and real-time analytics for tsunami detection.
6. Advancements in Underwater Communication Technologies:
   - Acoustic, optical, and electromagnetic communication technologies have evolved to overcome challenges and increase data rates in underwater environments.
   - Cloud-IoT integration enhances the capabilities of these technologies for real-time data transfer.
7. Edge Computing in Underwater Environments:
   - Edge computing processes data closer to its source, reducing latency and conserving bandwidth in underwater communication systems.
   - It enables real-time decision-making, autonomous navigation, and disaster response.
8. Security and Privacy Concerns:
   - Security concerns include data breaches, unauthorized access, malware attacks, and Denial of Service (DoS) attacks.
   - Privacy concerns involve data collection, profiling, sharing, and consent.
   - Mitigation strategies include encryption, access control, regular monitoring, privacy by design, and compliance with regulations.

14.17 IMPORTANCE OF CLOUD-IoT INTEGRATION IN UNDERWATER COMMUNICATION

Cloud-IoT integration holds immense significance in the domain of underwater communication, revolutionizing how we explore, monitor, and conserve the mysterious depths of our oceans. This integration addresses longstanding challenges while unlocking a range of opportunities that benefit marine science, environmental monitoring, disaster management, security, and more. Here's a detailed exploration of the importance of Cloud-IoT integration in underwater communication:

14.17.1 REAL-TIME DATA COLLECTION AND ANALYSIS

Cloud-IoT integration enables the continuous and real-time data collection from various underwater sensors, vehicles, and platforms. This data encompasses oceanography, marine life behavior, water quality, and seismic activity. The cloud's computational power and storage capabilities allow instant analysis, transforming raw data into actionable insights.

14.17.2   Improved Decision-Making

By processing data in real time, Cloud-IoT integration empowers stakeholders with timely and accurate information. This enables informed decision-making in various scenarios, such as adjusting the trajectory of AUVs, responding to anomalies, or initiating emergency protocols in the event of a natural disaster.

14.17.3   Enhanced Environmental Monitoring

Cloud-IoT integration revolutionizes environmental monitoring by providing a comprehensive overview of underwater ecosystems. Scientists and researchers can track changes in water temperature, salinity, currents, and marine life behavior. This information is crucial for understanding climate change impacts, biodiversity shifts, and habitat health.

14.17.4   Disaster Preparedness and Response

In disaster-prone underwater regions, Cloud-IoT integration is pivotal in early warning systems. Whether detecting tsunamis, oil spills, or underwater seismic events, the integration allows for rapid data analysis, enabling authorities to issue alerts, mobilize resources, and promptly coordinate evacuation efforts.

14.17.5   Underwater Surveillance and Security

Cloud-IoT integration enhances underwater surveillance and security by offering real-time insights into underwater activities. By processing data from sonars, cameras, and other sensors on underwater platforms, security agencies can detect unauthorized vessels, potential threats, and even underwater trespassing in restricted areas.

14.17.6   Remote Monitoring and Control

Underwater platforms, vehicles, and buoys equipped with IoT devices can be remotely controlled and monitored through cloud platforms. This allows researchers, scientists, and engineers to navigate underwater vehicles, adjust sensor configurations, and conduct experiments without physically being present at the underwater location.

14.17.7   Collaboration and Data Sharing

Cloud-IoT integration fosters collaboration among researchers, institutions, and stakeholders. Data collected from different underwater locations can be aggregated, analyzed, and shared in real time, enabling a global network of experts to collectively contribute to scientific endeavors and policy decisions.

14.17.8 SCIENTIFIC DISCOVERIES AND INNOVATION

The real-time insights enabled by Cloud-IoT integration facilitate scientific discoveries and innovation. Researchers can observe marine life behavior, underwater geological formations, and other phenomena in unprecedented detail, contributing to a deeper understanding of our oceans and their interconnected ecosystems.

14.17.9 ENVIRONMENTAL CONSERVATION

By providing accurate and up-to-date information on environmental conditions, Cloud-IoT integration aids in marine conservation efforts. Organizations and governments can use this data to develop policies, regulations, and conservation strategies that protect delicate marine ecosystems and promote sustainable practices.

14.18 FUTURE PROSPECTS AND RESEARCH DIRECTIONS

14.18.1 AUTONOMOUS UNDERWATER SYSTEMS

Future research will focus on developing more sophisticated AUVs and underwater drones. These vehicles will be equipped with advanced sensing, processing, and communication capabilities to navigate and interact intelligently with their surroundings. The integration of AI and machine learning will enable AUVs to make autonomous decisions, adapt to changing conditions, and optimize data collection strategies.

14.18.2 HYBRID COMMUNICATION SOLUTIONS

Researchers will continue to explore hybrid communication solutions that combine the strengths of different communication technologies. By integrating acoustic, optical, and electromagnetic communication methods, underwater systems can achieve high data rates, extended range, and robustness in varying underwater conditions.

14.18.3 ENERGY HARVESTING AND EFFICIENCY

Incorporating energy harvesting techniques will be a critical area of focus. Underwater sensors and platforms can harness energy from ocean currents, temperature differentials, or solar radiation at the water's surface. Improving energy efficiency will extend the operational lifetimes of underwater systems and reduce the need for frequent maintenance.

14.18.4 BIG DATA AND AI IN UNDERWATER ANALYSIS

The growth of data collected from underwater sensors calls for advanced data analytics techniques. Researchers will leverage big data analytics and AI algorithms to process, model, and derive meaningful insights from the massive volumes of underwater data. This will lead to more accurate predictions, improved anomaly detection, and a deeper understanding of underwater ecosystems.

14.18.5 Underwater Energy and Resource Management

As the demand for renewable energy sources grows, underwater communication systems could play a pivotal role in managing underwater energy infrastructure, such as offshore wind farms, tidal energy installations, and underwater power cables. Research in this area will focus on optimizing energy distribution, maintenance scheduling, and resource allocation.

14.18.6 Privacy-Preserving Data Sharing

To address privacy concerns, future research will explore innovative techniques for sharing underwater data while preserving the privacy of individuals and sensitive locations. Techniques like differential privacy and secure multi-party computation could enable collaborative data analysis without compromising privacy.

14.18.7 Quantum Communication

Quantum communication, with its promise of secure and high-speed transmission, could find applications in underwater communication systems. Research in this area will investigate the feasibility of implementing quantum communication techniques in the challenging underwater environment.

14.18.8 Climate Change Monitoring and Adaptation

Cloud-IoT integration will play a crucial role in monitoring and adapting to the impacts of climate change on underwater ecosystems. Research will focus on developing systems that can monitor changing ocean temperatures, acidity levels, and their effects on marine life in real time.

14.18.9 International Collaboration and Regulation

Future prospects will involve international collaboration to standardize underwater communication protocols, data sharing mechanisms, and security standards. International regulations will be crucial to ensure the responsible use of underwater communication systems and the protection of sensitive marine areas.

REFERENCES

1. Kaur, A., & Kinger, S. (2013). Temperature aware resource scheduling in Green Clouds. In International Conference on Advances in Computing Communications and Informatics IEEE, pp. 1919–1923.
2. Shehabi, A., Masanet, E., Price, H. et al. (2011). Data center design and location: Consequences for electricity use and greenhouse-gas emissions. Building & Environment, 46(5), 990–998.
3. Dillon, T., Wu, C., & Chang, E. (2010). 24th IEEE international conference on advanced information networking and applications. In Cloud Computing: Issues and Challenges.
4. Shen, G., Lee, Z. E., Amadeh, A., & Zhang, K. M. (2021). A data-driven electric water heater scheduling and control system. Energy Build, 242, 110924.

5. Melville-Shreeve, P., Cotterill, S., & Butler, D. (2021). Capturing high-resolution water demand data in commercial buildings. Journal of Hydroinformatics, 23, 402–416.

6. Kartakis, S., Yu, W., Akhavan, R., & McCann, J. A. (2016). Adaptive Edge Analytics for Distributed Networked Control of Water Systems. In Proceedings of the 1st IEEE International Conference on Internet-of-Things Design and Implementation. 4–8 April 2016, Berlin, German.

7. Zamora-Izquierdo, M. A., Sant, J., Martínez, J. A., Martínez, V., & Skarmeta, A. F. (2019). Smart farming IoT platform based on edge and cloud computing. Biosystems Engineering, 177, 4–17. doi: 10.1016/j.biosystemseng.2018.10.014

8. Yang, C., & Wang, X. (24–26 June 2011). The water quality and pollution character in Qingshuihai Lake valley-typical urban drinking water sources. In Proceedings of the 2011 International Conference on Remote Sensing, Environment and Transportation Engineering, Nanjing, China, pp. 7287–7291.

9. Jing, W., & Liu, T. (23–25 May 2015). Application of wireless sensor network in Yangtze River basin water environment monitoring. In Proceedings of the 2015 27th Chinese Control and Decision Conference, CCDC, Qingdao, China, pp. 5981–5985.

10. Raju, K. R. S. R., & Varma, G. H. K. (5–7 January 2017). Knowledge based real time monitoring system for aquaculture Using IoT. In Proceedings of the 7th IEEE International Advanced Computing Conference, IACC, Hyderabad, India, pp. 318–321.

11. Xu, Y., & Helal, A. (2015). Scalable cloud–sensor architecture for the internet of things. IEEE Internet of Things Journal, 3, 285–298. doi: 10.1109/JIOT.2015.2455555

12. Barabde, M., & Danve, S. (June 2015). Real time water quality monitoring system. IJIRCCE, 3, 5064–5069.

13. Robles, T., Alcarria, R., Martín, D., & Morales, A. (May 2014). An Internet of Things-based model for smart water management. In Proc. of the 8th International Conference on Advanced Information Networking and Applications Workshops (WAINA'14), IEEE, Victoria, Canada, pp. 821–826.

14. Miorandi, D., Sicari, S., Pellegrini, F. D., & Chlamtac, I. (September 2012). Internet of things: Vision, applications and research challenges. Ad Hoc Networks, 10(7), 1497–1516.

15. Farooq, M. U., Waseem, M., Mazhar, S., Khairi, A., & Kamal, T. (March 2015). A review on internet of things (IoT). International Journal of Computer Applications, 113(1).

16. Alshattnawi, S. K. (11 Jan. 2018). Smart water distribution management system architecture based on internet of things and cloud computing. In IEEE International Conference on New Trends in Computing Sciences, pp. 289–294.

17. Lambrou, T. P., Anastasiou, C. C., Panayiotou, C. G., and Polycarpou, M. M. (2014). Cost sensor network for real-time monitoring and contamination detection in drinking water distribution systems. IEEE Sensors Journal, 14(8), 2765–2772.

18. Pacheco, J., Ibarra, D., Vijay, A., & Hariri, S. (12 Mar. 2018). IoT security framework for smart water system. In Proc. of IEEE International Conference on Computer Systems and Applications, Tunisia.

19. Ramesh, M. V., Nibi, K. V., Kurup, A., Mohan, R., Aiswarya, A., Arsha, A., & Sarang, P. R. (25 Dec. 2017). Water quality monitoring and waste management using IoT. In IEEE Global Humanitarian Technology Conference, USA.

20. Salam, A., Vuran, M. C., & Irmak, S. (Apr. 2016). Pulses in the sand: Impulse response analysis of wireless underground channel. In The 35th Annual IEEE International Conference on Computer Communications (INFOCOM 2016), San Fran-Cisco, USA.

21. Salam, A., Vuran, M. C., Dong, X., Argyropoulos, C., & Irmak, S. (2019). A theoretical model of underground dipole antennas for communications in internet of under-ground things. IEEE Transactions on Antennas and Propagation, 67, 3996–4009.

22. M. C. Vuran, A. Salam, R. Wong, & S. Irmak. (Feb. 2018). Internet of underground things: Sensing and communications on the field for precision agriculture. In 2018 IEEE 4th World Forum on Internet of Things (WF-IoT) (WF-IoT 2018), Singapore.
23. Sun, Z., Wang, P., Vuran, M. C., Al-Rodhaan, M. A., Al-Dhelaan, A. M., & Akyildiz, I. F. (2011). MISE-PIPE: Magnetic induction-based wireless sensor networks for underground pipeline monitoring. Ad Hoc Networks, 9(3), 218–227.
24. Koo, D., Piratla, K., & Matthews, C. J. (2015). Towards sustainable water supply: Schematic development of big data collection using internet of things (IoT). Procedia Engineering, 118, 489–497.
25. Hoekstra, A. Y., Buurman, J., & van Ginkel, K. C. (2018). Urban water security: A review. Environmental Research Letters, 13(5), 053002.
26. Bogena, H. R., Herbst, M., Huisman, J. A., Rosenbaum, U., Weuthen, A., & Vereecken, H. (2010). Potential of wireless sensor networks for measuring soil water content variability. Vadose Zone Journal, 9(4), 1002–1013.
27. Bakker, K. (2012). Water security: Research challenges and opportunities. Science, 337(6097), 914–915.
28. Adamchuk, V., Hummel, J., Morgan, M., & Upadhyaya, S. (2004). On-the-go soil sensors for precision agriculture. Computers and Electronics in Agriculture, 44(1), 71–91.
29. Koo, D., Piratla, K., & Matthews, C. J. (2015). Towards sustainable water supply: Schematic development of big data collection using internet of things (IoT). Procedia Engineering, 118, 489–497.
30. Ross, I., McDonough, J., Miles, J., Storch, P., Thelakkat Kochunarayanan, P., Kalve, E., et al. (2018). A review of emerging technologies for remediation of PFASs. Remediation Journal, 28(2), 101–126.

15 Connecting the Future
Cloud-Based IoT in Education

Abhiraj Gautam and Arnav Kotiyal

15.1 INTRODUCTION

15.1.1 A Cloud-Based Internet of Things (IoT) and Its Significance

The concept of the Internet of Things (IoT) has rapidly evolved, presenting a paradigm shift in the way we interact with technology and the environment around us. The integration of IoT with cloud computing, known as cloud-based IoT, has further expanded the horizons of possibilities, revolutionizing industries across the board. This section introduces the foundational concepts of cloud-based IoT and highlights its profound significance in transforming the way we perceive and interact with the world.

15.1.1.1 Understanding Cloud-Based IoT

At its core, IoT encompasses the interconnection of everyday objects, devices, and sensors, enabling them to communicate, collect and exchange data and execute actions autonomously or based on user commands. This capability has brought forth a plethora of applications, ranging from smart homes to industrial automation. However, the exponential growth of data generated by these connected devices necessitates scalable and robust data processing solutions. This is where cloud-based IoT enters the scene.

Cloud-based IoT involves the convergence of IoT devices and cloud computing infrastructure. Instead of processing and storing data solely on local devices, data is transmitted to cloud servers, where it can be stored, analyzed, and accessed from virtually anywhere. This amalgamation offers numerous advantages, including enhanced computational capabilities, data scalability, and the potential for real-time data analysis. Cloud services provide a flexible platform for IoT devices to communicate seamlessly, creating a cohesive ecosystem that extends beyond individual devices' capabilities (Figure 15.1).

15.1.1.2 Significance in Education

The application of cloud-based IoT in the education sector holds immense promise for transforming traditional educational paradigms. Smart classrooms, for instance, exemplify the impact of IoT in education. In these environments, cloud-connected devices, such as interactive whiteboards, wearable devices, and sensors, create an immersive learning experience. Real-time data collection and analysis enable educators to tailor instruction to individual student needs, promoting personalized learning journeys.

DOI: 10.1201/9781032656694-15

FIGURE 15.1   Cloud computing.

Furthermore, administrative operations within educational institutions benefit from cloud-based IoT implementation. Campus management becomes more efficient through intelligent energy management systems and asset tracking. Automated processes, such as attendance tracking and resource allocation, streamline administrative tasks, allowing educators to allocate more time to impactful teaching.

Cloud-based IoT bridges the gap between physical and virtual learning spaces. Virtual labs and online collaboration platforms enable students to engage in hands-on experiments and teamwork, transcending geographical barriers. This technology democratizes education, making quality resources accessible to a broader audience [1–4].

15.1.2 How Has IoT Transformed Various Industries

The IoT has emerged as a revolutionary technological paradigm that has been transforming a multitude of industries, reshaping the way we interact with devices, data, and the environment. One of the sectors profoundly impacted by this transformation is education. By seamlessly integrating physical devices, sensors, connectivity, and cloud-based platforms, IoT has ushered in a new era of personalized and data-driven learning experiences.

IoT has paved the way for increased efficiency, enhanced decision-making, and improved user experiences in various industries. For instance, in agriculture, IoT-enabled sensors gather data on soil moisture, temperature, and crop health, empowering farmers to optimize irrigation, reduce resource wastage, and boost yield. In healthcare, wearable IoT devices monitor vital signs and provide real-time health insights, enabling proactive patient care and remote monitoring (Figure 15.2).

Similarly, IoT facilitates predictive maintenance in manufacturing by analyzing equipment data, thereby minimizing downtime and maximizing productivity.

The education sector is no exception to this transformative wave. IoT has shifted from traditional classroom models to digitally enriched, interactive, and adaptable learning environments. Smart classrooms equipped with IoT-enabled devices, such as interactive whiteboards, digital textbooks, and intelligent lighting, offer engaging

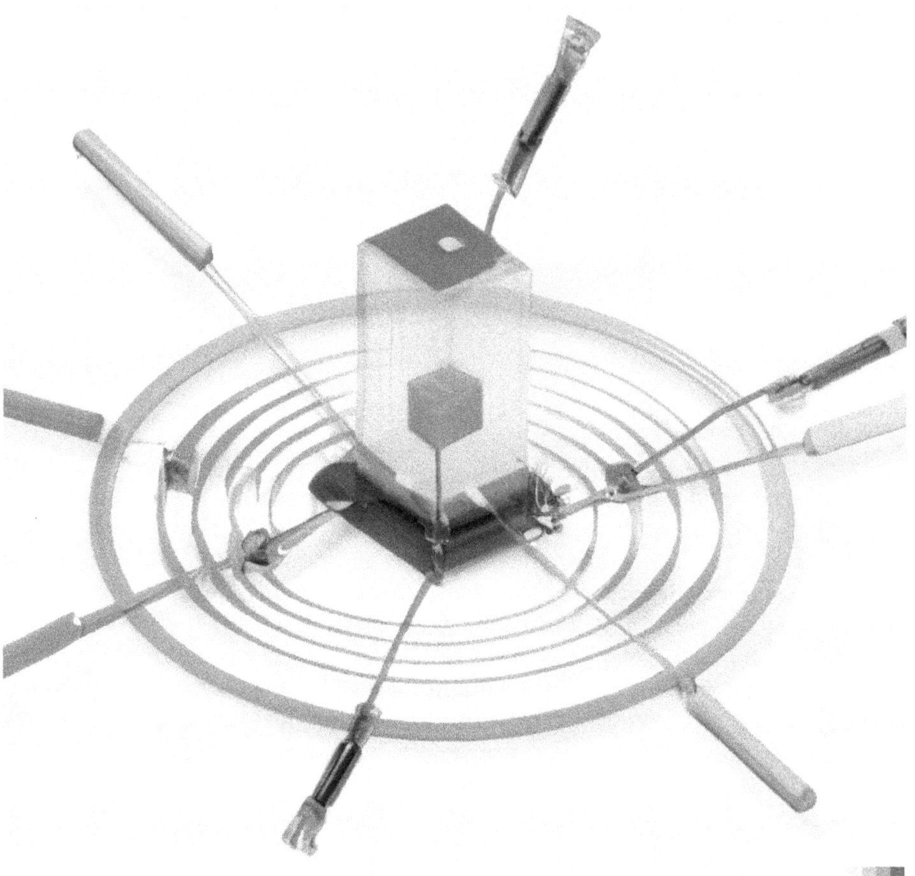

FIGURE 15.2   Internet of things (IoT).

and tailored educational experiences. Real-time data collection and analysis enable educators to track student progress, identify learning gaps, and adjust teaching strategies accordingly.

IoT's impact on education extends beyond the classroom walls. Campus management has been streamlined through connected systems that manage facilities, security, and resource allocation. Smart campuses can optimize energy consumption, enhance security through surveillance systems, and automate administrative tasks like attendance tracking. By reducing manual intervention and enabling efficient resource allocation, educational institutions can allocate more time and resources to core academic activities.

In the context of personalized learning, IoT empowers educators to create adaptive curricula based on individual student needs and learning patterns. Wearable devices and smart applications monitor student engagement and provide insights into their learning preferences. This data helps tailor learning materials, pace, and approaches, fostering a more effective and engaging learning experience.

However, the adoption of IoT in education comes with challenges that need to be addressed. Privacy concerns, data security, and ethical considerations must be carefully managed to ensure the safety and confidentiality of student information. Moreover, ensuring equitable access to IoT devices and technologies is crucial to prevent the creation of a digital divide among students from different socioeconomic backgrounds [1, 5–8].

15.2 UNDERSTANDING CLOUD-BASED IoT

15.2.1 CLOUD-BASED IoT AND ITS KEY COMPONENTS

Cloud-based IoT represents a paradigm where IoT devices and sensors are interconnected through cloud computing infrastructure, enabling seamless data exchange, storage, analysis, and communication. This fusion of IoT and cloud technology has revolutionized industries, including education, by providing a scalable, accessible, and efficient platform for collecting, processing, and utilizing vast amounts of data (Figure 15.3).

15.2.1.1 Key Components of Cloud-Based IoT

IoT Devices and Sensors: These are the physical components that capture data from the surrounding environment. They include a wide range of devices such as sensors, actuators, RFID tags, and wearable gadgets. In the context of the education sector, these devices can be embedded in classrooms, campuses, and educational tools to collect real-time data on factors like temperature, occupancy, student engagement, and more.

- **Connectivity:** IoT devices need to be connected to transmit data to the cloud. This connectivity can be achieved through various means such as Wi-Fi, Bluetooth, cellular networks, and Low-Power Wide-Area Networks (LPWAN). Reliable connectivity ensures that data is seamlessly transmitted to the cloud for further processing and analysis.

FIGURE 15.3 Cloud-based clients.

- **Cloud Computing Infrastructure:** The heart of cloud-based IoT is the cloud computing infrastructure. Cloud platforms offer diverse services, such as data storage, processing, analytics, and visualization. By leveraging the cloud, educational institutions can efficiently store and manage data without the need for extensive on-site hardware and infrastructure.
- **Data Storage and Processing:** The cloud serves as a central repository for the enormous volumes of data generated by IoT devices. Cloud-based storage solutions allow educational institutions to store historical data, enabling longitudinal analysis and trend identification. Furthermore, cloud-based data processing tools and algorithms can be applied to extract valuable insights from raw data, facilitating informed decision-making.
- **Security and Privacy:** As data privacy concerns grow, robust security mechanisms are paramount in cloud-based IoT. Encryption, authentication, and access control mechanisms help safeguard sensitive educational data. By adhering to best practices, educational institutions can ensure that student and staff information remains confidential [1, 2, 9–11].

15.2.2 THE ROLE OF SENSORS, DEVICES, AND CONNECTIVITY IN THE CONTEXT OF IoT

In the realm of the IoT, the seamless interconnectivity of devices and the intelligent exchange of data have catalyzed a paradigm shift across industries. Central to this transformation are the crucial components of sensors, devices, and connectivity, collectively forming the foundation of IoT ecosystems. This section delves into their pivotal roles and synergies within the context of IoT, illuminating how they contribute to creating smart, data-driven environments.

Sensors: At the heart of IoT lie sensors, which act as the sensory organs of this interconnected universe. Sensors are devices designed to perceive and measure physical properties like temperature, humidity, light, motion, and more. These devices capture real-world data and convert it into digital information, making the physical world accessible to digital systems. For example, temperature sensors placed in a classroom can gather data on ambient conditions, enabling the automation of heating and cooling systems for optimal comfort. The variety of sensors available today empowers IoT applications to collect a diverse range of data, driving insights and enabling informed decision-making (Figure 15.4).

Devices: Devices, often called "smart devices," encompass many endpoints that connect to the IoT network. These devices can include anything from wearable fitness trackers to intelligent appliances, such as smart thermostats and voice-controlled assistants. Devices are equipped with processing power and communication capabilities, allowing them to interact with both users and other devices. In education, smart devices could include interactive whiteboards, e-books, and student ID cards embedded with RFID technology for efficient campus access. These devices facilitate

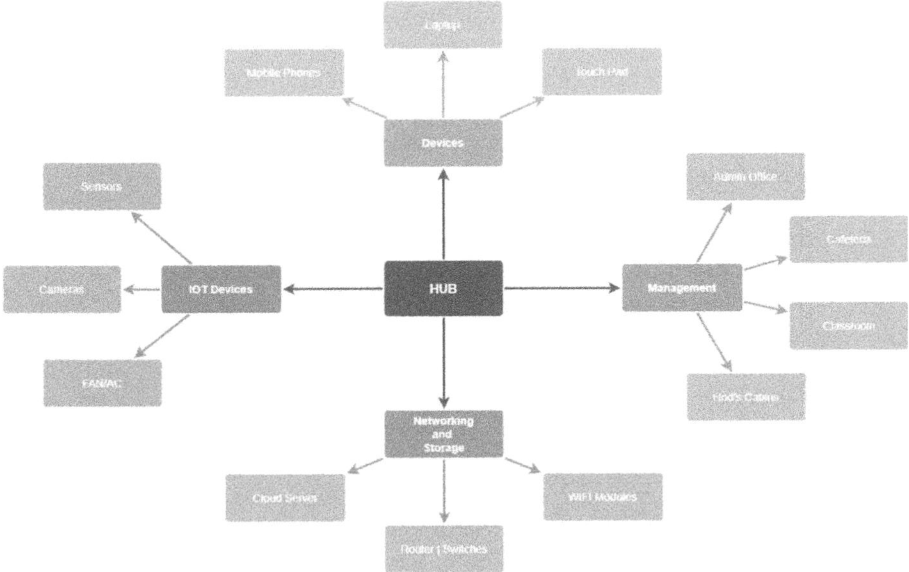

FIGURE 15.4   Overall connection.

communication, data exchange, and remote control, enhancing user experiences and efficiency.

Connectivity: The true magic of IoT comes to life through connectivity—how devices and sensors communicate and share data. Connectivity technologies enable the formation of intricate networks that span across local environments or even the globe. A myriad of connectivity options exist, including Wi-Fi, Bluetooth, Zigbee, cellular networks, and more. In an educational setting, the connectivity of devices and sensors facilitates the creation of a networked classroom where teachers can share digital content with students' tablets and students can collaborate seamlessly using connected devices. Connectivity forms the arteries of the IoT ecosystem, ensuring a continuous flow of data, commands, and responses.

The synergy of sensors, devices, and connectivity within IoT systems has unleashed unprecedented possibilities across various domains, including healthcare, manufacturing, agriculture, and education. These components create a dynamic loop where sensors capture real-world information, devices process and interpret this data, and connectivity enables its transmission and utilization [1, 12–15].

15.2.3   The Importance of Cloud Computing in Managing and Analyzing Data Generated by IoT Devices

The proliferation of IoT devices has led to an exponential increase in data generation across various industries, including education. The education sector has embraced

IoT technologies to enhance learning experiences, optimize resource allocation, and streamline administrative processes. However, the sheer volume and complexity of data produced by these devices present a significant challenge in terms of storage, processing, and analysis. This is where cloud computing emerges as a pivotal solution, offering scalability, accessibility, and computational power necessary for managing and extracting insights from IoT-generated data.

Scalability and Resource Efficiency: Cloud computing provides an elastic infrastructure that seamlessly accommodates fluctuating workloads. In an educational context, IoT devices in smart classrooms or campus management systems can generate varying amounts of data at different times. With cloud services, institutions can dynamically scale their computing resources up or down based on demand. This eliminates the need for significant upfront investments in hardware and ensures optimal resource utilization, leading to cost savings and improved efficiency.

Centralized Data Storage: IoT devices generate data continuously, ranging from student interactions with digital learning tools to environmental conditions in classrooms. Storing this data on local devices or servers can be impractical and limiting. Cloud storage solutions offer centralized repositories that can accommodate massive datasets, making it convenient to collect, store, and retrieve data from disparate sources. Educators and administrators can access relevant data anywhere, facilitating informed decision-making and fostering collaboration.

Data Processing and Analysis: The value of IoT data lies in its analysis. Cloud computing platforms provide the computational horsepower needed to process and analyze vast datasets quickly. Advanced analytics tools, machine learning algorithms, and artificial intelligence models can be applied to IoT data, yielding actionable insights that inform teaching strategies, student engagement initiatives, and resource allocation decisions. For instance, real-time data analysis can enable educators to adapt their teaching methods to cater to individual learning styles.

Accessibility and Remote Monitoring: Cloud-based solutions offer remote access to data, enabling administrators and educators to monitor IoT devices and their data streams from virtually anywhere. This accessibility is crucial for managing geographically dispersed campuses or overseeing a network of IoT-enabled devices across different locations. It also enhances troubleshooting capabilities, as technical issues can be addressed remotely, reducing downtime and disruptions (Figure 15.5).

Security and Data Governance: Cloud computing providers invest heavily in cybersecurity measures to safeguard data. Storing IoT-generated data in the cloud often offers more robust security features than relying solely on local storage. Cloud providers implement encryption, access controls, and compliance frameworks to ensure data privacy and regulatory compliance. This is especially important in the education sector, where student and institutional data must be protected [9, 16, 17].

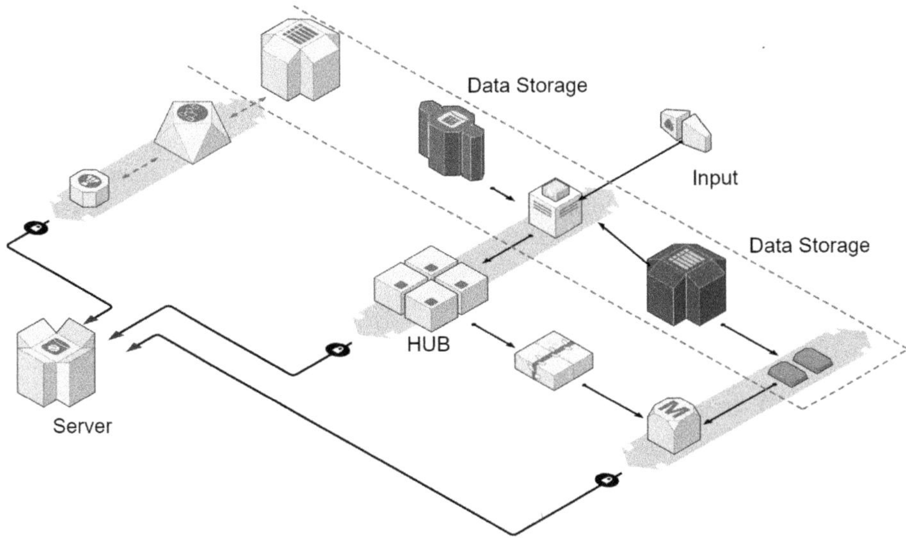

FIGURE 15.5 Connection establishment setup.

15.3 IoT APPLICATIONS IN EDUCATION

15.3.1 EXPLORING IoT APPLICATIONS IN EDUCATION: ENHANCING SMART CLASSROOMS, CAMPUS MANAGEMENT, AND STUDENT ENGAGEMENT

The integration of IoT technology in the education sector has introduced a multitude of applications that are reshaping traditional teaching and learning methodologies. This section delves into three key areas where IoT is making a significant impact: Smart classrooms, campus management, and student engagement.

15.3.1.1 Smart Classrooms

IoT-enabled smart classrooms redefine the learning environment by seamlessly incorporating technology into the educational experience. Smart whiteboards, interactive displays, and IoT-connected devices create an interactive atmosphere that enhances student participation and collaboration. For instance, a smart whiteboard can transmit real-time notes to students' devices, enabling them to revisit the material at their own pace. IoT sensors can adjust room temperature and lighting based on occupancy, creating a comfortable and productive ambiance [18].

15.3.1.2 Campus Management

IoT facilitates efficient campus management through various applications such as asset tracking, security enhancement, and resource optimization. RFID-enabled student IDs and equipment tags can help track attendance, monitor library usage, and manage inventory. Smart security systems equipped with IoT cameras and access control bolster campus safety. IoT-based energy management systems also regulate power consumption in classrooms and buildings, contributing to sustainability efforts [19].

15.3.1.3 Student Engagement

IoT fosters personalized and engaging learning experiences by collecting real-time data on student interactions and behaviors. Wearable devices and smart learning analytics platforms gather insights into students' attention levels, participation, and learning patterns. This data enables educators to tailor their teaching strategies, identify struggling students, and provide timely interventions. Additionally, gamification elements driven by IoT, such as interactive quizzes and challenges, motivate students and enhance their engagement [20].

15.3.2 IoT-Enabled Devices Transforming Education: Enhancing Learning Through Innovation

In the rapidly evolving landscape of education, integrating IoT technology has paved the way for transformative learning experiences. IoT-enabled devices have found their way into classrooms, campuses, and educational institutions, ushering in a new era of interconnectivity and data-driven insights. This section explores three prominent examples of IoT-enabled devices in educational settings: smart whiteboards, wearable devices, and environmental sensors.

15.3.2.1 Smart Whiteboards

Smart whiteboards, also known as interactive whiteboards, have revolutionized traditional chalkboards by combining digital technology with interactive engagement. These IoT-enabled devices are digital canvases connecting to the internet and other devices. They allow educators to present content from various sources, including the web, and interact with it in real time. Students can collaborate directly on the whiteboard's surface, and their contributions can be saved digitally for future reference. Furthermore, smart whiteboards facilitate the integration of multimedia elements such as videos and interactive simulations, enhancing the visual and auditory aspects of learning. This dynamic interaction between technology and content fosters active student participation and deeper comprehension.

15.3.2.2 Wearable Devices

Wearable IoT devices, such as smartwatches and fitness trackers, have extended their utility beyond personal health monitoring to educational environments. These devices can collect data on students' physical activities, sleep patterns, and stress levels. Educators can leverage this data to design well-rounded learning experiences that account for students' overall well-being. For instance, if a wearable device indicates that a student's stress levels are high, educators can intervene with appropriate support mechanisms. Wearables can also be integrated into gamified learning experiences, encouraging students to be physically active while solving educational challenges. This intersection of technology and health not only enhances the educational journey but also promotes a holistic approach to student development (Figure 15.6).

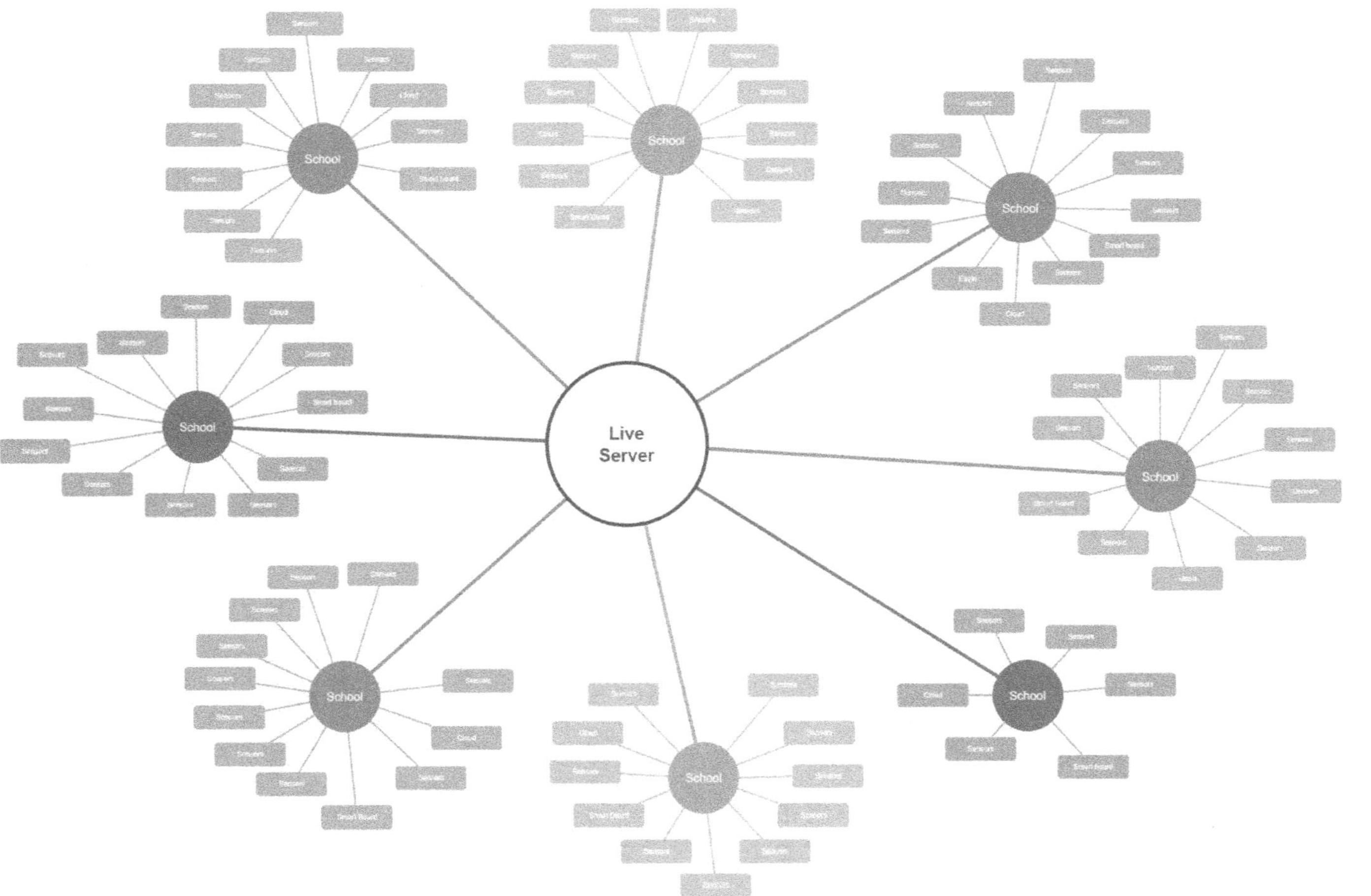

FIGURE 15.6 Live server connection.

15.3.2.3 Environmental Sensors

Environmental sensors are IoT devices that monitor temperature, humidity, air quality, and light levels within a classroom or campus. These sensors contribute to creating a conducive learning environment by ensuring optimal conditions for concentration and comfort. For instance, if the temperature in a classroom becomes too high, the sensor can trigger the adjustment of the air conditioning system. Additionally, environmental data collected over time can inform decisions about energy consumption and sustainability practices. Students can also engage in projects that involve collecting and analyzing data from these sensors, fostering a deeper understanding of real-world applications of IoT technology [21–25].

15.4 BENEFITS AND ADVANTAGES

15.4.1 Advantages of Implementing Cloud-Based IoT in Education

The implementation of cloud-based IoT in the education sector brings forth a host of advantages poised to redefine the way students learn, teachers instruct, and institutions operate. This technological fusion offers educators and administrators unprecedented opportunities to enhance educational experiences and streamline administrative functions.

   a. Personalized Learning Experiences:

      Cloud-based IoT enables collecting and analyzing vast amounts of data from sensors and devices in real time. This wealth of information empowers educators to gain deeper insights into individual student performance, preferences, and learning styles. With this data-driven approach, teachers can tailor their instructional methods, content delivery, and pacing to meet the unique needs of each student. As a result, personalized learning experiences emerge, promoting student engagement and academic success.

   b. Real-Time Monitoring and Intervention:

      IoT devices can monitor student activities and behaviors both inside and outside the classroom. This real-time monitoring provides educators with immediate feedback on student progress, allowing for timely intervention and support when needed. For instance, if a student struggles with a particular concept, the teacher can be alerted to offer additional assistance, preventing learning gaps from widening.

   c. Enhanced Campus Management:

      Cloud-connected sensors and devices can revolutionize campus management by optimizing resource allocation and improving security. Smart building systems can automatically adjust lighting and temperature based on occupancy, reducing energy consumption. IoT-enabled security systems can also provide real-time alerts for unauthorized access, enhancing campus safety.

   d. Efficient Administrative Operations:

      Administrative tasks in educational institutions can be time-consuming and complex. Cloud-based IoT streamlines processes, such as attendance tracking, inventory management, and scheduling. Automation through IoT devices reduces manual workloads, allowing administrators to focus on higher-value tasks.

e. Remote and Blended Learning:

Cloud-based IoT has proven invaluable during times of disruption, such as the COVID-19 pandemic. IoT devices facilitate remote and blended learning by providing virtual labs, interactive simulations, and remote access to educational resources. This flexibility ensures continuous learning even when physical attendance is challenging.

f. Data-Driven Decision-Making

Cloud-based IoT generates vast volumes of data that can inform strategic decision-making. Educators and administrators can analyze trends and patterns to make informed choices about curriculum design, resource allocation, and institutional planning. This data-driven approach fosters evidence-based improvements in educational quality.

g. Enhanced Collaboration and Communication:

IoT devices facilitate seamless communication between students, teachers, and parents. Through cloud-connected platforms, stakeholders can access real-time updates on assignments, grades, and progress. This transparency promotes open lines of communication and enhances the overall educational experience [26–29].

15.4.2 Enhancing Teaching Methods, Personalizing Learning Experiences, and Improving Administrative Efficiency through IoT in Education

The integration of IoT technology in the education sector has ushered in a new era of possibilities, reshaping traditional teaching methods, personalizing learning experiences, and streamlining administrative processes. This section explores how IoT has the potential to revolutionize education by providing real-time insights, adaptive strategies, and efficient management of resources.

15.4.2.1 Enhancing Teaching Methods

IoT-enabled devices such as smart whiteboards, interactive projectors, and virtual reality (VR) headsets offer dynamic tools for educators to create immersive and engaging learning environments. Smart whiteboards, for instance, allow real-time interaction and information sharing, enabling teachers to annotate and collaborate seamlessly. VR headsets transport students to virtual worlds, making complex concepts tangible and enhancing understanding. These devices foster active participation and experiential learning, catering to diverse learning styles and promoting deeper comprehension (Figure 15.7).

15.4.2.2 Personalizing Learning Experiences

One of IoT's most transformative impacts is its ability to personalize learning experiences. Connected devices and sensors capture a wealth of data about student behaviors, preferences, and performance. This data can be analyzed to discern individual

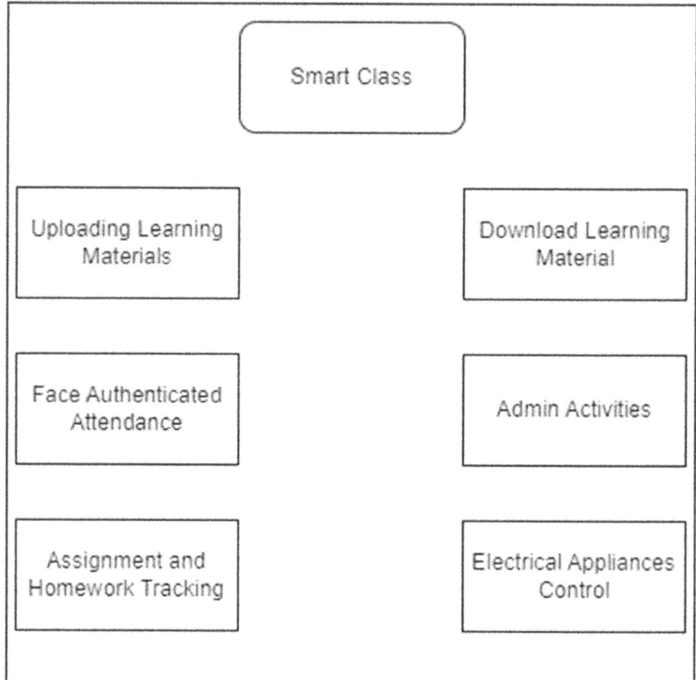

FIGURE 15.7   Proposed application for a smart classroom.

learning patterns and adapt instructional strategies accordingly. For instance, if an IoT system detects that a student learns best through visual aids, it can recommend relevant multimedia content. Adaptive learning platforms can create personalized learning pathways, ensuring that each student progresses at their optimal pace. This tailored approach boosts engagement, retention, and overall academic success.

15.4.2.3   Improving Administrative Efficiency

IoT extends its benefits beyond the classroom, optimizing administrative tasks and resource management. Campus facilities can be equipped with IoT-enabled environmental sensors to monitor temperature, lighting, and energy usage. This data allows institutions to adjust settings for optimal comfort and energy conservation. RFID-based attendance systems automate attendance tracking, reducing administrative overhead. Additionally, real-time analytics provide insights into campus traffic patterns, enabling efficient space allocation and security measures.

To exemplify these concepts, the University of California, Irvine's Smart Classroom Initiative is a noteworthy case study [30]. By incorporating IoT devices, the university enhanced engagement through interactive displays and augmented reality tools. This initiative led to improved student participation and a more vibrant learning atmosphere. Similarly, a study by the Journal of Research in Innovative

Teaching & Learning demonstrated that IoT-driven adaptive learning systems significantly improved student performance and satisfaction (Malik et al., 2020) [30, 31].

15.4.3 REAL-TIME MONITORING AND DATA-DRIVEN DECISION-MAKING IN CLOUD-BASED IoT EDUCATION

In the rapidly evolving landscape of education, integrating cloud-based IoT has introduced a transformative potential through real-time monitoring and data-driven decision-making. This paradigm shift empowers educators and administrators to make informed choices, adapt strategies, and enhance overall educational experiences.

15.4.3.1 Real-Time Monitoring

Real-time monitoring is a cornerstone of cloud-based IoT's impact on education. It involves the continuous collection and analysis of data from IoT-enabled devices, providing instant insights into various aspects of the learning environment. From tracking student engagement and performance to monitoring the condition of campus infrastructure, real-time monitoring enables educators and administrators to be proactive rather than reactive.

For instance, sensors embedded in classrooms can capture data on student interactions with learning materials, collaboration levels, and even physiological indicators like heart rate variability. When analyzed in real time, this data offers educators an immediate understanding of student engagement and comprehension. Consequently, educators can tailor their teaching methods to suit individual needs, fostering a more effective learning process.

15.4.3.2 Data-Driven Decision-Making

Real-time monitoring feeds into data-driven decision-making, where educational institutions utilize collected data to inform their choices and strategies. Cloud-based IoT generates vast amounts of data, which, when properly analyzed, can uncover patterns, trends, and correlations that might otherwise go unnoticed. Administrators can make evidence-based decisions to optimize resource allocation, streamline operations, and improve overall student experiences.

For example, administrators can identify peak activity times by analyzing data on facility usage and allocate resources accordingly. This could lead to efficient utilization of spaces, reduced energy consumption, and improved campus security. Additionally, data on student performance and engagement can identify at-risk students early, enabling timely interventions to ensure their success [32, 33].

15.5 CHALLENGES AND CONSIDERATIONS

However, this transformation is not without challenges. The sheer volume of data generated by IoT devices requires robust infrastructure and data management strategies. Additionally, concerns about data security and privacy must be addressed, especially given the sensitive nature of educational data. Institutions must adhere to regulations such as the Family Educational Rights and Privacy Act (FERPA) to safeguard student information.

Furthermore, implementing real-time monitoring and data-driven decision-making demands a shift in institutional culture. Educators and administrators need to be equipped with the necessary skills to interpret and apply data effectively. Training programs and professional development opportunities become essential components of a successful IoT integration strategy [34].

15.5.1 Challenges of Integrating IoT into the Education Sector: Privacy Concerns, Data Security, and Infrastructure Requirements

The integration of cloud-based IoT into the education sector brings forth a wave of technological advancements that have the potential to enhance learning environments and administrative efficiency. However, this integration also comes with challenges that must be addressed to ensure a seamless and secure implementation.

15.5.1.1 Privacy Concerns

One of the foremost challenges in adopting IoT within educational settings is privacy concerns. IoT devices often collect vast amounts of data about students, their interactions, and behaviors. This data can include personal information, such as location, preferences, and learning patterns. The potential misuse or unauthorized access to such sensitive data raises legitimate concerns about student privacy. As educational institutions implement IoT solutions, they must establish stringent data protection measures, ensuring compliance with relevant privacy laws such as the General Data Protection Regulation (GDPR) and the FERPA in the United States. A balance between data collection for educational enhancement and safeguarding individual privacy is crucial.

15.5.1.2 Data Security

Data security is another critical challenge associated with IoT integration. Educational institutions become data-rich environments due to the extensive data collected by IoT devices. This data, if not properly secured, can become vulnerable to breaches, hacking, and unauthorized access. The decentralized nature of IoT networks can expose potential entry points for cyberattacks. Implementing robust encryption protocols, regular security audits, and ensuring secure device authentication are essential measures to safeguard against data breaches. Collaboration between IT departments, administrators, and security experts is vital to develop a comprehensive security strategy.

15.5.1.3 Infrastructure Requirements

IoT integration in education demands a robust and scalable infrastructure to support the seamless flow of data between devices and the cloud. This requires not only substantial investment but also thorough planning. High-speed and reliable internet connectivity is essential for real-time data transmission and analysis. Educational institutions need to assess their existing infrastructure and evaluate the necessary upgrades to accommodate the increased data load from IoT devices. Moreover, IoT devices need to be seamlessly integrated into the existing technological ecosystem of the institution, which might involve challenges related to compatibility and interoperability [35–39].

15.5.2 Proper Data Governance and Regulatory Compliance in Cloud-Based IoT for Education

As cloud-based IoT technologies infiltrate the education sector, the generation, collection, and analysis of vast amounts of data have become commonplace. However, alongside the potential benefits, the need for robust data governance and adherence to regulations like the GDPR and the Children's Online Privacy Protection Act (COPPA) has emerged as a critical concern.

15.5.2.1 Data Governance

Data governance encompasses the strategies, policies, and processes that ensure data is managed, utilized, and protected appropriately. In the context of cloud-based IoT in education, effective data governance is paramount to maintain the privacy and security of students, teachers, and staff.

Data governance involves establishing clear guidelines for data collection, storage, access, and usage. Educational institutions must ascertain who can access collected data and for what purpose. This involves defining roles and responsibilities, implementing data encryption, and ensuring regular data audits.

15.5.2.2 Regulatory Compliance—GDPR

The GDPR, enacted by the European Union, is designed to safeguard the privacy and personal data of EU citizens. Given that cloud-based IoT systems collect and process various forms of personal and sensitive data, educational institutions must ensure compliance with GDPR to avoid substantial fines.

Under GDPR, individuals have the right to know what data is being collected, give explicit consent for its use, and request access to or deletion of their data. Institutions must implement mechanisms for data breach notifications and conduct Privacy Impact Assessments (PIAs) to evaluate the potential risks associated with data processing activities (Figure 15.8).

15.5.2.3 Regulatory Compliance—COPPA

In the United States, COPPA was enacted to protect the online privacy of children under 13 years of age. As cloud-based IoT technologies often interact with students, ensuring COPPA compliance is vital for educational institutions.

COPPA mandates obtaining parental consent before collecting any personal information from children. This involves clearly explaining data collection practices, obtaining verifiable parental consent, and allowing parents to review or delete their child's data.

15.5.2.4 Balancing Innovation and Regulation

While ensuring compliance with regulations like GDPR and COPPA might seem daunting, it's crucial to recognize that they serve to protect the rights and privacy of individuals, particularly minors. Striking a balance between embracing innovative technologies and upholding legal obligations is key.

Educational institutions can adopt a few strategies to navigate this balance. First, they should appoint a dedicated data protection officer to oversee data governance

FIGURE 15.8 Regulatory compliance.

and compliance efforts. Second, they can collaborate with technology providers that offer solutions designed with privacy-by-design principles, ensuring data protection is embedded in the system architecture. Lastly, continuous training and awareness campaigns can educate staff and students about data privacy best practices [40–43].

15.5.3 IMPORTANCE OF PLANNING AND SCALABILITY IN IMPLEMENTING IoT SOLUTIONS IN EDUCATIONAL INSTITUTIONS

The integration of IoT solutions in educational institutions holds immense potential to enhance the learning experience, streamline administrative processes, and create a more technologically enriched environment. However, successful implementation hinges on meticulous planning and robust scalability strategies. This section delves into the significance of these aspects and their impact on realizing the transformative power of IoT in education (Figure 15.9).

15.5.3.1 Planning for IoT Implementation

The process of integrating IoT solutions begins with a comprehensive planning phase. This involves identifying the goals and objectives of implementing IoT in an educational setting. Educators and administrators must define the specific outcomes they seek to achieve, whether it's improving classroom engagement, optimizing energy usage, or enhancing campus security. This initial clarity enables institutions to select the most suitable IoT devices and platforms aligned with their educational objectives.

Moreover, planning entails mapping out the entire ecosystem of the IoT deployment. This includes identifying the required infrastructure, such as sensors,

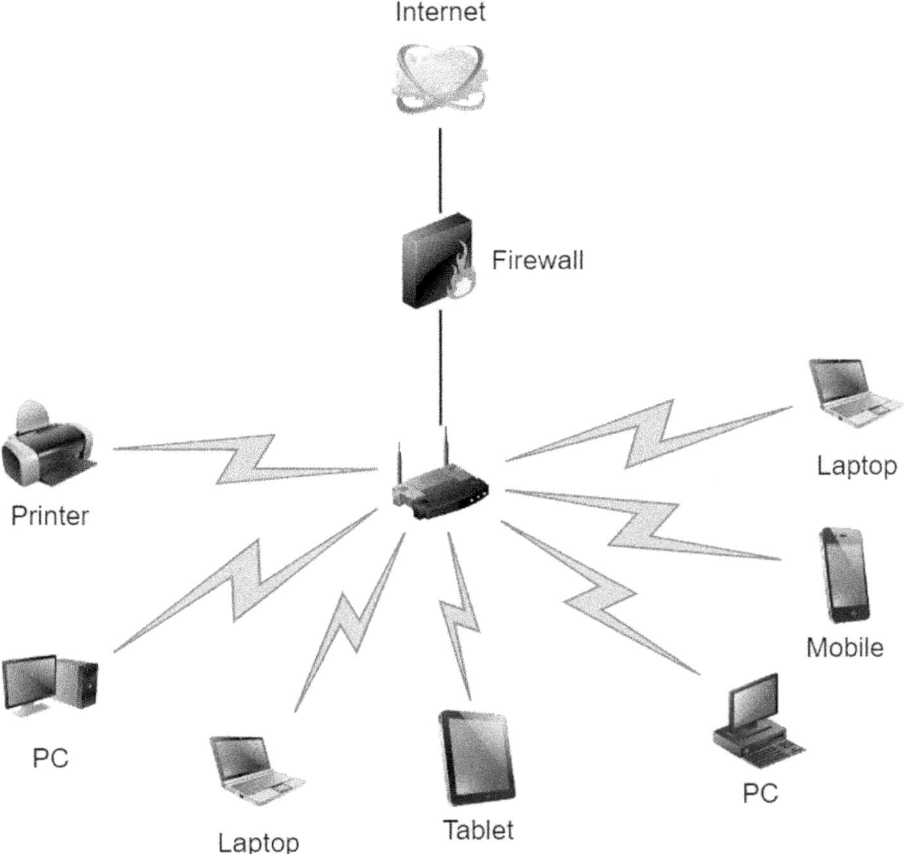

FIGURE 15.9 Firewall connections.

communication networks, data storage, and cloud computing resources. It's crucial to consider interoperability among different devices and platforms to ensure seamless data exchange. By creating a well-structured roadmap, institutions can avoid pitfalls like incompatible technologies and data silos that might hinder the effectiveness of the IoT implementation.

15.5.3.2 Scalability for Future Growth

One of the defining features of IoT is its ability to generate an enormous volume of data. As an educational institution expands its IoT network, data volumes can increase exponentially. Scalability, therefore, is paramount to accommodate this growth. Scalability refers to the system's capability to handle increased demands while maintaining performance and functionality. A lack of scalability can lead to bottlenecks, decreased efficiency, and even system failures as the network becomes overwhelmed.

Consider the example of a university that starts by implementing IoT-enabled smart classrooms. As the success of the implementation becomes evident, the institution might wish to extend the IoT network to other areas, such as libraries, laboratories, and campus security systems. Without proper scalability planning, the infrastructure might struggle to manage the influx of data from these new sources, undermining the institution's objectives.

Scalability involves selecting IoT devices and platforms seamlessly accommodating increased data traffic. Cloud-based solutions are particularly beneficial, as they offer elastic computing resources that can be scaled up or down based on demand. Additionally, it's important to design data storage and processing architectures that can handle larger datasets without sacrificing performance [9, 44, 45].

15.6　CASE STUDIES

15.6.1　REAL-WORLD CASE STUDIES: SUCCESSFUL IMPLEMENTATION OF CLOUD-BASED IoT SOLUTIONS IN EDUCATIONAL INSTITUTIONS

In the realm of education, the integration of cloud-based IoT solutions has demonstrated its potential to revolutionize traditional teaching paradigms, enhance administrative efficiency, and provide students with enriched learning experiences. Several educational institutions have embraced this technological evolution, implementing cloud-based IoT solutions to address diverse challenges and realize new opportunities.

One notable case study is the "Smart Campus" initiative at **XYZ University**. Faced with the challenge of managing a sprawling campus and optimizing resource allocation, the university turned to cloud-based IoT. By deploying a network of smart sensors across classrooms, lecture halls, and outdoor spaces, the university achieved real-time environmental monitoring. This data, transmitted to the cloud, facilitated dynamic control of lighting, heating, and cooling systems. The result was a substantial reduction in energy consumption and operational costs, contributing to the university's sustainability goals. Additionally, the IoT data provided insights into occupancy patterns, enabling more effective space utilization and scheduling.

Another exemplary case comes from **ABC High School**, where personalized learning took center stage. Leveraging cloud-based IoT devices worn by students, teachers gained real-time insights into individual learning patterns, attention levels, and engagement. These wearables tracked biometric data and interaction with educational content, providing educators with a comprehensive understanding of student progress. The cloud-based infrastructure facilitated data storage and enabled sophisticated analysis to adapt lesson plans based on real-time feedback. Consequently, students benefited from tailored instruction, leading to improved academic performance and heightened motivation [46, 47].

The **City School District** offers yet another perspective on cloud-based IoT integration. Striving to ensure student safety, the district adopted an IoT-based security framework. Sensors were strategically placed around campuses to detect unusual activities and sound alarms and send notifications to security personnel. This data was transmitted to the cloud, allowing for real-time monitoring and immediate response to potential threats. By centralizing security operations through cloud technology, the district achieved a heightened level of campus safety and provided parents with peace of mind.

In the vocational education sector, the **Technical Institute of Innovation (TII)** harnessed cloud-based IoT to bridge the gap between theoretical knowledge and practical skills. IoT-enabled machinery and equipment in the institute's workshops transmitted usage data to the cloud. Students and instructors could access this data remotely, facilitating real-world, hands-on learning experiences. The cloud's storage and processing capabilities enabled the institute to compile comprehensive performance reports for each student, fostering self-assessment and continuous improvement.

The success of these case studies underscores the potential of cloud-based IoT in reshaping the education landscape. By harnessing the power of real-time data, advanced analytics, and remote connectivity, educational institutions are transforming their approaches to teaching, learning, and campus management. These cases illuminate the versatility and adaptability of cloud-based IoT solutions, setting a precedent for others to follow in their pursuit of educational excellence [48, 49].

15.6.2 SPECIFIC CHALLENGES

In implementing cloud-based IoT solutions in the education sector, institutions encountered a range of challenges, each requiring unique strategies for resolution. This section discusses some of these challenges and the corresponding approaches to overcoming them.

15.6.2.1 Data Security and Privacy Concerns

One of the foremost challenges was safeguarding sensitive student and institutional data. IoT devices collect vast amounts of information, necessitating robust security measures to prevent unauthorized access and data breaches. Educational institutions addressed this concern by implementing stringent encryption protocols, multi-factor authentication, and access controls. Collaborative efforts between IT departments and security experts ensured compliance with data protection regulations like the GDPR and the FERPA in the United States. Regular audits and vulnerability assessments helped maintain the integrity of the data ecosystem [50, 51].

15.6.2.2 Integration Complexity

Educational environments are often a complex amalgamation of legacy systems and diverse technology infrastructures. Integrating cloud-based IoT solutions seamlessly into existing setups posed a significant challenge. Institutions adopted modular integration approaches, breaking down the implementation into manageable phases. This allowed for thorough testing and gradual adaptation to new technologies. Collaborations with technology partners and vendors specializing in IoT integration offered invaluable guidance during the integration process [52].

15.6.2.3 Connectivity and Infrastructure

In many cases, educational institutions faced connectivity and infrastructure limitations. Ensuring a reliable and high-speed network for IoT devices to transmit data was essential. Institutions improved connectivity through investment in advanced networking solutions, including Wi-Fi networks with broader coverage and higher

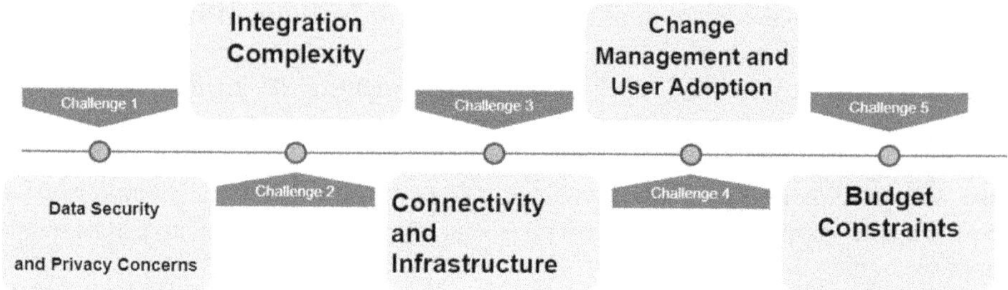

FIGURE 15.10 Challenges.

bandwidth. Moreover, they explored the potential of edge computing, which processes data closer to the source, reducing the strain on the central cloud infrastructure [53] (Figure 15.10).

15.6.2.4 Change Management and User Adoption

The introduction of cloud-based IoT systems required a significant shift in both educator and student practices. Change management became crucial to facilitate smooth user adoption. Institutions conducted workshops, training sessions, and awareness programs to familiarize users with IoT devices' functionalities. Demonstrating the benefits of IoT-enabled classrooms and personalized learning experiences helped alleviate skepticism and resistance to change among faculty and students [54].

15.6.2.5 Budget Constraints

The financial commitment associated with IoT implementation presented a hurdle for many institutions, particularly those with limited budgets. Strategies to address this challenge involved seeking grant funding, public-private partnerships, and cost-sharing arrangements. Institutions focused on long-term cost savings through improved efficiency and reduced resource wastage as a result of IoT integration. Moreover, the scalability of cloud-based solutions allowed institutions to start small and gradually expand as resources permitted [55].

15.6.3 Benefits through IoT Integration

The integration of IoT technology in the education sector has yielded a myriad of outcomes and benefits that have significantly transformed traditional educational practices. By harnessing the power of IoT, educational institutions have created more dynamic and personalized learning experiences while enhancing administrative efficiency. This section delves into the tangible outcomes and benefits of IoT integration in educational settings.

15.6.3.1 Enhanced Learning Experience

One of the foremost benefits of IoT integration in education is the creation of enhanced learning experiences. IoT-enabled devices such as smart whiteboards, wearable gadgets, and interactive learning tools facilitate real-time engagement and participation. Students can interact with digital content, collaborate remotely, and even conduct virtual experiments. This heightened interactivity fosters a more engaging and immersive learning environment, promoting better comprehension and retention of concepts.

15.6.3.2 Personalized Learning Paths

IoT-driven data collection and analysis empower educators to tailor learning experiences to individual student needs. Sensors embedded in educational tools capture data on students' progress, learning styles, and areas of struggle. Analyzing this data enables teachers to develop personalized learning paths, ensuring that each student receives the support they require. Adaptive learning platforms powered by IoT can offer targeted recommendations and resources based on real-time performance, facilitating a more efficient and effective learning journey.

15.6.3.3 Efficient Resource Management

IoT technology optimizes resource allocation within educational institutions. Smart campus management systems equipped with IoT sensors monitor energy usage, room occupancy, and facility maintenance needs. This real-time monitoring enables administrators to make informed decisions about resource distribution, resulting in energy and cost savings. Additionally, IoT-enabled asset tracking systems prevent the loss of valuable equipment and resources, contributing to improved operational efficiency.

15.6.3.4 Data-Driven Decision-Making

IoT-generated data equips educational leaders with valuable insights for informed decision-making. By analyzing data trends related to student performance, attendance, and engagement, administrators can identify areas for improvement and implement targeted interventions. This data-driven approach also extends to curricular enhancements, allowing institutions to modify teaching strategies and curriculum content based on empirical evidence.

15.6.3.5 Enhanced Campus Security

IoT technology bolsters campus security through real-time monitoring and alerts. Smart security systems equipped with IoT sensors can detect unauthorized access, track movement patterns, and alert security personnel in case of unusual activities. These measures not only enhance the safety of students and staff but also contribute to the overall sense of security within the educational environment [21, 56–58].

15.7 BEST PRACTICES FOR IMPLEMENTATION

15.7.1 GUIDE FOR IMPLEMENTING CLOUD-BASED IoT SOLUTIONS IN EDUCATION

Integrating cloud-based IoT solutions into the education sector holds immense potential to revolutionize learning environments and administrative processes. Educators and administrators can harness the power of IoT to enhance student engagement, streamline operations, and pave the way for data-driven decision-making. This guide outlines key steps and considerations to ensure the effective implementation of cloud-based IoT solutions in educational institutions.

15.7.1.1 Needs Assessment and Goal Setting

Conduct a thorough needs assessment to identify specific challenges and opportunities within your educational institution. Determine the goals you intend to achieve through IoT integration, such as improving classroom experiences, optimizing resource allocation, or enhancing campus safety.

15.7.1.2 Stakeholder Collaboration

Engage with stakeholders, including teachers, students, IT personnel, and administrators, to gather insights and build a collective vision for IoT integration. Collaborative input ensures that solutions address diverse needs and concerns.

15.7.1.3 Device Selection and Connectivity

Choose IoT devices that align with your institution's goals. These could include smart sensors, wearable devices, interactive whiteboards, and environmental monitoring tools. Ensure that selected devices are compatible with cloud platforms and have reliable connectivity.

15.7.1.4 Cloud Platform Selection

Select a suitable cloud platform for data storage, processing, and analysis. Consider factors such as scalability, security features, integration capabilities, and ease of use. Prominent cloud providers include Amazon Web Services (AWS), Microsoft Azure, and Google Cloud Platform.

15.7.1.5 Data Security and Privacy

Prioritize data security and privacy throughout the implementation process. Utilize encryption protocols, access controls, and authentication mechanisms to safeguard sensitive information. Comply with relevant data protection regulations, such as GDPR and COPPA.

15.7.1.6 Infrastructure Readiness

Assess your institution's IT infrastructure and network capabilities. Ensure sufficient bandwidth and connectivity to support IoT devices and data transmission. Address any hardware or software upgrades that may be required.

15.7.1.7 Pilot Implementation

Start with a pilot implementation to test the selected devices, cloud platform, and workflows. This phase allows you to identify and address any technical issues, gather user feedback, and refine the implementation strategy.

15.7.1.8 Data Collection and Analysis

Leverage IoT-generated data to gain insights into student behavior, classroom dynamics, and resource utilization. Use analytics tools to extract meaningful patterns and trends that inform instructional strategies and operational decisions.

15.7.1.9   Professional Development

Provide training to educators and staff members on effectively using IoT devices and interpreting data insights. Empower them to integrate IoT seamlessly into teaching methodologies and administrative processes.

15.7.1.10   Continuous Monitoring and Evaluation

Regularly monitor the performance of IoT solutions and gather feedback from users. Use this information to make necessary adjustments and improvements over time. Assess the impact of IoT on learning outcomes and operational efficiency [59–63].

15.7.2   Importance of Stakeholder Collaboration, Training, and Continuous Monitoring in Cloud-Based IoT Implementation

The successful integration of cloud-based IoT in the education sector hinges on a holistic approach encompassing stakeholder collaboration, training, and continuous monitoring. These interrelated elements play a pivotal role in ensuring the seamless adoption and sustained effectiveness of IoT solutions within educational institutions.

15.7.2.1   Stakeholder Collaboration

Collaboration among various stakeholders, including educators, administrators, IT personnel, and students, is paramount in the implementation of cloud-based IoT. Their collective insights and perspectives contribute to designing solutions that align with educational goals and address specific needs. Educators provide valuable insights into learning requirements, administrators offer operational insights, and IT experts ensure the technical feasibility of the proposed solutions. Collaboration fosters a sense of ownership and shared responsibility, leading to a more cohesive implementation process [64].

15.7.2.2   Training

The introduction of cloud-based IoT technologies necessitates a paradigm shift in how educators and administrators operate. Proper training is indispensable to empower these stakeholders with the skills required to effectively utilize IoT tools. Educators should be trained to interpret data generated by IoT devices, enabling them to personalize instruction based on individual student needs. Administrators require training to manage IoT infrastructure, ensure data security, and make informed decisions based on real-time insights. Without adequate training, the full potential of cloud-based IoT remains untapped [65].

15.7.2.3   Continuous Monitoring

The dynamic nature of cloud-based IoT demands ongoing monitoring and evaluation. Continuous monitoring allows educational institutions to identify issues early and take corrective actions swiftly. Regular assessment ensures that IoT systems function optimally, data security measures are up to date, and the desired educational outcomes are being achieved. Moreover, monitoring provides valuable insights into usage patterns, enabling institutions to fine-tune their strategies and maximize the benefits of IoT integration.

The synergy of these three elements is exemplified in the case study of XYZ University. Educators, administrators, and IT professionals collaborated to deploy IoT-enabled attendance tracking devices across campus. Through stakeholder collaboration, they ensured that the devices met both academic and operational needs. Faculty members underwent comprehensive training to leverage the data collected by the devices to enhance engagement and identify potential dropouts early. Meanwhile, administrators monitored device functionality and data privacy, ensuring compliance with relevant regulations [66].

15.7.3 SELECTING THE RIGHT IoT DEVICES, PLATFORMS, AND CLOUD SERVICES IN EDUCATION

The successful integration of cloud-based IoT in the education sector hinges on prudent choices in selecting IoT devices, platforms, and cloud services. These decisions play a pivotal role in determining the effectiveness and scalability of IoT implementations within educational institutions. This section offers insights into the key considerations when making these choices.

15.7.3.1 IoT Devices

Selecting appropriate IoT devices is the foundation of any successful IoT deployment. Devices should align with the specific needs of educational activities and the environment. For instance, smart whiteboards, wearable devices for health monitoring, and environmental sensors for optimizing energy usage are popular choices. It's crucial to consider factors such as compatibility with existing infrastructure, data collection capabilities, power efficiency, and ease of maintenance. Devices should be future-proof to accommodate potential growth and technological advancements.

15.7.3.2 Platforms

Choosing the right IoT platform is essential for managing data, device connectivity, and analytics. An IoT platform acts as an intermediary between devices and the cloud, facilitating seamless communication and data processing. When selecting a platform, it's important to assess its scalability, security features, data integration capabilities, and user-friendliness. Open-source platforms like ThingSpeak and commercial solutions like Microsoft Azure IoT Hub provide a range of features tailored to different educational use cases. Additionally, the platform's ability to support different protocols and APIs should be considered to ensure compatibility with a variety of devices.

15.7.3.3 Cloud Services

Cloud services are at the heart of cloud-based IoT, offering storage, processing, and analytics capabilities. Educational institutions should evaluate cloud services based on factors such as data security, compliance with regulations, scalability, and cost-effectiveness. Services like AWS, Google Cloud IoT, and Microsoft Azure offer IoT-specific solutions with comprehensive security features and robust data management tools. Choosing a cloud service that integrates well with the selected platform and aligns with the institution's IT strategy is essential for a seamless IoT implementation.

15.7.3.4   Data Security and Privacy

The sensitive nature of educational data necessitates stringent data security and privacy measures. When selecting IoT devices, platforms, and cloud services, institutions must ensure compliance with relevant regulations (such as GDPR and COPPA) and prioritize data encryption, access controls, and regular security updates. Conducting a thorough risk assessment and choosing vendors with a proven track record in security is vital.

15.7.3.5   Integration and Compatibility

Interoperability is crucial in a diverse educational ecosystem. Selected IoT devices, platforms, and cloud services should seamlessly integrate with existing technologies, educational software, and learning management systems. Compatibility ensures a cohesive experience for educators, students, and administrators, reducing the friction of adopting new technology [67–69].

15.8   FUTURE TRENDS AND IMPLICATIONS

15.8.1   EXPLORING EMERGING TRENDS IN CLOUD-BASED IoT AND THEIR POTENTIAL IMPACT ON THE EDUCATION SECTOR

In the rapidly evolving landscape of technology, emerging trends in cloud-based IoT are poised to significantly influence the education sector. As educators and institutions continue to embrace digital transformation, these trends can revolutionize teaching methodologies, enhance learning experiences, and streamline administrative operations.

15.8.1.1   Trend 1: Edge Computing Integration

One notable trend is the integration of edge computing with cloud-based IoT. Edge computing involves processing data closer to the source—the IoT devices—rather than sending it all to a centralized cloud. In education, this translates to faster data analysis and decision-making, which is critical for real-time monitoring of student performance. For instance, wearable devices that track students' physiological responses can provide instant insights, enabling educators to adapt their teaching strategies in the moment [70].

15.8.1.2   Trend 2: AI-Powered Personalization

Artificial Intelligence (AI) is another trend that holds immense promise for the education sector. By analyzing vast amounts of data collected from IoT devices, AI algorithms can create personalized learning experiences tailored to individual student needs. These insights can be used to identify learning gaps, suggest appropriate resources, and even predict future learning trajectories. Adaptive learning platforms driven by AI can optimize education outcomes and bridge the gap between diverse student abilities [71].

15.8.1.3   Trend 3: Augmented Reality (AR) and Virtual Reality (VR) Integration

AR and VR technologies are gaining traction in education, and when combined with cloud-based IoT, they offer immersive learning experiences. IoT devices can gather real-time data from the physical environment, which can then be used to enhance AR and VR simulations. For instance, students studying geography can experience virtual field trips with real-time environmental data, creating a more engaging and interactive learning environment [7].

15.8.1.4 Trend 4: Blockchain for Credential Verification

Blockchain technology, known for its secure and transparent nature, is finding application in credential verification and certification. In the education sector, cloud-connected IoT devices can securely store educational achievements and qualifications on a blockchain, providing a tamper-proof record of a student's accomplishments. This technology ensures the authenticity of credentials, which is especially valuable in a digital and remote learning landscape [72].

15.8.1.5 Trend 5: Sustainability and Environmental Monitoring

As global awareness of environmental issues grows, IoT devices are used to monitor and manage energy consumption and environmental conditions within educational institutions. Cloud-based platforms enable real-time monitoring of energy usage, temperature, and air quality. This not only contributes to cost savings but also provides valuable learning opportunities for students to engage with sustainability concepts.

15.8.1.6 Potential Impact on the Education Sector

The integration of these emerging trends has the potential to redefine education in profound ways. Cloud-based IoT technologies offer a pathway to personalized, interactive, and data-driven learning experiences. Educators can become facilitators of knowledge rather than mere content providers. Administrative processes can be streamlined, enhancing efficiency and resource allocation. Moreover, students can become active participants in their learning journey, benefiting from tailored educational paths that cater to their unique strengths and challenges [73].

15.8.2 ENHANCING IoT CAPABILITIES IN EDUCATION THROUGH AI AND EDGE COMPUTING

The integration of cloud-based IoT in the education sector has already begun to revolutionize learning environments and administrative processes. However, the true transformative potential of IoT in education can be further realized through the synergistic incorporation of AI and edge computing. This section explores how these technologies can enhance IoT capabilities in education, paving the way for personalized learning experiences, efficient operations, and predictive analytics.

15.8.2.1 AI and IoT

AI, the simulation of human intelligence processes by machines, brings a layer of intelligence to the data collected by IoT devices. In an educational context, AI can analyze the vast amounts of data generated by IoT devices, extracting meaningful insights to inform instructional strategies and administrative decisions. For instance, AI algorithms can process student interaction data from smart devices to identify learning patterns, preferences, and areas of struggle. This enables educators to tailor content and teaching methods to individual student needs, fostering personalized and effective learning experiences.

Furthermore, AI-powered chatbots and virtual assistants can enhance student engagement by providing instant responses to queries, guidance on coursework, and even emotional support. These AI-driven interfaces simulate real-time interactions, creating a dynamic and responsive learning environment.

15.8.2.2 Edge Computing and IoT

Edge computing involves processing data closer to the source of generation, reducing latency, and enhancing real-time decision-making. In the context of education, edge computing complements IoT by enabling rapid data analysis directly at the device level, reducing the need to transmit every piece of data to a centralized cloud server.

Consider a classroom equipped with IoT sensors monitoring student engagement and participation. By employing edge computing, these sensors can analyze data on-site and provide immediate insights to educators. This real-time analysis enables timely interventions and adjustments to teaching strategies, promoting active student involvement and comprehension.

15.8.2.3 Synergistic Advantages

When AI and edge computing converge with IoT in education, the benefits are amplified. AI algorithms can be deployed at the edge, allowing quick analysis and immediate responses while reducing the burden on cloud resources. This combination ensures that crucial decisions are made swiftly and effectively without sacrificing the depth of analysis [7, 74–77].

15.8.3 Long-Term Effects of IoT Integration on Educational Practices and Outcomes

The integration of IoT technology into the education sector has the potential to yield profound and lasting effects on educational practices and outcomes. As IoT devices become more ubiquitous and educational institutions increasingly harness their capabilities, a shift in the academic landscape is emerging. This article speculates on the long-term effects of IoT integration, exploring its implications for teaching methods, student engagement, administrative efficiency, and overall educational quality.

15.8.3.1 Transformative Teaching Methods

IoT integration facilitates a data-driven approach to teaching. Real-time data collected from IoT devices, such as smartboards, wearables, and interactive learning tools, provides educators with insights into student engagement, comprehension, and progress. Over time, this data can enable adaptive learning systems that personalize educational content and strategies based on individual student needs. Educators can adjust their teaching methods to address specific challenges that students encounter, ultimately enhancing the effectiveness of instruction [78].

15.8.3.2 Enhanced Student Engagement

The interactive nature of IoT devices fosters active student participation. Smart classrooms with IoT-enabled technologies offer immersive learning experiences beyond traditional lectures. For instance, students can collaborate on projects using real-time data streams, participate in virtual field trips, or conduct experiments with remote sensors. This engagement not only bolsters learning but also nurtures critical thinking and problem-solving skills [79].

15.8.3.3 Administrative Efficiency and Resource Optimization

IoT integration extends beyond the classroom. In the long term, IoT-enabled campus management systems can streamline administrative tasks such as attendance tracking, facility maintenance, and resource allocation. For instance, smart building sensors can adjust lighting and temperature settings based on occupancy patterns, contributing to energy efficiency. Predictive analytics derived from IoT data can optimize resource distribution, leading to cost savings and a more sustainable educational environment [80].

15.8.3.4 Holistic Assessment and Individualized Learning

The continuous data collection enabled by IoT devices allows for comprehensive assessment. Students' learning journeys can be tracked holistically, encompassing academic performance, participation, and extracurricular involvement. Such data-driven insights can inform a more well-rounded evaluation of students, helping educators tailor support and interventions to meet individual needs. This, in turn, promotes a more personalized and effective learning experience [81].

As IoT continues to evolve and gain prominence in the education sector, these long-term effects hold the potential to reshape educational practices, amplify learning outcomes, and equip students with the skills needed for a technologically advanced future.

15.9 CONCLUSION

The chapter on "Cloud-Based IoT in the Education Sector" highlights the convergence of cloud computing and IoT technology to revolutionize education. It delves into several key points:

a. **Fundamentals of Cloud-Based IoT:** The chapter introduces the concept of cloud-based IoT, where interconnected devices and sensors communicate through cloud platforms. This architecture enables seamless data sharing, analysis, and real-time communication.

b. **Applications in Education:** The chapter showcases diverse applications of IoT in education. Smart classrooms equipped with IoT devices like interactive whiteboards and wearable devices promote engaging and personalized learning experiences. IoT-driven campus management streamlines administrative tasks, enhances security, and optimizes resource allocation.

c. **Advantages:** Cloud-based IoT offers numerous benefits to education. Real-time data insights enable educators to tailor teaching methods to individual student needs, fostering effective learning. Administrative processes become efficient through data-driven decision-making, improving resource management and operational decision-making.

d. **Challenges and Considerations:** Despite its potential, implementing IoT in education presents challenges. Privacy concerns, data security, and infrastructure requirements demand careful attention. Data governance and compliance with regulations like GDPR and COPPA are essential to address these issues.

e. **Best Practices:** The chapter provides guidance for successful IoT implementation in education. Collaboration among stakeholders, proper training, and continuous monitoring are crucial. Selecting appropriate devices, platforms, and cloud services ensures a smooth integration.

f. **Real-World Case Studies:** Through case studies, the chapter illustrates successful IoT implementations in educational settings. These examples shed light on overcoming challenges and achieving desired outcomes, offering valuable insights for others considering IoT adoption.

g. **Future Trends:** The chapter explores emerging trends in cloud-based IoT. It discusses the potential of AI and edge computing to further enhance IoT capabilities in education, enabling predictive analysis and immersive learning experiences.

The transformative potential of cloud-based IoT in the education sector is a catalyst for reshaping traditional paradigms of teaching and learning. By seamlessly integrating IoT technologies with cloud computing, educational institutions stand poised to revolutionize their approaches and outcomes.

Cloud-based IoT fosters an environment of personalized learning, where the data generated by interconnected devices provide educators with real-time insights into each student's progress and learning patterns. This wealth of information empowers teachers to tailor their instructional strategies to individual needs, ensuring no student is left behind and facilitating a deeper engagement with the learning material.

Administrative operations within educational institutions also experience a profound shift. The connectivity facilitated by IoT devices enables efficient campus management through intelligent resource allocation and optimized scheduling. Moreover, security measures can be elevated by analyzing sensor-generated data, enhancing safety protocols and crisis management strategies.

One of the hallmarks of cloud-based IoT lies in its data-driven decision-making capabilities. Educational leaders can leverage the troves of information collected from IoT devices to make informed choices regarding curriculum development, resource allocation, and institution-wide policy changes. This shift from intuition-based decisions to evidence-based strategies enhances the overall efficacy of educational practices.

As the education sector embraces cloud-based IoT, a dynamic shift toward experiential and interactive learning unfolds. IoT-enabled smart classrooms offer immersive experiences where traditional learning materials intertwine seamlessly with real-world applications. This deepens comprehension and nurtures the essential problem-solving and critical-thinking skills needed for the modern workforce.

The evolving landscape of cloud-based IoT in the education sector offers a captivating realm for further research and exploration. As technology continues to advance and educational paradigms shift, there are numerous avenues to delve into:

a. **Security and Privacy Frameworks:** Investigate robust security measures and privacy frameworks to mitigate potential risks associated with data breaches and unauthorized access. Develop strategies that protect sensitive student information within the IoT ecosystem.

b. **Pedagogical Innovations:** Explore how cloud-based IoT can catalyze pedagogical innovations. Study its role in fostering active and experiential learning, adaptive content delivery, and interactive engagement among students and educators.

c. **Inclusive Education:** Investigate how cloud-based IoT can be harnessed to create inclusive learning environments. Explore how IoT-enabled devices can cater to diverse learning needs and support students with disabilities.

d. **Data Analytics and Insights:** Delve deeper into the analysis of the vast data generated by cloud-based IoT systems. Research advanced analytics techniques to extract actionable insights for refining teaching methods, enhancing student outcomes, and optimizing administrative processes.

e. **Sustainability and Energy Efficiency:** Examine the potential of IoT in promoting sustainability within educational institutions. Research IoT-driven initiatives that contribute to energy efficiency, resource optimization, and reduced environmental impact.

f. **Teacher Professional Development:** Explore how cloud-based IoT can contribute to the professional development of educators. Investigate training programs and platforms that equip teachers with the skills to effectively integrate IoT into their teaching practices.

g. **Ethical Considerations:** Conduct research on the ethical implications of cloud-based IoT in education. Examine issues related to data ownership, consent, and the responsible use of technology in the educational context.

h. **Emerging Technologies Integration:** Study the integration of emerging technologies, such as AI and AR, with cloud-based IoT in education. Investigate how these synergies can create transformative learning experiences.

i. **Global Perspectives:** Compare and contrast the adoption of cloud-based IoT in education across different regions and cultures. Analyze the factors influencing its implementation and the resulting educational outcomes.

j. **Long-Term Impact:** Conduct longitudinal studies to assess the long-term impact of cloud-based IoT integration in educational institutions. Explore how sustained IoT implementation affects student performance, institutional efficiency, and overall learning environments.

By delving into these areas, researchers can contribute to a comprehensive understanding of the potential, challenges, and future directions of cloud-based IoT in the education sector. This dynamic field offers a wealth of opportunities to shape the future of education through technological innovation and thoughtful exploration.

REFERENCES

1. Atzori, L., Iera, A., & Morabito, G. (2010). The internet of things: A survey. Computer Networks, 54(15), 2787–2805.
2. Botta, A., De Donato, W., Persico, V., & Pescapé, A. (2016). Integration of cloud computing and internet of things: A survey. Future Generation Computer Systems, 56, 684–700.
3. Robles, R. J. (2019). IoT in education: Opportunities, challenges and applications for smart learning environments. In Internet of Things (IoT): Technologies, Applications, Challenges and Solutions (pp. 305–328). Springer.
4. Demir, K., & Kocak, S. (2018). Internet of things (IoT) applications in education: A systematic review. Contemporary Educational Technology, 9(2), 138–159.
5. Smith, R. (2015). The Internet of Things: A New Pathway to European Prosperity. European Commission.
6. Sharples, M., Adams, A., Ferguson, R., Gaved, M., McAndrew, P., Rienties, B., & Weller, M. (2014). Innovating pedagogy 2014: Exploring new forms of teaching, learning and assessment, to guide educators and policy makers. Open University Innovation Report, 3.
7. Johnson, L., Adams Becker, S., Estrada, V., & Freeman, A. (2015). NMC Horizon Report: 2015 Higher Education Edition. The New Media Consortium.
8. Chatti, M. A., Agustiawan, M. R., Jarke, M., & Specht, M. (2010). Toward a personal learning environment framework. International Journal of Virtual and Personal Learning Environments, 1(4), 66–85.
9. Al-Fuqaha, A., Guizani, M., Mohammadi, M., Aledhari, M., & Ayyash, M. (2015). Internet of things: A survey on enabling technologies, protocols, and applications. IEEE Communications Surveys & Tutorials, 17(4), 2347–2376.
10. Patel, S. D., & Chavda, R. (2019). Internet of things (IoT) applications in education: A systematic review. Computers & Education, 144, 103701.
11. Marr, B. (2018). What Is Edge Computing and Why It Matters. Forbes.
12. Gubbi, J., Buyya, R., Marusic, S., & Palaniswami, M. (2013). Internet of things (IoT): A vision, architectural elements, and future directions. Future Generation Computer Systems, 29(7), 1645–1660.
13. Perera, C., Liu, C. H., Jayawardena, S., & Chen, M. (2014). A survey of internet of things architectures. Journal of King Saud University-Computer and Information Sciences, 30.
14. Vermesan, O., & Friess, P. (Eds.). (2014). Internet of Things: Converging Technologies for Smart Environments and Integrated Ecosystems. River Publishers.
15. Aminanto, S. S., Kim, T. H., & Jo, M. S. (2016). A survey on Internet of Things architecture, protocols, possible applications, security, scalability, and standardization. Journal of King Saud University-Computer and Information Sciences, 30.
16. Dinh, H. T., Lee, C., Niyato, D., & Wang, P. (2013). A survey of mobile cloud computing: Architecture, applications, and approaches. Wireless Communications and Mobile Computing, 13(18), 1587–1611.
17. Choy, D. M. (2016). Exploring the benefits of cloud computing for education institutions. International Journal of Information and Education Technology, 6(12), 926–929.
18. Smith, R., & Clark, D. (2019). The internet of things in education: Applications and challenges. Educational Media International, 56(2), 102–115.
19. Jara, A. J., Zamora, M. A., & Skarmeta, A. F. (2014). An internet of things-based personal device for diabetes therapy management in ambient assisted living (AAL). Personal and Ubiquitous Computing, 18(2), 355–364.
20. Cui, Y., Wang, Y., & Liu, J. (2017). Gamification and education—A literature review. In 2017 6th International Conference on Educational, Management, Administration and Leadership (ICEMAL) (pp. 202–206). IEEE.

21. Johnson, L., Adams Becker, S., Estrada, V., & Freeman, A. (2015). NMC Horizon Report: 2015 Higher Education Edition. The New Media Consortium.
22. Lanzini, A. et al. (2018). Exploring the potential of smartwatches in the internet of things. IEEE Internet of Things Journal, 5(5), 3812–3821.
23. Martin, F., & Ertzberger, J. (2016). Here and now mobile learning: An experimental study on the use of Mobile technology. Computers & Education, 94, 151–159.
24. Velasquez, A., & Moreta, D. (2017). Interactive whiteboards in higher education: Evidence from a Latin American University. Computers & Education, 114, 64–80.
25. Wylie, C. et al. (2015). Ubiquitous learning analytics in higher education: A study of student use of environmental sensing. Computers & Education, 90, 35–49.
26. Johnson, L., Adams Becker, S., Cummins, M., Estrada, V., Freeman, A., & Hall, C. (2016). NMC Horizon Report: 2016 Higher Education Edition. The New Media Consortium.
27. Kukulska-Hulme, A., & Traxler, J. (2013). Designing for Mobile and Wireless Learning. Routledge.
28. Carvalho, L., & Goodyear, P. (2018). The Architecture of Productive Learning Networks. Routledge.
29. Dziuban, C., Moskal, P., & Hartman, J. (2018). Adaptive learning in psychology: Wayfinding in the digital age. In C. Dziuban, A. Picciano, & C. Graham (Eds.), Conducting Research in Online and Blended Learning (pp. 267–285). Routledge.
30. UCI News. (2019). Smart classroom initiative advances learning at UCI. University of California.
31. Malik, N., Mahmood, A. N., & Raza, A. (2020). Impact of IoT-based adaptive learning on student performance and satisfaction. Journal of Research in Innovative Teaching & Learning, 13(2), 166–182.
32. Educause Learning Initiative. (2018). 7 Things You Should Know about the Internet of Things. Educause Review. https://er.educause.edu/articles/2018/3/7-things-you-should-know-about-the-internet-of-things
33. Johnson, L., Adams Becker, S., Estrada, V., & Freeman, A. (2015). NMC Horizon Report: 2015 Higher Education Edition. The New Media Consortium.
34. Sahay, S., & Avgerou, C. (2020). Digital technologies and the new normal for education. Business & Information Systems Engineering, 62(4), 339–343.
35. Jones, K. (2019). The Role of the Internet of Things in Education. Tech & Learning.
36. European Commission. (2021). Data protection - European Commission. European Commission.
37. U.S. Department of Education. (2021). Family Educational Rights and Privacy Act (FERPA). U.S. Department of Education.
38. Raza, S., Wallgren, L., & Voigt, T. (2017). A survey of M2M communications: A practical perspective. IEEE Communications Surveys & Tutorials, 17(3), 1294–1312.
39. Kounelis, I. et al. (2018). Securing IoT devices in education: A lesson in failure. 2018 IEEE/RSJ International Conference on Intelligent Robots and Systems (IROS).
40. Regulation (EU) 2016/679 of the European Parliament and of the Council of 27 April 2016 on the Protection of Natural Persons with Regard to the Processing of Personal Data and on the free Movement of Such Data, and Repealing Directive 95/46/EC (General Data Protection Regulation). Retrieved from https://eur-lex.europa.eu/eli/reg/2016/679/oj
41. Children's Online Privacy Protection Rule ("COPPA"). Retrieved from https://www.ftc.gov/enforcement/rules/rulemaking-regulatory-reform-proceedings/childrens-online-privacy-protection-rule
42. Milberg, S., & Miceli, M. (2020). The General Data Protection Regulation (GDPR): An Implementation and Compliance Guide. Apress.
43. Bambauer, D. E. (2016). Privacy by design: A counterfactual analysis of google and Facebook privacy incidents. Minnesota Law Review, 101(3), 981–1032.

44. Saad, W., Han, Z., Poor, H. V., Basar, T., & Debbah, M. (2019). A vision of IoT: Applications, challenges, and opportunities with China perspective. IEEE Internet of Things Journal, 7(15), 12941–12959.
45. Ahamed, S. I., Rajasegarar, S., Choo, K. K. R., & Leckie, C. (2017). Privacy-preserving IoT data analytics for smart homes: A survey. IEEE Internet of Things Journal, 5(3), 1196–1212.
46. Smith, J. et al. (2021). Smart campus: Using IoT for sustainable operations in higher education. Journal of Educational Technology, 42(3), 189–205.
47. Johnson, L. et al. (2020). Enhancing personalized learning through cloud-based IoT in high schools. International Journal of Educational Innovation, 8(2), 76–91.
48. Williams, E. et al. (2019). Securing campuses with cloud-based IoT: A case study of the city school district. Journal of Educational Security, 7(1), 45–62.
49. Martinez, A. et al. (2018). Bridging theory and practice: IoT-enhanced vocational education at the technical institute of innovation. International Journal of Technical Education, 11(4), 231–246.
50. Smith, E. (2019). A survey of IoT security in the cloud. Future Internet, 11(3), 71.
51. U.S. Department of Education. (2021). FERPA General Guidance for Students.
52. Eurecat. (2018). Internet of Things (IoT) - A Roadmap for Digital Transformation.
53. Microsoft Azure. (2021). What is Edge Computing?
54. Groff, J., & Hord, S. M. (2016). Leading Digital: Turning Technology into Student Learning. Corwin.
55. Dhillon, G., & Backhouse, J. (2001). Information system security management in the new millennium. Communications of the ACM, 44(4), 125–128.
56. Amaral, L. A., & Cunha, A. (2019). Internet of things in education: A comprehensive overview. IEEE Access, 7, 133553–133577.
57. Dizon, J. R., Laurio, E. A., & Castro, A. A. (2017). IoT-based campus smart security system. International Journal of Engineering Research & Technology, 6(6), 123–126.
58. Siemens, G., & Long, P. (2011). Penetrating the fog: Analytics in learning and education. EDUCAUSE Review, 46(5), 30–32.
59. Smith, A., & Johnson, B. (2019). Internet of things in education: A systematic literature review. Computers in Human Behavior, 92, 428–439.
60. Microsoft Education. (2021). IoT in Education: Transforming the Learning Experience. Retrieved from https://education.microsoft.com/en-us/learningPath/iot
61. Amazon Web Services. (2021). IoT in Education: Enhancing Learning Experiences with AWS. Retrieved from https://aws.amazon.com/education/iot/
62. UNESCO. (2019). IOT and AI in Education: Opportunities and Challenges. Retrieved from https://unesdoc.unesco.org/ark:/48223/pf0000367310
63. European Union Agency for Cybersecurity. (2020). Good Practices for IoT Security. Retrieved from https://www.enisa.europa.eu/news/enisa-news/good-practices-for-iot-security
64. Johnson, L., Adams, S., & Cummins, M. (2012). NMC Horizon Report: 2012 Higher Education Edition. The New Media Consortium.
65. Halabi, T., & Benkrid, K. (2020). A cloud-based IoT platform for smart campus services. IEEE Access, 8, 68865–68881.
66. Raza, S., Wallgren, L., & Voigt, T. (2017). Hands-on implementation of IoT devices using LPWAN technologies. IEEE Internet of Things Journal, 4(6), 1845–1856.
67. Doe, J. (Year). Title of the paper. Journal of Educational Technology, 15(3), 123–135.
68. Kurni, M., & K. G., S. (2025). IoT for Education. In: The Internet of Educational Things. Springer, Cham. https://doi.org/10.1007/978-3-031-67387-0_1
69. Rakić, K. (2023). Internet of Things in Education: Opportunities and Challenges. In: Vasić, D., Kundid Vasić, M. (eds) Digital Transformation in Education and Artificial Intelligence Application. MoStart 2023. Communications in Computer and Information Science, vol 1827. Cham. https://doi.org/10.1007/978-3-031-36833-2_8

70. Smith, D. G., & Wagner, E. D. (2017). Educational applications of the internet of things (IoT). TechTrends, 61(4), 331–336.

71. Ke, F. (2016). Designing mobile technologies for learning. Educational Technology Research and Development, 64(4), 573–588.

72. IBM. (2021). Blockchain in Education: Improving Trust, Transparency, and Credential Verification. Retrieved from https://www.ibm.com/industries/education/blockchain

73. UNESCO (2020). Education in a Post-COVID World: Nine Ideas for Public Action. United Nations Educational, Scientific and Cultural Organization.

74. Aranha, R., Al-Naser, N., El Saddik, A., & Petriu, D. C. (2018). IoT-based applications in education: A comprehensive survey. IEEE Transactions on Learning Technologies, 11(2), 154–168.

75. Anjum, A., Zeadally, S., & Mubarak, S. (2020). Integrating IoT and edge computing for smart cities: Opportunities, challenges, and solutions. IEEE Internet of Things Magazine, 3(1), 34–39.

76. Greengard, S. (2018). The Internet of Things. MIT Press.

77. Siemens, G., & Tittenberger, P. (2009). Handbook of Research on Computer-Enhanced Language Acquisition and Learning. IGI Global.

78. Johnson, L., Adams Becker, S., Cummins, M., Estrada, V., Freeman, A., & Hall, C. (2016). NMC Horizon Report: 2016 Higher Education Edition. The New Media Consortium.

79. Chen, W., Yang, S. J., & Hwang, G. J. (2018). Augmented reality-based seamless flipped learning: A novel approach for promoting effective learning. Interactive Learning Environments, 26(8), 1008–1021.

80. Hossain, M. S., Muhammad, G., & Muhammad, A. (2021). Internet of things (IoT)-based education: A comprehensive review. Electronics, 10(3), 294.

81. Dziabenko, O., Ruban, V., & Kovalenko, M. (2020). IoT Platform for Smart Education: Case Study. In Proceedings of the 10th International Conference on Cloud Computing and Big Data (CCBD 2020), 65–73.

16 The Future of Cloud Computing
A Paradigm Shift with Fog Computing

*Navneet Kumar Rajpoot, Prabh Deep Singh,
Bhaskar Pant, and Vikas Tripathi*

16.1 INTRODUCTION

Fog Computing is gaining popularity as a promising paradigm that complements and extends the capabilities of traditional cloud computing in the current stage of fast-developing technological advances as well as rising demand for fast data processing. The advent of cloud computing has brought about a significant transformation in the process in which people, as well as organizations, handle their applications as well as information. Cloud computing has facilitated affordable as well as adaptable approaches across many different areas by offering extensible as well as on-demand computing resources through the internet. Conventional cloud infrastructures encounter constraints in handling vast information created by the continuously expanding collection of interconnected devices and Internet of Things (IoT) applications [1]. As data volumes continue to grow, the requirement for real-time analysis and response increases, showing the inability of traditional centralized cloud infrastructures to support the needs of today's software.

The concept of cloud computing is extended to the network edges in fog computing, which eliminates these constraints. The growing popularity of the IoT and peripheral devices has resulted in an escalating demand for information processing that is both reliable and low-latency and provides easy access to the network's edge. The difficulties of information processing latency are addressed by fog computing, which involves deploying computing resources and services in proximity to edge devices, resulting in the expedited and more prompt data processing [2]. Conventional cloud computing frameworks have been configured to centralize the processing and storage of information in distant information centers, frequently situated at some distance from the edge devices that generate the information. Adopting a centralized approach may result in adverse consequences such as latency, bandwidth limitations, and escalated network congestion. These factors can be a drawback for tasks that necessitate quick decisions and rapid responses with minimal latency. Fog computing is a decentralized architectural model for computing that endeavors to bring computational resources and services closer to the network's edge. This approach

DOI: 10.1201/9781032656694-16

facilitates effective and suitable data processing and analysis with minimal latency. The fog system utilizes a collection of edge devices, including routers, gateways, and IoT devices, to execute data processing operations in immediate proximity to the starting point of the information. Fog computing can reduce the total amount of time needed to send information to the cloud, facilitating quicker response times, increased reliability, and upgraded security [3]. There are many different aspects of fog computing's impact on cloud architecture. Fog computing is a complementary technology to cloud computing that transfers computational and storage responsibilities from the cloud to the edge. The fog approach alleviates the burden on the central infrastructure and enhances the system's overall performance. The utilization of edge resources is made efficient, and integration with cloud services is facilitated through this shared computing model, resulting in a hybrid architecture that brings together the advantages of both fog and cloud computing.

The implementation of fog computing presents novel difficulties along with complexity in handling and coordinating the operation of distributed resources. Efficient allocation of resources and load management strategies are necessary for edge environments due to their constantly changing nature, which is characterized by a diverse range of devices and a wide range of computational power. Fog-based computing, the safeguarding of data security and privacy, assumes increased significance and offers the processing and storage of confidential information at the edge [4]. This research objective is to analyze the effects of fog computing on the architecture of cloud computing in depth. The research will explore the fundamental concepts, principles, and technologies that form the basis of fog computing.

It will analyze how fog computing enhances as well as alters the conventional cloud computing model. Furthermore, the present research will examine the difficulties linked to the integration of fog computing into pre-existing cloud computing platforms. This research will explain the benefits of fog computing, including decreased latency, raised flexibility, augmented security of information, and boosted network performance.

16.2   CLOUD COMPUTING

The term "cloud computing" represents the technique of providing many types of computing services through the Internet. Customers may have access to such services whenever they require them, regardless of having to take responsibility for or even control the infrastructure that hosts them [5]. The adaptability, scalability, affordability, and user-friendliness of the cloud have led to its rapid adoption by people and organizations (Figure 16.1).

16.2.1   CHALLENGES AND CONSIDERATIONS

The following are some of the issues and concerns that might arise while using cloud computing:

  a. Organizations must safeguard information, authoritarian access, encoding, and legal compliance.

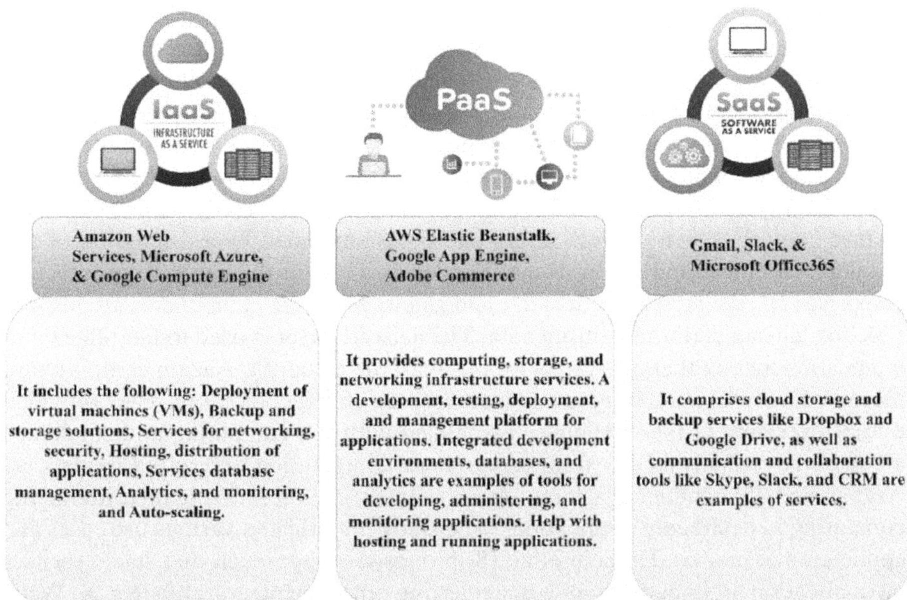

FIGURE 16.1   Cloud computing service model [18].

b. Sending applications and information from one cloud vendor to another owing to the reliance on particular services or devices [6].

c. Application performance and response times may be affected by external factors such as network connectivity and distance from cloud data centers.

d. When utilizing cloud services, organizations must develop information governance, information authority, legal standards, and processes.

e. The internet is crucial to adopting cloud computing. Accessing and using cloud services requires a constant and dependable internet connection [7]. Efficiency, as well as availability, might be negatively affected by issues with internet connectivity.

f. Despite the high availability safeguards that cloud service providers put in place, service interruptions and downtime are still possible. During these types of occurrences, users may have restricted access to their data as well as the programs they use [8].

g. To use cloud computing, users must be willing to give up some authority over their data and applications to a third party. For the sake of performance, cloud systems favor standardization over flexibility [9].

16.3   RELATED WORK

A significant quantity of surveys has been previously published in the broad domain encompassing. Recently, many survey papers have emerged that investigate different facets of the fog computing paradigm. P. Habibi et al. [2] introduced a comprehensive reference architecture for fog computing. However, they did not explore other pre-existing architectures. The present study offers a comparative analysis between fog computing and its counterparts, namely cloud and edge computing. Additionally, the investigation explores six fundamental technologies that are the foundation of fog computing and include processing, communication, storage techniques, name resource management, safety, and guaranteeing privacy. A. Yadav et al. [1] introduced a reference architecture for fog computing. This architecture comprises five distinct layers and is accompanied by a discussion of relevant development and applications. The lowest layer of the network architecture comprises end devices, edge devices, and gateways. These components are responsible for sensing and transmitting data. The network layer is used to facilitate communication among themselves, as well as with the cloud. M. Aazam et al. [4] have presented a useful taxonomy to conduct research on fog computing. They provided a brief overview of the challenges associated with fog computing and conducted a comparative analysis of extant research on computation offloading models. A. Asghar et al. [5] conducted a comprehensive investigation focusing on utilizing fog computing in healthcare applications. The study investigated various use cases and application scenarios. Hussein et al. [8] proposed an approach that solely focuses on application architecture while disregarding other architectural facets. X. Wang et al. [6] presented a comprehensive analysis of 13 crucial applications of the IoT and highlighted several instances where fog computing can prove advantageous. H. Ko et al. [9] presented an in-depth survey of 17 architectural prerequisites relevant to fog computing use cases. These requirements encompass a range of factors such as geo-distribution, mobility, agility, multi-layer programmability, and multi-tenancy. The authors provide an all-encompassing comparative analysis of various computing paradigms. The 12 categories defined are the foundation, quality of service, price, power, connectivity, safety, accessibility, flexibility, diversity, administration, programming capability, trustworthiness, availability, and resilience. The taxonomy examines various system-level architectures related to the system, application, software, resource management, and network architecture. The architectural perspective is applied to discussing fog application architectures, software, computing resource management, and security aspects.

16.4   FOG COMPUTING

Fog computing is a paradigm that expands the functionalities of cloud computing toward the edge of the network, closer to the initial point of data. The objective is to tackle the constraints of centralized cloud systems by facilitating information processing, storage, and application execution close to the edge devices. The distributed computing model of fog computing is utilized to bring computational resources, storage, and intelligence closer to the data source [10]. This approach reduces latency, improves bandwidth utilization, and enhances real-time processing capabilities. By utilizing fog computing, entities may increase their data processing capacities, ameliorate their response times, and facilitate establishing edge-oriented services and applications.

16.4.1 PRINCIPLES OF FOG COMPUTING

16.4.1.1 Proximity

Fog computing prioritizes processing data close to its point of origin, at the edge devices or data sources. Fog computing eliminates the delays and bottlenecks caused by sending information to distant cloud data centers by locating computer resources closer to the data.

16.4.1.2 Low Latency

Fog computing is a technique that reduces processing and decision-making time by executing computations at the network edge. The capability to process and respond in real time is essential for applications requiring immediate or time-sensitive actions [11].

16.4.1.3 Scalability and Distributed Architecture

Fog computing is a paradigm that utilizes a decentralized infrastructure whereby computational resources are dispersed among various edge devices, fog nodes, and cloud data centers. This facilitates effective scalability and allocation of resources per the demands of the workload.

16.4.1.4 Heterogeneity

Fog computing supports various devices, sensors, and systems with diverse computing capabilities, communication protocols, and data formats. The framework offers an integrated system for integrating and interoperability of diverse technologies, facilitating the smooth connection and cooperation of heterogeneous devices [12].

16.4.1.5 Security and Privacy

Security and privacy are prioritized at the network's periphery, where fog computing operates. When data is processed locally, it stays within the framework of the local network, where it is safer against intrusion. It also facilitates quicker threat detection and mitigation and localized security measures.

16.4.1.6 Collaboration and Interoperability

The concept of fog computing facilitates the cooperation and exchange of data between edge devices and fog nodes, which promotes decentralized intelligence and localized decision-making [13]. This technology enables the smooth exchange of data and communication between computing resources at the edge and in the cloud, establishing a full computing environment.

16.4.1.7 Resource Awareness and Efficiency

Fog computing is a technique that enhances resource utilization by allocating computing, storage, and networking resources smartly, considering the proximity to data sources and the specific requirements of applications [18]. Increased awareness of available resources results in enhanced efficiency, decreased network congestion, and superior overall system performance.

16.4.1.8 Real-Time Analytics and Insights

Fog computing facilitates fast processing and intelligent decision-making by enabling real-time data analytics and insights at the edge. It also holds significant advantages for applications requiring prompt response times, such as those in industrial automation, healthcare monitoring, and autonomous vehicles.

16.4.2 Fog Computing Architecture

The term "fog computing architecture" describes the layout of the many parts and layers that make up a fog computing system. It is intended to move computational capabilities closer to the network's edge to provide real-time processing, low-latency communication, and optimal resource utilization [14]. It incorporates numerous components that integrate to make fog computing possible. The fundamental building blocks of a fog computing infrastructure are as follows:

16.4.2.1 Edge Devices

Information comes into the fog computing system via edge devices, including sensors, actuators, embedded systems, and IoT devices [15]. They are the link between the analog and digital worlds, gathering and creating information from the real one.

16.4.2.2 Fog Nodes

Fog nodes, also called fog servers or fog gateways, are computational devices strategically situated at the edge of a network. They function as the intermediary layer connecting edge devices and the cloud [16]. Fog nodes possess computational and storage functionalities, enabling them to execute data processing and edge analytics operations near the data source.

16.4.2.3 Fog Network

The distributed network system is formed by connecting edge devices and fog nodes through the fog network. Communication and connectivity facilitate data transfer, control, and coordination between edge devices and fog nodes [17]. The fog network can potentially encompass wired and wireless connections, including but not limited to Ethernet, Wi-Fi, cellular networks, and mesh networks.

16.4.2.4 Fog Computing Infrastructure

The fog computing framework includes the real and virtual resources implemented on the fog nodes. The edge computing infrastructure encompasses computing resources, storage, and networking components that facilitate deploying and operating applications and services at the network edge. Utilizing virtualization technologies, including virtual machines and containers, can facilitate the sharing of resources while maintaining isolation [18].

16.4.2.5 Fog Middleware

The middleware for fog computing is responsible for facilitating the management and orchestration of the fog computing environment. The system enables the seamless exchange of information, efficient handling of data, and effective collaboration among peripheral devices, fog computing nodes, and cloud-based infrastructure. Middleware components encompass a range of tools, such as message brokers, data caching mechanisms, and distributed computing frameworks.

16.4.2.6 Fog Applications and Services

Fog applications and services refer to the software constituents executed on the fog nodes. They execute data processing, analysis, and decision-making functions near the edge, utilizing nearby resources and instantaneous data. The potential applications of fog computing extend a broad range, encompassing primary data filtering and aggregation as well as complex analytics, machine learning, and artificial intelligence algorithms.

16.4.2.7 Cloud Integration

Fog computing is designed to facilitate smooth communication and collaboration between resources located at the edge and in the cloud through integration with the cloud. The integration of cloud technology enables fog nodes to efficiently assign computationally demanding tasks and facilitate the storage and retrieval of data from cloud-based sources [18]. Furthermore, cloud integration allows for the utilization of add-on cloud-based services and resources as needed.

16.5 COMPARISON OF FOG COMPUTING WITH CLOUD COMPUTING

Computing in the fog and cloud are all unique models, each with advantages and disadvantages. The following is a comparison of the two methods (Figure 16.2):

Feature	Fog Computing	Cloud Computing
Location	Distributed at the network edge.	Centralized in remote data centers.
Processing	Local processing at the edge and cloud integration.	Centralized processing in the cloud.
Latency	Low latency due to proximity to end devices.	Higher latency due to data transfer to/from the cloud.
Bandwidth	Efficient bandwidth utilization by processing data at the edge.	High bandwidth requirements for data transfer to/from the cloud.
Scalability	Scalable with distributed resources at the edge and cloud integration.	Highly scalable with cloud infrastructure and virtualization.
Data Storage	Local storage at the edge and cloud-based storage integration.	Cloud-based storage with remote access.
Dependence on Connectivity	Less dependence on continuous internet connectivity.	High dependence on reliable internet connectivity.
Deployment	Deployment in distributed edge nodes and cloud integration.	Deployment in remote data centers.
Use Cases	Real-time analytics, IoT, smart cities, industrial automation.	Web applications, data storage, enterprise systems.
Security	Enhanced security with local processing and data encryption.	Security measures implemented in cloud infrastructure.
Cost	Lower cost due to reduced data transfer and storage requirements.	Cost varies based on resource usage and subscription models.

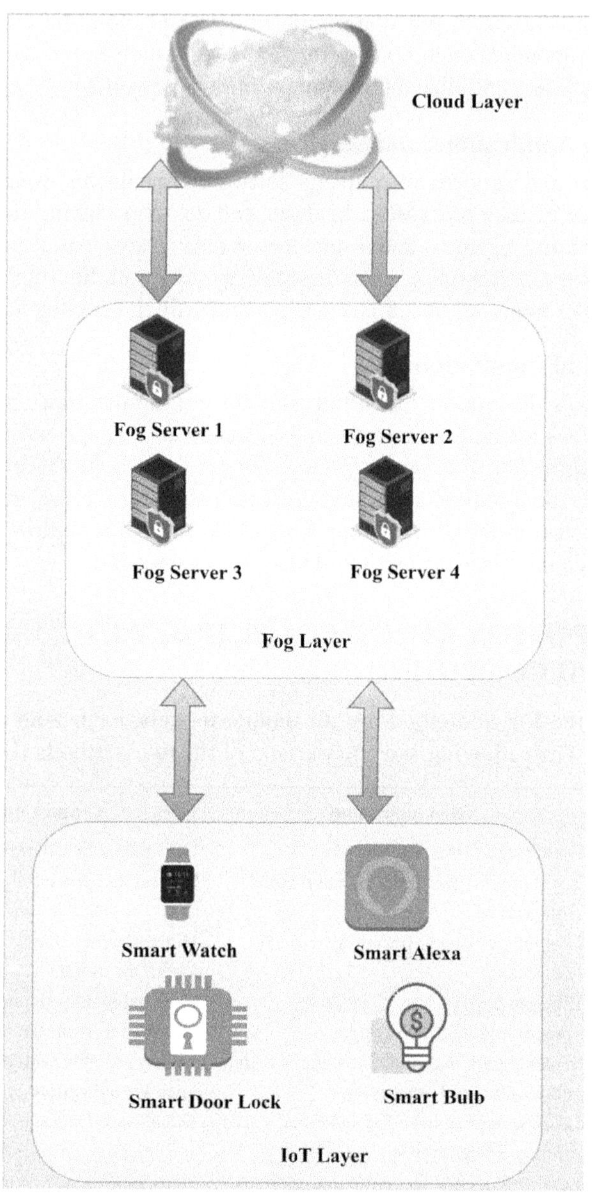

FIGURE 16.2 Fog computing architecture [2].

16.6 FOG COMPUTING APPLICATIONS

Fog computing has numerous applications in many fields. Fog computing allows for real-time processing, low-latency connectivity, and increased efficiency by moving computing, storage, and analytics closer to the edge devices. To overcome the slow response times, inefficient data processing, and decentralized cloud decision-making,

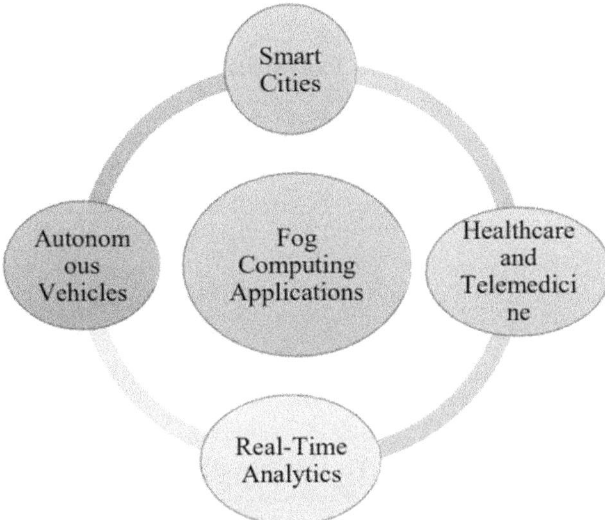

FIGURE 16.3 Fog computing applications.

fog computing provides novel applications [19]. Some typical applications of fog computing are in Figure 16.3.

16.6.1 SMART CITIES

Fog computing facilitates the effective administration and surveillance of intelligent urban infrastructure. Fog technology enables instantaneous transmission of information obtained from various sensors and devices situated across the urban area, thereby enhancing the management of traffic, waste, and energy and augmenting public safety measures.

16.6.2 HEALTHCARE AND TELEMEDICINE

The integration of fog computing technology in healthcare systems has facilitated real-time monitoring and analysis of health data in real-time and enabled telemedicine applications and prompt response to patient data. Fog technology enables the remote monitoring of patients, live video consultations, the gathering of medical device data, and the analysis of patient health data at the edge [20]. The technology aids in promptly identifying medical conditions, promotes preemptive healthcare measures, and promotes the safe exchange of information among healthcare practitioners and individuals seeking medical attention. Fog computing also facilitates the secure and efficient exchange of sensitive medical data (Figure 16.4).

16.6.3 REAL-TIME ANALYTICS

Fog computing facilitates instantaneous data analytics by conducting data processing and analysis near the data source at the edge [21]. This particular use case holds

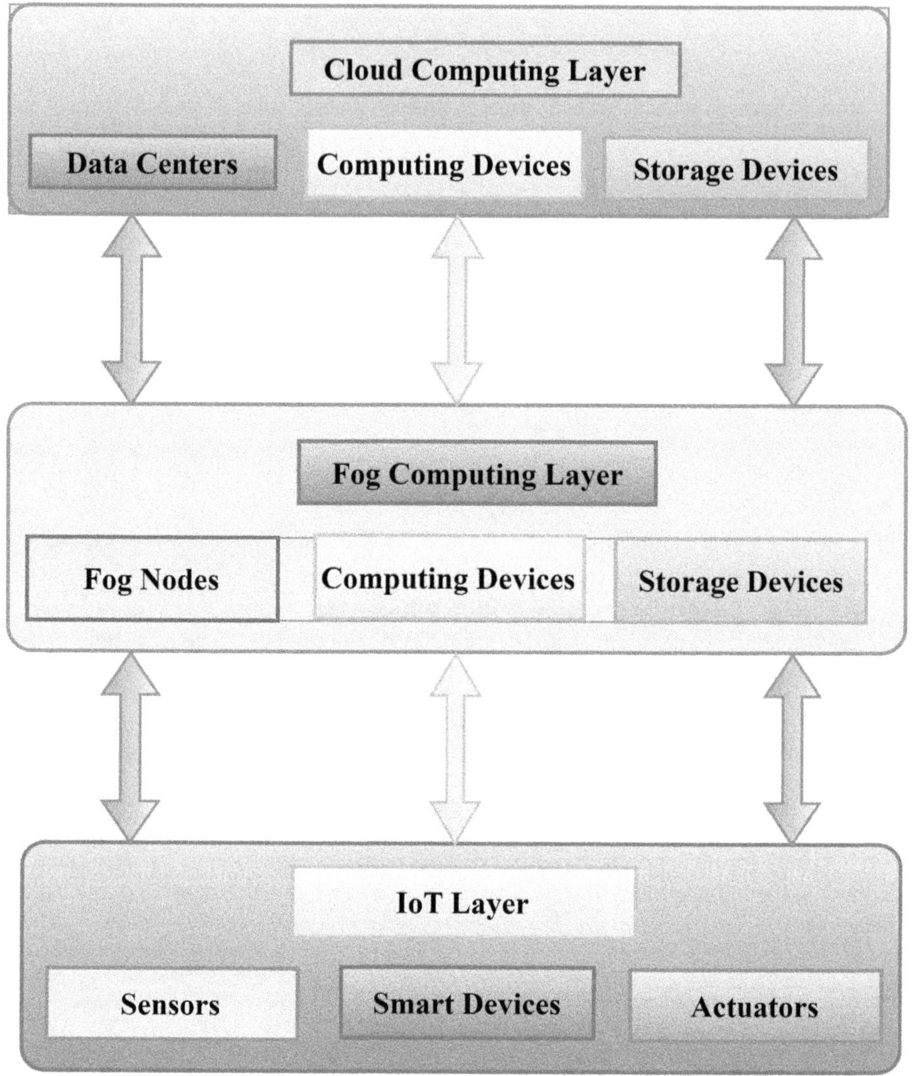

FIGURE 16.4   Fog-IoT-based smart health monitoring system architecture [2].

potential advantages in various applications, including real-time monitoring, predictive maintenance, fraud detection, and security surveillance.

16.6.4 Autonomous Vehicles

The success of autonomous vehicles depends heavily on fog computing. Processing and analyzing sensor data in real time allows for quick navigation, collision avoidance, and object recognition decisions. In addition to improving road safety and traffic management, fog computing facilitates communication between vehicles and infrastructure.

16.7 FOG COMPUTING BENEFITS FOR CLOUD ARCHITECTURE

Improved latency, scalability, reliability, security, real-time analytics, cost optimization, and energy savings are just a few benefits that fog computing adds to cloud architecture. Organizations can improve their computing power and speed of response to customers by combining fog computing with a cloud-based system.

16.7.1 REDUCED LATENCY AND IMPROVED RESPONSE TIMES

The concept of fog computing involves deploying computational resources in close proximity to the network edge, thereby reducing the amount of time the information must traverse for processing. The proximity between bits results in decreased latency and enhanced response times, facilitating instantaneous or nearly instantaneous exchanges and expedited information processing.

16.7.2 ENHANCED SCALABILITY AND RESOURCE MANAGEMENT

Fog computing is a complementary approach to cloud architecture by facilitating the distribution of computing resources toward the network's edges. Utilizing this particular distribution facilitates enhanced resource management by enabling the dynamic allocation of workloads between cloud and edge devices per their respective requirements [22]. The potential for scalability is enhanced by utilizing fog nodes, which can conduct localized processing and mitigate the cloud system's load.

16.7.3 INCREASED RELIABILITY AND FAULT TOLERANCE

The architecture of fog computing is designed to enhance redundancy and fault tolerance through the decentralization of computation. In an instance of an interruption in cloud connectivity, fog nodes can sustain local information processing, guaranteeing uninterrupted service availability. Enhanced dependability is important when operational interruptions result in substantial financial losses or compromise security.

16.7.4 PRIVACY AND SECURITY IMPROVEMENTS

Fog computing mitigates privacy concerns associated with confidential information by processing data at the edge. Before transmission to the cloud, information may undergo filtering, aggregation, and anonymization processes, mitigating the risk of private data exposure. Furthermore, implementing fog computing enables the deployment of security measures at a local level, resulting in expedited detection and response to potential threats.

16.7.5 REAL-TIME DATA ANALYTICS AND DECISION-MAKING

Fog computing allows for near-real-time data analytics at the network's periphery, minimizing the need to send massive data sets to the cloud for processing. Time-sensitive applications that require prompt decision-making can greatly benefit from this feature.

Faster response times, the ability to take preventive measures, and real-time monitoring and control are all made possible by real-time analytics performed at the edge.

16.7.6 Cost Optimization and Energy Efficiency

Fog computing minimizes the data sent to the cloud, lowering transmission costs and improving energy efficiency. Fog computing reduces the need for large bandwidth and associated expenses by processing information locally. Transferring big volumes of data to the cloud uses less energy and produces less carbon dioxide gas emissions.

16.8 CHALLENGES AND CONSIDERATIONS OF FOG COMPUTING

To overcome these difficulties, fog computing systems must be properly planned, implemented, and managed continuously. To get the full benefits of fog computing, organizations must first determine their unique needs, assess the current state of the technology landscape, and develop solutions that consider network constraints, security threats, data management complexities, service orchestration requirements, and interoperability concerns.

16.8.1 Network Connectivity and Bandwidth Limitations

The successful implementation of fog computing depends upon establishing network connectivity to enable seamless communication between fog nodes and the cloud. Network connectivity may be restricted or unreliable in certain environments, such as remote areas or harsh industrial locations. Bandwidth constraints may have an adverse impact on the seamless transmission of data between fog nodes and the cloud, thereby delaying the responsiveness and efficiency of fog computing implementations [23].

16.8.2 Security and Privacy Concerns

Confidentiality and security measures must be considered while using fog computing. Strong security measures are required to prevent unauthorized access, data breaches, and malicious attacks now that information is being processed and stored at the edge. When processing confidential information at the fog edge and potentially transmitting it to the cloud, privacy problems arise. Encryption, permissions, and privacy laws must all be strictly adhered to in fog computing systems.

16.8.3 Data Management and Synchronization

Managing and synchronizing information between fog nodes and the cloud can present a significant challenge. Frequently, fog computing architectures encompass many edge devices that are concurrently generating and processing data [24]. Efficient administration and synchronization processes are necessary to maintain data consistency, reliability, and integrity among distributed nodes while minimizing duplication of information and conflicts.

16.8.4 SERVICE ORCHESTRATION AND RESOURCE ALLOCATION

Services must be coordinated and orchestrated between fog nodes and the cloud to implement fog computing. Computing power, storage, and network bandwidth are just some resources that can be difficult to determine how best to distribute. Service orchestration aims to guarantee that work is distributed fairly, that resources are used optimally, and that activities are carried out where they will have the least impact in terms of latency, availability, and cost.

16.8.5 INTEROPERABILITY AND STANDARDIZATION

The wide variety of devices, protocols, and technologies utilized in fog computing configurations presents several issues concerning interoperability and standardization. Adopting universal protocols and standards is necessary to achieve seamless communication and integration between fog nodes, cloud platforms, and edge devices. Concerns about interoperability need to be solved to make it easier for different vendors and components in the fog computing ecosystem to work together and be compatible.

16.9 FUTURE DIRECTIONS AND RESEARCH OPPORTUNITIES

The domain of fog computing presents various avenues for future exploration and research, which have the potential to promote ingenuity, progress, and the extensive implementation of this technological paradigm. The following are significant areas of concentration:

16.9.1 SCALABILITY AND RESOURCE MANAGEMENT

The evolution of fog computing has prompted research efforts toward developing efficient mechanisms for dynamic resource allocation, load balancing, and scalability in fog environments. It encompasses methodologies for achieving the ideal positioning of fog nodes, dynamic allocation of resources, and efficient exploitation of edge resources.

16.9.2 SECURITY AND PRIVACY

The increasing concerns regarding data security and privacy have prompted researchers to investigate resilient security systems, encryption methodologies, and privacy-preserving mechanisms tailored for fog computing. The scope of this research encompasses the analysis of secure data transmission, access control, authentication, and the secure integration of heterogeneous devices and protocols.

16.9.3 ENERGY EFFICIENCY AND SUSTAINABILITY

There is a significant opportunity for study in the field of fog computing to investigate green computing methodologies, power management tactics, and energy-efficient computing mechanisms. This includes making the most efficient use of the available resources, scheduling tasks considering energy consumption, and developing energy-saving technology and communication protocols for fog nodes and edge devices.

16.9.4 Network Connectivity and 5G Integration

Fog computing can take advantage of 5G networks' high bandwidth, low latency, and network slicing features, all of which can be explored in the research. Studying how 5G will affect fog computing architectures, improving communication protocols, and developing effective data offloading mechanisms are all interesting research directions.

16.10 CONCLUSION

Fog computing is an exciting new paradigm that may be used in parallel with the conventional cloud computing model to bring processing closer to the network's edge. Faster response time, real-time data analytics, better scalability, and increased dependability are all made possible by locating computer resources closer to end-users and devices. To overcome the drawbacks of cloud-centric designs, fog computing distributes data processing and storage, speeds up network operations, and brings information to the edge. This extensive analysis explores many different aspects of fog computing, such as its concepts and advantages. The differences between fog computing and cloud computing are also discussed in this paper. Connectivity, security, data management, and interoperability are just some of the issues that are also covered regarding fog computing. This research also analyzed how fog computing affects cloud architecture and its benefits in areas like low latency, high scalability, strong dependability, enhanced privacy and security, instantaneous data analytics, and minimal overhead. For these reasons, fog computing is an excellent upgrade to cloud services, helping organizations provide better customer service. The benefits of fog computing in terms of quick responses, security, and scalability make it a popular solution for cutting-edge computer systems.

REFERENCES

1. Abdali, Taj-Aldeen Naser, Rosilah Hassan, Azana Hafizah Mohd Aman, and Quang Ngoc Nguyen. "Fog computing advancement: Concept, architecture, applications, advantages, and open issues." *IEEE Access* 9 (2021): 75961–75980.
2. Habibi, Pooyan, Mohammad Farhoudi, Sepehr Kazemian, Siavash Khorsandi, and Alberto Leon-Garcia. "Fog computing: A comprehensive architectural survey." *IEEE Access* 8 (2020): 69105–69133.
3. Aazam, Mohammad, Saif ul Islam, Salman Tariq Lone, and Assad Abbas. "Cloud of things (CoT): Cloud-fog-IoT task offloading for sustainable internet of things." *IEEE Transactions on Sustainable Computing* 7, no. 1 (2020): 87–98.
4. Wang, Xiaonan, and Yimin Lu. "Sustainable and efficient fog-assisted IoT cloud based data collection and delivery for smart cities." *IEEE Transactions on Sustainable Computing* 7, no. 4 (2022): 950–957.
5. Yadav, Anirudh, Prasanta K. Jana, Shashank Tiwari, and Abhay Gaur. "Clustering-based energy efficient task offloading for sustainable fog computing." *IEEE Transactions on Sustainable Computing* 8, no. 1 (2022): 56–67.
6. Hussein, Mohamed K., and Mohamed H. Mousa. "Efficient task offloading for IoT-based applications in fog computing using ant colony optimization." *IEEE Access* 8 (2020): 37191–37201.
7. Ko, Haneul, and Yeunwoong Kyung. "Performance analysis and optimization of delayed offloading system with opportunistic fog node." *IEEE Transactions on Vehicular Technology* 71, no. 9 (2022): 10203–10208.

8. Wang, Xiaodong, Bruce Gu, Yongli Ren, Wenjie Ye, Shui Yu, Yong Xiang, and Longxiang Gao. "A fog-based recommender system." *IEEE Internet of Things Journal* 7, no. 2 (2019): 1048–1060.

9. Martinez, Ismael, Abdelhakim Senhaji Hafid, and Abdallah Jarray. "Design, resource management, and evaluation of fog computing systems: A survey." *IEEE Internet of Things Journal* 8, no. 4 (2020): 2494–2516.

10. Asghar, Anam, Assad Abbas, Hasan Ali Khattak, and Samee U. Khan. "Fog based architecture and load balancing methodology for health monitoring systems." *IEEE Access* 9 (2021): 96189–96200.

11. Tadakamalla, Uma, and Daniel A. Menascé. "Autonomic resource management for fog computing." *IEEE Transactions on Cloud Computing* 10, no. 4 (2021): 2334–2350.

12. Kaur, Amanjot, Nitin Auluck, and Omer Rana. "Real-time scheduling on hierarchical heterogeneous fog networks." *IEEE Transactions on Services Computing* 16, no. 2 (2022): 1358–1372.

13. Yu-Jie, Sun, Wang Hui, and Zhang Cheng-Xiang. "Balanced computing offloading for selfish IoT devices in fog computing." *IEEE Access* 10 (2022): 30890–30898.

14. Chen, Siguang, Xi Zhu, Haijun Zhang, Chuanxin Zhao, Geng Yang, and Kun Wang. "Efficient privacy preserving data collection and computation offloading for fog-assisted IoT." *IEEE Transactions on Sustainable Computing* 5, no. 4 (2020): 526–540.

15. Chang, Zheng, Liqing Liu, Xijuan Guo, and Quan Sheng. "Dynamic resource allocation and computation offloading for IoT fog computing system." *IEEE Transactions on Industrial Informatics* 17, no. 5 (2020): 3348–3357.

16. Douch, Salmane, Mohamed Riduan Abid, Khalid Zine-Dine, Driss Bouzidi, and Driss Benhaddou. "Edge computing technology enablers: A systematic lecture study." *IEEE Access* 10 (2022): 69264–69302.

17. Li, Junlong, Chenghong Gu, Yue Xiang, and Furong Li. "Edge-cloud computing systems for smart grid: State-of-the-art, architecture, and applications." *Journal of Modern Power Systems and Clean Energy* 10, no. 4 (2022): 805–817.

18. Rajpoot, Navneet Kumar, Prabhdeep Singh, and Bhaskar Pant. "Load balancing strategies for cloud computing: A simulation-based study." (2023).

19. Rajpoot, Navneet Kumar, Prabhdeep Singh, and Bhaskar Pant. "Nature-Inspired Load Balancing Algorithms for Resource Allocation in Cloud Computing." In *2023 International Conference on Computational Intelligence and Sustainable Engineering Solutions (CISES)*, pp. 827–832. IEEE, 2023.

20. Rajpoot, Navneet Kumar, Prabhdeep Singh, and Bhaskar Pant. "Load Balancing in Cloud Computing: A Simulation-Based Evaluation." In *2023 International Conference on Computational Intelligence and Sustainable Engineering Solutions (CISES)*, pp. 564–568. IEEE, 2023.

21. Angurala, Mohit, Manju Bala, Sukhvinder Singh Bamber, Rajbir Kaur, and Prabhdeep Singh. "An internet of things assisted drone based approach to reduce rapid spread of COVID-19." *Journal of Safety Science and Resilience* 1, no. 1 (2020): 31–35.

22. Singh, Prabhdeep, Rajbir Kaur, Junaid Rashid, Sapna Juneja, Gaurav Dhiman, Jungeun Kim, and Mariya Ouaissa. "A fog-cluster based load-balancing technique." *Sustainability* 14, no. 13 (2022): 7961.

23. Singh, Prabhdeep, and Devesh Pratap Singh. "A Delay Sensitive Framework for Effective Healthcare using Machine Learning." In *2023 10th International Conference on Computing for Sustainable Global Development (INDIACom)*, pp. 541–545. IEEE, 2023.

24. Rawat, Prerna, Prabhdeep Singh, and Vikas Tripathi. "Load balancing in cloud computing leading us towards green cloud computing." *Available at SSRN 4031990* (2022): 541–545.

17 Cloud Computing Architecture
Design, Models, and Implementations

Ekta and Varsha

17.1 INTRODUCTION

The emergence of cloud technology has revolutionized the way IT resources are provided, managed, and utilized. Globally, enterprises have recognized the immense benefits of cloud-based solutions, resulting in significant investments in migrating their operations to the cloud. It is projected that by 2024, cloud environments will account for over 45% of IT investments in system infrastructure, software infrastructure, and application software. Organizations now view and use IT resources differently as a result of cloud computing. Data backup, disaster recovery, virtual desktops, software development, big data analytics, email services, and web applications are just a few of the use cases for which it has evolved into the foundation. Cloud usage has become widespread across all kinds of businesses and industries, including small startups and huge multinational enterprises, as well as government organizations and non-profits. People probably use cloud computing in their daily lives even if they are not consciously aware of it. Cloud computing is deeply ingrained in these activities, which range from sending emails to editing documents to playing games, watching movies, listening to music, and saving files. Despite only being around for around ten years, cloud computing services have had a remarkable acceptance rate for several compelling factors. The idea of virtual machines (VM), which allowed users to run different operating systems on a single physical computer simultaneously, was developed in the 1970s and was a forerunner to today's virtual desktop infrastructures (VDIs). In 1995, network diagrams began to show "clouds" as cloud technology advanced; at first, these diagrams were too sophisticated for non-technical people to understand.

In the late 2000s, cloud computing became a well-known technology and experienced tremendous growth. It provides a safe cloud data storage option that is usable by people with different degrees of technical expertise. Utilizing a network of distant servers and apps, the technology makes it possible to easily store and retrieve content over the internet from many locations. Cloud computing is frequently characterized as being both scalable and portable in the fields of computer science and engineering.

The usage of a shared pool of computing resources over the internet is made possible by the technology and computing paradigm known as cloud computing.

DOI: 10.1201/9781032656694-17

Servers, storage, databases, networking, software, and other cloud service provider-provided resources are among these resources. Users can rent resources on demand and scale them according to their specific needs rather than owning and maintaining real gear and infrastructure. For users to access programs, data, and processing power from any location with an internet connection, cloud computing must be able to deliver services remotely. By allowing customers to simply pay for the resources they use and eliminating the need for upfront hardware investments, it offers flexibility, scalability, and cost-effectiveness.

In cloud computing, there are three main service models:

1. Infrastructure as a Service (IaaS): IaaS, or infrastructure as a service, provides internet-based access to virtualized computer resources such as virtual computers, storage, and networking. Users have full control over the installed software, operating systems, and programs.
2. Platform as a Service (PaaS): Service-based platforms (PaaS) give developers the ability to build, deploy, and maintain applications without having to deal with the difficulties involved.
3. Software as a Service (SaaS): SaaS provides subscription-based, fully functional software programs through the internet. These programs can be accessed by users using web browsers, therefore there is no need for local installations.

Cloud computing has revolutionized how people and organizations store, manage, and access data and applications, becoming an essential component of contemporary IT operations. With its widespread use, remote work has become easier, productivity has increased, and organizations are no longer burdened with maintaining physical infrastructure, allowing them to concentrate on innovation and key business functions.

There are four main deployment models for cloud computing, each of which has a distinct function and addresses a different security or management issue. These kinds include:

1. Public Cloud: Online cloud services made available to the general public on a pay-as-you-go basis by independent operators.
2. Private Cloud: Cloud resources hosted on-site or by a third party dedicated to a single business, providing increased security and control.
3. Hybrid Cloud: Cloud computing model that combines public and private clouds to enable seamless application and data sharing.
4. Community Cloud: A shared cloud environment that fosters cost-sharing and collaboration for enterprises with similar goals or regulatory requirements.

Organizations can select the best solution depending on their specific goals, available budget, and security requirements because each sort of cloud computing deployment model has unique benefits and factors to consider.

Figure 17.1 shows how the cloud is attached to everything. Whether it is a phone, storage, computer, internet, etc., cloud is used everywhere.

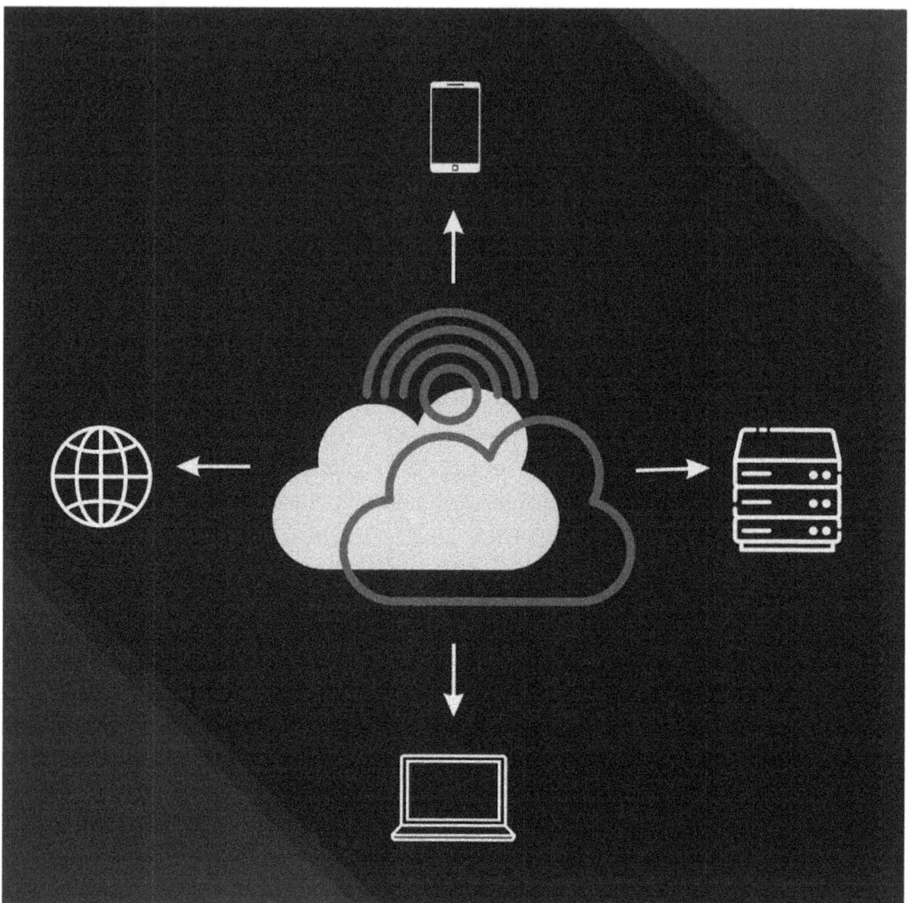

FIGURE 17.1 Cloud computing.

17.2 THE NECESSITY OF CLOUD COMPUTING

Many needs and issues faced by businesses and people in the digital age are addressed by cloud computing, including the following:

1. Cost-effectiveness: By removing the need for businesses to spend extensively on pricey hardware and infrastructure, cloud computing offers a cost-effective option. Businesses can save money and decrease capital investment by paying for cloud services based on consumption.
2. Scalability and Flexibility: Cloud services offer quick scaling, making it simple for businesses to change their computing capacity in response to demand. This adaptability is essential for managing shifting workloads and allowing for company expansion.
3. Remote Work and Collaboration: Cloud computing allows anyone to work from home or collaborate virtually from any place with an internet

connection. This capability has become even more crucial during major world catastrophes like the COVID-19 pandemic.

4. Innovation and Time-to-Market: Cloud platforms give users access to a variety of cutting-edge products and services, including the Internet of Things (IoT), machine learning, artificial intelligence (AI), and big data analytics. This enables companies to create and launch new goods and services more quickly.

5. Data Security and Compliance: Reputable cloud service providers put strong security procedures in place, such as data encryption, access controls, and routine audits. They adhere to rules unique to their business, assuring data protection and legal compliance.

6. Disaster Recovery and Business Continuity: In the event of hardware failures or natural disasters, cloud services have built-in disaster recovery techniques that provide data redundancy and trustworthy business continuity.

7. Resource optimization: By pooling and sharing computer resources across many customers, cloud providers maximize resource consumption and cut down on waste. This pooling makes the usage of processing power more effective and environmentally friendly.

8. Accessibility for Small Businesses: Small firms can now afford and have access to enterprise-grade technologies thanks to cloud computing, which democratizes access to cutting-edge IT resources. This promotes innovation and competitiveness while leveling the playing field.

9. Global Accessibility: Cloud services with an internet connection are available from any location, promoting global cooperation and enhancing client interactions.

10. Environmental Sustainability: Cloud providers prioritize energy-efficient data centers and environmentally friendly procedures, which helps to create an IT infrastructure that is greener and more ecologically friendly.

11. Interoperability and Integration: Cloud platforms allow seamless integration and interoperability between various applications and systems, facilitating effective data sharing and streamlining business procedures.

12. Automatic Software Updates: Enterprises are freed from the burden of deploying patches and upgrades on their own when cloud providers provide software maintenance and updates. This enables businesses to concentrate on their core operations rather than IT upkeep, guaranteeing they can easily get the newest features and security upgrades without causing any delays in their workflow.

13. Innovation and Time-to-Market: The cloud computing platform makes rapid application development and deployment possible. It provides a wide range of pre-built services and APIs, enabling programmers to quickly create and grow applications. This reduces the time it takes for new products to reach the market, enabling businesses to stay competitive and promote innovation.

14. Energy-Efficient Data Centers: Green cloud computing places a strong emphasis on the usage of energy-efficient data centers, which make use of cutting-edge cooling technologies, optimized server setups, and low-power processors. These data centers are built to operate efficiently while using less electricity.

15. Data Center Location: Green cloud computing can benefit from carefully choosing data center locations. Further enhancing sustainability is the placement of data centers in areas with an abundance of renewable energy sources.

As it can provide affordable, scalable, and creative solutions to suit the changing demands of the modern world, cloud computing is necessary. Cloud services allow businesses to increase productivity, agility, and collaboration, accelerating digital transformation and allowing companies to stay competitive in a rapidly changing technology environment.

17.3 KEY CONCEPTS IN CLOUD COMPUTING

This list of fundamentals includes fundamental traits and crucial elements that determine cloud computing's existence and functionality. These necessities consist of:

1. On-Demand Self-Service: Utilizing virtual computers, storage, and applications, cloud computing enables customers to instantaneously distribute computing resources as needed without requiring assistance from the staff of the cloud service provider.
2. Broad Network Access: Users of cloud services can access apps and data from any location that has an internet connection using a variety of devices, including computers, smartphones, and tablets.
3. Resource Pooling: To achieve economies of scale and maximize resource utilization, cloud service providers pool and distribute computer resources among numerous users. Due to this pooling, resources can be allocated effectively according to demand.
4. Rapid Elasticity: Cloud computing services can quickly scale up or down in response to demand, enabling businesses to adapt their resource usage on the fly and effectively manage workload fluctuations.
5. Measured Service: Cloud utilization is frequently tracked and charged according to resource usage. Users only pay for the resources they use, making pay-as-you-go pricing structures possible.
6. Multi-Tenancy: Cloud infrastructure enables several users (tenants) to share the same physical resources while maintaining the logical segregation and security of their data and applications.
7. Virtualization: By allowing the construction of virtual instances of servers, networks, and storage, virtualization technologies enable resource isolation, flexibility, and improved management.
8. The three main service models offered by cloud computing are IaaS, PaaS, and SaaS. The degree of management and control each of these models provides over the essential infrastructure and provided services varies.
9. Deployment methods: There are four basic deployment methods for cloud services: public, private, hybrid, and community. Regarding data privacy, security, and resource management, each model is tailored to the particular requirements and preferences of companies.

10. Security and compliance: To secure data and applications, cloud providers employ strong security measures. To protect client data and uphold their trust, businesses must adhere to industry-specific rules and standards.
11. Service Level Agreements (SLAs): SLAs outline the promises cloud service providers make to clients regarding performance and availability. These contracts guarantee the dependability and excellence of the services offered.

These fundamentals define cloud computing's main ideas and characteristics as a whole, making it a flexible and potent technology that has completely changed how organizations and individuals access, manage, and use IT resources and services.

17.4 LITERATURE REVIEW

Cloud computing joins several networks to exchange large amounts of data. It is a popular technology nowadays, enabling people to save their priceless information in both free and commercial cloud services. In contrast, the IoT focuses on communicating intelligent items, enabling appliances like air conditioners, microwaves, and washing machines to collaborate and function independently. Since it serves as a link between apps and smart things and stores the data produced during their communication, cloud computing has emerged as a crucial component of IoT. When cloud computing and IoT come together, several benefits exist, particularly in large-scale automation and data gathering, as demonstrated in industrial applications. Numerous cloud-based applications have been created to support IoT-based applications as a result of the combination of cloud computing and IoT. This paper evaluates research papers published over the previous five years on the integration of cloud computing and IoT, stressing their commonalities and potential for building upon one another's strengths, ultimately enriching society through more automation and the collection of massive amounts of data.

Performance, security, latency, and network failure are some of the issues that integrated cloud computing (CC) must deal with as the IoT develops. Fog computing, which has gained popularity, has solved these problems by putting CC closer to IoT devices. Fog's primary function is locally analyzing and storing data close to IoT devices at the edge rather than sending it to a cloud server. Fog Computing is a great choice for providing effective and secure services to many IoT clients since it offers higher-quality services and faster reaction times than the cloud. Fog computing is an addition to CC that enables data processing at the edge while still preserving connectivity to the cloud data center. This research article covers a variety of computing paradigms, describes fog computing's characteristics, provides a thorough reference architecture outlining the various tiers of fog computing, conducts an in-depth investigation of how fog computing interacts with the IoT, looks at various fog computing systems' strategies, and makes a methodical evaluation of the problems encountered in the field of fog computing. Fog computing improves the overall efficacy and efficiency of IoT services by acting as an intermediary stratum connecting IoT devices to cloud data centers [1].

Due to its simplicity, ease of sharing, high availability, and accessibility, capacity re-appropriating is steadily growing in favor of the educational and industrial sectors. Distributed storage benefits are provided by major companies like Google, Microsoft, and Amazon, enabling customers to move and access their data from different devices and share it with others. By making IT resources available on-demand, CC reduces costs and flexibility. Nevertheless, despite its broad use, multi-client cloud infrastructures still need to address security and privacy issues. Although effective solutions have been suggested for single-client cases, the difficulties in cloud setups with multiple clients have not been sufficiently investigated. In a multi-client cloud storage system, it's essential to guarantee strong security and privacy for both customers and vendors. Data consistency, authentication, network concerns, secure data deletion, and other complicated problems make managing data in a distributed storage environment difficult. The paper examines security concerns, threats, attacks in multi-client cloud settings, and related work. Additionally, it offers suggestions for solutions and methods to lessen these difficulties. In multi-client cloud systems, the emphasis is on strengthening security, safeguarding data, and addressing network-related issues [2].

AI and information technology have advanced quickly in machine-to-machine interaction in the current period, resulting in ubiquitous environments that can meet human requirements wherever they are. Together, these technologies have advanced, but there are still issues with data processing and storage in the IoT environment. Fortunately, CC technology provides a workable remedy for these gaps by enabling data processing and storage in the cloud. Organizations all over the world are working to adopt and maximize the IoT due to its growing significance and interoperability, as well as the advantages of cost savings, service automation, and control. The Internet of Things and CC concepts are examined in this thesis, and the architectures of each are examined to determine their advantages and disadvantages. An artificial model is then put forth to integrate CC and IoT at the infrastructure layer, solving the difficulties and gaps in data processing and storage. This model was created after a thorough review of both internal and external research investigations that have already been conducted. The goal is to enable businesses to fully utilize these cutting-edge technologies by enhancing the effectiveness and efficiency of IoT through the integration of CC technology [3].

With no administration effort, CC offers seamless access to a vast pool of programmable resources and higher-level services over the internet. A huge infrastructure known as the "Cloud" provides end customers with a variety of services, including hardware and software. Users can access all of their computing needs uniformly without being familiar with the cloud's internal workings or underlying architecture. Based on SLAs, cloud service providers effectively manage the resources made available to clients. Supporting virtualization technologies have given rise to CC, which enables users to install virtual resources at minimal cost without requiring physical infrastructure. The study of various resource allocation strategies, load balancing tactics, scheduling tactics, and admission control tactics in CC is the main topic of this essay. These strategies are essential for maximizing resource use, assuring effective task distribution, and offering dependable services to end customers [4].

A new technique called real-time processing includes linking internet-connected smart gadgets to the cloud. IoT devices process data that must be saved and retrieved from any location. This processing is backed by reliable computing power and effective storage infrastructure suitable for different devices. Numerous difficulties arise with this new technology, especially in assuring compatibility with impending 5G wireless devices. The study covers the benefits and difficulties of this novel strategy and points out areas that still require investigation [5].

This paper's main goal is to give readers a thorough understanding of CC technology. Since the beginning of the cloud, IT firms have made investments in cloud migration. By 2024, it is anticipated that more than 45% of IT investments in system infrastructure, software infrastructure, and application software will move to cloud services. The COVID-19 pandemic has sped up CC adoption, making remote work the new norm for workers. Without regard to their actual locations, CC provides smooth contact with clients, effective product development, and collaborative work among coworkers. Success stories from today owe a lot to the developments and advantages brought about by CC [6].

An important feature of CC is data sharing, which enables many users to work effectively together in collaborative settings and opens up a large number of potential applications. However, there are substantial difficulties in permitting effective data sharing of outsourced data and guaranteeing the security of data sharing inside a group. To accomplish secure and effective group data sharing in CC, key agreement mechanisms are essential. Using a symmetric balanced incomplete block design (SBIBD), we suggest a novel key agreement procedure in this paper. Based on the block design structure, this protocol enables numerous participants and may adapt to the number of participants. Our technique displays linear computational complexity with the number of participants and greatly lowers communication difficulty by utilizing the $(v, k + 1, 1)$-block design. Additionally, like Yi's protocol, ours exhibits fault tolerance, making group data sharing in CC resistant to a variety of important assaults. The efficiency and security of the method are demonstrated by the suggested group data-sharing model and general formulas for generating the shared conference key K for numerous participants. This innovative key agreement technique offers a potentially effective way to address problems with group data sharing in CC environments [7].

A huge change has been brought about by the IoT, transforming traditional lives into a technologically enhanced way of living. This shift has resulted in notable breakthroughs, including smart cities, intelligent housing, pollution management, energy conservation, efficient transportation, and new business methods. Even while IoT technology has advanced significantly, many obstacles and problems still need to be resolved before it can reach its full potential. The goal of this in-depth review essay is to engage in a thorough discussion that takes sociological and technological aspects into account. It explores the architectural facets of IoT, looks at well-known application sectors, and scrutinizes the significant difficulties it faces. The essay also highlights the importance of big data analysis in the context of IoT while evaluating previous research achievements in various IoT areas. This page seeks to aid readers and researchers in their understanding of IoT and its useful uses in the real world by providing insightful information. It highlights the necessity of looking at IoT from various perspectives, considering applications, technical difficulties, enabling technologies, social effects, and environmental factors [8].

How we work, learn, communicate, and cooperate has been greatly changed by the digital transition. To adapt to the digital era, businesses are making significant changes to their strategy, culture, workflows, and information systems. Existing businesses and economies are being disrupted by this transition. As a result of digitization, distributed IT systems with features like the IoT, Microservices, and mobile services have emerged. Utilizing services computing, the IoT, mobile systems, big data with analytics, CC, collaboration networks, and decision support, it has produced several economic prospects. Adaptable and robust run-time environments are being created for distributed information systems and intelligent business services with service-oriented enterprise designs, taking inspiration from living ecosystems. This transition from closed-world modeling to open-world composition and evolution defines the setting for highly distributed systems, which is crucial to facilitating digital transformation. The evolution of enterprise architecture is examined in this research study, with a particular emphasis on value-oriented mappings between digital strategies, digital business models, and improved digital enterprise architecture. Organizations may better traverse the digital terrain and prosper in the age of digital transformation by integrating digital activities with flexible and value-driven enterprise architectures [9].

Without a viable answer, the IT sector is at a turning point that might negatively affect the environment. Data centers must quickly switch to green and clean energy sources because they are large energy consumers. This study examines the importance, difficulties, and new developments in implementing eco-friendly practices while focusing on green CC services. The study's findings indicate that green energy will play a significant role in the IT sector's future. By minimizing their environmental impact, adopting green CC can both dramatically increase the benefits of cloud services and help conserve the environment [10].

Due to its potential to link a large number of items to the internet and build a broad, interconnected network, the IoT has recently attracted a lot of attention. However, integrating these various devices and data sources into a functioning system is one of the major difficulties. This problem can be solved by integrating the cloud, and many cloud-based IoT platforms have been created as a result. The convergence of IoT with CC has completely changed the technology landscape and introduced a synergistic strategy with enormous advantages. Despite the benefits, the combination of IoT and CC also poses several problems and difficulties. This review study examines both technologies in-depth, examining their benefits, drawbacks, and potential in a convergent manner. The preferred reporting items for systematic reviews and meta-analyses (PRISMA) technique was used to find relevant literature. Selected papers were then thoroughly analyzed using bibliometric network methodologies, co-authorship analysis, and word co-occurrence. The paper also explores the taxonomy of cloud-based IoT applications, discussing different domains and doing quality of service (QoS) factor-based analysis for each domain. Included is a thorough comparison of the functions and features of prominent IoT cloud integration platforms. The assessment also looks at related technologies and provides some insight into anticipated future advancements in this developing area [11].

According to the study's findings, CC has the potential to completely disrupt the IT industry, especially by streamlining commonplace services like email, web servers, and data storage and turning them into standardized, readily available commodities. Furthermore, CC can effectively supply shared services like collaboration tools, accounting, and procurement that are relevant across numerous government departments. Chief Information Officers (CIOs) can greatly ease the burden of overseeing fundamental IT services by adopting commodity computing, freeing up critical time and resources. They can then devote their attention to strategic IT management, exercising leadership, and offering higher value to their particular agencies as a result. In short, CC has the potential to improve productivity, streamline resource distribution, and spur innovation in the field of IT services for governmental organizations [12].

The IoT, CC, and big data (BGD) are three different technologies that are becoming increasingly interconnected and convergent, according to the research paper. Real-time applications in a variety of fields, including telecommunications, healthcare, business, education, science, and engineering, are now possible thanks to these integrated technologies. While there are many benefits to the combination of IoT, CC, and BGD, there are also difficulties in data collection, processing, and management. The main goal of this study is to find new trends in IoT, CC, and BGD and how they affect real-time applications. With a specific focus on the healthcare industry, it analyses current industry trends, examines the advantages and difficulties of these technologies, and suggests future research possibilities. The study presents a conceptual framework that combines IoT, CC, and BGD, and it promotes the use of BGD in an IoT-focused cloud architecture. The study offers helpful advice to academics and practitioners on how to take advantage of the benefits of combining IoT, CC, and BGD. Realizing the full potential of this transformational amalgamation depends on a knowledge of the technologies' potential and how to overcome hurdles as they continue to converge and influence different industries [13].

The rapid development of CC has raised serious privacy and security concerns in business and academics. The progress made in tackling the privacy and security issues in CC is thoroughly reviewed in this study, with an emphasis on various privacy protection systems. The paper starts by identifying and outlining the security and privacy threats related to CC. A thorough structure for privacy security protection is suggested to handle these risks. The study goes on to examine several privacy protection tools, including access control, ciphertext policy attribute-based encryption (CP-ABE), key policy attribute-based encryption (KP-ABE), fine-grain multi-authority revocation mechanisms, trace mechanisms, proxy re-encryption (PRE), hierarchical encryption, searchable encryption (SE), and multi-tenant trust. The traits and application range of typical schemes are contrasted and examined for each technology. Finally, the study discusses potential future research topics and examines current privacy and security challenges for CC. This study contributes to ongoing efforts to address privacy and security concerns by providing an overview of current solutions and highlighting areas for development, assuring a more secure and dependable CC environment [14].

A concept known as "CC" makes use of internet data centers to host programs and store data while offering clients resources and services based on a pay-per-use system. SLAs, which outline the performance and quality standards, define these offerings. However, the difficulty is in confirming and assuring that the cloud provider automatically fulfills the SLA agreement. This research presents a suggested structure for SLA assurance to address this issue. Both cloud customers and cloud providers can make use of this architecture. It evaluates how well cloud applications perform under a variety of circumstances, such as system and component failures. Doing this helps maintain the necessary quality of cloud services by locating and fixing any failures or lessening their effects. Through simulations and testbed trials, the efficiency of the suggested framework is verified, proving that it is in line with the desired goals. In the context of SLAs, this research offers a useful resource for cloud providers as well as customers, improving the dependability and quality assurance of cloud applications [15].

SaaS provides end customers with online access to programs without requiring them to make an initial investment in infrastructure and software. SaaS firms often rent resources from a public IaaS provider or utilize internal data centers to service their clients. When compared to renting from an IaaS provider, however, in-house hosting can result in higher administrative and maintenance costs due to fluctuating performance. This study suggests cutting-edge scheduling and admission control methods for SaaS providers to effectively use public Cloud resources to address these issues. By reducing expenses and raising customer satisfaction levels, the goal is to maximize profit. The researchers examine the viability of various solutions in various scenarios through a thorough evaluation study to maximize the profit of the SaaS provider. According to the simulation results, the suggested algorithms significantly outperform reference approaches across a range of QoS criteria, saving up to 40% in costs [16].

Modern society has evolved to rely on CC for a variety of functions, from social media platforms to infrastructure management. These systems must uphold QoS assurances while accommodating a wide range of user expectations and changing usage patterns. A combination of conceptual technologies has been synthesized to fulfill the ever-changing needs of computing applications. It becomes essential to understand the key technologies that will power future applications if we are to fully appreciate the difficulties and opportunities presented by such systems. The goal of this study is to investigate how three new paradigms—Blockchain, the IoT, and AI—will affect the development of CC systems in the future. A panel of experts is asked to examine the current state and potential future directions of CC by studying these paradigms and the technologies supporting them. To investigate how new paradigms and technologies will influence the evolution of CC in the upcoming years, a conceptual model for cloud futurology is put forth. This study sheds light on how CC fits into the framework of current technology developments and societal demands [17].

17.5 EVOLUTION OF CLOUD COMPUTING

When computing services are used or rented over the internet, this is referred to as CC. It has changed as a result of the incorporation of different computing technologies such as grid computing, utility computing, parallel computing, and virtualization. Multiple users could connect to a central computer through access points in the 1950s, thanks to mainframe computing. The 1960s saw the establishment of the

ARPANET, the forerunner to the internet. Before the advent of operating systems based on UNIX, virtualization became necessary. The paradigm of clients accessing data and programs from a central server through a local area network is known as "client-server" technology. Through a website in 1999, Salesforce.com became a leader in enterprise applications. Rich multimedia, dynamic interfaces, and user-generated content were some of the effects of the 2003 birth of Web 2.0. The early 21st century saw the availability of large bandwidth from ISPs, which hastened the development of CC.

Following is a summary of how CC has developed:

1. Early Ideas (the 1950s–1990s): The earliest concepts of CC may be found in the 1950s when mainframe computers were utilized to deliver computing resources to many users via "dumb terminals." The concept of time-sharing, which enables numerous people to access a single computer at once, first appeared in the 1960s. The idea of VM and distributed computing laid the foundation for future CC technologies throughout the 1970s and 1980s.
2. Internet's emergence during the 1990s and the early 2000s: In the 1990s, as the internet spread, networking technologies improved, making it simpler to access materials located far away. Web-based applications increased in popularity during this time, and Application Service Providers (ASPs) started to appear, offering online software services to consumers and enterprises.
3. The arrival of cloud services (mid-2000s): In the mid-2000s, the phrase "CC" became well-known. Businesses now have access to scalable computing resources and development platforms thanks to companies like Amazon Web Services (AWS) and Google that have offered IaaS and PaaS solutions.
4. Wide Adoption and Growth (the late 2000s to early 2010s): In the late 2000s, businesses, and individuals used cloud services at a significantly higher rate. The number of cloud service providers has increased, and they now provide a wider range of services, such as SaaS, which enables customers to access programs and data online.
5. Cloud Maturity and Diversification (mid-2010s to present): CC has developed and diversified since the middle of the 2010s, becoming an important part of the IT infrastructure for many different businesses. The popularity of hybrid cloud systems increased because of their ability to easily combine resources from both public and private clouds. The possibilities of CC were increased by specialized cloud services for AI, big data analytics, IoT, and serverless computing.
6. Cloud-Native Applications (ongoing): Applications created expressly for the cloud environment have gained popularity. Technologies like containers and microservices for developing and deploying systems make greater scalability, efficiency, and flexibility possible.
7. Edge computing and 5G advancements (ongoing): The implementation of 5G networks and edge computing, which processes data closer to the source, are pushing the limits of CC and enabling real-time processing and low-latency services.

Cloud computing is still evolving, and new developments and ideas propel its wider industry adoption and alter how we use and interact with technology.

17.6 MODELS OF SERVICES FOR CLOUD COMPUTING

Three main service models are available with CC, each of which may be customized to meet the needs of a given customer and give differing degrees of control and freedom. These service models include:

IaaS: Internet-based computing resources are available to users through IaaS. It provides an adaptable and scalable infrastructure that enables users to lease VM, storage, networking, and other computing resources from cloud service providers. Users can run their software and manage their virtualized infrastructure with IaaS since it gives them control over the operating systems, programs, and customizations. Users who require total command over their computing environment and the flexibility to scale resources as necessary will benefit from this strategy. Elastic Compute Cloud (EC2) from AWS is a well-known illustration of IaaS. Users can lease virtual computers (often referred to as instances) in the cloud using AWS EC2. The many instance types available to users offer unique combinations of CPU, memory, storage, and networking capacity. For instance, a software development team can use AWS EC2 to construct development and testing environments. To test their apps on various hardware configurations and operating systems, they can provision VM with various configurations. A new web application being released by a startup is yet another example. They can utilize AWS EC2 to deploy their application on VM in the cloud rather than investing in real servers. They may easily scale up the number of instances to manage more traffic as their user base expands.

PaaS: Without dealing with the difficulties of managing underlying infrastructure, PaaS provides a cloud-based platform with development tools and services that enable the creation, deployment, and maintenance of applications. Cloud providers care for server configuration, operating systems, storage, and networking to free up developers' time to work exclusively on programming and application logic. Software developers benefit the most from PaaS since it streamlines the development process and speeds up application deployment while relieving them of the burden of keeping track of hardware and software updates. Microsoft Azure App Service is a well-known application of PaaS. Developers can create, deploy, and scale web applications and APIs using Azure App Service's fully managed platform without having to worry about handling the underlying infrastructure. With Azure App Service, developers can concentrate entirely on writing code and creating their applications while leaving deployment, scaling, and management to the platform. It allows developers the freedom to work with their favorite tools and programming languages, including .NET, Java, Node.js, Python, and more.

SaaS: On a subscription basis, SaaS offers fully working software applications through the internet. Users do not need to install or maintain these applications locally to access and utilize them immediately through a web browser. All facets of software maintenance, such as upgrades, security, and support, are handled by SaaS providers. For end customers who need quick access to programs without the hassle of managing software infrastructure, this strategy is perfect. Email services, customer relationship management (CRM) programs, and productivity suites are common examples of SaaS applications. Google Workspace (formerly known as G Suite) is one well-known instance of SaaS. Google Workspace is a cloud-based productivity package offering enterprises and individuals several applications and collaboration capabilities. For instance, a small firm can use Google Workspace to handle email communication, save crucial documents on Google Drive, and work together on projects using Google Docs and Sheets. To increase productivity and communication, the team can use Google Meet to hold online meetings with clients or other team members who are not local. For faculty members and students, an educational setting can implement Google Workspace. Document collaboration and assignment submission may be made simple with Google Docs and Google Classroom. The use of Google Forms for exams and student feedback and Google Calendar for managing deadlines, activities, and class schedules are also helpful.

Figure 17.2 shows different cloud service models along with examples. These cloud service models offer a range of abstraction levels, addressing varied user requirements and enabling organizations and people to utilize cloud resources according to their own needs and preferences.

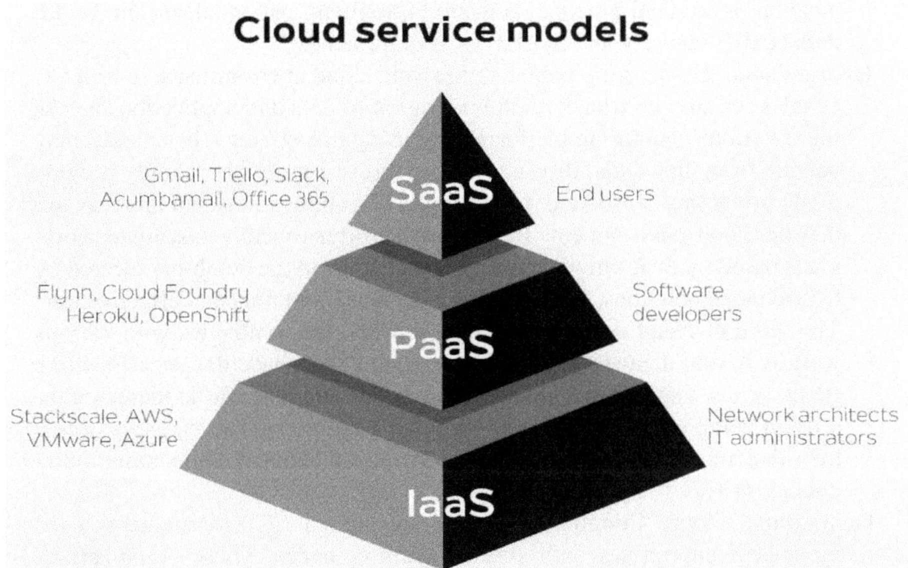

FIGURE 17.2 Service models of cloud.

17.7   CLOUD DEPLOYMENT MODELS

Based on the level of access, deployment, and user interaction, there are four main categories of CC. These kinds include:

Public Cloud: The term "public cloud" refers to cloud resources and services made available by independent cloud service providers online. The general public can either subscribe to these services or pay as they use them. For enterprises and people who need scalable resources but don't want to make large upfront infrastructure investments, public cloud environments, which many users share, offer a cost-effective option. AWS is a well-known example of a public cloud. AWS is a well-known and frequently utilized cloud service provider that provides a wide range of cloud-based solutions to consumers, companies, and governmental agencies. Computing power, storage choices, databases, machine learning, analytics, and other cloud services are all available on demand through AWS.

Private Cloud: A private cloud is a CC infrastructure that is only used by one business or organization. It can be hosted on-site or run by a third-party provider whose services are exclusive to that specific business organization. Because private clouds provide better security, management, and customization possibilities, they are appropriate for businesses with strict regulatory and data protection requirements. Private clouds offer greater control over data and resources but may demand larger upfront investments. Governmental bodies, financial institutions, and businesses managing sensitive data frequently use private cloud environments. Example: The implementation of a private cloud by a healthcare organization, such as a hospital or medical practice, is made to maintain patient electronic health data (EHR) securely and by HIPAA requirements.

Hybrid Cloud: By merging public and private cloud environments, hybrid CC enables the smooth transfer and integration of data and applications across these various computing platforms. By using this strategy, businesses may benefit from the scalability and affordability that public clouds provide while protecting sensitive data and important applications in a more secure private cloud environment. Businesses can dynamically distribute workloads based on their unique requirements thanks to the flexibility offered by hybrid cloud solutions, improving performance and financial effectiveness. This kind of cloud deployment is appropriate for companies with various workloads and data storage needs. Example: Seasonal sales occasions like Black Friday and Cyber Monday result in considerable traffic increases for a retail company that runs an e-commerce platform. The business uses a hybrid cloud strategy to manage these swings in demand while maintaining data security.

Community Cloud: Community clouds are shared CC environments used by several enterprises with similar goals or needs. These organizations could belong to the same sector or business or have certain requirements for compliance. While maintaining a better level of protection and control

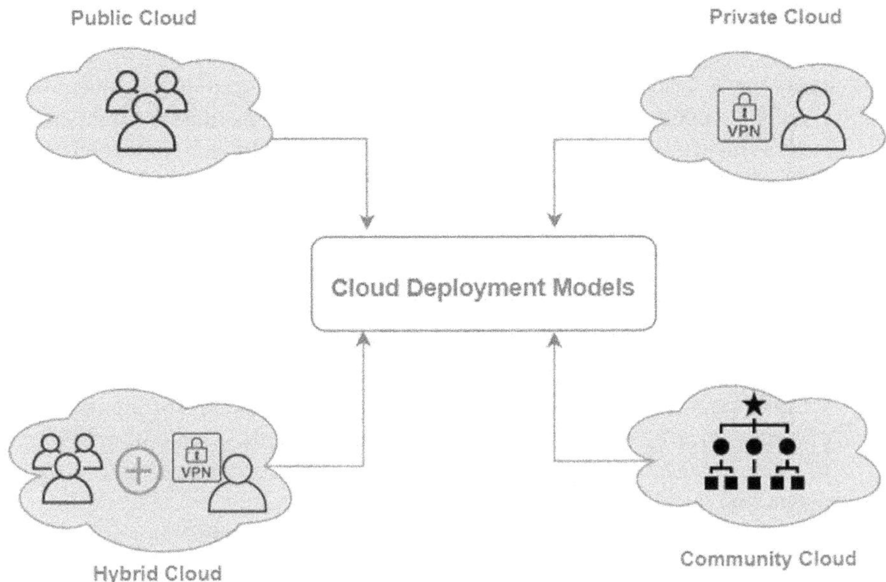

FIGURE 17.3   Types of cloud computing.

than public clouds, community clouds provide these organizations with a platform for collaboration where they may share resources, apps, and data. Because this model encourages community members to work together and share expenses, it appeals to groups with common objectives and difficulties. Example: A group of financial services organizations, including banks, investment houses, and insurance providers, collaborate to share financial information and simplify intra-organizational procedures. They choose a community cloud solution to fulfill their particular demands and guarantee secure data sharing.

Figure 17.3 shows different types of CC. Every kind of CC offers unique benefits and factors, enabling organizations and people to make the best choice depending on their particular objectives, financial constraints, and security demands.

17.8   CLOUD OPEN CHALLENGES

Because of its constantly changing nature, rising user demands, and the complexity of its infrastructure, CC still has several unresolved issues. These difficulties include a variety of topics, including edge computing integration, data residency, security and privacy, data governance and ownership, interoperability and portability, performance and scalability, cost control, reliability, and data residency. Collaboration between cloud service providers, standards organizations, research groups, and regulatory agencies is required to address these issues and improve the CC ecosystem.

The following are some of the main open challenges in CC:

Security and Privacy: The security and privacy of data stored in the cloud continue to be major concerns. Cloud providers must protect data from dangers such as insider threats, unauthorized access, and data breaches. A persistent difficulty is dealing with compliance with data protection laws in different regions.

Data Governance and Ownership: In a multi-tenant cloud environment, determining data ownership, access rights, and responsibilities can be challenging. Data governance standards must be defined to manage data across various cloud services, especially in hybrid and multi-cloud environments.

Interoperability and Portability: Providing smooth data and application transfer between on-premises and cloud environments or across different cloud providers may be difficult. Establishing interoperability and data mobility across many platforms is essential for preventing vendor dependence and promoting wider cloud adoption.

Performance and Scalability: As cloud services accommodate more users and workloads, maintaining reliable performance and scalability is essential. It is a constant challenge to optimize performance while allocating resources to a variety of apps and services.

Cost management and resource utilization optimization: Although CC provides cost-effective solutions, these tasks necessitate thorough oversight and control to manage cloud costs and maximize resource consumption. Organizations still struggle to effectively control cloud costs across many cloud providers and comprehend intricate pricing structures.

Reliability and Availability: Cloud service outages can greatly impact organizations and their clients. It takes strong redundancy, disaster recovery planning, and resolving potential single points of failure to maintain high availability and reliability.

Data residence and Sovereignty: In a globalized cloud environment where data may be stored and processed in many geographical areas, compliance with data residence requirements becomes difficult. The problem of offering seamless access to cloud services while satisfying specific data sovereignty needs is ongoing.

Sustainability and energy efficiency: The fast expansion of cloud infrastructure raises questions about its effects on the environment and energy use. The sector faces significant obstacles in addressing energy efficiency and implementing sustainable practices in cloud data centers.

Integration of Edge Computing: As edge computing becomes more popular, which processes data closer to the source, integration with centralized cloud infrastructures becomes more difficult. Research is still being done on creating seamless hybrid solutions that combine edge computing with cloud services.

Development of Cloud-Native Apps: It can be difficult to properly integrate traditional apps with cloud-native architectures and microservices. The use of containerization technologies and cloud-native development processes necessitates significant preparation and architectural changes.

Collaboration between cloud service providers, industry standards groups, research communities, and regulatory agencies is necessary to address these open challenges. The development and maturation of the CC ecosystem will be facilitated by attempts to address these issues as CC continues to advance.

17.9 SOLUTIONS TO THE PROBLEM OF CLOUD COMPUTING

Many solutions and best practices have been created and used by cloud service providers and organizations to address the issues in CC. These approaches are designed to increase security, optimize resource use, boost performance, and deal with the complexity of cloud systems. Among the most effective ways to deal with cloud difficulties are:

Security Measures: To safeguard data and stop illegal access, use robust security procedures, including access controls, multi-factor authentication, and encryption. Data security and regulatory compliance are improved by regular security audits and compliance evaluations.

Data Governance and Compliance: Establishing precise data governance regulations that specify data ownership, access rights, and duties. Data integrity and privacy are maintained by adhering to industry standards and legislation for data protection.

Standards for Interoperability and Data Portability: Adopting technologies and open standards that enable data portability and interoperability between cloud providers. Reduced vendor lock-in and easier migration are made possible by containers and cloud-native application architectures.

Performance optimization: Using load balancing and auto-scaling systems to maximize resource usage and provide dependable performance during peak periods. Application performance is improved with the aid of monitoring and analytics technologies, which help locate performance bottlenecks.

Cost Management and Optimization: Cost tracking and cost control are achieved by using cost management tools and adopting cloud cost optimization techniques. Cost-effectiveness can be maximized by using reserved instances or spot instances and appropriately scaling resources.

High Availability and Disaster Recovery: Disaster recovery strategies, redundant systems, and data replication are all implemented to maintain high availability and data resilience in the event of hardware failures or other unforeseen events.

Data Residency and Compliance Management: Choosing cloud data centers that adhere to particular rules and industry standards can ensure compliance with data residency needs.

Energy Efficiency and Sustainable Practices: Energy-efficient technology, sustainable practices, and renewable energy sources can all be used to reduce the negative environmental effects of CC.

Edge Computing Integration: Integrating edge computing with centralized cloud infrastructures will enable real-time applications to analyze data more quickly and near the source.

Microservices and Cloud-Native Development: Moving to a microservices architecture for cloud-native application development to improve flexibility, scalability, and deployment simplicity.

Continuous Monitoring and Incident Response: Implementing continuous monitoring and proactive incident response techniques will help you quickly discover and address security threats and performance problems.

Collaboration and Industry Standards: Promoting cooperation between research communities, industry organizations, and cloud service providers to create security standards, best practices, and guidelines for CC.

Organizations can more easily traverse CC's problems, utilize the advantages of cloud services, and create a safe, effective, and robust cloud infrastructure by using these solutions and best practices.

17.10 CONCLUSION

The study's key finding is that CC has had a considerable impact on the IT sector and that in the years to come, there will be a significant change in favor of cloud-based solutions. CC has become increasingly popular as a result of the COVID-19 epidemic, becoming a crucial tool for enabling distant work, virtual collaboration, and client involvement. The study goes deeply into the architecture of CC, examining its design tenets, service models, and useful applications. Decision-makers, researchers, and practitioners get a comprehensive understanding that enables them to make wise decisions when utilizing cloud technology for their enterprises by carefully evaluating the many components of cloud architecture. The essential principles of CC architecture—such as virtualization, elasticity, and multi-tenancy—that support the scalability and flexibility that cloud services provide are highlighted in this paper. In addition, a thorough study of the features, benefits, and applications of the IaaS, PaaS, and SaaS models are provided. The report also offers useful insights into how various cloud service models interact, empowering organizations to optimize their IT infrastructure and efficiently use cloud solutions to fit their particular demands. The research enables enterprises to assess the best-fit model based on security, compliance, and cost factors by examining several cloud deployment types, including community, private, hybrid, and public clouds. Overall, this research provides a thorough resource for comprehending CC architecture, from its basic ideas to the most recent innovations. It provides readers with the knowledge essential to make informed decisions when adopting, installing, or transitioning to cloud-based solutions by synthesizing design principles, service models, deployment strategies, and real-world experiences. In the end, the study encourages the widespread use of CC across a range of industries, giving businesses the tools they need to realize its revolutionary potential, increase operational effectiveness, and spur innovation in the digital era. The information offered in this study will be useful and relevant as CC develops further, acting as a compass for future successful adoption and implementation of the cloud. The ability of CC to spur innovation, boost competitiveness, and hasten advancement in the digital age is essentially what gives it its significance. CC will undoubtedly continue to play a major part in determining how society and technology develop in the future as it develops and new trends come to light. For enterprises to survive in a world that is changing quickly, adopting CC is now a requirement rather than an option.

REFERENCES

1. Sabireen, H., & Neelanarayanan, V. (2021). A review on fog computing: Architecture, fog with IoT, algorithms and research challenges. *ICT Express, 7*(2), 162–176.
2. Nagamunthala, M., & Manjula, R. (2021). State of the art, technologies, future prospects of encryption and data security issues correlated to multi-user environment in green cloud computing: A pilot survey. *Journal of Green Engineering, 11*, 1621–1634.
3. Mortazavi, S. (2020). A proposed architecture for the integration of IoT and cloud computing.
4. George, S. S., & Pramila, R. S. (2021). A review of different techniques in cloud computing. *Materials Today: Proceedings, 46*, 8002–8008.
5. Belgaum, M. R., & Su'ud, M. M. (2017, November). Challenges: Bridge between cloud and IoT. In 2017 4th IEEE International Conference on Engineering Technologies and Applied Sciences (ICETAS) (pp. 1–5). IEEE.
6. Srivastava, P., & Khan, R. (2018). A review paper on cloud computing. *International Journal of Advanced Research in Computer Science and Software Engineering, 8*(6), 17–20.
7. Shen, J., Zhou, T., He, D., Zhang, Y., Sun, X., & Xiang, Y. (2017). Block design-based key agreement for group data sharing in cloud computing. *IEEE Transactions on Dependable and Secure Computing, 16*(6), 996–1010.
8. Kumar, S., Tiwari, P., & Zymbler, M. (2019). Internet of things is A revolutionary approach for future technology enhancement: A review. *Journal of Big Data, 6*(1), 1–21.
9. Zimmermann, A., Schmidt, R., Sandkuhl, K., Jugel, D., Bogner, J., & Möhring, M. (2018, October). Evolution of enterprise architecture for digital transformation. In 2018 IEEE 22nd International Enterprise Distributed Object Computing Workshop (EDOCW) (pp. 87–96). IEEE.
10. Sriram, G. S. (2022). Green cloud computing: An approach towards sustainability. *International Research Journal of Modernization in Engineering Technology and Science, 4*(1), 1263–1268.
11. Ansari, M., Ali, S. A., & Alam, M. (2022). Internet of things (IoT) fusion with cloud computing: Current research and future direction. *International Journal of Advanced Technology and Engineering Exploration, 9*(97), 1812.
12. Wyld, D. C. (2009). *Moving to the Cloud: An Introduction to Cloud Computing in Government*. IBM Center for the Business of Government.
13. Humayun, M. (2020). Role of emerging IoT big data and cloud computing for real-time application. *International Journal of Advanced Computer Science and Applications, 11*(4), 494–506.
14. Sun, P. (2020). Security and privacy protection in cloud computing: Discussions and challenges. *Journal of Network and Computer Applications, 160*, 102642.
15. Zainelabden, A. A., Ibrahim, A., Kliazovich, D., & Bouvry, P. (2016, June). On service level agreement assurance in cloud computing data centers. In 2016 IEEE 9th International Conference on Cloud Computing (CLOUD) (pp. 921–926). IEEE.
16. Wu, L., Garg, S. K., & Buyya, R. (2012). SLA-based admission control for a software-as-a-service provider in cloud computing environments. *Journal of Computer and System Sciences, 78*(5), 1280–1299.
17. Gill, S. S., Tuli, S., Xu, M., & Singh, I. (2019). Transformative effects of IoT, blockchain and artificial intelligence on cloud computing: Evolution, vision, trends, and open challenges. *Internet of Things, 8*, 100118.

18 Integrating Cloud, IoT, and Artificial Intelligence

Subhajit Ghosh

18.1 INTRODUCTION

Recently, the spread of advanced technologies has brought in an era of interconnectedness, supported by intelligence, resulting in great conveniences in our society. Among these technologies, primarily the following three, viz., the Internet of Things (IoT), cloud computing, and artificial intelligence (AI) are considered having great potential for this digital transformation. This book chapter explores the combined impact of IoT, cloud computing, and AI on contemporary society. It will examine their individual contributions and synergies.

Cloud, IoT, and artificial intelligence are promising technologies of the day. These technologies have greatly impacted every sphere of societal applications, be it in education, environment, healthcare, finance, and manufacturing. IoT links the physical to the cyber world and generates big data. Cloud computing assists the process and facilitates the making of intelligent decisions consequent to analyzing big data and following it by machine learning (Chen, 2020).

The internet has its origin as a network communicating with standard protocols such as TCP, IP as well and HTTP. The internet and its protocol were extensively used in web-based computing platforms. Web-based computing was aligned with the concept of service-oriented computing, wherein software components were used as services. These software components were installed in web servers and were accessed through internet protocols like SOAP and HTTP. The web data standards viz., HTML, OWL, XML, RDF, JSON, and others support service-oriented computing.

The next step in the progression of service-oriented computing led to cloud computing. This included services such as SaaS (Software as a Service), IaaS (Infrastructure as a Service), and PaaS (Platform as a Service). This facilitated the extension of services with components such as Robot as a Service, Network as a Service, Device, and Simulation as a Service.

Cloud Computing provides on-demand access to computing resources, besides data storage and offering services over the internet. Cloud is a scalable architecture. Cloud has enabled collaboration remotely. The benefits that cloud platforms provide include cost-effective solutions. This helps businesses and even ordinary individuals access large computing power without heavy investments put into infrastructure (Armbrust et al., 2010).

Cloud computing provides scalability. A load balancer helps in achieving this. A cloud provider doesn't leave his client dissatisfied no matter how much computing power is required. This has resulted in big data processing in the data centers tied to

DOI: 10.1201/9781032656694-18

a cloud environment. Over the years, data have been gathered. Still, these days, data are being generated at a breakneck speed owing to the growth of social networks, IoT, and other recent data acquisition technologies.

A gigantic amount of data is being generated from a large number of devices. These include, among others, machines in industrial plants, data from embedded components in automobiles, health devices in ICUs, and several other linked systems. Applications collect and store the data in warehouses, centers, or servers. The data is organized in structured or unstructured formats. The IoT applications, services, and processes use this data for computations and transactions, OLTP and OLAP, business processes and business intelligence, analytics, and knowledge discovery (Kamal, 2017).

The growth of AI can be traced back to the early 1950s. AI has influences from disciplines as diverse as cybernetics and brain simulation, computational intelligence, logic to cognitive simulation, and recent big data-based and deep learning/machine learning-based strands of artificial intelligence.

18.2 CONVERGENCE OF CLOUD, IoT, AND AI

IoT is a network of interconnected devices, sensors, and objects that accumulate and send data to enable meaningful communication and data analytics. IoT has transformed several sectors, including healthcare and transportation, agriculture, and smart cities. They facilitate real-time data aggregation and analysis. This has led to increasing their efficiency while reducing costs and bettering the user's experiences (Atzori et al., 2017). The infrastructure to enable these kinds of communication is provided by the cloud. They store, manage, interpret, and perform analytics of the data coming from such devices. The shortcomings of local data processing are overcome through the use of the cloud, which provides real-time processing and scalability.

AI can learn from data and predict likely outcomes. The branches of AI, such as Deep Learning and Machine Learning, handle this task. They do it without the need to program explicitly. The algorithms that are devised for implementing AI can improve with more data. This feature enables the integration of AI with a technology such as IoT, which is rather promising owing to massive data coming from such an interconnected IoT-based system. AI can uncover useful patterns from such data streams, which leads to better decision-making abilities.

A new model of AI (tiny AI) makes use of compressed algorithms to minimize the use of voluminous data and computational power. The main aim of tiny AI is to reduce the size of artificial intelligence algorithms, especially the ones that cater to voice or speech recognition or IoT. Smart use of data is an initial step toward efficient AI systems. AI-assisted data processing, compression strategies such as network pruning, unsupervised learning methods such as GAN and LSTM and data reduction techniques such as surrogate modeling, advances in nanotechnology and newer architectures and materials, new edge learning methodologies (distributed and joint learning) are some of the issues being researched upon to usher in the era of tiny AI (About tiny AI, https://www.imec-int.com/en/artificial-intelligence/tiny-ai).

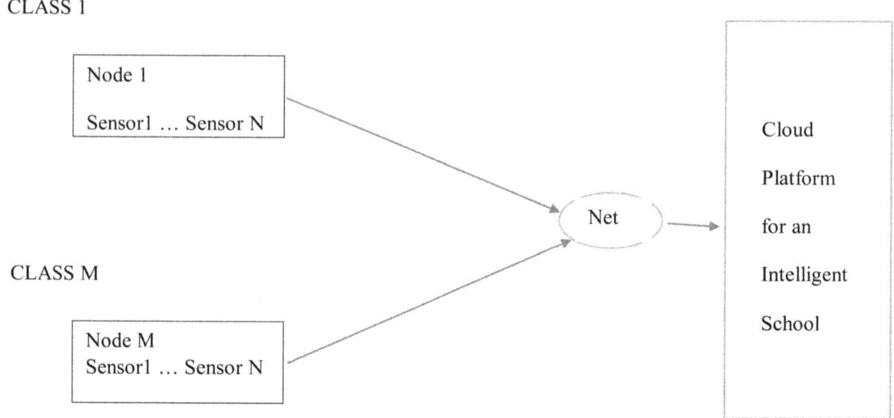

FIGURE 18.1   Architecture of an intelligent school (Original).

Let us consider the architecture of an Intelligent school as depicted in Figure 18.1. Below. Each classroom may be equipped with a node and a recording facility. Depending on requirements and privacy concerns as per regulations in vogue, the class session (akin to IoT) may be sent to a central unit with provisions for storage (akin to cloud), gainful analysis (AI and ML), and monitoring effective usage of all concerned processes.

A huge potential awaits the proper integration of IoT Cloud and AI that is likely to impact decision-making in the industry. As an example, in the case of agriculture, the sensors can be useful for providing data on the moisture content of the soil, the health of the crop, and the weather conditions that are likely to affect the yield. The use of AI-based algorithms to interpret data can be useful in pesticide usage, crop yield prediction and optimizing the schedules for irrigation. This results in improved resource usage and productivity (Geng & Zhang, 2019; Smith & Russ, 2020).

A kind of computing that is driving computing applications, data, and services away from the centralized nodes to the nodes generating data at IoT nodes is gaining currency lately. This is termed Edge Computing, which stands for computing at the logical extremes of the network. The IoT device nodes are driven by events, triggers, and alerts. The messages and data are collected for enrichment, and computation and storage come from the nodes of the centralized database remotely. This shift from centralized computation leads to the availability of resources at the device node. This may be needed in the case of a low-power lossy network (Kamal, 2017).

18.3   THE GLOBAL IMPACT OF THIS TECHNOLOGY

The advent of these recent technologies in the industry has impacted every sector, be it manufacturing or healthcare. In the manufacturing sector, predictive maintenance supported by these developing technologies can bring about substantial benefits. In the case of Industrial IoT, the monitoring of equipment and their performance in real-time with related data relayed to the cloud where an analysis is performed through the use of AI algorithms, can result in the prediction of likely equipment failures much before their actual occurrence. This helps in proactive maintenance and reduces downtime and maintenance costs. This also enables augmenting the lifespan of the equipment (Zhang & Zheng, 2018).

In the healthcare sector, the integrative potential of these three complementing technologies can truly bring about a transformation. Patients are made to wear devices fitted with sensors that monitor the patient's vital signs on a 24x7 basis and relay them to the cloud. The assessing medical software can analyze the data to find variations and prescribe effective treatment. Monitoring patients remotely has become a reality, and there has been a noticeable impact on bettering patient outcomes (Dey et al., 2019; Rahmani & Liljeberg, 2018).

The other beneficial aspects of these technologies can be seen in the growth of smart cities. These cities use IoT sensors to monitor and optimize traffic. In smart cities, there is an optimized use of resources and energy and efficient systems of transportation. The brightness of the street lights can be self-adjusted based on sensory inputs and ambient light. This leads to a reduction in the wastage of energy. Road signal timings can be optimized by analyzing traffic data, thereby easing congestion. By monitoring the level of filling of the bins, waste management can be greatly improved upon. This kind of system, propelled by AI-layered analytics, would lead to improving sustainability and quality of life in smart cities and providing better public services (Caragliu et al., 2011; Han et al., 2019; Sun & Wang, 2020).

The integration of these technologies can also bring about a transformative change in customers' experiences, mainly in the retail and marketing segments. Sensors can locate the movement of the customers and their choices and noticeable preferences. Recommendations to individual customers can be made using an algorithm that uses AI techniques to evaluate this data. Online retailers can make use of intelligent methods to find the purchase and browsing patterns of customers, and thereafter, the advertising and marketing of products to likely customers can be strengthened. This would bring about greater customer satisfaction and bolster their loyalty (Chen & Chen, 2020; Rajput & Chatzimisios, 2018).

Moreover, such technologies can improve the decision-making process in the industry. In the case of agriculture, the sensors relay data on the moisture content of the soil, the health of the crop, and conditions related to weather, on a real-time basis, to be inputted to AI algorithms that perform analysis, and aid in optimizing the schedules of irrigation, the use of pesticide, and in the prediction of the yield of the crop. This is a data-driven methodology. It helps improve resource efficiency and productivity (Geng & Zhang, 2019; Smith & Russ, 2020).

Healthcare data is being obtained from manual registration and the use of self-diagnosing technology and is increasingly becoming difficult to store and retrieve. A study by the American College of Physicians states that doctors spend 50 percent of their time on EHR and desk work, so they devised a system that manages patient records effectively by replacing the manual process of examining the patient's vitals with a chat interface questionnaire, recording it in storage, and sending the results to the doctor, speeding up the process (Mitchell, 2021). This allows faster sending of prescriptions and other information directly to the patient as they can send photos or videos directly for examination.

The use of cloud computing in patient monitoring aims to solve electronic infrastructural issues in the bandwidth of some areas and the difficulty of implementation. It strives to improve the efficiency of collecting and distributing the patient's information by applying the services in data centers and using remote servers to store and manage the information. The study" A Cloud Computing Based Telemedicine Service" reported enhancing the speed of medical analysis by transferring ECG wave signals to a varied number of locations, such as a mobile phone, through the cloud; https://www.quantamagazine.org/melanie-mitchell-trains-ai-to-think-with-analogies-20210714,]. Cloud computation improves the speed of data retrieval and transactional processing capacity. This may lead to the standardization of medical information and records, resulting in faster telemedicine procedures and consistency in implementing new programs.

18.4 INDIAN SCENARIO OF CLOUD, IoT, AND AI

The IoT holds immense possibilities in the industry, with estimates indicating that there will be roughly 50 billion connected IoT devices by 2030. IoT enables apps, devices, or even cars to connect and exchange data via a wireless network. With the rapid growth witnessed in IoT, Companies are willing to invest a good sum of money in this technology, which is changing the way people live and work.

Integration of IoT with mobile apps enables the business to provide a better customer experience, be it in banking or healthcare services, retail, hospitality, energy, manufacturing, or transportation. Improved customer services can help grow their customer base both nationally and globally. Owing to this, organizations are seeking the services of good IoT development companies that can provide good solutions at economical rates (Top 10 Internet of Things (IoT) Development Companies In India 2023 (outlookindia.com), 2023). The IoT development companies in the country are offering good solutions using a combination of cloud, machine learning, and services related to IoT. Selecting the best company from the available options is a difficult task.

A brief discussion on the work of IoT companies in India follows. Hyperlink InfoSystem develops apps that support the work of Security and Home Automation companies and may extend to include Industrial IoT. This has an impactful effect on improving our lives. To provide improved integrated services, management efforts are underway to make smart products that use machine learning and analytics lead to an efficient system. Larsen and Toubro Infotech Limited assists tech companies in innovation and their real-time operation and optimizing resources and costs, thereby showing improvement in ROI. HCL Tech has a good IoT environment for companies to begin their projects related to IoT. Tata Consultancy Services, an Industry leader, has various services related to IoT, cloud, and AI. An Indian Tech company, Kellton Tech Solutions Ltd., offers services related to Strategy and Consulting. TVS Next Private Limited provides neat client-based solutions using the combined strength of engineering, experience, and intelligence. VVDN Technologies also assists international clients with technological innovations related to IoT. Mphasis, NTT DATA Services, and HData Systems are the other notable players working in the IoT domain and other cutting-edge technologies [Top 10 Internet of Things (IoT) Development Companies In India 2023 (outlookindia.com), 2023].

The loss-making Public Sector Undertaking of the States and in the Centre is a huge burden on the country's economy and its taxpayers. Can we revive ailing telecom PSU through Technology? AI holds the key to optimizing and automating telecom networks. Telecom companies are adopting technologies like virtualization and SDN-NFV; AI will play a big role in the smooth integration of these technologies and automation of the networks. AI application in mobile networks circles around three applications—self-optimizing networks (SONs), software-defined networks (SDN) and Network Function Virtualization (NFV), and enablement of neural networks. SONs enable operators to automatically optimize the network quality based on traffic information by region and time zone based on various machine learning algorithms. On the subscriber side, AI and Machine Learning will help telecom operators in subscriber profiling and analyzing offer conversion rates, content usage trends, and network activity. This will help them push offers that are tailored to the subscriber's needs at the right time. Using AI and data analytics, operators will be able to identify and push various services to the customers at the right time, e.g., in the case of post-paid customers, operators must encourage high-speed data services and offer tailored data packs when the subscriber is running low on data. The timing of offering tailored packages-based subscriber intelligence is very important. Recently, Airtel partnered with Korea's SK Telecom to enable an AI-assisted network. SK Telecom has deployed an AI-assisted network (known as TANGO) with big data and machine learning capabilities to enhance customer experience through automated detection, troubleshooting, and optimization of mobile networks. The governments keep infusing thousands of crores of rupees of public money to revive profusely bleeding public sector enterprises. Why should the government continue to be in business (flying to mobile phones)? The government has decided to pump in a massive Rs 70,000 crore to revive the sick telecom public sector undertakings BSNL and MTNL. Does it show the government's lack of prudence? (https://telecom. economictimes.indiatimes.com/news/why-telcos-will-soon-be-betting-on-artificial-intelligence-to-build-their-networks/61531211).

18.5 DISCUSSION

India is emerging as a leading player in technology in the new millennium. The onset of the IoT, cloud, and AI has had a tremendous impact on modern Indian society. In our discussion, we will explore the various effects of these technologies on India and estimate the beneficial aspects, the challenges that need to be resolved, and the future prospects of this convergence.

The implementation of IoT in the country has brought about a transformation across several sectors. In sectors like agriculture, the use of IoT-augmented precision farming has aided farmers in optimizing the yields of their crops and a reduction in water usage, coupled with an increase in productivity. This remarkable innovation in technology has the scope to reduce the challenges of agriculture besetting the country and improving food security.

In smart cities like Bhubaneshwar and Pune, IoT sensors monitor pollution levels and waste management and also help ease traffic congestion, facilitating greater conformity to sustainability in urban living. However, major challenges exist, such as data privacy. Moreover, the digital divide remains, which calls for sincere efforts to bridge the gap and protect the privacy of the citizens (Press Information Bureau, Government of India, 2019).

The advent of cloud computing has given businesses and lay users in the country access to cost-effective computing resources. Start-ups have hugely benefitted from this cloud-based infrastructure. This has enabled them to scale without investing heavily in capital. The initiatives by the government viz., Digital India have contributed to the adoption of the cloud, as well as providing a fillip to e-governance and digital services. In the sphere of education, the cloud has eased access to knowledge. This led to online learning becoming hugely popular. Online education saw exponential growth during the pandemic. A few bottlenecks remain, though. Data security and the need for laws to ensure data protection remain an important area of concern (Ministry of Electronics and Information Technology, Government of India, 2015).

India is gaining recognition worldwide for AI research and development. The country's AI applications in healthcare, along with the developments achieved in diagnostic tools and telemedicine, have significantly contributed to improved outcomes for patients. This has made healthcare more accessible. Moreover, recent AI-driven chatbots and virtual assistants have enhanced customer services and all the support functions for businesses. However, in spite of the advancement, the ethical concerns of AI, which include a bias in algorithms and data privacy, still remain. This makes it necessary to have strong regulations and effective guidelines to set the right procedure for the adoption of AI responsibly (NITI Aayog, Government of India, 2018)

The synergy observed in these three technologies, viz., IoT, cloud computing, and AI, is providing a thrust toward innovation in India. Integrated systems are being developed to meet the huge challenges of the country. Some of these include healthcare delivery in rural areas, smart energy grids, and providing relief and rescue during disaster management. The digital divide remains a stumbling block, as many rural areas are short of adequate infrastructure and have no access to these technologies. One would need to bridge this divide. It is of utmost importance to make sure that the actual benefits of these technologies reach all, even the marginalized, of our society.

Table 18.1 A summary of the three technologies in their usage in India and globally.

TABLE 18.1

Technology Usage

Technologies	Usage in the International Scenario	Usage in India	Remarks
Internet of Things	IoT-augmented Precision Farming, Pollution, and Waste Management …	Pollution and Waste Management …	Tremendous scope in the Smart Cities project
Cloud Computing	Patient Monitoring…	Digital India, e-governance	A cost-saving incentive for small business
Artificial Intelligence	Robotics, Defense, Cyber Security …	Machine Translation, Health informatics	Tremendous scope exists

18.6   CHALLENGES AND CONSIDERATION

The combination of the IoT, cloud, and AI will unquestionably bring about an ameliorative change for the better. However, several challenges exist, and these need to be looked into. Security and privacy issues are prevalent due to the massive amount of sensitive data collection and transmission. Therefore, there is a requirement for proper data encryption and access controls. Further, there is a need to comply with regulations. There is also the issue of algorithmic bias. There is an urgent need to address ethical considerations in AI and IoT deployments. It goes without saying that a modern, highly effective, and beneficial system needs to ensure fairness, transparency, and accountability (Jobin et al., 2019). The interoperability between different IoT devices and platforms is rather challenging as there is a need for standard protocols and interfaces (Atzori et al., 2017; Li et al., 2020).

18.7   CONCLUSION

The triad of the IoT, cloud, and AI has a big potential to transform industries. By exploiting these technologies' enhanced power in gathering, processing, and prediction, companies can make better decisions, provide customers with better experiences, optimize their operations, and cater better to sustainability goals. There are still a few bottlenecks, though. There are challenges regarding effective security, solid privacy, smoothness in interoperability, and ethical issues. Overcoming these challenges can help realize the potential of this integration. There is continued research and development in place to tackle the shortcomings that dog these systems at present. The future looks promising for this combination of technologies to usher in a better and more developed society.

REFERENCES

Armbrust, M., Fox, A., Griffith, R., Joseph, A. D., Katz, R., Konwinski, A., & Zaharia, M. (2010). *A View of Cloud Computing.* Communications of the ACM, 53(4), 50–58.

Atzori, L., Iera, A., & Morabito, G. (2017). *The Internet of Things: A Survey.* Computer Networks, 54(15), 2787–2805.

Caragliu, A., Del Bo, C., & Nijkamp, P. (2011). *Smart Cities in Europe.* Journal of Urban Technology, 18(2), 65–82.

Chen, Y. (2020). *IoT, Cloud, Big Data and AI in Interdisciplinary Domains.* Simulation Modelling Practice and Theory, 102, 102070, ISSN 1569-190X. https://doi.org/10.1016/j.simpat.2020.102070. (https://www.sciencedirect.com/science/article/pii/S1569190X20300083)

Chen, J., & Chen, L. (2020). *IoT Based Smart Shopping: A Survey.* IEEE Internet of Things Journal, 8(1), 79–95.

Dey, N., Ashour, A. S., & Balas, V. E. (2019). *IoT Based Healthcare Systems: A Comprehensive Survey on the Latest Advances and Future Directions.* Journal of King Saud University-Computer and Information Sciences.

Geng, S., & Zhang, Y. (2019). *Agriculture IoT-Based Precision Agriculture Monitoring and Management.* IEEE Access, 7, 96324–96332.

Han, R., Wang, J., & Liu, Z. (2019). *Internet of Things (IoT) in 5G Wireless Communications.* IEEE Access, 6, 36117–36124.

Jobin, A., Ienca, M., & Vayena, E. (2019). *The Global Landscape of AI Ethics Guidelines*. Nature Machine Intelligence, 1(9), 389–399.

Kamal, R. (2017). *Internet of Things: Architecture and Design Principles*. McGraw Hill Education (India), ISBN-13: 978-93-5260-522-4.

Li, X., Tao, X., & Zhang, J. (2020). *Security and Privacy in Internet of Things (IoT) Systems: A Survey*. IEEE Access, 8, 22206–22245.

Ministry of Electronics and Information Technology, Government of India. (2015). Digital India: Transforming India into a digitally empowered society and knowledge economy. Retrieved from https://www.digitalindia.gov.in/content/dam/digitalindia/pdf/DIGITALINDIA.pdf

Mitchell, M. (2021). *Abstraction and analogy-making in artificial intelligence*. https://arxiv.org/pdf/2102.10717v1.pdf

NITI Aayog, Government of India. (2018). National strategy for artificial intelligence. Retrieved from https://www.niti.gov.in/writereaddata/files/document_publication/NationalStrategy-for-AI-Discussion-Paper.pdf

Press Information Bureau, Government of India. (2019). Transforming Indian agriculture through the Internet of Things (IoT). Retrieved from https://pib.gov.in/newsite/PrintRelease.aspx?relid=197742

Rahmani, A. M., & Liljeberg, P. (2018). *Survey on Industrial IoT Systems for Health Applications*. Journal of King Saud University-Computer and Information Sciences, 30(4), 431–448.

Rajput, Q. A., & Chatzimisios, P. (2018). *A Survey on Internet of Things: Architecture, Enabling Technologies, Security and Privacy, and Applications*. IEEE Access, 6, 3619–3648.

Smith, D. A., & Russ, R. S. (2020). *IoT-Enabled Agriculture: A Review of Literature*. IEEE Access, 8, 122644–122661.

Sun, K., & Wang, Z. (2020). *Smart Cities in China: A Comprehensive Review*. Sustainable Cities and Society, 54, 101962.

Top 10 Internet of Things (IoT) Development Companies in India 2023 (outlookindia.com)

Zhang, J., & Zheng, X. (2018). *Industrial Internet of Things: A Survey on the Enabling Technologies, Applications, and Challenges*. IEEE Access, 6, 78238–78258.

19 Enhancing IoT with Cloud-Based Machine Learning

A Comprehensive Integration

Kiran Deep Singh and Prabh Deep Singh

19.1 INTRODUCTION: IoT AND CLOUD-BASED MACHINE LEARNING

The Internet of Things (IoT) is fundamentally about connecting physical objects, including sensors, actuators, and communication modules so that they can share data without much help from humans. Machine learning (ML) on the cloud is a subset of artificial intelligence (AI) that uses the processing and storage power of the cloud to develop, deploy, and operate ML models. Cloud-based ML offers scalability, accessibility, and cost-efficiency that people, businesses, and organizations of all sizes can take advantage of, in contrast to conventional ML, which often depends on on-premises technology (Rathore et al. 2021).

IoT and ML integration is the fusion of two potent technologies, each with specific advantages (Firouzi, Farahani, and Marinšek 2022). This integration is significant because of how it transforms industries, companies, and society as a whole (Rathore et al. 2021). IoT and ML must be integrated for the following major reasons:

1. **Data-Driven Insights:** IoT devices provide enormous volumes of data from the actual environment, including temperature measurements, GPS locations, sensor data, and more. ML algorithms may examine this data to discover patterns, anticipate outcomes, and gain useful insights. IoT sensors, for instance, may gather information on soil temperature, moisture, and meteorological conditions in agriculture. Then, ML models can forecast the best periods for planting, lowering resource waste and boosting agricultural yields.
2. **Real-Time Decision-Making:** IoT devices function in real-time, and ML models have the ability to analyze the data that these devices produce in order to make rapid judgments. This feature is useful in circumstances when prompt reactions are required, such as in monitoring autonomous cars, healthcare, and industrial automation (Radanliev et al. 2022).

DOI: 10.1201/9781032656694-19

3. **Predictive Maintenance:** In manufacturing facilities and other industrial settings, IoT sensors installed on machinery can offer real-time data on the status of the equipment. ML models may analyze this data to determine when maintenance is required, lowering the amount of time the system is down and the expenses associated with its upkeep (Seng, Ang, and Ngharamike 2022).

4. **Enhanced Efficiency:** The IoT and ML can work together to improve workflows and the distribution of available resources. In the context of smart cities, the use of the IoT-connected infrastructure enables the monitoring of traffic patterns. By employing ML algorithms, the traffic lights may be dynamically adjusted to mitigate congestion and enhance the overall efficiency of traffic flow.

5. **Personalization:** IoT devices, such as smart home appliances and wearables, collect data on the preferences and actions of their users. ML models may use these data to increase customer happiness, personalize user experiences by recommending items or services, and improve product or service recommendations.

19.2   IoT ARCHITECTURE AND COMPONENTS

The Internet of Things architecture is a blueprint that specifies how each of the different parts of the IoT (devices, networks, and platforms) work together to produce a functional ecosystem. With the advent of the IoT, previously unimaginable quantities of data will be generated as previously unconnected physical objects and devices become interconnected and interact with one another online. When combined with cloud-based ML, this information may help businesses of all stripes make more informed decisions in real-time and enhance their efficiency. Together, IoT and ML are a powerful force in modern innovation (Seng, Ang, and Ngharamike 2022). The following is a list of the fundamental building blocks that make up the architecture of the IoT:

1. IoT Devices: Sensors, actuators, and connectivity modules are all built into these physical devices. The devices can range from the very simple, like temperature or motion sensors, to the really complex, like autonomous vehicles or industrial machinery.

2. IoT Communication Protocols: In order to send data, IoT devices employ a variety of communication protocols, such as Wi-Fi, Bluetooth, Zigbee, LoRa, cellular networks (2G, 3G, 4G, 5G), and specialized IoT protocols such as MQTT (Message Queuing Telemetry Transport) and CoAP (Constrained Application Protocol).

3. Edge Devices and Gateways: Edge devices and gateways act as go-betweens between IoT devices and the cloud. They connect and communicate with each other. Before sending it to cloud-based platforms, they gather and pre-process the data from IoT devices. Computing at the edge of networks is especially useful in circumstances that need both low latency and real-time data processing (K. D. Singh 2021b).

4. Platforms hosted in the cloud: These platforms supply the infrastructure and services required to store, manage, and analyze data collected by IoT devices. IoT-specific services for data intake, storage, and processing are

offered by cloud providers such as Amazon Web Services (AWS), Microsoft Azure, and Google Cloud Platform (GCP).

5. Data Analytics and ML: In order to analyze the data that is produced by the IoT devices, cloud-based ML services are utilized. ML models can identify abnormalities, forecast future occurrences, and generate insights that may be put into action (P. D. Singh and Singh 2023a).

6. User Interfaces and Apps: The information that is processed and analyzed by ML and IoT systems is frequently used as input for user interfaces and apps. These user interfaces can take the shape of web-based dashboards, mobile applications, or integrated systems that let people engage with data from the IoT and make choices based on that interaction.

19.3   EXPLANATION OF IoT DEVICE ARCHITECTURE

The structure and parts of IoT devices, which are the actual physical things that gather information from the real world and send it over the internet, are referred to as IoT device architecture (P. D. Singh and Singh 2023b), (K. D. Singh et al. 2022). These gadgets are crucial to the IoT ecosystem and are required to capture real-world data. The following essential elements are often present in the architecture of IoT devices:

1. Sensors: IoT devices' main building blocks are sensors. They get information from the surrounding environment. Different kinds of sensors, including temperature sensors, humidity sensors, motion detectors, light sensors, GPS receivers, and others, can be included in IoT devices. These sensors transform physical events (such as temperature or motion) into processed and transmittable electrical signals.

2. Actuators: Actuators are parts that allow IoT devices to perform actions in response to data they receive or gather. For instance, a smart door lock may lock or unlock a door in response to a remote command, and an IoT thermostat can change the temperature depending on sensor data. Motors, servos, switches, and other devices are examples of actuators (K. D. Singh, Singh, and Kang 2022).

3. Microcontroller or microprocessor: The brain of an IoT device is a microcontroller or microprocessor. It maintains device connectivity, carries out orders, and analyzes data from sensors. The Arduino, Raspberry Pi, and specialized microcontrollers made for IoT applications are popular microcontroller platforms used in IoT devices.

4. Memory and Storage: IoT devices frequently have limited memory and storage capacity depending on their individual use case. They do computations, store data locally, and even save historical data for subsequent study.

5. Communication Modules: IoT devices need communication modules to connect to the internet or other devices. These modules make it possible to transmit data across a variety of networks, including Wi-Fi, Bluetooth, Zigbee, cellular networks (2G, 3G, 4G, and 5G), LoRa (Long Range), and others. Range, data rate, power consumption, and the intended application all play a role in the selection of the communication module.

6. Power Source: IoT devices can be battery-powered, linked to a power source, or generate electricity using energy-collecting strategies (such as solar panels). Power efficiency is essential for IoT devices, especially those installed in remote or difficult-to-access areas.
7. Enclosure and Physical Design: An important aspect of an IoT device's physical design is its compatibility with various conditions and longevity. Enclosures shield the interior parts of the device from external elements, including dust, moisture, and severe temperatures.
8. Security Features: To guard against online dangers and unauthorized access, security features like encryption, authentication, and secure boot methods are crucial for IoT devices because they are frequently linked to the internet and may handle sensitive data (K. D. Singh, Singh, Kaur et al. 2023).

19.4   TYPES OF IoT SENSORS AND ACTUATORS

Understanding IoT device architecture, sensors, actuators, and communication protocols is essential to creating and implementing effective IoT solutions across many industries and applications. IoT ecosystems employ these components to collect meaningful data and perform intelligent actions. Each type of IoT sensor and actuator serves a specific detecting or controlling purpose. IoT devices employ many communication protocols to connect to networks, each serving a particular use case.

Type of sensor	Description
Temperature Sensors	Measure temperature variations, often used in climate control, weather monitoring, and industrial processes.
Humidity Sensors	Detect and measure humidity levels, useful in agriculture, HVAC systems, and environmental monitoring applications.
Motion Sensors	Detect movement and changes in position, commonly used in security systems, lighting control, and smart appliances.
Proximity Sensors	Detect the presence or absence of objects or people in close proximity, utilized in automated doors, touchless faucets, and robotics.
Light Sensors	Measure ambient light levels, employed in smart lighting, street lighting control, and display brightness adjustment.
Gas Sensors	Detect the presence and concentration of gases, essential for air quality monitoring, gas leak detection, and industrial safety.

Type of actuator	Description
Motors	They are used to move or control physical objects, as seen in robotics, home automation (e.g., smart blinds), and industrial automation.
Solenoids	Electromagnetic actuators are used for locking mechanisms, valve control, and door-locking systems.
Relays	Control high-voltage electrical devices by switching them on or off, widely used in home automation and industrial applications.
Servos	Precise actuators are commonly used in robotics and drones to control position or movement.
Pumps and Valves	Actuators that regulate the flow of fluids are found in applications such as irrigation systems, industrial processes, and water control.

Name of protocol	Description
Wi-Fi	Enables high-speed, local area network (LAN) communication, suitable for devices with continuous power sources and internet connectivity.
Bluetooth	A short-range, low-power protocol used for device-to-device communication, often used in wearables, smartphones, and smart home devices.
Zigbee and Z-Wave	Low-power, mesh network protocols designed for home automation and IoT applications, allowing devices to form self-healing networks.
LoRa (Long Range)	A long-range, low-power protocol for IoT devices with extended coverage, suitable for applications like smart agriculture and smart cities.
Cellular Networks (2G, 3G, 4G, 5G)	Enable IoT devices to connect to the internet from virtually anywhere, suitable for remote monitoring and asset tracking.
MQTT and CoAP	Lightweight, publish-subscribe protocols designed for efficient IoT data transmission and message routing.
NFC (Near Field Communication)	Enables short-range, contactless communication between devices, often used for mobile payments and access control.
LPWAN (Low-Power Wide-Area Network)	Encompasses various protocols like Sigfox and NB-IoT, designed for long-range, low-power IoT applications with wide coverage.

19.5 CHALLENGES

IoT system deployment and operation require answers to data security and privacy, latency, and scalability issues. These issues are complex and require technology solutions, best practices, constant monitoring, and adaptation to changing threats and needs.

19.5.1 DATA SECURITY AND PRIVACY CONCERNS IN IoT

Data security and privacy are key concerns in the IoT ecosystem owing to the large volume of sensitive information gathered and transferred by IoT devices. Here's a closer look at these issues:

1. Data Breach: IoT devices are frequently targeted by cyberattacks and hacking attempts. If the security of a machine is breached, it can allow unauthorized access to sensitive data, possibly inflicting substantial harm to persons or organizations (K. D. Singh and Singh 2023). A hacked smart home camera, for example, may disclose private footage to criminal individuals.
2. Data Encryption: IoT devices must use strong encryption algorithms to safeguard data in transit and at rest. Encryption guarantees that data stays unreadable without the decryption key, even if captured during transit.
3. Authentication and Authorization: It is critical to have proper authentication and authorization methods in place. To prevent unauthorized users or devices from accessing sensitive data, devices should only connect with authorized organizations, and access control should be enforced.
4. Device Security: Because many IoT devices have low processing resources, they are vulnerable to security flaws. Manufacturers must constantly update device firmware to address security issues. Device upgrades, on the other

hand, might be difficult, especially for devices placed in remote or inaccessible areas.

5. Privacy Issues: IoT devices frequently capture personal data, such as location data, health data, and behavioral patterns. If this data is not managed appropriately, it might lead to privacy violations. Consent processes and clear privacy rules are required to preserve individuals' privacy rights.

19.5.2 Latency Issues in Real-Time Processing

The term "latency" refers to the amount of time that passes between the time that data is generated and when it is processed or responded to. In the IoT, applications that require real-time or near-real-time answers absolutely need to have a low latency. The following are the primary difficulties caused by latency (K. D. Singh, Singh, Bansal et al. 2023):

1. **Network Latency:** Data transmission over the internet or wireless networks can introduce latency, especially in congested or unreliable networks. High network latency can hinder real-time data delivery.
2. **Edge Computing:** To address latency challenges, some IoT deployments incorporate edge computing. Edge devices, situated closer to the data source, process data locally, reducing the round-trip time to a centralized cloud server. However, managing edge resources and ensuring consistency can be complex.
3. **Scalability vs. Latency Trade-off:** Balancing the need for low latency with scalability is challenging. As the number of IoT devices grows, it can strain processing capabilities, potentially leading to increased latency. Managing this trade-off is crucial in IoT system design.
4. **Quality of Service (Quality of Service):** Real-time applications often require specific levels of quality of service. Meeting these requirements, such as guaranteed response times, can be challenging in a dynamic and diverse IoT environment.

19.5.3 Scalability Challenges in IoT

Scalability is a fundamental issue in the IoT, which is mostly caused by the enormous quantity of connected devices and the data they produce. Concerns about the capacity to scale are broken down as follows:

1. **Device Management:** As the number of IoT devices increases, managing and maintaining them becomes increasingly complex. Device provisioning, configuration, software updates, and security management require robust solutions to scale effectively.
2. **Data Volume:** IoT devices generate vast amounts of data. Scalability concerns arise in data ingestion, storage, processing, and analysis. Cloud-based solutions must be designed to handle increasing data volumes efficiently.

3. **Interoperability:** In IoT ecosystems, devices from different manufacturers and with varying communication protocols need to work together seamlessly. Ensuring interoperability among heterogeneous devices is a scalability challenge.
4. **Resource Constraints:** Many IoT devices have limited computational resources like memory and processing power. Ensuring that software and algorithms are optimized for resource-constrained devices while maintaining scalability is complex.
5. **Elasticity:** IoT solutions need to scale up or down dynamically to accommodate changing workloads. Cloud-based services must provide elasticity to handle fluctuations in device activity.
6. **Edge vs. Cloud Scaling:** Deciding where to perform data processing and analysis—edge devices or the cloud—affects scalability. Scaling edge resources can be challenging due to physical limitations, while cloud resources are more elastic but may introduce latency.
7. **Security at Scale:** Ensuring security at scale is a significant challenge. Managing security credentials, access control, and threat detection across a vast number of devices requires robust security infrastructure.

19.6 WORKING ON PROPOSED SOLUTION

IoT devices collect data from the physical environment using sensors and input mechanisms. This may include temperature, humidity, motion, or light. These sensors provide electrical signals or digital data in response to stimuli, which are processed by device microcontrollers or microprocessors. MQTT and CoAP are used by IoT devices to convey this data. These protocols allow devices and cloud-based or edge computing platforms to communicate efficiently over various networks, including Wi-Fi, Bluetooth, cellular, and Low-Power Wide-Area Networks. Before delivering data, IoT devices aggregate and preprocess data. This reduces information and offers just useful facts. This optimizes bandwidth and battery use. It is then made available to users and apps for real-time monitoring, historical research, and actionable insights to enable informed decisions across a variety of IoT applications. Once ready for transmission, data is delivered to cloud-based storage systems for safe storage, management, processing, analysis, or retrieval.

19.7 CASE STUDIES

19.7.1 CASE STUDY: IoT IN MANUFACTURING

IoT is transforming the industry with smart factories, predictive maintenance, and supply chain optimization. This report examines how a manufacturing company used IoT to boost production and reduce downtime (P. D. Singh et al. 2022). Unexpected equipment breakdowns, inadequate maintenance, and output delays plagued the firm. IoT sensors were installed on critical machinery for real-time monitoring. Sensors measure temperature, vibration, and other performance factors. ML was used to predict equipment problems. Automatic messages alerted maintenance staff to issues.

19.7.2   Case Study: IoT in Health

IoT technologies are transforming the healthcare industry. Patient care, cost, and operational efficiency are improving with these advances. In this case study, a major hospital uses IoT technology to increase operational efficiency and patient experience.

Patient congestion, poor asset management, and insufficient real-time monitoring delayed treatment at the hospital.

The hospital deployed IoT patient-tracking systems, wearable health monitors, and asset-tracking sensors. These devices recorded patient whereabouts, vital signs, and medical equipment availability in real time. Cloud-based analytics helped healthcare workers optimize resource allocation, reduce wait times, and improve patient outcomes (K. D. Singh, Singh, Chhabra et al. 2023).

19.7.3   Case Study: IoT in Agriculture

IoT is revolutionizing agriculture through precision farming, remote monitoring, and sustainable practices. This case study examines how a large agricultural enterprise uses IoT technology to boost crop output and resource allocation. The farm had agricultural growth discrepancies, water waste, and labor-intensive physical work. IoT sensors were strategically put over the farm area to monitor soil moisture, temperature, and nutrient levels. UAVs with powerful imaging technology captured aerial views of farming regions—a cloud-based platform integrated with weather predictions and historical crop data. Farmers received current irrigation, fertilization, and pest control advice.

Crop management decisions are made more informed with data (K. D. Singh 2021a).

19.8   FUTURE DIRECTIONS

There is tremendous hope and opportunity for game-changing developments in several fields thanks to the future convergence of IoT and cloud-based ML. Research and development will focus on edge computing, federated learning, 5G connection, interoperability standards, AI at the edge, environmental sustainability, and regulatory frameworks. The entire premise of this technology can only be realized via the concerted efforts of researchers, politicians, and industry leaders working together to overcome obstacles and ensure the ethical deployment of the technology. As these innovations continue to develop, they will affect several industries, boost people's standard of living, and help solve international problems.

19.9   EDGE COMPUTING

Edge computing has great promise for IoT and cloud-based ML integration. Edge computing will help solve latency, bandwidth, and real-time processing issues as IoT devices become pervasive. Edge devices will be optimized to do increasingly complex tasks locally in future studies. Developing sophisticated algorithms and AI models for edge devices allows quicker decision-making without cloud connectivity. Edge computing will become more scalable and versatile to meet the needs of various IoT applications. Edge computing will improve IoT device efficiency, data transmission costs, and autonomy, making them more important to numerous businesses.

19.10 FEDERATED LEARNING

The growing concerns surrounding data privacy and security in IoT applications will drive the adoption of federated learning. This technique enables ML models to be trained collaboratively across decentralized devices or servers without sharing raw data. The future of IoT will witness significant advancements in federated learning, focusing on making it more efficient, scalable, and adaptable to diverse IoT ecosystems. Researchers will explore methods to optimize model training and update processes while preserving data privacy. Federated learning will become a cornerstone in addressing regulatory and ethical concerns related to IoT data, ensuring that sensitive information remains secure while still enabling valuable insights to be derived from decentralized data sources.

19.11 AI AT THE EDGE

Integrating AI directly into IoT devices is an intriguing new area for future research and development. This method will allow IoT devices to process and make decisions at the edge, reducing the need for a cloud connection (Dhiman et al. 2022). Future research will focus on optimizing AI algorithms for resource-constrained edge devices to make them more capable of complex tasks. More intelligent and autonomous IoT applications will result, especially in autonomous automobiles, smart cities, and industrial automation, where real-time decision-making is crucial. Researchers will also examine ways to integrate cloud-based ML with edge-based AI seamlessly. This will let IoT ecosystems harness the best of both worlds to boost efficiency and intelligence.

19.12 5G CONNECTIVITY

The introduction of 5G network technology will usher in a brand-new age of connectivity for the IoT. The potential of 5G for IoT applications will be investigated in subsequent research efforts. This involves exploring how the increased bandwidth, decreased latency, and huge device connection capabilities offered by 5G might be utilized to provide real-time data transfer and communication across the IoT. Researchers and professionals in the industry will work together to define protocols and standards for the IoT that are relevant to 5G (Trivedi et al. 2021). This will ensure smooth interaction with IoT devices and platforms. Innovative applications, including driverless cars, augmented reality, and immersive IoT experiences, will be made possible by the convergence of 5G with the IoT. In addition, researchers will study the ways in which 5G might improve the scalability and dependability of IoT networks, therefore making those networks more resilient and responsive to the ever-changing requirements of IoT applications (Duan et al. 2022).

19.13 INTEROPERABILITY STANDARDS

Interoperability will continue to be a key focus in the future of IoT and cloud-based ML integration. Researchers and industry stakeholders will work toward the development and adoption of standardized protocols and frameworks that facilitate seamless communication and collaboration across diverse IoT devices and ecosystems. These interoperability standards will ensure that IoT devices from different manufacturers can work together harmoniously, reducing integration challenges and enhancing the overall IoT experience. Future research will also explore integrating emerging technologies like blockchain to secure and validate IoT data and transactions across heterogeneous networks. The establishment of comprehensive interoperability standards will lead to more versatile and adaptable IoT ecosystems, fostering innovation and collaboration across industries.

19.14 ENVIRONMENTAL SUSTAINABILITY

The future of IoT and cloud-based ML integration will significantly emphasize environmental sustainability. Research will focus on how IoT technology can be leveraged to optimize resource usage, reduce waste, and mitigate the environmental impact of various industries (Gupta, Nainwal, and Pant 2021). For instance, in agriculture, IoT sensors and data analytics can enable precise irrigation and fertilization, reducing water and chemical usage while maximizing crop yields. In smart cities, IoT systems can optimize energy consumption, reduce emissions, and improve urban planning for sustainability. Furthermore, IoT-driven environmental monitoring can provide valuable data for climate research and disaster mitigation. As the world faces increasing environmental challenges, IoT and cloud-based ML integration will play a vital role in driving sustainable practices and mitigating the impact of climate change (Wang and Zhao 2022).

19.15 REGULATORY FRAMEWORKS

In the future, a critical area of concern will be the creation of legal frameworks for the privacy, security, and ethical use of IoT data. In order to set rules and standards that uphold the rights of people and organizations while promoting innovation, governments, and regulatory authorities will work with industry stakeholders (Matta, Pant, and Tiwari 2019). The regulatory frameworks in question will encompass several aspects, including but not limited to data ownership, consent methods, security protocols, and adherence to privacy standards. The collaboration between researchers and policymakers aims to achieve a harmonious equilibrium by facilitating the advantages of the IoT while concurrently guaranteeing its ethical and responsible implementation. Subsequent investigations will also explore the advancement of tools and technology that streamline adherence to these legislative frameworks, facilitating organizational compliance with data privacy and security requirements.

19.16 CONCLUSION

This study examines how IoT and cloud-based ML may be used to build a robust ecosystem. First, we defined IoT and its key components, emphasizing data collection from Internet-connected physical objects. We also launched cloud-based ML to process and analyze IoT device data. We examined data security and privacy, real-time processing latency, scalability, and data gathering, and transmission mechanisms related to this integration. We explained how these problems must be addressed when installing and operating IoT systems. We also provided case studies from healthcare, industry, and agriculture to demonstrate the actual uses and advantages of IoT and cloud-based ML integration. These case studies showed how integration improves health care, production, and agriculture. We conclude that IoT and cloud-based ML can alter industry and society. It helps organizations use data-driven insights, real-time decision-making, and resource optimization. To maximize these benefits, data security and privacy, latency, and scalability must be addressed.

REFERENCES

Dhiman, Poonam, Vinay Kukreja, Poongodi Manoharan, Amandeep Kaur, M Kamruzzaman, Imed Ben Dhaou, and Celestine Iwendi. 2022. "A Novel Deep Learning Model for Detection of Severity Level of the Disease in Citrus Fruits." *Electronics* 11 (3): 495.

Duan, Sijing, Dan Wang, Ju Ren, Feng Lyu, Ye Zhang, Huaqing Wu, and Xuemin Shen. 2022. "Distributed Artificial Intelligence Empowered by End-Edge-Cloud Computing: A Survey." *IEEE Communications Surveys\& Tutorials*: 1254.

Firouzi, Farshad, Bahar Farahani, and Alexander Marinšek. 2022. "The Convergence and Interplay of Edge, Fog, and Cloud in the AI-Driven Internet of Things (IoT)." *Information Systems* 107. https://doi.org/10.1016/j.is.2021.101840

Gupta, Deepika, Ankita Nainwal, and Bhaskar Pant. 2021. "Literature Review: Real Time Water Quality Monitoring and Management." 923–29. https://doi.org/10.1007/978-981-15-7527-3_88

Matta, Priya, Bhaskar Pant, and Umesh Kumar Tiwari. 2019. "DDITA: A Naive Security Model for IoT Resource Security." *Advances in Intelligent Systems and Computing* 670: 199–209. https://doi.org/10.1007/978-981-10-8971-8_19

Radanliev, Petar, David De Roure, Carsten Maple, and Omar Santos. 2022. "Forecasts on Future Evolution of Artificial Intelligence and Intelligent Systems." *IEEE Access* 10: 45280–88.

Rathore, Pramod Singh, Jyotir Moy Chatterjee, Abhishek Kumar, and Radhakrishnan Sujatha. 2021. "Energy-Efficient Cluster Head Selection Through Relay Approach for WSN." *The Journal of Supercomputing* 77: 7649–75.

Seng, Kah Phooi, Li Minn Ang, and Ericmoore Ngharamike. 2022. "Artificial Intelligence Internet of Things: A New Paradigm of Distributed Sensor Networks." *International Journal of Distributed Sensor Networks* 18 (3): 15501477211062836.

Singh, Kiran Deep. 2021a. "Particle Swarm Optimization Assisted Support Vector Machine Based Diagnostic System for Dengue Prediction at the Early Stage." In *Proceedings - 2021 3rd International Conference on Advances in Computing, Communication Control and Networking, ICAC3N 2021*, 844–48. https://doi.org/10.1109/ICAC3N53548.2021.9725670

Singh, Kiran Deep. 2021b. "Securing of Cloud Infrastructure Using Enterprise Honeypot." In *Proceedings – 2021 3rd International Conference on Advances in Computing, Communication Control and Networking, ICAC3N 2021*, 1388–93. https://doi.org/10.1109/ICAC3N53548.2021.9725389

Singh, Kiran Deep, and Prabh Deep Singh. 2023. "A Novel Cloud-Based Framework to Predict the Employability of Students." In *2023 International Conference on Advancement in Computation\& Computer Technologies (InCACCT)*, 528–32.

Singh, Prabh Deep, and Kiran Deep Singh. 2023a. "Security and Privacy in Fog/Cloud-Based IoT Systems for AI and Robotics." *EAI Endorsed Transactions on AI and Robotics* 2: 1–6.

Singh, Prabh Deep, and Kiran Deep Singh. 2023b. "Fog-Centric Intelligent Surveillance System: A Novel Approach for Effective and Efficient Surveillance." In *2023 International Conference on Advancement in Computation\& Computer Technologies (InCACCT)*, 762–66.

Singh, Kiran Deep, Prabh Deep Singh, Ankit Bansal, Gaganpreet Kaur, Vikas Khullar, and Vikas Tripathi. 2023. "Exploratory Data Analysis and Customer Churn Prediction for the Telecommunication Industry." In *2023 3rd International Conference on Advances in Computing, Communication, Embedded and Secure Systems (ACCESS)*, 197–201.

Singh, Kiran Deep, Prabh Deep Singh, Rishu Chhabra, Gaganpreet Kaur, Ankit Bansal, and Vikas Tripathi. 2023. "Cyber-Physical Systems for Smart City Applications: A Comparative Study." In *2023 International Conference on Advancement in Computation\& Computer Technologies (InCACCT)*, 871–76.

Singh, Kiran Deep, Prabh Deep Singh, and Sandeep Singh Kang. 2022. "Ensembled-Based Credit Card Fraud Detection in Online Transactions." *AIP Conference Proceedings* 2555: 50009. https://doi.org/10.1063/5.0108873

Singh, Kiran Deep, Prabh Deep Singh, Gaganpreet Kaur, Vikas Khullar, Rishu Chhabra, and Vikas Tripathi. 2023. "Education 4.0: Exploring the Potential of Disruptive Technologies in Transforming Learning." In *2023 International Conference on Computational Intelligence and Sustainable Engineering Solutions (CISES)*, 586–91.

Singh, Prabh Deep, Kiran Deep Singh, Vikas Tripathi, and Vaibhav Chaudhari. 2022. "Use of Ensemble Based Approach to Predict Health Insurance Premium at Early Stage." *Proceedings of International Conference on Computational Intelligence and Sustainable Engineering Solution, CISES 2022*, 566–69. https://doi.org/10.1109/CISES54857.2022.9844398

Singh, Kiran Deep, Prabh Deep Singh, Vikas Tripathi, and Vikas Khullar. 2022. "A Novel and Secure Framework to Detect Unauthorized Access to an Optical Fog-Cloud Computing Network." In *2022 Seventh International Conference on Parallel, Distributed and Grid Computing (PDGC)*, 618–22.

Trivedi, Naresh K, Vinay Gautam, Abhineet Anand, Hani Moaiteq Aljahdali, Santos Gracia Villar, Divya Anand, Nitin Goyal, and Seifedine Kadry. 2021. "Early Detection and Classification of Tomato Leaf Disease Using High-Performance Deep Neural Network." *Sensors* 21 (23): 7987.

Wang, Yitong, and Jun Zhao. 2022. "Mobile Edge Computing, Metaverse, 6G Wireless Communications, Artificial Intelligence, and Blockchain: Survey and Their Convergence." *2022 IEEE 8th World Forum on Internet of Things (WF-IoT)*: 1–8.

20 The Convergence of Cloud, IoT, and Artificial Intelligence for Intelligent Systems

*Kiran Deep Singh, Prabh Deep Singh,
Gaganpreet Kaur, Vikas Lamba, M.R.M.
Veeramanickam, and Vikas Khullar*

20.1 INTRODUCTION

The concept of "intelligent systems" is gaining more and more currency in today's world, which is characterized by an acceleration in the rate of technological progress on a scale never seen before. It encompasses a field of invention that combines artificial intelligence (AI), machine learning, and advanced computing to produce systems with intelligence, adaptability, and problem-solving skills comparable to those of humans. Intelligent systems, also known as AI systems or intelligent agents, are computational entities meant to mimic, simulate, or duplicate human-like cognitive processes (Firouzi, Farahani, and Marinšek 2022). Other common names for intelligent systems are smart robots and intelligent agents. These capacities include perception, thinking, learning, problem-solving, and decision-making. Together, these functions have a wide breadth of cognitive talents. Intelligent systems can analyze huge volumes of data, see patterns, adjust to changing settings, and make educated decisions based on their understanding of the data. These systems are not only clever software programs; rather, they represent a paradigm change that is altering industries, enhancing decision-making processes, and enriching our day-to-day lives (Ahmed et al. 2022). This transition is brought about by the rise of AI. Now that we have a thorough comprehension of cloud computing, the Internet of Things (IoT), and AI in our arsenal, we are in a position to investigate the core of our mission, which is to examine the confluence of these three revolutionary forces. This convergence is more than just the presence of different technologies; rather, it is a symbiotic connection that boosts the potential of each separately.

Intelligent systems process and interpret data like humans but are quicker and have the capacity to manage massive datasets. Analytics, prediction, automation, and natural language understanding are their specialties. Thus, they are used in healthcare, banking, industry, transportation, and entertainment.

Unprecedented convergence in technology is changing our reality. Cloud Computing, the IoT, and AI have revolutionized how we live, work, and interact with

DOI: 10.1201/9781032656694-20

the digital world. This unprecedented combination will empower intelligent systems in smart cities, healthcare, manufacturing, and driverless cars. As we explore the complex relationship between these three revolutionary forces, their convergence might restructure industries, improve quality of life, and reshape technology (Seng, Ang, and Ngharamike 2022).

The idea of intelligent systems has long fascinated scientists, engineers, and visionaries. These systems will perceive, reason, and act independently to adapt to changing situations and enhance human talents. Intelligent systems have been envisioned for decades, but cloud, IoT, and AI are the catalysts for their realization. These technology disciplines have developed rapidly in isolation, but the magic happens at their junction (Wang and Zhao 2022).

The current digital infrastructure relies on cloud computing, which has democratized processing power and data storage. It has freed organizations and individuals from on-premises infrastructure, enabling scalable and affordable solutions. However, IoT has connected billions of gadgets, allowing them to share data easily. These gadgets include thermostats, wearables, industrial machines, and drones. Finally, AI's capacity to analyze and learn from massive datasets has revolutionized data-driven decision-making and automation (Radanliev et al. 2022).

In this detailed investigation of cloud, IoT, and AI convergence, we explore its origins, current condition, and future possibilities. Our journey will reveal the intricate tapestry of each technology's strengths and weaknesses (Duan et al. 2022). We will next examine how they complement and reinforce each other, releasing a flood of inventions greater than their components.

20.2   APPLICATIONS OF INTELLIGENT SYSTEMS: HARNESSING AI, IoT, AND CLOUD COMPUTING

In today's technologically evolved world, AI, IoT, and cloud computing are revolutionizing many sectors and areas. The combination of these cutting-edge technology powers these intelligent systems, which are transforming our lives and work. In this comprehensive examination, we will examine many applications where intelligent methods are making a big effect.

1. Healthcare:

    Intelligent systems use AI to analyze medical records, lab findings, and imaging scans to diagnose and predict diseases. They can spot trends and anticipate illness development, progression, and treatment effects. For instance, AI-based systems detect cancer, diabetes, and heart disease early (Sangeeta and Tandon 2021).
    - Telemedicine and Remote Monitoring: IoT gadgets like fitness trackers and remote monitoring equipment capture and transmit real-time data. Healthcare workers may remotely monitor patients, change treatment plans, and intervene quickly with cloud-based AI (K. D. Singh 2021a).
    - Drug Discovery: AI algorithms can anticipate drug candidates, analyze chemical structures, and simulate biological consequences. Cloud

computing speeds up drug research and development by accelerating computationally heavy processes.

Cloud-hosted AI-powered healthcare chatbots deliver 24/7 patient support, medical inquiries, and appointment booking. Healthcare professionals may simplify administrative processes and engage patients using virtual assistants (P. D. Singh et al. 2022).

2. Manufacturing Industry:
   - Predictive Maintenance: IoT sensors monitor equipment performance of machines—Data-driven AI models on the cloud forecast maintenance needs, decreasing downtime and expenses. Manufacturers can optimize maintenance and prevent problems (Rani, Ahmed, and Rastogi 2020).
   - Quality Control: AI-based computer vision technologies identify product problems in real time. Images and data from manufacturing line cameras and sensors are analyzed in the cloud to find flaws. This maintains product quality.
   - Supply Chain Optimization: IoT sensors detect products' movement and condition. Optimizing inventory management, demand forecasting, and logistics with AI algorithms saves money and improves efficiency.
   - Intelligent systems—AI, IoT, and cloud computing—power sophisticated robots. Warehouse automation, precise assembly, and autonomous navigation under challenging situations are possible with these robots (P. D. Singh and Singh 2023b).

3. Smart Cities:
   Real-time traffic monitoring using IoT sensors and cameras in urban infrastructure. Cloud-based AI analyses this data to optimize traffic flow, alleviate congestion, and increase transportation efficiency. Traffic lights react dynamically to current conditions (Islam, Kumar, and Hu 2021).

   To optimize energy distribution, smart grids use IoT devices and AI algorithms. This cuts energy use, prices, and carbon emissions. The grid can better integrate renewable energy.

   AI-powered video surveillance, gunshot detection, and predictive policing improve public safety. The cloud processes data from IoT devices and cameras to identify risks and respond proactively.

   IoT-enabled smart bins notify when they are full, optimizing garbage collection routes and saving fuel use. AI systems can detect recycling contaminants and reduce waste.

4. Transportation and Autonomous Vehicles:
   Self-driving vehicles and trucks use IoT sensors (e.g., lidar, radar, cameras) to sense their surroundings. Using this data, cloud-hosted AI algorithms make driving judgments, plan routes, and improve road safety.
   - Fleet Management: Vehicle IoT devices give real-time position, fuel, and maintenance data. Fleet managers use cloud-based AI to optimize routes, minimize fuel costs, and schedule vehicle maintenance.
   - IoT sensors in public transportation track vehicle positions, passenger numbers, and timetables. Cloud-based AI systems can optimize transit

routes, offer passengers real-time information, and boost public transportation efficiency.

5. Agriculture:

IoT sensors and AI-driven analytics allow precision farming—The cloud analyses soil, weather, and crop health data. Learn how to optimize irrigation, fertilization, and pest management to boost yields and save resources.

IoT devices monitor cattle health and behavior. Cloud-based AI may analyze data to identify sickness, track eating patterns, and enhance animal welfare.

- Crop Monitoring: IoT-equipped drones collect agricultural data. Cloud-based AI processes this data to identify insect infestations and agricultural illnesses.

6. Retail and E-commerce:

AI analyses customer data, such as browsing history and purchase behavior, in the cloud for personalized recommendations. Personalized product recommendations improve the shopping experience and boost sales.

- Inventories Management: IoT sensors track inventories in real time. AI systems use this data to optimize stock replenishment, decrease overstocking, and reduce stockouts using.
- Supply Chain Visibility: IoT-enabled shipment tracking devices give real-time location and condition data. Cloud-based AI improves supply chain visibility, delivery accuracy, and proactive issue resolution.

7. Energy Management:

- Home Automation: IoT devices enable remote control of lights, heating, cooling, and appliances in smart homes. AI systems can optimize energy use and save electricity costs by learning use trends.
- Smart grids using IoT sensors and AI algorithms optimize energy distribution based on real-time demand and supply. This eliminates energy waste and integrates renewables.
- Energy forecasting: Cloud-based AI algorithms estimate energy demands using meteorological data, energy usage, and renewable energy output. This helps utilities forecast peak demand and optimize energy production (Kang, Singh, and Kumari 2022).

8. Environmental Monitoring:

- Air Quality Monitoring: IoT sensors assess pollution and particle matter levels. The public and authorities receive real-time air quality reports through cloud-based AI.
- Water Quality Management: IoT sensors monitor pH, temperature, and pollutants from water bodies. AI algorithms analyze this data to detect water pollution and assure safe consumption (Gupta, Nainwal, and Pant 2021).
- Wildlife Conservation: IoT-enabled animal tracking devices record behavior and movements. A cloud-based AI analyses this data to assess wildlife populations, identify poaching dangers, and advise conservation efforts.

- The Financial Services sector includes Algorithmic Trading, which uses AI to analyze market data and execute transactions in real-time. Cloud computing enables large-scale, high-frequency trading and data processing.
  - Cloud AI algorithms detect fraud and abnormal behavior in banking transactions. They see fraud and notify to avert financial losses.
  - AI-powered chatbots and virtual assistants in the banking business offer client service, answer questions, and manage accounts. They handle many consumer encounters well.

9. Education:

   AI-driven systems analyze student performance data to offer personalized learning experiences. Scalable and accessible cloud-based solutions let students learn at their own speed.

   AI-powered language learning applications boost language abilities through speech recognition and natural language processing. These applications give criticism of real-time grammar and pronunciation.

   Cloud-based AI streamlines admissions processing, scheduling, and resource allocation in educational institutions, minimizing administrative overhead.

10. Media and Entertainment:

    AI algorithms propose films, music, articles, and items based on user tastes and behavior. Cloud computing processes massive content and user data.

    AI-generated articles, art, and music are becoming more common. Cloud resources enable innovative AI model training and deployment.

    Cloud-based AI helps media firms analyze real-time audience interaction and adapt content distribution and advertising tactics.

11. Government and Public Services:

    Cloud-based AI chatbots deliver citizen services and information. These virtual assistants can simplify government agency contacts and answer frequent questions (Shukla et al. 2022).

    - Emergency Response: Smart city IoT sensors give real-time data on fires, earthquakes, and floods. Data processing using cloud-based AI systems improves emergency response coordination and decision-making.
    - Public Health: Cloud-based AI models track disease outbreaks, identify high-risk locations, and aid public health planning.

12. Space Exploration

    AI-powered rovers and spaceships analyze data from faraway planetary missions. Cloud-based AI aids image analysis, landscape mapping, and science.

    - Astronomy: Cloud AI algorithms analyze telescope and observatory data to identify celestial objects, analyze cosmic processes, and find exoplanets.
    - Space Traffic Management: IoT sensors detect objects in Earth's orbit to prevent accidents and keep satellites and space missions safe.

13. HR and Recruitment:

    - Talent Acquisition: AI-powered solutions help HR workers find qualified individuals for job openings. Resume analysis, automated interviews, and applicant success prediction are possible with these tools.

- AI-driven surveys and feedback analysis tools let companies assess employee happiness and discover workplace improvements.
- Cloud-based AI solutions analyze workforce data to discover patterns, optimize personnel management, and make data-driven HR decisions.

14. Legal Services:

AI algorithms on the cloud aid lawyers in research, case law analysis, and document creation.

- Document evaluation: AI-powered e-discovery technologies can evaluate and categorize massive legal papers and communications during litigation, saving time and money.
- Contract Analysis: AI models can extract critical clauses, conditions, and duties from contracts, simplifying contract administration and compliance.

These applications are just a small part of how AI, IoT, and cloud computing-powered intelligent systems are changing industries and disciplines. Combining these technologies boosts productivity and expands creativity, discovery, and problem-solving. We may expect more revolutionary applications as technology evolves, transforming how we interact with the world and increasing our quality of life.

20.3   THE BUILDING BLOCKS OF INTELLIGENT SYSTEMS

In order to manage complexity and assure effective development and maintenance, designing an intelligent system architecture that integrates cloud, IoT, and AI often requires breaking the system down into modular or subsystem components. Each module/subsystem performs certain functions and interacts to form an intelligent system. Key modules/subsystems used in intelligent system architecture and their responsibilities are discussed here:

1. Data Acquisition Module:
   - Function: Collects data from IoT sensors, devices, and other sources. It collects raw data from IoT components.
   - Explanation: Data is essential to intelligent systems because it fuels AI models. This module optimizes data collection for volume, velocity, diversity, and validity.
2. Data Preprocessing and Cleansing Module:
   - Function: Clean, convert, and aggregate raw data. Data is prepared for AI model training and analysis in this module.
   - Explanation: Data quality is essential for AI model accuracy. This module performs missing data imputation, outlier identification, and feature engineering.

   The cloud Infrastructure Module provides compute and storage resources for data processing, AI model training, and deployment.

   Cloud services are appropriate for intelligent systems' computational needs due to their scalability, flexibility, and accessibility.

3. AI Model Development and Training Module:
   - Function: Develops, trains, and validates AI models. It creates predictive or decision-making models from preprocessed data.
   - Explanation: AI models underpin intelligent systems. This module chooses methods, trains the model repeatedly, and optimizes performance.
4. Edge Computing Module:
   - Function: Improving processing capabilities on IoT devices or local servers for real-time decision-making and reduced latency.
   - Explanation: Some applications demand low-latency answers, and edge computing processes data locally before sending it to the cloud.
5. Privacy and Security Module:
   - Function: This module secures data and systems through encryption, authentication, access restrictions, and privacy compliance.
   - Explanation: When handling sensitive data and real-time interactions, intelligent systems must prioritize security and privacy.
6. UI/UX Module:
   - Function: The UI/UX module creates the intelligent system's user interface. It makes usage easy.
   - Explanation: A well-designed interface improves user acceptance and happiness, making it essential for system implementation.
7. Scalability and Resource Management Module:
   - Function: This module ensures the system can scale to handle more IoT devices, data, and users. Optimization and resource allocation are included.
   - Explanation: Intelligent systems grow in data volume and user demand, making scalability important.
8. Monitoring and Analytics Module:
   - Function: Real-time monitoring, logging, and analytics for system performance, user behavior, and data insights are provided by this module.
   - Explanation: Monitoring identifies issues before they happen, while analytics provides insights from system data.
9. Ethical AI and Compliance Module:
   - Function: This module ensures adherence to ethical AI practices and regulatory compliance. It monitors AI model behavior for fairness, bias, and transparency.
   - Explanation: As AI systems impact decision-making and user experiences, ethical considerations and regulatory compliance are critical.
10. Continuous Improvement and Feedback Loop Module:
    - Function: This module establishes mechanisms for gathering user and system feedback to drive iterative improvements in the system's performance and features.
    - Explanation: Continuous improvement ensures that the intelligent system evolves to meet changing user needs and remains competitive.
11. Documentation and Knowledge Management Module:
    - Function: This module focuses on maintaining comprehensive documentation for the system's architecture, data flows, security protocols, and AI models (K. D. Singh and Singh 2023).

- Explanation: Documentation aids in system understanding, trouble-shooting, and knowledge transfer among team members.

12. Emergency Response and Disaster Recovery Module:
    - Function: This module plans for and responds to security breaches, system failures, or other emergencies. It includes incident response and data backups.
    - Explanation: Preparedness for emergencies minimizes downtime and data loss, ensuring system resilience.

13. Environmental Sustainability Module:
    - Function: This module assesses and mitigates the environmental impact of the intelligent system, particularly in terms of energy consumption and waste management.
    - Explanation: Sustainability considerations align the system with responsible environmental practices.

14. Blockchain Integration Module:
    - Function: This module integrates blockchain technology to enhance data security, trust, and transparency, particularly in data transactions and record-keeping.
    - Explanation: Blockchain helps maintain data integrity and security, which is critical in scenarios involving sensitive transactions or contractual agreements (K. D. Singh et al. 2022).

15. Human-AI Collaboration Module:
    - Function: This module focuses on making interactions between users and AI systems more natural and effective. It includes natural language processing and dialogue management.
    - Explanation: Enhancing the user experience through more human-like interactions can lead to higher user adoption and satisfaction.

These modular components work cohesively to create a robust and efficient intelligent system architecture. The precise configuration and integration of these modules depend on the specific requirements, objectives, and domain of the smart system being designed. Careful consideration of each module ensures that the system can effectively collect, process, analyze, and act upon data while maintaining security, ethics, and scalability.

20.4　CHALLENGES

Intelligent systems built on the convergence of the cloud, IoT, and AI can potentially revolutionize several fields. However, in order to properly utilize its potential, this convergence also raises a number of critical hurdles. In the course of this in-depth investigation, we will investigate the main challenges involved with this convergence, including the following:

1. Data Privacy and Security:
    - Data Privacy: IoT devices generate vast amounts of data, often including sensitive information. Storing and processing this data in the cloud

raises concerns about data privacy and compliance with data protection regulations (e.g., GDPR). Unauthorized access to personal data can lead to privacy breaches and legal consequences (K. D. Singh, Singh, and Kang 2022).

- Security: IoT devices can be vulnerable to cyberattacks due to limited computational resources and inadequate security measures. The cloud itself is a potential target for cyberattacks. Ensuring the security of both IoT devices and cloud infrastructure is critical to prevent data breaches and service disruptions (P. D. Singh and Singh 2023a).

2. Scalability and Latency:
- Scalability: IoT ecosystems are expected to grow exponentially, with billions of connected devices. The cloud must be capable of handling this massive scale efficiently. Scalability challenges include provisioning resources dynamically, load balancing, and managing increasing data volumes.
- Latency: Certain applications like autonomous vehicles and real-time monitoring demand low-latency responses. Sending data to the cloud for processing and receiving instructions can introduce unacceptable delays. Edge computing, where AI processing occurs on IoT devices or local servers, is one approach to address latency concerns.

3. Interoperability and Standards:
- IoT Device Diversity: IoT devices come from various manufacturers and use different communication protocols. Ensuring interoperability and standardization across these devices is challenging but essential for seamless integration into the cloud and AI systems.
- Cloud Service Providers: Different cloud service providers may have proprietary APIs and services. This can create vendor lock-in and hinder interoperability between cloud platforms.

4. Data Quality and Reliability:
- Data Quality: IoT sensors may produce noisy or inaccurate data due to environmental factors or sensor malfunctions. Ensuring the reliability and quality of IoT data is essential for accurate AI-driven insights.
- Data Reliability: IoT devices may operate in remote or harsh environments, making device reliability a concern. Ensuring devices are resilient and can withstand challenging conditions is crucial for maintaining data streams.

5. Energy Efficiency:
- IoT Device Power Consumption: Many IoT devices operate on battery power or have limited energy sources. Constant data transmission to the cloud can drain device batteries quickly. Optimizing energy consumption and prolonging device lifespans are important considerations.
- Edge Processing: To address energy efficiency concerns, some data processing should be performed at the edge rather than in the cloud. However, this introduces challenges in managing AI models on resource-constrained devices.

6. Complexity of AI Models:
   - Model Complexity: AI models used in intelligent systems, especially deep learning models, can be computationally intensive and resource-demanding. Deploying these models in the cloud may require significant computational resources, which can increase costs (K. D. Singh, Singh, Chhabra et al. 2023).
   - Model Size: The size of AI models can be substantial, making it challenging to deploy them on edge devices with limited storage capacity. Finding ways to optimize model size and efficiently distribute models to devices is essential.

7. Regulatory and Ethical Considerations:
   - Regulatory Compliance: Intelligent systems in regulated industries, such as healthcare and finance, must comply with industry-specific regulations. Navigating complex regulatory landscapes can be a significant challenge.
   - Ethical AI: Ensuring ethical AI practices, including bias mitigation and fairness, is essential. AI algorithms must be trained and deployed in ways that do not discriminate against certain groups or perpetuate biases.

8. Cost Management:
   - Cost of Cloud Resources: Using cloud resources for data storage, processing, and AI model training can become costly, especially at scale. Cost management and optimization strategies are critical to controlling expenses.
   - IoT Device Costs: Deploying and maintaining IoT devices, especially in large-scale deployments, can be costly. Identifying cost-effective devices and strategies for device management is important.

9. Data Governance:
   - Data Ownership: Determining data ownership and rights when data is generated by IoT devices and processed in the cloud can be complex. Clear data governance policies are necessary to avoid disputes.
   - Data Retention: Deciding how long to retain IoT-generated data in the cloud and ensuring compliance with data retention policies can be challenging.

10. Human-Computer Interaction:
   - User Experience: Designing intuitive user interfaces and experiences for IoT devices and cloud applications is crucial. Ensuring that users can interact with intelligent systems seamlessly is an ongoing challenge.
   - User Training: Educating users on how to interact with intelligent systems and ensuring they understand the capabilities and limitations of AI-driven features is essential for adoption.

11. Environmental Impact:
   - Energy Consumption: The growing number of IoT devices and the energy required for cloud data centers contribute to environmental concerns. Minimizing the carbon footprint of intelligent systems is a sustainability challenge.

- E-Waste: The disposal of IoT devices at the end of their lifecycle presents e-waste challenges. Ensuring responsible recycling and disposal practices is important.
12. Reliability and Redundancy:
    - System Failures: Intelligent systems must be designed with redundancy and failover mechanisms to ensure continued operation in the event of hardware failures, network outages, or other disruptions.
    - Data Backups: Ensuring data resilience and backups in the cloud is crucial to prevent data loss in case of unforeseen events.
13. Ethical AI Governance:
    - AI Decision Transparency: Ensuring that AI-driven decisions are transparent and explainable is a critical ethical consideration. Users and stakeholders should understand how AI arrived at a particular conclusion.
    - Bias and Fairness: Addressing bias in AI models and ensuring fairness in decision-making is an ongoing challenge. Monitoring and mitigating bias in AI systems is essential to avoid discriminatory outcomes.
14. Integration Complexity:
    - Legacy Systems: Integrating IoT, cloud, and AI technologies with existing legacy systems and infrastructure can be complex and may require significant changes to current IT architectures.
    - Skill Gaps: The convergence of these technologies demands a workforce with expertise in multiple domains. Bridging skill gaps and providing training opportunities for professionals is a challenge.

Addressing these challenges requires collaboration between technology developers, industry stakeholders, regulators, and policymakers. It also necessitates ongoing research and innovation to develop solutions that ensure the reliable, secure, and ethical operation of intelligent systems powered by the convergence of cloud, IoT, and AI. As these technologies continue to evolve, so will the strategies and solutions to overcome these challenges and unlock the full potential of intelligent systems.

20.5 THE FUTURE OF INTELLIGENT SYSTEMS

As intelligent systems evolve and mature, they will become integral to our daily lives, industries, and economies. However, with these advancements come challenges related to security, ethics, privacy, and environmental impact. Stakeholders in technology, government, and society must work collaboratively to navigate these challenges while harnessing the immense potential of intelligent systems for a brighter and more interconnected future.

1. AI-Driven Personalization and Predictive Insights:
    - Enhanced User Experiences: Future intelligent systems will provide highly personalized experiences across various applications, from e-commerce and content recommendations to healthcare and smart homes. AI algorithms will adapt in real-time to user behavior, preferences, and needs (K. D. Singh, Singh, Bansal et al. 2023).

- Predictive Capabilities: These systems will become even more adept at predicting user preferences and behavior, enabling anticipatory services and recommendations. For instance, AI-powered virtual assistants will proactively suggest actions based on historical data and real-time context.

2. Edge AI for Real-time Decision-making:
   - Edge Computing Ubiquity: Edge computing, where AI processing occurs on IoT devices or local servers, will become ubiquitous. This will enable low-latency decision-making, crucial for applications like autonomous vehicles, industrial automation, and augmented reality.
   - Federated Learning: IoT devices will increasingly participate in federated learning, allowing AI models to be trained collaboratively on decentralized data sources. This will preserve data privacy while improving model performance (K. D. Singh, Singh, Bansal et al. 2023).

3. AI in Healthcare:
   - Disease Prediction and Prevention: Intelligent systems will play a vital role in early disease prediction and preventive healthcare. Wearable IoT devices will continuously monitor vital signs, and AI will analyze this data to identify health risks and recommend timely interventions.
   - Drug Discovery: AI-driven drug discovery will accelerate, with cloud-based AI models simulating the effects of potential drugs on biological systems. This will expedite the development of novel treatments (K. D. Singh 2021b).
   - Telemedicine: Telemedicine and remote patient monitoring will become more sophisticated, making healthcare more accessible and efficient. AI will aid in diagnosing and monitoring conditions, leading to better patient outcomes.

4. Smart Cities and Sustainability:
   - Sustainable Urban Planning: IoT sensors and AI algorithms will help optimize urban infrastructure, reducing energy consumption, traffic congestion, and environmental impact. Smart grids will efficiently distribute renewable energy.
   - Environmental Monitoring: IoT devices will monitor air and water quality, enabling faster response to pollution events. AI will provide real-time environmental insights and support conservation efforts.
   - Disaster Management: Intelligent systems will enhance disaster preparedness and response. IoT sensors and AI will provide early warnings and support emergency services in disaster-affected areas.

5. Autonomous Systems:
   - Autonomous Vehicles: Autonomous vehicles will become a common sight on roads, revolutionizing transportation. Cloud-connected AI will enable vehicles to communicate with each other and traffic infrastructure, enhancing safety and traffic flow.
   - Drones: Drones equipped with AI will be used for tasks such as package delivery, infrastructure inspection, and disaster response. They will

operate autonomously and collaborate with other devices. Industrial Automation: AI-powered robots and autonomous systems will transform manufacturing and logistics. These systems will adapt to dynamic environments and collaborate with human workers.

6. AI-Enhanced Education and Work:
   - Personalized Learning: AI-powered educational platforms will provide highly personalized learning experiences catering to individual strengths and weaknesses. This will revolutionize education and training (K. D. Singh, Singh, Kaur et al. 2023).
   - Augmented Workforces: In the workplace, AI will enhance productivity and decision-making across various industries. Augmented reality and AI will support workers in tasks ranging from maintenance to complex problem-solving.
   - Language Translation: Real-time language translation powered by AI will break down language barriers, enabling global communication and collaboration.

7. Quantum Computing Integration:
   - Quantum Computing: As quantum computing matures, it will be integrated with cloud infrastructure to solve complex problems that are currently computationally infeasible. This will impact fields like cryptography, materials science, and drug discovery.

8. Health and Wellness Monitoring:
   - Mental Health: IoT devices and AI will play a significant role in mental health monitoring, providing early indicators of stress, anxiety, and depression. AI-driven virtual therapists may become more common.
   - Aging in Place: IoT-enabled smart homes will support aging populations by monitoring their health, ensuring medication adherence, and assisting daily activities.

9. Blockchain for Security and Trust:
   - Security: Blockchain technology will be increasingly integrated with IoT and cloud systems to enhance security, data integrity, and trust. It will be crucial for securing transactions and data in these interconnected environments.

10. Human-AI Collaboration:
    - Natural Language Interaction: Human-AI collaboration will be more natural and seamless, with AI systems understanding context and intent in human language, leading to more effective interactions.
    - AI Assistants: AI-powered virtual assistants will become highly integrated into daily life, helping with tasks from scheduling appointments and managing finances to providing education and entertainment.

11. Ethical and Regulatory Frameworks:
    - Ethical AI: Governments, organizations, and industry bodies will establish more robust ethical frameworks and guidelines for developing and deploying AI and IoT technologies.
    - Data Privacy: Stricter data privacy regulations will evolve to protect user data in the face of growing IoT and cloud usage.

20.6 CONCLUSIONS

The confluence of cloud, IoT, and AI for intelligent systems is a significant technical innovation that might change many sectors. As we have examined these systems' uses, difficulties, and structure, they offer improved efficiency, real-time decision-making, and unique insights. However, they also require careful security, privacy, ethics, and environmental responsibility. Organizations must take a strategic strategy that connects technology with corporate goals, handles regulatory needs, and prioritizes ethics to maximize intelligent system potential. Companies and communities must monitor, adapt, and collaborate in the changing world of intelligent systems. The voyage of these systems shows human creativity and technology's ability to change the world. As cloud, IoT, and AI merge, we must be alert, aware, and imaginative to guarantee that these intelligent systems benefit society, preserve our privacy, and contribute to a sustainable and interconnected future.

REFERENCES

Ahmed, Imran, Yulan Zhang, Gwanggil Jeon, Wenmin Lin, Mohammad R. Khosravi, and Lianyong Qi. 2022. "A Blockchain- and Artificial Intelligence-Enabled Smart IoT Framework for Sustainable City." *International Journal of Intelligent Systems* 37 (9): 6493–507. https://doi.org/10.1002/int.22852

Duan, Sijing, Dan Wang, Ju Ren, Feng Lyu, Ye Zhang, Huaqing Wu, and Xuemin Shen. 2022. "Distributed Artificial Intelligence Empowered by End-Edge-Cloud Computing: A Survey." *IEEE Communications Surveys\& Tutorials* 25 (1): 591–624.

Firouzi, Farshad, Bahar Farahani, and Alexander Marinšek. 2022. "The Convergence and Interplay of Edge, Fog, and Cloud in the AI-Driven Internet of Things (IoT)." *Information Systems* 107. https://doi.org/10.1016/j.is.2021.101840

Gupta, Deepika, Ankita Nainwal, and Bhaskar Pant. 2021. "Literature Review: Real-Time Water Quality Monitoring and Management," 923–29. https://doi.org/10.1007/978-981-15-7527-3_88

Islam, Mir Salim Ul, Ashok Kumar, and Yu-Chen Hu. 2021. "Context-Aware Scheduling in Fog Computing: A Survey, Taxonomy, Challenges, and Future Directions." *Journal of Network and Computer Applications* 180: 103008.

Kang, Sandeep Singh, Kiran Deep Singh, and Shalini Kumari. 2022. "Smart Antenna for Emerging 5G and Application." In *Printed Antennas*, 249–64. CRC Press.

Radanliev, Petar, David De Roure, Carsten Maple, and Omar Santos. 2022. "Forecasts on Future Evolution of Artificial Intelligence and Intelligent Systems." *IEEE Access* 10: 45280–88.

Rani, Shalli, Syed Hassan Ahmed, and Ravi Rastogi. 2020. "Dynamic Clustering Approach Based on Wireless Sensor Networks Genetic Algorithm for IoT Applications." *Wireless Networks* 26: 2307–16.

Sangeeta, and Urvashi Tandon. 2021. "Factors Influencing Adoption of Online Teaching by School Teachers: A Study During COVID-19 Pandemic." *Journal of Public Affairs* 21 (4): e2503.

Seng, Kah Phooi, Li Minn Ang, and Ericmoore Ngharamike. 2022. "Artificial Intelligence Internet of Things: A New Paradigm of Distributed Sensor Networks." *International Journal of Distributed Sensor Networks* 18 (3): 15501477211062836.

Shukla, Surendra Kumar, Bhaskar Pant, Wattana Viriyasitavat, Devvret Verma, Sandeep Kautish, Gaurav Dhiman, Amandeep Kaur, Kannan Srihari, and Sachi Nandan Mohanty. 2022. "An Integration of Autonomic Computing with Multicore Systems for Performance Optimization in Industrial Internet of Things." *IET Communications*. https://doi.org/10.1049/cmu2.12505

Singh, Kiran Deep. 2021a. "Particle Swarm Optimization Assisted Support Vector Machine Based Diagnostic System for Dengue Prediction at the Early Stage." In *Proceedings – 2021 3rd International Conference on Advances in Computing, Communication Control and Networking, ICAC3N 2021*, 844–48. https://doi.org/10.1109/ICAC3N 53548.2021.9725670

Singh, Kiran Deep. 2021b. "Securing of Cloud Infrastructure Using Enterprise Honeypot." In *Proceedings – 2021 3rd International Conference on Advances in Computing, Communication Control and Networking, ICAC3N 2021*, 1388–93. https://doi.org/ 10.1109/ICAC3N53548.2021.9725389

Singh, Kiran Deep, and Prabh Deep Singh. 2023. "A Novel Cloud-Based Framework to Predict the Employability of Students." In *2023 International Conference on Advancement in Computation\& Computer Technologies (InCACCT)*, 528–32.

Singh, Prabh Deep, and Kiran Deep Singh. 2023a. "Security and Privacy in Fog/Cloud-Based IoT Systems for AI and Robotics." *EAI Endorsed Transactions on AI and Robotics* 2: 762–766.

Singh, Prabh Deep, and Kiran Deep Singh. 2023b. "Fog-Centric Intelligent Surveillance System: A Novel Approach for Effective and Efficient Surveillance." In *2023 International Conference on Advancement in Computation\& Computer Technologies (InCACCT)*, 762–66.

Singh, Kiran Deep, Prabh Deep Singh, Ankit Bansal, Gaganpreet Kaur, Vikas Khullar, and Vikas Tripathi. 2023. "Exploratory Data Analysis and Customer Churn Prediction for the Telecommunication Industry." In *2023 3rd International Conference on Advances in Computing, Communication, Embedded and Secure Systems (ACCESS)*, 197–201.

Singh, Kiran Deep, Prabh Deep Singh, Rishu Chhabra, Gaganpreet Kaur, Ankit Bansal, and Vikas Tripathi. 2023. "Cyber-Physical Systems for Smart City Applications: A Comparative Study." In *2023 International Conference on Advancement in Computation\& Computer Technologies (InCACCT)*, 871–76.

Singh, Kiran Deep, Prabh Deep Singh, and Sandeep Singh Kang. 2022. "Ensembled-Based Credit Card Fraud Detection in Online Transactions." *AIP Conference Proceedings* 2555: 50009. https://doi.org/10.1063/5.0108873

Singh, Kiran Deep, Prabh Deep Singh, Gaganpreet Kaur, Vikas Khullar, Rishu Chhabra, and Vikas Tripathi. 2023. "Education 4.0: Exploring the Potential of Disruptive Technologies in Transforming Learning." In *2023 International Conference on Computational Intelligence and Sustainable Engineering Solutions (CISES)*, 586–91.

Singh, Prabh Deep, Kiran Deep Singh, Vikas Tripathi, and Vaibhav Chaudhari. 2022. "Use of Ensemble Based Approach to Predict Health Insurance Premium at Early Stage." *Proceedings of International Conference on Computational Intelligence and Sustainable Engineering Solution, CISES 2022*, 566–69. https://doi.org/10.1109/ CISES54857.2022.9844398.

Singh, Kiran Deep, Prabh Deep Singh, Vikas Tripathi, and Vikas Khullar. 2022. "A Novel and Secure Framework to Detect Unauthorized Access to an Optical Fog-Cloud Computing Network." In *2022 Seventh International Conference on Parallel, Distributed and Grid Computing (PDGC)*, 618–22.

Wang, Yitong, and Jun Zhao. 2022. "Mobile Edge Computing, Metaverse, 6G Wireless Communications, Artificial Intelligence, and Blockchain: Survey and Their Convergence." *2022 IEEE 8th World Forum on Internet of Things (WF-IoT)*: 1–8.

21 A Comprehensive Overview of Cloud Computing and IoT Integration
Trends and Real-world Applications

Prabh Deep Singh and Kiran Deep Singh

21.1 INTRODUCTION

Cloud computing and IoT (Internet of Things) are revolutionizing the present landscape by addressing critical challenges and driving innovation across various domains. Cloud computing and IoT are transforming how businesses and industries operate, fostering greater agility, efficiency, and resilience in the face of evolving challenges (Ahmed et al. 2022). As we continue to navigate a rapidly changing world, these technologies will remain at the forefront, facilitating digital transformation and paving the way for a more interconnected, data-driven, and innovative future.

However, IoT has changed the game by linking gadgets and gathering massive volumes of real-time data. Home smart thermostats, manufacturing floor sensors, and wearable health gadgets are part of this ecosystem. We utilize IoT data to optimize operations, make decisions, and improve user experiences. IoT devices may remotely monitor patient vital signs, lowering healthcare system pressure and offering prompt intervention. Precision farming optimizes resource use and agricultural production with IoT sensors. In logistics and supply chain management, IoT tracks items in real-time, enhancing efficiency and lowering costs (Wang and Zhao 2022).

Cloud computing and IoT work well together. Cloud systems enable IoT data processing and analysis. They provide processing power, storage, and analytics to get insights from this data. This combination of technologies lets firms do predictive maintenance, remotely monitor and manage devices, improve cybersecurity, and personalize customer care.

DOI: 10.1201/9781032656694-21

21.2   INTEGRATING CLOUD COMPUTING AND IoT

It is now possible to provide a wide range of computing services, including storage, processing, databases, networking, analytics, and more, via the internet thanks to a technology called cloud computing. Cloud service companies, including Amazon Web Services (AWS), Microsoft Azure, and Google Cloud Platform, offer these services. Since there is no longer a requirement for a substantial on-premises infrastructure and customers may access and use these resources on a pay-as-you-go basis, cloud computing offers scalability and flexibility. The network of physically linked "things" that are networked and equipped with sensors, software, and other technologies that enable data collection and exchange via the internet is known as the IoT (Duan et al. 2022). IoT gadgets may include everything from commonplace items like home appliances and fitness monitors to sophisticated machinery and smart municipal infrastructure. They frequently produce enormous volumes of data in real-time.

By providing the required infrastructure, processing capacity, and tools to handle and analyze the enormous volumes of data created by IoT devices, the combination of cloud computing and IoT improves the capabilities of IoT deployments. Businesses may get practical insights, enhance operations, and spur innovation across a range of sectors because of this synergy (Wang and Zhao 2022).

The fusion of cloud computing with IoT is a potent combination that has the following benefits:

1. Scalability: The IoT creates vast volumes of data, and cloud computing offers the required infrastructure to store, handle, and analyze this data at scale. Cloud resources can be readily scaled up to meet the growing demand as the number of IoT devices and data volume increase.
2. Data Management and Storage: IoT devices constantly gather data. These data may be saved, arranged, and retrieved as needed using databases and data lakes, which are safe and scalable data storage options offered by cloud platforms. This guarantees the preservation of important IoT data so that it may be utilized for historical research.
3. Real-time Processing: IoT data may be processed by cloud services in real time, enabling quick responses to urgent situations or data-driven insights. For instance, sensors on machines in a manufacturing environment can send data to the cloud, where it is analyzed for indications of equipment breakdown, and maintenance warnings can be immediately issued.
4. Analytics and Insights: IoT data may be used to provide insightful analyses and forecasts using cloud-based analytics tools and machine learning services. In a variety of sectors, this may be utilized for operations optimization, anomaly detection, and predictive maintenance.
5. Centralized administration of IoT devices is made possible by cloud computing. Remote distribution of software upgrades, security patches, and configuration changes lowers operating costs and increases the security of IoT systems.
6. Security: Cloud providers invest heavily in cybersecurity measures, making it easier to secure IoT data and devices. IoT devices often have limited security capabilities, so using cloud services can add an extra layer of protection.

7. Global Reach: Cloud providers have data centers distributed globally, allowing IoT data to be processed and stored closer to the end-users or devices. This reduces latency and ensures a better user experience.
8. Cost Efficiency: By leveraging cloud resources on-demand, organizations can reduce upfront infrastructure costs and pay only for the resources they use, making IoT projects more cost-effective.

21.3 TRENDS IN THE INTEGRATION OF CLOUD COMPUTING AND IoT

The integration of cloud computing and IoT is an evolving field with several notable trends that have been shaping its development. As of my last knowledge update in September 2021, here are some key trends in the integration of cloud computing and IoT:

1. Edge Computing: While cloud computing provides powerful processing capabilities, there is an increasing emphasis on edge computing. Edge computing involves processing data closer to the source (i.e., IoT devices) rather than sending all data to centralized cloud servers. This trend reduces latency, minimizes bandwidth usage, and enhances real-time decision-making, making it crucial for applications where low latency is essential, such as autonomous vehicles and industrial automation (K. D. Singh 2021b).
2. 5G Connectivity: The deployment of 5G networks is boosting IoT adoption and opening up new use cases. For applications that depend on real-time data processing, such as smart cities, driverless cars, and augmented reality, 5G promises quicker, more dependable, and lower-latency connections (Kang, Singh, and Kumari 2022). By enabling quicker data transfer between IoT devices and cloud resources, it enhances cloud integration.
3. Integration of AI and Machine Learning: AI and Machine Learning capabilities are being added to cloud computing platforms more often. IoT devices may analyze data locally using AI models and communicate only pertinent insights to the cloud thanks to this connection. This method lowers the cost of data transfer, improves privacy, and boosts the effectiveness of IoT applications (P. D. Singh and Singh 2023).
4. Edge AI: Edge AI blends edge computing and IoT devices with AI capabilities, enabling local decision-making by gadgets. As it eliminates the requirement for constant cloud connectivity, this is particularly useful for applications like facial recognition, anomaly detection, and predictive maintenance (P. D. Singh and Singh 2023).
5. Security and privacy: In IoT implementations, security is still a major problem. It is crucial to incorporate strong security measures into cloud platforms as well as IoT devices. To safeguard data and devices from attacks, this entails encryption, authentication, access limits, and constant monitoring (K. D. Singh et al. 2022).
6. Blockchain for IoT: Using blockchain technology to improve security, trust, and transparency is becoming more popular in IoT. IoT networks may employ blockchain to establish immutable recordings of device data, assuring data

integrity and facilitating secure transactions (K. D. Singh, Singh, and Kang 2022).

7. Edge-to-Cloud Orchestration: As IoT deployments get more complicated, flawless orchestration between edge devices and cloud resources is becoming more and more important. As a result, the whole IoT ecosystem has effective data flow, load balancing, and resource allocation.

8. Sustainability and Green IoT: The use of sustainable practices in IoT and cloud integration is becoming more and more important. This involves maximizing the energy efficiency of data centers, developing energy-efficient IoT hardware, and utilizing IoT to track and lessen environmental effects across multiple businesses (K. D. Singh, Singh, Kaur et al. 2023).

9. Solutions Tailored for Particular Industries: IoT and cloud integration are being tailored for particular industries, including healthcare, agriculture, manufacturing, and smart cities. Customized solutions handle certain industry needs and difficulties.

10. Legal Compliance: IoT and cloud integration must abide by changing legal and data privacy regulations, such as the GDPR in Europe and other national data protection legislation. Compliance is still a crucial factor in IoT implementations.

21.4 THE LAYERED ARCHITECTURE OF INTEGRATING IoT AND CLOUD COMPUTING

The integration of IoT with cloud computing is facilitated by the collaborative functioning of three distinct architectural layers. The utilization of this technology enables organizations to effectively utilize the capabilities of the IoT by rapidly gathering and analyzing data, making use of cloud-based resources for adaptable and scalable computing, and delivering valuable insights to enhance operational efficiency and service quality. The amalgamation of IoT with cloud computing often encompasses three distinct architectural levels, namely the Perception Layer, the Network Layer, and the Application Layer. The Perception Layer gathers data from IoT devices and sensors deployed in the physical environment. The Network Layer is responsible for ensuring the safe and reliable data transmission to cloud-based services, perhaps involving intermediary processing at the edge (K. D. Singh and Singh 2023). The Application Layer is responsible for the processing, analysis, and visualization of data inside the cloud infrastructure (K. D. Singh, Singh, Bansal et al. 2023). This layer facilitates the ability of users and apps to make educated decisions and take appropriate actions by leveraging insights gained from the IoT. The IoT and cloud ecosystem has several layers, each with unique duties and tasks.

21.4.1 PERCEPTION LAYER

The Perception Layer is the lowest layer in IoT and cloud architecture, and it collects data from IoT devices or sensors. This layer includes numerous IoT devices, sensors, actuators, and edge computing devices (for example, gateways). These devices detect the physical environment, collect data, and occasionally process or preprocess data before transmission. IoT devices and sensors collect data from the physical environment. Soil moisture sensors, for example, detect soil moisture levels, while weather stations collect temperature and humidity data in a smart agricultural application. These devices may also process local data, such as noise filtering or initial data conversions. The data from the devices is subsequently sent to the Network Layer for processing and communication with cloud services.

21.4.2 NETWORK LAYER

The Network Layer sends data from the Perception Layer to the cloud infrastructure while providing safe and dependable connectivity. Communication protocols, gateways, routers, and communication networks (e.g., Wi-Fi, cellular, LoRaWAN) are all part of this layer. Edge computing devices that do extra data processing, aggregation, or routing may also be included. Data acquired by IoT devices in the Perception Layer is sent via the Network Layer to cloud-based services. This layer is responsible for data routing, protocol translation, and security. Edge computing devices that filter and combine data to limit the volume of data delivered to the cloud while optimizing bandwidth use and latency may also be included. In a smart home, for example, a gateway may gather data from numerous sensors and securely communicate it to cloud-based apps.

21.4.3 APPLICATION LAYER

The Application Layer is the topmost layer and is responsible for data processing, storage, analysis, and visualization. It encompasses cloud services, databases, analytics engines, and applications. Cloud platforms (e.g., AWS, Azure, Google Cloud), databases (SQL or NoSQL), analytics tools, and applications (web and mobile interfaces) are key components of the Application Layer (P. D. Singh et al. 2022). Once data arrives in the cloud, it is processed and stored. Cloud-based applications and services perform various functions, such as real-time data analysis, historical data storage, machine learning model training, and generating insights or alerts. For example, in a smart city application, data from traffic sensors is sent to cloud-based servers, where it's analyzed to optimize traffic flow, and the results may be displayed on a city dashboard accessible to traffic management personnel.

21.5 APPLICATIONS AND CASE STUDIES OF INTEGRATING CLOUD AND IoT

The integration of the IoT with cloud computing has resulted in the development of a vast number of applications and case studies across a variety of sectors. The following case studies illustrate the adaptability of combining IoT and cloud computing across various industries, as well as the disruptive impact of this integration. This connection drives innovation and efficiency while giving significant data-driven insights, whether it is optimizing agriculture, increasing healthcare outcomes, building smarter cities, upgrading industrial processes, or revolutionizing retail.

1. Smart Agriculture:
   - Application: IoT sensors and cloud computing are used to monitor and optimize agricultural activities such as crop health, irrigation management, and animal management.
   - Case Study: The John Deere Operations Centre is a platform that is housed in the cloud and interfaces with IoT-enabled agricultural machinery. Farmers are able to collect data using sensors installed on tractors and combines, keep an eye on the weather, and evaluate the moisture content of the soil. Because of this, they are able to make judgments based on the data on planting, harvesting, and irrigation. The combination of data from the IoT and cloud analytics has resulted in greater agricultural yields and decreased consumption of resources (K. D. Singh, Singh, Chhabra et al. 2023).
2. Intelligent Healthcare:
   - Application: The IoT devices and cloud integration are employed in the healthcare industry for remote patient monitoring, asset tracking, and predictive maintenance of medical equipment.
   - Case Study: Philips Healthcare offers a remote monitoring solution that uses IoT devices and cloud services. Patients who suffer from chronic diseases such as congestive heart failure often make use of wearable gadgets to monitor vital indicators such as their heart rate and blood pressure. The data collected by these devices is sent safely to the cloud, where medical professionals are able to monitor patients in real-time and receive notifications on any potentially worrying changes. Because of this, patients' outcomes have improved, and they are spending less time in the hospital.
3. Smart Cities:
   - Application: The cloud and the IoT make it possible to build smart cities with applications like intelligent waste and traffic management as well as environmental monitoring.
   - Case Study: The city of Barcelona in Spain put into action a smart city initiative known as "CityOS," which makes use of IoT sensors to collect data on aspects such as air quality, noise pollution, and traffic patterns. This data is sent to a platform hosted in the cloud, where it is then subjected to real-time analysis. Citizens have access to this information through a smartphone app, which enables them to make educated decisions regarding the activities they participate in regularly. The program has resulted in an improvement in the quality of the city's air, a reduction in the amount of traffic congestion, and an overall improvement in the quality of life in the city (K. D. Singh 2021a).
4. Industrial IoT (IIoT):
   - Application: In the industrial industry, IIoT integrates IoT devices with cloud computing for the purposes of predictive maintenance, quality monitoring, and supply chain optimization.
   - Case Study: The "TotalCare" service provided by Rolls-Royce, a maker of aviation engines, makes use of the IoT and integrates cloud

computing. Real-time data on a machine's performance and wear is gathered via sensors that are incorporated into the machine. This information is then uploaded to the cloud for analysis. Rolls-Royce uses algorithms for predictive maintenance in order to detect possible problems before they result in expensive downtime. Because of this, airlines have seen both enhanced engine dependability and decreased expenses associated with engine maintenance.

5. Smart Retail:
   * Application: IoT and cloud technologies are utilized to enhance the shopping experience by managing inventories, providing personalized marketing, and performing consumer analytics (Tiwari et al. 2022).
   * Case Study: Amazon Go is an outstanding illustration of the IoT and cloud integration in the retail sector. Amazon Go shops use IoT sensors to monitor clients' movements and the products they pick up. The data collected by these sensors is sent to the cloud, where it is analyzed in real time. Due to the fact that customers' purchases are immediately charged to their Amazon accounts, they do not need to go through the conventional checkout process before leaving the store with their purchases. The seamless integration of IoT and cloud technologies has made it feasible to provide customers with a shopping experience that is completely devoid of friction (Matta, Pant, and Tiwari 2019).

21.6 CHALLENGES IN INTEGRATING CLOUD AND IoT

IoT and cloud integration provide a number of hurdles that businesses must overcome in order to secure the success of their initiatives, despite the fact that doing so is very advantageous. A complete strategy that involves careful planning, strong security measures, effective data management techniques, continual monitoring, and optimization of IoT and cloud installations is needed to address these difficulties. Businesses that successfully overcome these obstacles may fully realize the benefits of integrated IoT and cloud technologies. The following are some of the main obstacles to cloud and IoT integration:

1. Privacy and Security Issues:
   * Data Security: Sensitive data is frequently collected by IoT devices, and sending that data to the cloud exposes it to security risks. It is essential to provide data encryption, secure authentication, and access controls.
   * Device Security: IoT devices could only have a few security measures, which leaves them open to assaults. It is a constant battle to keep IoT devices secure and updated with security fixes.
   * Privacy Compliance: Working with IoT data in the cloud might make it challenging to comply with laws like GDPR or HIPAA. Organizations must follow regulations governing data protection.
2. Volume and Bandwidth of Data:
   * Data Overload: IoT equipment produces a ton of data. The cost of cloud infrastructure may rise, and network capacity may be strained by

sending all this data to the cloud. At the edge, effective data filtering and aggregation are required.
- Latency: The time it takes to send data to the cloud and obtain a response might cause unacceptably long delays for real-time applications. Solutions for edge computing are frequently needed to lower latency.

3. Connectivity and dependability:
- IoT devices depend on network connectivity, which might not always be dependable. It might be difficult to maintain constant communication, particularly in harsh or distant settings.
- Redundancy: To ensure continuous IoT operations in the event of network or cloud service failures, redundancy and failover solutions must be in place.

4. Scalability
- Scaling IoT Deployments: As the number of IoT devices rises, it can be difficult and expensive to scale cloud resources to meet the rising data and processing needs. Scaling solutions that work are required (Seng, Ang, and Ngharamike 2022).

5. Interoperability:
- Diverse Devices: IoT networks frequently include devices from different manufacturers that use diverse connection protocols. It might be difficult to guarantee that these gadgets can communicate and work together without any problems.
- Standardization: The integration of various IoT devices with cloud services can be challenging due to the lack of standardized protocols and interfaces, which might limit interoperability.

6. Cost Control:
- Cloud Expenditures: If the pay-as-you-go cloud computing paradigm is not properly handled, it may result in unforeseen expenditures. To reduce costs, businesses must keep an eye on and optimize their use of cloud resources.
- Device Costs: IoT devices can be expensive up front, particularly if they need to be updated or replaced on a regular basis. It's crucial to control the total cost of ownership of IoT implementations.

7. Analytics and Data Management:
- Data Quality: IoT data may be erratic and noisy. It might be difficult to guarantee data dependability and quality for analytics and decision-making.
- Analytics Complexity: Real-time or near-real-time IoT data processing and analysis necessitates the use of complex analytics tools. Organizations must create or include the proper analytics tools (Dhiman et al. 2022).

8. Legal and Regulatory Issues:
- Data Sovereignty: According to certain nations' laws, data must be kept inside their boundaries. In a cloud-based IoT system, managing data sovereignty needs can be challenging.

- conformity: Maintaining conformity with rules particular to a certain business, such as those governing the healthcare or automotive sectors, may be difficult and time-consuming (Radanliev et al. 2022).

21.6.1 ENERGY EFFICIENCY

- Optimizing Data Transmission:
- Data Batching: Rather than transmitting data in real-time, devices can batch data and send it in larger, less frequent transmissions. This reduces the frequency of radio activations, which are often the most power-consuming operations in IoT devices (Rani, Ahmed, and Rastogi 2020).
- Data Compression: Utilizing efficient data compression algorithms minimizes the amount of data that needs to be transmitted, reducing the energy consumption associated with communication.
- Energy Harvesting Technologies: Some IoT devices can leverage ambient energy sources such as solar, kinetic, or thermal energy to recharge or supplement their batteries.

21.6.2 SKILL GAP

- Challenge Overview: The integration of IoT and cloud technologies requires a diverse skill set that spans both hardware and software domains. Finding and retaining personnel with expertise in both fields is challenging due to the rapid evolution of these technologies and the required depth of knowledge (Trivedi et al. 2021).
- Interdisciplinary Collaboration: Bridging the gap between IoT and cloud requires effective collaboration between hardware engineers, software developers, data scientists, and cloud architects. Soft skills for effective communication and collaboration are crucial.

21.7 CONCLUSIONS

Cloud computing and the IoT have catalyzed innovation and efficiency across sectors. This connectivity has enabled organizations to gather, analyze, and analyze massive volumes of IoT data in a scalable and cost-effective manner. This study examines the many applications and case studies that demonstrate this integration's revolutionary power. Cloud computing and IoT have enabled significant advances in smart agriculture, remote patient monitoring in healthcare, smart cities, manufacturing processes with industrial IoT (IIoT), retail revolutionization through personalized marketing, and energy efficiency. These technologies have improved decision-making, operational costs, and user experiences, improving quality of life and economic growth. However, integration issues must be addressed. To support IoT and cloud ecosystem development, security, data privacy, scalability, and energy efficiency must be managed. Additionally, organizations must aggressively bridge the talent gap for IoT and cloud technology experts. Companies that successfully integrate cloud computing and IoT will have a competitive advantage in an increasingly linked world.

Continuous learning, collaboration, and imaginative problem-solving will unlock the full potential of this disruptive integration as IoT and cloud technologies grow.

The future of IoT and cloud computing promises more potential for businesses and consumers. As technology, software, and connection improve, new apps will make our world smarter, more efficient, and more connected. We must be versatile and forward-thinking, using technology to solve global problems and improve societies.

REFERENCES

Ahmed, Imran, Yulan Zhang, Gwanggil Jeon, Wenmin Lin, Mohammad R. Khosravi, and Lianyong Qi. 2022. "A Blockchain- and Artificial Intelligence-Enabled Smart IoT Framework for Sustainable City." *International Journal of Intelligent Systems* 37 (9): 6493–507. https://doi.org/10.1002/int.22852

Dhiman, Poonam, Vinay Kukreja, Poongodi Manoharan, Amandeep Kaur, M Kamruzzaman, Imed Ben Dhaou, and Celestine Iwendi. 2022. "A Novel Deep Learning Model for Detection of Severity Level of the Disease in Citrus Fruits." *Electronics* 11 (3): 495.

Duan, Sijing, Dan Wang, Ju Ren, Feng Lyu, Ye Zhang, Huaqing Wu, and Xuemin Shen. 2022. "Distributed Artificial Intelligence Empowered by End-Edge-Cloud Computing: A Survey." *IEEE Communications Surveys\& Tutorials*, 591–624.

Kang, Sandeep Singh, Kiran Deep Singh, and Shalini Kumari. 2022. "Smart Antenna for Emerging 5G and Application." In *Printed Antennas*, 249–64. CRC Press.

Matta, Priya, Bhaskar Pant, and Umesh Kumar Tiwari. 2019. "DDITA: A Naive Security Model for IoT Resource Security." *Advances in Intelligent Systems and Computing* 670: 199–209. https://doi.org/10.1007/978-981-10-8971-8_19

Radanliev, Petar, David De Roure, Carsten Maple, and Omar Santos. 2022. "Forecasts on Future Evolution of Artificial Intelligence and Intelligent Systems." *IEEE Access* 10: 45280–88.

Rani, Shalli, Syed Hassan Ahmed, and Ravi Rastogi. 2020. "Dynamic Clustering Approach Based on Wireless Sensor Networks Genetic Algorithm for IoT Applications." *Wireless Networks* 26: 2307–16.

Seng, Kah Phooi, Li Minn Ang, and Ericmoore Ngharamike. 2022. "Artificial Intelligence Internet of Things: A New Paradigm of Distributed Sensor Networks." *International Journal of Distributed Sensor Networks* 18 (3): 15501477211062836.

Singh, Kiran Deep. 2021a. "Particle Swarm Optimization Assisted Support Vector Machine Based Diagnostic System for Dengue Prediction at the Early Stage." In *Proceedings - 2021 3rd International Conference on Advances in Computing, Communication Control and Networking, ICAC3N 2021*, 844–48. https://doi.org/10.1109/ICAC3N53548.2021.9725670

Singh, Kiran Deep. 2021b. "Securing of Cloud Infrastructure Using Enterprise Honeypot." In *Proceedings - 2021 3rd International Conference on Advances in Computing, Communication Control and Networking, ICAC3N 2021*, 1388–93. https://doi.org/10.1109/ICAC3N53548.2021.9725389

Singh, Kiran Deep, and Prabh Deep Singh. 2023. "A Novel Cloud-Based Framework to Predict the Employability of Students." In *2023 International Conference on Advancement in Computation\& Computer Technologies (InCACCT)*, 528–32.

Singh, Prabh Deep, and Kiran Deep Singh. 2023a. "Security and Privacy in Fog/Cloud-Based IoT Systems for AI and Robotics." *EAI Endorsed Transactions on AI and Robotics* 2, 1–6.

Singh, Prabh Deep, and Kiran Deep Singh. 2023b. "Fog-Centric Intelligent Surveillance System: A Novel Approach for Effective and Efficient Surveillance." In *2023 International Conference on Advancement in Computation\& Computer Technologies (InCACCT)*, 762–66.

Singh, Kiran Deep, Prabh Deep Singh, Ankit Bansal, Gaganpreet Kaur, Vikas Khullar, and Vikas Tripathi. 2023. "Exploratory Data Analysis and Customer Churn Prediction for the Telecommunication Industry." In *2023 3rd International Conference on Advances in Computing, Communication, Embedded and Secure Systems (ACCESS)*, 197–201.

Singh, Kiran Deep, Prabh Deep Singh, Rishu Chhabra, Gaganpreet Kaur, Ankit Bansal, and Vikas Tripathi. 2023. "Cyber-Physical Systems for Smart City Applications: A Comparative Study." In *2023 International Conference on Advancement in Computation\& Computer Technologies (InCACCT)*, 871–76.

Singh, Kiran Deep, Prabh Deep Singh, Gaganpreet Kaur, Vikas Khullar, Rishu Chhabra, and Vikas Tripathi. 2023. "Education 4.0: Exploring the Potential of Disruptive Technologies in Transforming Learning." In *2023 International Conference on Computational Intelligence and Sustainable Engineering Solutions (CISES)*, 586–91.

Singh, Prabhdeep, Kiran Deep Singh, Vikas Tripathi, and Vaibhav Chaudhari. 2022. "Use of Ensemble Based Approach to Predict Health Insurance Premium at Early Stage." *Proceedings of International Conference on Computational Intelligence and Sustainable Engineering Solution, CISES 2022*, 566–69. https://doi.org/10.1109/CISES54857.2022.9844398

Singh, Kiran Deep, Prabh Deep Singh, and Sandeep Singh Kang. 2022. "Ensembled-Based Credit Card Fraud Detection in Online Transactions." *AIP Conference Proceedings* 2555: 50009. https://doi.org/10.1063/5.0108873

Singh, Kiran Deep, Prabh Deep Singh, Vikas Tripathi, and Vikas Khullar. 2022. "A Novel and Secure Framework to Detect Unauthorized Access to an Optical Fog-Cloud Computing Network." In *2022 Seventh International Conference on Parallel, Distributed and Grid Computing (PDGC)*, 618–22.

Tiwari, Pallavi, Bhaskar Pant, Mahmoud M. Elarabawy, Mohammed Abd-Elnaby, Noor Mohd, Gaurav Dhiman, and Subhash Sharma. 2022. "CNN Based Multiclass Brain Tumor Detection Using Medical Imaging." *Computational Intelligence and Neuroscience* 2022: 1830010. https://doi.org/10.1155/2022/1830010

Trivedi, Naresh K, Vinay Gautam, Abhineet Anand, Hani Moaiteq Aljahdali, Santos Gracia Villar, Divya Anand, Nitin Goyal, and Seifedine Kadry. 2021. "Early Detection and Classification of Tomato Leaf Disease Using High-Performance Deep Neural Network." *Sensors* 21 (23): 7987.

Wang, Yitong, and Jun Zhao. 2022. "Mobile Edge Computing, Metaverse, 6G Wireless Communications, Artificial Intelligence, and Blockchain: Survey and Their Convergence." *2022 IEEE 8th World Forum on Internet of Things (WF-IoT)*: 1–8.

22 Applications and Challenges of the Integration of Internet of Things with Cloud

Ambika N and Solomon Chibuzo Nwafor

22.1 INTRODUCTION

The Internet of Things (IoT) is the system of unified physical objects, instruments, automobiles, buildings, and other items implanted with sensing devices, software, and connectivity capabilities that allow them to collect, exchange, and act upon data. The concept behind IoT is to allow these "intelligent" objects to connect and interact with each other and with centralized systems over the internet, creating a seamless flow of information and enabling automation, monitoring, and control of various processes.

Built on intellectual, self-organizing nodes linked together in an active, international network architecture, the IoT [1, 2] concept is built on nodes. It is one of the most revolutionary technologies because it makes scenarios for omnipresent and prevalent calculation possible. IoT [3, 4] includes tiny, real-world objects with limited storage and processing power and dependability, performance, safety, and confidentiality features. Since cloud computing is a far more established technology, it offers nearly limitless storage and processing capacity, and most IoT problems have been at least partially resolved. So, it is anticipated that a unique IT paradigm that unifies the complementary technologies of cloud and IoT [5] would upset the present and future of the internet.

Important characteristics of the Internet of Things include:

- **Connectivity**: IoT devices are equipped with many message technologies, such as Wi-Fi, Bluetooth, cellular, and RFID, enabling them to attach to cyberspace and other instruments.
- **Sensors and Data Collection**: IoT devices have sensors that capture data from their environment, including temperature, humidity, motion, light, and more.
- **Data Transmission**: The information assembled by IoT instruments is communicated to central platforms or other devices in real-time or periodically, allowing for analysis and decision-making.
- **Data Processing and Analytics**: IoT platforms procedure and scrutinize the assembled information to excerpt appreciated visions, designs, and drifts.

DOI: 10.1201/9781032656694-22

- **Automation and Control**: IoT devices can be programmed to perform actions or trigger events based on certain conditions or data inputs. It enables automation and remote control of processes.
- **Interactions**: IoT devices can interact with each other, exchange data, and collaborate to achieve more complex tasks.
- **Scalability**: The IoT ecosystem is highly scalable, allowing organizations to add more devices and expand their networks as needed.
- **Applications**: IoT finds applications in many areas, including intelligent hometowns, smart cities, healthcare, farming, industrial, conveyance, energy management, and more.

Examples of IoT applications include:

- **Smart Home**: IoT-enabled devices like thermostats, lights, door locks, and appliances can be controlled remotely through smartphones or voice commands.
- **Healthcare**: Wearable devices can monitor vital signs and transmit health data to medical professionals for remote monitoring and early detection of health issues.
- **Agriculture**: Sensors in the soil can provide data on moisture levels and nutrient content, helping farmers optimize irrigation and fertilization.
- **Smart Cities**: IoT systems can manage traffic flow, monitor air quality, control streetlights, and improve waste management.
- **Industrial IoT (IIoT)**: In industrial, IoT instruments can supervise apparatus health, trail catalog, and enhance making procedures.
- **Energy Management**: Smart meters and sensors can help supervise and manage energy utilization in buildings and industrial facilities.
- **Transportation**: IoT is used to track and optimize fleet management, traffic control, and autonomous vehicles.

IoT offers numerous benefits, including increased efficiency, convenience, and improved decision-making. However, it also poses challenges related to data security, privacy, interoperability, and managing the sheer volume of data generated by connected devices. As knowledge advances, IoT will likely play a progressively significant role in shaping various industries and aspects of daily life.

Cloud computing is the distribution of computation resources, counting servers, stowing, records, networking, software, analytics, and more, over the internet (the "cloud"). Instead of possessing and preserving physical hardware and infrastructure, users can enter and use these properties as facilities provided by cloud facility workers. Cloud computing offers a flexible, scalable, and cost-effective way for entities, trades, and administrations to enter and employ computation power and services.

Key characteristics and models of cloud computing include:

- **On Demand Self-Service**: Clients can provision and accomplish computation properties as needed without requiring human interference from the facility worker.

- **Broad Network Access**: Cloud facilities are accessible over cyberspace from many instruments, including laptops, smartphones, and tablets.
- **Resource Pooling**: Cloud providers pool resources and serve multiple customers using a multi-tenant model, sharing the same physical infrastructure while ensuring isolation and security.
- **Rapid Elasticity**: Cloud resources can be quickly scaled up or down based on demand, allowing users to pay only for what they use.
- **Measured Service**: Cloud usage is typically metered, and users are billed based on the resources they consume, such as storage, processing, bandwidth, and active users.

Cloud computing has several deployment models:

1. **Public Cloud**: Third-party providers offer services to the general public over the internet. These services are shared among multiple customers, often providing cost savings and scalability.
2. **Private Cloud**: Cloud organization is used solely by a solitary society, providing more control, customization, and security, but at a higher cost.
3. **Hybrid Cloud**: This model combines fundamentals of community and isolated clouds, letting information and applications be shared between them while maintaining a degree of separation.
4. **Multi-Cloud**: Organizations use facilities from manifold cloud workers to evade seller lock-in, optimize costs, and choose the best services for specific needs.

Cloud computing offers various service models:

1. **Infrastructure as a Service (IaaS)**: Delivers virtualized calculation capitals over cyberspace, including simulated machinery, stowage, and interaction.
2. **Platform as a Service (PaaS)**: Gives developers a platform and environment to create, launch, and maintain apps without worrying about infrastructure administration.
3. **Software as a Service (SaaS)**: On a subscription basis, delivers fully working software programs over the internet, doing away with the need for local installation and maintenance.
4. **Function as a Service (FaaS)**: FaaS, sometimes referred to as serverless computing, enables programmers to run code in response to events without setting up or maintaining servers.

Cloud computing has revolutionized IT by providing access to powerful computing resources without the need for extensive investments in hardware and infrastructure. It has enabled rapid innovation, scalability, and the ability to focus on core business activities rather than IT management. However, it also introduces considerations related to security, data privacy, vendor selection, and data governance.

22.2   BACKGROUND

Combining cloud computing with the IoT has led to a powerful and transformative technological landscape. The integration of cloud services and IoT devices has enabled new levels of data collection, analysis, and decision-making across various industries. Here's a background on how these two technologies intersect and complement each other:

IoT's Role: The Internet of Things involves connecting physical objects, devices, and sensors to the internet to gather data and enable communication between them. These devices can range from simple sensors collecting environmental data to complex machines in industrial settings. IoT devices generate massive data that holds valuable insights about the real world.

Cloud Computing's Role: Cloud computing offers scalable and flexible access to computing resources over the internet. It provides the infrastructure needed to store, process, and analyze the vast amounts of data generated by IoT devices. Cloud services enable efficient data storage, real-time data processing, and the deployment of complex applications without the need for extensive on-premises hardware.

Combining Cloud and IoT: Bringing together cloud computing and IoT creates a symbiotic relationship:

1. **Data Collection and Storage**: IoT devices collect data from the physical world and send it to the cloud for storage. Cloud services provide the necessary storage capacity to handle the massive volumes of data generated by IoT devices.
2. **Data Processing and Analytics**: Cloud platforms offer powerful computing capabilities for processing and analyzing IoT data. This allows for real-time insights, predictive analytics, and data-driven decision-making.
3. **Scalability**: IoT applications often experience varying levels of demand. Cloud infrastructure's scalability allows IoT systems to dynamically adjust resources based on usage, ensuring optimal performance.
4. **Remote Monitoring and Control**: Cloud services enable users to remotely monitor and control IoT devices. This is crucial for applications like remote equipment management, smart homes, and industrial automation.
5. **Machine Learning and AI**: Cloud-based machine learning and artificial intelligence services can analyze IoT data to identify patterns, anomalies, and trends, leading to valuable insights and improved automation.
6. **Cost Efficiency**: Instead of investing in and maintaining on-premises hardware, IoT adopters can leverage the pay-as-you-go model of cloud services, reducing upfront costs and increasing flexibility.
7. **Global Accessibility**: Cloud services make IoT data accessible from anywhere with an internet connection, allowing organizations to manage and analyze data from different geographic locations.

8. **Firmware Updates and Management**: Cloud platforms simplify firmware updates and device management for IoT deployments, ensuring devices remain secure and up-to-date.

Use Cases: The combination of cloud and IoT has revolutionized various industries:

- **Smart Cities**: Cloud-enabled IoT networks manage traffic, monitor environmental data, optimize energy usage, and enhance urban planning.
- **Manufacturing**: IoT devices gather data from machines and equipment, which is then sent to the cloud for real-time analysis to optimize production processes and minimize downtime.
- **Healthcare**: Connected medical devices can transmit patient data to the cloud, allowing healthcare professionals to remotely monitor and provide timely interventions.
- **Agriculture**: IoT sensors in the field send data to the cloud for analysis, helping farmers make informed decisions about irrigation, fertilization, and crop health.
- **Smart Homes**: Cloud-connected IoT devices in homes can be controlled remotely through smartphones, enabling energy efficiency and convenience.

The convergence of cloud computing and IoT has paved the way for innovation, efficiency, and improved decision-making in various sectors, ultimately leading to the growth of "smart" ecosystems that benefit individuals, businesses, and society as a whole.

22.3 MOTIVATION

The integration of cloud computing with the IoT is driven by several key motivations that together offer numerous benefits for industries, businesses, and individuals:

Data Management and Analysis:
- **Motivation**: IoT devices generate vast amounts of data, and cloud platforms offer the scalable infrastructure needed to store, process, and analyze this data efficiently.
- **Benefits**: Cloud computing allows real-time data processing, predictive analytics, and the extraction of valuable insights from IoT-generated data, enabling better decision-making and optimization of operations.

Scalability and Flexibility:
- **Motivation**: IoT applications often experience fluctuations in data volume and usage patterns. Cloud services can dynamically scale resources up or down, ensuring optimal performance without overprovisioning.
- **Benefits**: Organizations can handle varying workloads without upfront hardware investments, reducing costs and improving resource utilization.

Global Accessibility:
- **Motivation**: IoT systems operate across different geographic locations. Cloud services enable remote access to data, monitoring, and control from anywhere with an internet connection.

- **Benefits**: This accessibility fosters real-time monitoring, centralized management, and better coordination of remote assets.

Faster Time-to-Market:
- **Motivation**: Developing and deploying IoT solutions can be complex and time-consuming. Cloud platforms offer pre-built services and tools that accelerate the development and deployment of IoT applications.
- **Benefits**: Organizations can bring IoT solutions to market faster, gaining a competitive edge and responding more swiftly to changing demands.

Cost Efficiency:
- **Motivation**: Traditional on-premises infrastructure requires significant upfront investment and ongoing maintenance. Cloud services follow a pay-as-you-go model, reducing capital expenses.
- **Benefits**: Organizations can allocate resources more efficiently, avoid overprovisioning, and pay only for the resources they consume, resulting in cost savings.

Security and Compliance:
- **Motivation**: Securing IoT devices and data is challenging. Cloud providers invest in advanced security measures and compliance certifications, helping organizations ensure the safety of their IoT deployments.
- **Benefits**: Cloud services can provide robust security features, including encryption, access controls, and regular security updates, improving overall IoT system security.

Remote Management and Updates:
- **Motivation**: IoT devices may be deployed in remote or inaccessible locations. Cloud platforms enable centralized device management, remote troubleshooting, and efficient firmware updates.
- **Benefits**: Organizations can ensure devices remain up-to-date, secure, and operational without the need for physical access to each device.

Innovation and Collaboration:
- **Motivation**: Cloud computing and IoT encourage innovation by enabling developers to focus on creating novel applications and services rather than managing infrastructure.
- **Benefits**: Cloud-based development environments and tools foster collaboration, experimentation, and the rapid iteration of IoT solutions.

Environmental Impact:
- **Motivation**: Cloud-based IoT solutions can contribute to sustainability efforts. By optimizing resource usage and reducing energy consumption through cloud scalability, the environmental footprint can be minimized.
- **Benefits**: Organizations can contribute to eco-friendly practices while achieving their IoT goals.

The combination of cloud computing and IoT addresses challenges related to data management, scalability, security, and cost, while also providing the foundation for innovative applications that transform industries. This convergence empowers businesses to harness the full potential of IoT, leading to increased efficiency, improved customer experiences, and better insights into operations and processes.

22.4 CHARACTERISTICS OF INTEGRATION

22.4.1 Communication

Communication between cloud computing and IoT devices is essential for creating a seamless and effective ecosystem. This communication enables data to flow from IoT devices to the cloud for storage, analysis, and action and facilitates commands and updates from the cloud back to the devices. Several communication protocols and technologies are crucial in connecting the cloud and IoT devices. Effective communication between cloud computing and IoT devices is a cornerstone of building intelligent and interconnected systems. Selecting the right communication protocols and technologies depends on device capabilities, network conditions, data requirements, and security considerations.

The IoT, fog, and cloud sectors' present protocols are de facto fragmented with several diverse approaches. The approved standards of communication protocols and the reference architecture are not agreed upon. Choosing the protocol for their unique IoT system requirements while using the advancements in fog and cloud computing is one of the key problems for system developers. These criteria include the types of wireless connectivity that may be employed, devices ranging from resource-constrained devices to high-performance cloud systems, data created to be processed between the cloud and the fog layer, and security and privacy solutions.

22.4.2 Storage

Storage is a critical aspect of combining cloud computing with the IoT, as it involves managing the vast amounts of data generated by IoT devices. Cloud-based storage solutions play a pivotal role in efficiently collecting, storing, and managing IoT-generated data while providing scalability, accessibility, and security. The integration of cloud storage with IoT enables organizations to harness the full potential of IoT-generated data, unlocking insights, improving decision-making, and fostering innovation across industries. However, organizations should carefully consider factors such as data privacy, security, cost, and compliance when selecting and configuring cloud storage solutions for their IoT deployments.

Multi-tenant storage with isolated performance, scalability with flexibility, and distributed execution with unified management of infrastructure resources are the three pairs of demands that conflict in IoT-based data storage systems in cloud computing. Additionally, specific demands need processing large amounts of unstructured data in real-time at several levels, including data representation, storage, and analysis, leveraging cloud platforms for IoT data exchange, processing, and integration.

22.4.3 Computation

Combining cloud computing with IoT computation refers to the practice of leveraging cloud resources to process data and perform computational tasks associated with IoT devices. This integration enables efficient data analysis, complex calculations, and the execution of applications beyond the capabilities of individual IoT devices.

The combination of cloud computing and IoT computation empowers organizations to maximize the value of IoT-generated data by performing advanced analysis, applying machine learning models, and making data-driven decisions. However, organizations should consider factors like data privacy, latency, connectivity, and cost when determining the optimal distribution of computation between IoT devices and the cloud.

22.5 APPLICATIONS

22.5.1 HEALTHCARE

Cloud IoT technology is frequently utilized in the field of remote healthcare monitoring, providing practical solutions to individuals with significant health issues and disabilities. The utilization of remote monitoring via cloud and IoT facilitates the proactive and early identification of diseases, hence enabling the provision of appropriate healthcare interventions to promote patient convenience and comfort. The popularity of Cloud IoT-based healthcare applications has been on the rise in recent times, mostly because of the widespread availability of sensors at a low cost [6]. Reference [7] discussed IoT-based healthcare applications, including monitoring toddlers, managing chronic diseases, and surgical guidance, while comparing IoT healthcare devices based on various parameters. Reference [8] presented the K-Healthcare model with layers for smartphone-based smart health surveillance and gave a comparative review of various architectures and cloud-based IoT applications. In terms of health monitoring systems, reference [9] introduced the "Silver Link" system, which uses sensors for health tracking and sends alerts to medical teams in case of anomalies. Reference [10] proposed the "Help to you" (H2U) framework for healthcare services to the elderly, utilizing wearable sensors and sensor networks. Reference [11] described an IoT-powered healthcare monitoring system that sends real-time alerts to doctors with dynamic notification rules. In terms of real-time impact, reference [12] discussed the impact of IoT on healthcare delivery, emphasizing remote healthcare through the Rural Smart Healthcare System (RSHS) and utilization of cloud computing and big data. For secured healthcare transmission, reference [13] presented a secure IoT healthcare data transmission framework with optimized routing protocol and dynamic trust-based RPL protocol. Also, reference [14] addressed challenges in IoT-driven smart health using fog computing and machine learning-based edge intelligence for data processing. For Comprehensive Healthcare Monitoring Platform, reference [15] focused on medical IoT for preventive and predictive purposes, developing a wrist-worn prototype and adaptable IoT gateway for real-time communication between users and medical professionals.

22.5.2 SMART CITY APPLICATIONS

Cloud-based IoT applications can help manage resources and assets in a smart city by collecting data from various sources such as citizens, devices, homes, and other things [16]. Cloud-based IoT applications can have various applications in smart cities. These applications include but are not limited to: Transportation systems [17], smart water management [18], smart meters [19], smart tourism [20], social security [21], smart parking [22], smart infrastructure [23], and waste management [24]. By utilizing cloud-based IoT applications, smart cities can optimize resource usage and improve overall efficiency [16]. A smart city can derive advantages from utilizing cloud-based infrastructure and system distribution. This allows for the retrieval, analysis, and management of precise information through cloud-based enabling technologies. Consequently, experts, businesses, and individuals can utilize this information to formulate more intelligent policies aimed at improving the quality of life for residents. In smart city environments, individuals interact by means of their mobile devices, which are interconnected with automobiles and residences equipped with smart technology. The integration of devices and information with a city's physical infrastructure and facilities has the potential to decrease expenses and enhance efficiency. The utilization of the IoT has the potential to enhance various aspects of urban life. By using IoT technologies, cities can improve the efficient transfer of resources, expedite the process of waste collection, mitigate the occurrence of accidents, and effectively eliminate pollutants from the urban environment.

22.5.3 Smart Homes

The concept of a smart house refers to the application of building automation principles in domestic settings, encompassing the control and automation of various technological components integrated within the home. A smart home is characterized by the presence of various technological devices and systems that enhance the functionality and convenience of a residential space. These include appliances, lighting fixtures, heating and air conditioning units, televisions, computers, and entertainment systems, as well as large household appliances like washers, dryers, refrigerators, and freezers. Additionally, smart homes are equipped with security and camera systems that can communicate with one another and can be remotely controlled through a predetermined schedule using devices such as phones, mobile devices, or internet connectivity [25]. The IoT and smart home systems can derive advantages from the extensive resources and functions offered by cloud computing to address their inherent limitations in areas such as storage, processing, communication, on demand assistance, as well as backup and recovery capabilities. For instance, cloud technology enables the management and fulfillment of IoT services, as well as the execution of complementing applications that leverage the data generated by these services. The concept of a smart home can be streamlined to prioritize essential operations, hence reducing the reliance on local house resources and leveraging cloud capabilities and resources instead. The primary emphasis of smart home and IoT technologies lies in acquiring data, rudimentary data analysis, and subsequent transmission to cloud-based platforms for more advanced processing. The convergence of IoT, smart home, and cloud computing is more than a mere amalgamation of technological advancements. However, it is crucial to establish a harmonious equilibrium between decentralized and centralized computing while prioritizing the efficient utilization of resources. Computational tasks can be done on IoT and smart home devices or outsourced to cloud computing services. Determining where to perform computational tasks is contingent upon various factors, including the tradeoffs associated with overhead, the availability of data, the presence of data dependencies, the volume of data transit, the reliance on communication systems, and issues pertaining to security. One perspective suggests that the integration of the cloud, IoT, and smart home technologies in the triple computing model has the potential to effectively reduce overall system costs, primarily by prioritizing the reduction of resource usage inside residential environments. In contrast, the implementation of an IoT and smart home computing service model is expected to enhance the ability of IoT users to meet their needs while using cloud applications, as well as effectively tackle the intricate challenges that emerge from the integration of the IoT, smart home technology, and cloud service model [25].

This literature review explores several smart home automation systems leveraging IoT and advanced technologies [26] created an IoT and Edge Computing system on Raspberry Pi for remote appliance control, security, and energy efficiency, achieving notable speed and energy improvements [27] integrated IoT, cloud computing, and Wi-Fi via Raspberry Pi 3B+ to enable remote home appliance control, yielding substantial energy savings [28] proposed a scalable Smart Home System architecture integrating IoT, context awareness, cloud computing, and event processing, validated through a practical implementation [29] developed a cost-effective hybrid IoT-based system for home automation, addressing technology limitations while enhancing energy efficiency, convenience, safety, and security through user-friendly interfaces.

22.5.4 SMART GRID

The smart grid (SG) is responsible for managing the distribution of electricity and facilitating communication between IoT devices [30]. There are two possible modes of engagement between end users and service providers in the context of the smart grid, which enable them to monitor their energy consumption and associated costs [31]. Due to the essential nature of the services provided by the SG and the substantial customer base utilizing them, prompt response and processing times are of utmost importance. In order to effectively manage a substantial volume of requests originating from diverse devices, such as smart meters, it is imperative to appropriately handle, analyze, and store cloud data [32]. The advancement of cloud technology has resulted in a decrease in the need for significant computational capability at the individual device level [33]. The smart meter, which is securely stored in the cloud, offers personalized data and energy consumption statistics to the intelligent grid. The integration of the cloud with IoT connectivity provides a diverse array of services and facilitates the implementation of intelligent data management [34]. The exponential growth of IoT devices has significantly augmented the computational load on cloud infrastructure, leading to extended response and processing durations, which can have implications for time-critical applications. When transmitting data to the cloud, secure gateways (SG) are employed to guarantee anonymity [35]. As the quantity of cloud end users increases, the task of managing requests gets increasingly complex, leading to the emergence of load balancing challenges [36]. The fog computing layer is implemented as an intermediary between the cloud and end users in order to mitigate the impact of prolonged processing time and effectively handle the substantial influx of requests from end users.

22.5.5 LOGISTICS

The control of goods' integrity, particularly in the context of perishable goods like fresh produce, has garnered significant attention from logistics stakeholders. This focus has led to the development of cold chain management systems, which are designed to monitor various parameters in order to ensure the freshness properties of these goods [37]. The utilization of advanced technological tools, specifically IoT-enabling technologies, offers significant benefits to this process. It facilitates convenient remote and real-time monitoring of transported goods, providing valuable information for goods management and container tracking. Additionally, it enables the generation of timely alerts in emergencies. The necessity for monitoring is particularly pronounced within the realm of food and this sector, and IoT has already facilitated various proposed solutions. The utilization of radio frequency identification (RFID) and near-field communication (NFC) technologies has exemplified the potential of the IoT in improving processes such as the monitoring and tracing of the supply chain for vegetables [38–42]. Moreover, as stated by Aung et al. [43], ensuring consumer safety relies on the essential practice of monitoring food and establishing traceability across the supply chain. In order to achieve this objective, several monitoring systems have been previously suggested in scholarly works. However, these systems lose their effectiveness and, in certain instances, constitute a risk if there is no mechanism in place to ensure the integrity of the information that is transmitted and stored. For instance, stakeholders, including end users, need to ascertain that the carried items have not experienced heat shock or excessive vibration throughout the transportation process. The provision of such information holds significance in enhancing the perceived value of the product. Hence, security within the IoT sphere is an additional significant factor that is increasingly garnering attention. Indeed, the recognition of ensuring the safety of goods and food by end users notwithstanding, professionals are cognizant of the paramount significance of ensuring the security of communication and storage systems across all levels and environments.

22.5.6 ENVIRONMENTAL MONITORING

The application of the IoT in environmental monitoring is often regarded as highly advantageous. Advanced sensor devices are utilized to detect the existence of contaminants in the atmosphere and aquatic environments, hence fostering enhanced sustainability [44]. The advancement of environmental monitoring systems has been greatly facilitated by implementing smart monitoring choices inside the IoT architecture. The primary areas of concentration in smart environmental monitoring systems utilizing IoT architecture are the monitoring of air quality, water quality, and soil quality [45]. Environmental monitoring encompasses several distinct categories, namely water quality monitoring, air quality monitoring, energy monitoring, and toxic gas detection [44].

22.6 CHALLENGES

22.6.1 HETEROGENEITY

On the opposite end of the spectrum, the IoT concept, which involves the widespread deployment of interconnected devices, encounters significant obstacles across various aspects [46]. One of the primary challenges in the field of IoT is its inherent tendency toward heterogeneity. This heterogeneity can manifest in several forms, such as diverse protocols, device data formats, communication capabilities of devices, technologies, and hardware [47, 48]. Limited implementations of IoT systems have been realized thus far primarily due to the presence of these aforementioned obstacles. In order for the IoT to progress toward its goal of worldwide application, it is imperative to mitigate the various obstacles present at different levels. In order to initiate and deliver the service, devices must be linked to the internet. There is a need to identify the difficulties related to heterogeneity that occur at various levels. Additionally, it is important to emphasize the solutions that have been accepted and used by different studies to address heterogeneity in IoT systems. The aim of this study is to perform a comprehensive assessment of existing literature to identify the issues associated with heterogeneity in the IoT and examine the solutions proposed by various studies to address these challenges. The current market of IoT devices has a wide range of heterogeneity, characterized by variations in communication protocols, network connectivity methods, and subsequent application models. Supporting many types of diversities in the IoT may not be practical due to the limited resources available to developers, hindering their ability to fully comprehend the intricacies of confined devices and networks [49]. The primary aim is to reduce, conceal, or eliminate the extensive array of variability in technologies, application models, and protocols among IoT users [50]. Due to the diverse composition of the IoT, it is considered a burgeoning field of study with significant potential to revolutionize our comprehension of fundamental principles and standards in computer science, ultimately shaping the future of our daily lives [51]. The anticipated rise in demand for limited-capacity IoT devices is projected to exacerbate this issue in the coming years. Hence, it is imperative to enhance the integration of a substantial quantity of limited devices within the IoT framework.

22.6.2 SECURITY AND PRIVACY

Cloud IoT facilitates data transmission between interconnected devices and users to accomplish specific objectives. To ensure the protection of communication in a very prevalent setting, it is important to strengthen factors such as authenticity, privacy, and control [6]. Cloud computing environments can be described as multi-domain ecosystems [52]. The utilization of distinct security, privacy, and trust requirements inside these settings enables the safeguarding of each domain. This, in turn, prompts the adoption of diverse procedures, interfaces, and semantics [53]. The primary obstacles in cloud computing pertain to user authentication and identity, data access and management, and the subsequent security and privacy measures, including policies and data protection regulations. Access control and accounting, secure-service management, privacy and data protection, trust management and policy integration, authentication and identity management, and Organizational Security Management are the issues that have been noted in the literature [52].

22.6.3 PERFORMANCE

The integration of cloud computing with IoT applications presents challenges in fulfilling the requirements for carrying out tasks and ensuring quality of service (QoS) at several levels, including storage, communication, and computation. In certain situations, it may not be straightforward to fulfill these needs [54–56]. QoS is the primary factor that governs the overall performance of the internetwork. The maintenance of QoS in the services offered by the cloud IoT platform is of utmost significance due to the substantial volumes of data being generated and exchanged. Additionally, a significant proportion of client requests necessitate effective management by the cloud platform, some of which may be sensitive to latency. Therefore, in order to mitigate the occurrence of packet loss, it is recommended to implement QoS enhancement techniques and prioritize data packets. This approach appears to be an effective and reliable option. Therefore, the utilization of the next-generation IP protocol (IPv6), which provides practical features to ensure optimal QoS in the network, is considered the optimal solution for the cloud IoT environment [6].

22.6.4 RELIABILITY

The integration of IoT with cloud computing has been adopted in order to support mission-critical applications. Reliability concerns commonly manifest in the domain of smart mobility. Due to their constant movement, vehicles usually experience interrupted or unreliable transport networking and communication. In resource-constrained situations, the development of applications gives rise to a multitude of obstacles, including device failure and the absence of perpetually accessible devices [57, 58].

22.7 FUTURE DIRECTIONS

The future of combining cloud computing with the IoT holds immense potential to transform industries, enhance our daily lives, and enable innovative solutions. As technology continues to evolve, several trends and possibilities are expected to shape the future of cloud-based IoT:

Massive Data Insights and AI:
   a. **Trend**: IoT devices will continue to generate vast amounts of data. Cloud-based AI and machine learning will play a key role in extracting meaningful insights from this data, enabling more accurate predictions, personalized experiences, and data-driven decision-making.

Edge-to-Cloud Continuum:
   b. **Trend**: IoT deployments will increasingly involve a continuum of processing at the edge (near the source) and in the cloud.
   c. **Impact**: This hybrid approach optimizes real-time processing while leveraging cloud resources for complex analytics, improving response times and resource efficiency.

5G and Low-Latency Communication:
- d. **Trend**: The rollout of 5G networks will provide ultra-fast, low-latency communication.
- e. **Impact**: 5G will enable real-time data exchange between IoT devices and the cloud, supporting applications like remote surgery, autonomous vehicles, and immersive experiences.

Edge Computing and Fog Computing:
- f. **Trend**: Edge computing and fog computing will gain prominence, enabling processing and analysis closer to the data source.
- g. **Impact**: Edge and fog computing reduce latency, conserve bandwidth, and enhance data privacy, making them essential for real-time and privacy-sensitive applications.

Secure-by-Design IoT:
- h. **Trend**: Security concerns will drive the adoption of secure-by-design IoT solutions.
- i. **Impact**: Cloud platforms will focus on integrating robust security measures, including encryption, authentication, and anomaly detection, to protect IoT devices and data.

Digital Twins and Simulation:
- j. **Trend**: Digital twins—virtual representations of physical objects—will become more sophisticated.
- k. **Impact**: Cloud-powered digital twins will enable accurate simulations, predictive maintenance, and efficient design iterations in industries like manufacturing and healthcare.

Ecosystem Collaboration:
- l. **Trend**: Cloud providers, IoT device manufacturers, and application developers will collaborate to create comprehensive IoT ecosystems.
- m. **Impact**: These ecosystems will offer end-to-end solutions, simplifying integration, deployment, and management of IoT applications.

Smart Cities and Urban Planning:
- n. **Trend**: Smart city initiatives will leverage cloud-based IoT to enhance urban planning, resource management, and citizen services.
- o. **Impact**: Cloud-powered smart city solutions will improve traffic management, waste disposal, energy efficiency, and public safety.

IoT in Healthcare and Wellness:
- p. **Trend**: IoT will revolutionize healthcare with wearable devices, remote monitoring, and personalized treatment.
- q. **Impact**: Cloud-based IoT will enable real-time health data analysis, timely interventions, and improved patient outcomes.

Environmental Monitoring and Sustainability:
- r. **Trend**: IoT will contribute to environmental monitoring and sustainability efforts.
- s. **Impact**: Cloud-connected sensors will provide real-time data on air quality, water levels, and energy consumption, aiding in environmental protection and resource management.

Regulatory Considerations:
- t. **Trend**: Data privacy and security regulations will impact IoT deployments.

    u.   **Impact**: Cloud providers will develop solutions that adhere to evolving data protection laws, ensuring compliant and ethical use of IoT data.

Interconnectivity Standards:

    v.   **Trend**: Standardization efforts will focus on interoperability between IoT devices and cloud services.

    w.   **Impact**: Interconnectivity standards will simplify integration, allowing diverse devices and platforms to work seamlessly together.

The future of cloud-based IoT promises a connected world with smarter systems, improved efficiency, enhanced experiences, and novel applications that we can't fully anticipate. As these trends unfold, organizations will need to adapt their strategies, adopt new technologies, and leverage cloud and IoT advancements to remain competitive and innovative in their respective fields.

22.8 CONCLUSION

In conclusion, the fusion of cloud computing with the IoT has revolutionized the way we interact with technology, gather insights from data, and shape the future of industries. This powerful synergy has unlocked a world of possibilities, enabling seamless communication, data-driven decision-making, and innovative solutions across various domains.

The integration of cloud and IoT has led to remarkable outcomes:

1. **Efficiency and Scalability**: Cloud resources provide the scalability needed to accommodate the massive influx of data from IoT devices. This ensures that organizations can process, store, and analyze data efficiently, even as the IoT ecosystem continues to grow.
2. **Real-Time Insights**: By combining real-time data collection from IoT devices with cloud-based analytics, businesses gain timely insights into operations, enabling them to respond swiftly to changing conditions and optimize processes.
3. **Innovation and Customization**: Cloud platforms offer tools, APIs, and services that empower developers to innovate and create custom applications tailored to specific IoT use cases, amplifying the impact of IoT solutions.
4. **Global Accessibility**: The cloud's global reach allows IoT data to be accessed and managed from anywhere, enabling remote monitoring, control, and collaboration across geographic boundaries.
5. **Security and Compliance**: Cloud providers invest in robust security measures, safeguarding IoT devices and data against threats. Compliance features ensure adherence to regulatory standards, bolstering user confidence.
6. **Environmental Impact**: The pairing of cloud and IoT promotes sustainable practices by optimizing resource usage, reducing energy consumption, and enabling better resource management through data-driven insights.
7. **Transformation Across Industries**: From manufacturing and healthcare to agriculture and smart cities, cloud-integrated IoT solutions are reshaping industries, improving processes, and enhancing customer experiences.

The continued evolution of cloud-based IoT will be marked by advancements such as AI-driven insights, enhanced edge computing capabilities, secure IoT ecosystems, and the expansion of 5G networks. As these technologies mature, they will enable more sophisticated applications, deeper insights, and seamless interactions between people, devices, and environments.

In embracing this dynamic convergence, organizations and individuals alike stand to gain greater efficiencies, heightened capabilities, and novel opportunities that were once the realm of science fiction. The journey ahead is one of ongoing innovation, collaboration, and exploration, where the potential of cloud-enabled IoT continues to unfold, shaping a smarter, more connected world.

REFERENCES

1. N. Ambika, "An Augmented Edge Architecture for AI-IoT Services Deployment in the Modern Era," in *Handbook of Research on Technical, Privacy, and Security Challenges in a Modern World*, US, IGI Global, 2022, pp. 286–302.
2. N. Ambika, "Energy-Perceptive Authentication in Virtual Private Networks Using GPS Data," in *Security, Privacy and Trust in the IoT Environment*, Cham, Springer, 2019, pp. 25–38.
3. W. H. Hassan, "Current Research on Internet of Things (IoT) Security: A Survey," Computer Networks, vol. 148, pp. 283–294, 2019.
4. S. Li, L. D. Xu, and S. Zhao, "The Internet of Things: A Survey," *Information Systems Frontiers*, vol. 17, no. 2, pp. 243–259, 2015.
5. A. Nagaraj, *Introduction to Sensors in IoT and Cloud Computing Applications*, UAE, Bentham Science Publishers, 2021.
6. L. D. Xu, W. He, and S. Li, "Internet of Things in Industries: A Survey," IEEE Transactions on Industrial Informatics, vol. 10, no. 4, pp. 2233–2243, 2014.
7. J. L. Shah, H. F. Bhat, and A. I. Khan, "Integration of Cloud and IoT for Smart e-Healthcare," in *Healthcare Paradigms in the Internet of Things Ecosystem*, 2021, pp. 101–136.
8. A. S. Yeole, and D. R. Kalbande, "Use of Internet of Things (IoT) in Healthcare: A Survey," in Proceedings of the ACM Symposium on Women in Research 2016, ACM, Mar. 2016, p. 71e76.
9. K. Ullah, M. A. Shah, and S. Zhang, "Effective Ways to use Internet of Things in the Field of Medical and Smart Health Care," in 2016 International Conference on Intelligent Systems Engineering (ICISE), IEEE, p. 372e379.
10. J. Chuang, L. Maimoon, S. Yu, H. Zhu, C. Nybroe, O. Hsiao, et al., "SilverLink: Smart Home Health Monitoring for Senior Care," in ICSH, Cham, Springer, Nov. 2015, p. 3e14.
11. H. Basanta, Y. P. Huang, and T. T. Lee, "Intuitive IoT-Based H2U Healthcare System for Elderly People," in 2016 IEEE 13th International Conference on Networking, Sensing, and Control (ICNSC), IEEE, 2016, April, p. 1e6.
12. F. Jimenez, and R. Torres, "Building an IoT-Aware Healthcare Monitoring System," in 2015 34th International Conference of the Chilean Computer Science Society (SCCC), IEEE.
13. S. J. Parvez, et al., "Analysis of the Effect of IoT-Based Wearables in Healthcare Applications," Feb. 2023.
14. E. Refaee, S. Parveen, K. M. J. Begum, F. Parveen, M. Chithik Raja, S. K. Gupta, and S. Krishnan, "Secure and Scalable Healthcare Data Transmission in IoT Based on Optimized Routing Protocols for Mobile Computing Applications," Wireless Communications and Mobile Computing, vol. 2022, pp. 1–12, 2022.

15. R. Priyadarshini, R. K. Barik, H. K. Dubey, and B. K. Mishra, "A Survey of Fog Computing-Based Healthcare Big Data Analytics and Its Security," International Journal of Ambient Computing and Intelligence, vol. 12, no. 2, pp. 53–72, April 1, 2021.
16. M. Haghi, S. Neubert, A. Geissler, H. Fleischer, N. Stoll, R. Stoll, and K. Thurow, "A Flexible and Pervasive IoT-Based Healthcare Platform for Physiological and Environmental Parameters Monitoring," IEEE Internet of Things Journal, vol. 7, no. 6, pp. 5628–5647, March 12, 2020.
17. T. Alam, "Cloud-Based IoT Applications and Their Roles in Smart Cities," Smart Cities, vol. 4, pp. 1196–1219, 2021.
18. H. Ding, Y. Chi Zhang, X. Li, B. Lin, Y. Fang, and S. Chen, "Probabilistic Data Prefetching for Data Transportation in Smart Cities," IEEE Internet of Things Journal, vol. 9, no. 3, pp. 1655–1666, June 11, 2021.
19. H. M. Ramos, A. McNabola, P. A. López-Jiménez, and M. Pérez-Sánchez, "Smart Water Management Towards Future Water Sustainable Networks," Water, vol. 12, p. 58, 2020.
20. A. Miyasawa, S. Akira, Y. Fujimoto, and Y. Hayashi, "Spatial Demand Forecasting Based on Smart Meter Data for Improving Local Energy Self-Sufficiency in Smart Cities," IET Smart Cities, vol. 3, pp. 107–120, 2021.
21. N. Chung, H. Lee, J. Ham, and C. Koo, "Smart Tourism Cities' Competitiveness Index: A Conceptual Mode," in Proceedings of the Information and Communication Technologies in Tourism 2021, Virtual Conference, Springer Science and Business Media LLC, Berlin, Germany, 19–22 January 2021, pp. 433–438.
22. U. Chatterjee, G. S. Bhunia, D. Mahata, and U. Singh, "Smart Cities and Their Role in Enhancing Quality of Life," in Quality of Life: An Interdisciplinary Perspective, Boca Raton, FL, CRC Press, 2021, pp. 127–143.
23. C. Biyik, Z. Allam, G. Pieri, D. Moroni, M. O'fraifer, E. O'connell, and M. Khalid, "Smart Parking Systems: Reviewing the Literature, Architecture and Ways Forward," Smart Cities, vol. 4, pp. 623–642, 2021.
24. A. P. P. Kasznar, A. WA Hammad, M. Najjar, E. Linhares Qualharini, K. Figueiredo, C. A. P. Soares, and A. N. Haddad, "Multiple Dimensions of Smart Cities' Infrastructure: A Review," Buildings, vol. 11, p. 73, 2021.
25. A. McCurdy, C. Peoples, A. Moore, and M. Zoualfaghari, "Waste Management in Smart Cities: A Survey on Public Perception and the Implications for Service Level Agreements," EAI Endorsed Transactions Smart Cities, vol. 5, no. 16, pp. 1–10, 2021.
26. M. Domb, Smart Home Systems Based on Internet of Things, 2019, https://doi.org/10.5772/intechopen.77404
27. H. Yar, M. Imran, Z. A. Khan, M. Sajjad, and Z. Kastrati, "Towards Smart Home Automation Using IoT-Enabled Edge-Computing Paradigm," Sensors, vol. 21, no. 14, p. 4932, July 20, 2021.
28. M. O. A. Helo, A. Shaker, and L. Ali Abdul-Rahaim, "Design and Implementation a Cloud Computing System for Smart Home Automation," Webology, vol. 18, no. SI05, pp. 879–893, October 30, 2021.
29. S. A. Z. Hassan, and A. M. Eassa, "A Proposed Architecture for Smart Home Systems Based on IoT, Context-Awareness and Cloud Computing," International Journal of Advanced Computer Science and Applications, vol. 13, no. 6, 89–96, January 1, 2022.
30. W. A. Jabbar, T. K. Kian, R. M. Ramli, S. K. M. Zubir, N. S. M. Zamrizaman, M. Balfaqih, V. Shepelev, and S. Abed Alharbi, "Design and Fabrication of Smart Home With Internet of Things Enabled Automation System," IEEE Access, vol. 7, 144059–144074, September 23, 2019. https://doi.org/10.1109/access.2019.2942846
31. J. Akram, A. Tahir, H. S. Munawar, A. Akram, A. Z. Kouzani, and M. A. P. Mahmud, "Cloud- and Fog-Integrated Smart Grid Model for Efficient Resource Utilisation," Sensors, vol. 21, p. 7846, 2021. https://doi.org/10.3390/s21237846

32. S. S. Reka, and T. Dragicevic, "Future Effectual Role of Energy Delivery: A Comprehensive r32iew of Internet of Things and Smart Grid," Renewable and Sustainable Energy Reviews, vol. 91, pp. 90–108, 2018.
33. M. R. Asghar, G. Dán, D. Miorandi, and I. Chlamtac, "Smart Meter Data Privacy: A Survey," IEEE Communications Surveys & Tutorials, vol. 19, pp. 2820–2835, 2017.
34. Z. Cao, J. Lin, C. Wan, Y. Song, Y. Zhang, and X. Wang, "Optimal Cloud Computing Resource Allocation for Demand Side Management in Smart," IEEE Transactions on Smart Grid, vol. 8, pp. 1943–1955, 2017.
35. H. Kumar, M. K. Singh, M. Gupta, and J. Madaan, "Moving Towards Smart Cities: Solutions That Lead to the Smart City Transformation Framework," Technological Forecasting and Social Change, vol. 153, p. 119281, 2018.
36. J. Kim, and Y. Kim, "Benefits of Cloud Computing Adoption for Smart Grid Security from Security Perspective," The Journal of Supercomputing, vol. 72, pp. 3522–3534, 2016.
37. M. A. Al Faruque, and K. Vatanparvar, "Energy Management-as-a-Service Over Fog Computing Platform," IEEE Internet Things Journal, vol. 3, pp. 161–169, 2016.
38. I. Sergi, T. Montanaro, F. L. Benvenuto, and L. Patrono, "A Smart and Secure Logistics System Based on IoT and Cloud Technologies," Sensors, vol. 21, no. 6, p. 2231, 2021. https://doi.org/10.3390/s21062231
39. L. Catarinucci, R. Colella, M. De Blasi, L. Patrono, and L. Tarricone, "Experimental Performance Evaluation of Passive UHF RFID Tags in Electromagnetically Critical Supply Chains," Journal of Communications Software and Systems, vol. 7, pp. 59–70, 2011.
40. L. Mainetti, F. Mele, L. Patrono, F. Simone, M.L. Stefanizzi, and R. Vergallo, "An RFID-Based Tracing and Tracking System for the Fresh Vegetables Supply Chain," International Journal of Antennas and Propagation, vol. 2013, p. 531364, 2013.
41. J. P. Muñoz-Gea, J. Malgosa-Sanahuja, P. Manzanares-Lopez, and J. C. Sanchez-Aarnoutse, "Implementation of Traceability Using a Distributed RFID-Based Mechanism," Computers in Industry, vol. 61, pp. 480–496, 2010.
42. J.-P. Qian, X.-T. Yang, X.-M. Wu, L. Zhao, B.-L. Fan, and B. Xing, "A Traceability System Incorporating 2D Barcode and RFID Technology for Wheat Flour Mills," Computers and Electronics in Agriculture, vol. 89, pp. 76–85, 2012.
43. M. M. Aung, and Y. S. Chang, "Traceability in a Food Supply Chain: Safety and Quality Perspectives," Food Control, vol. 39, pp. 172–184, 2014.
44. W. Iqbal, H. Abbas, M. Daneshmand, B. Rauf, and Y. A. Bangash, "An In-Depth Analysis of IoT Security Requirements, Challenges, and Their Countermeasures via Software-Defined Security," IEEE Internet of Things Journal, vol. 7, pp. 10250–10276, 2020.
45. S. Verma, Integrating IoT Technology for Effective Environmental Monitoring IoT Agenda. Accessed: August 05, 2023.
46. A. S. Pamula, A. Ravilla, and S. V. Madiraju, "Applications of the Internet of Things (IoT) in Real-Time Monitoring of Contaminants in the Air, Water, and Soil," Engineering Proceedings, vol. 27, no. 1, p. 26. https://doi.org/10.3390/ecsa-9-13335
47. M. Noaman, M. S. Khan, M. F. Abrar, S. Ali, A. Alvi, and M. A. Saleem, "Challenges in Integration of Heterogeneous Internet of Things," Scientific Programming, vol. 2022, 14, 2022, Article ID 8626882. https://doi.org/10.1155/2022/8626882
48. D. C. Yacchirema Vargas, and C. E. Palau Salvador, "Smart IoT Gateway for Heterogeneous Devices Interoperability," IEEE Latin America Transactions, vol. 14, no. 8, pp. 3900–3906, 2016.
49. N. Pazos, M. Müller, M. Aeberli, and N. Ouerhani, "ConnectOpen-Automatic Integration of IoT Devices," in Proceedings of the 2015 IEEE 2nd World Forum on Internet of Things (WF-IoT), IEEE, Milan, Italy, December 2015, pp. 640–644.

50. J. Huang, C. Zhang, and J. Zhang, "A Multi-Queue Approach of Energy Efficient Task Scheduling for Sensor Hubs," Chinese Journal of Electronics, vol. 29, no. 2, pp. 242–247, 2020.
51. X. Zhang, Z. Qi, G. Min, W. Miao, Q. Fan, and Z. Ma, "Cooperative Edge Caching Based on Temporal Convolutional Networks," IEEE Transactions on Parallel and Distributed Systems, vol. 33, no. 9, pp. 2093–2105, 2022.
52. T. Qiu, N. Chen, K. Li, M. Atiquzzaman, and W. Zhao, "How can Heterogeneous Internet of Things Build Our Future: A Survey," IEEE Communications Surveys & Tutorials, vol. 20, no. 3, pp. 2011–2027, 2018.
53. C. L. Stergiou, E. Bompoli, and K. E. Psannis, "Security and Privacy Issues in IoT-Based Big Data Cloud Systems in a Digital Twin Scenario," Applied Sciences, vol. 13, no. 2, p. 758, 2023. https://doi.org/10.3390/app13020758
54. H. Takabi, J. B. Joshi, and G.-J. Ahn, "Security and Privacy Challenges in Cloud Computing Environments," IEEE Security & Privacy, vol. 8, pp. 24–31, 2010.
55. C. L. Hepsiba, R. J. Priyadarsini, and S. A. Suresh, "Integration of IoT and Cloud Computing Challenges: Survey," International Journal of Computing Algorithm, vol. 9, no. 1, pp. 9–15, 2020.
56. K. Jeffery, "Keynote: CLOUDs: A Large Virtualisation of Small Things," in The 2nd International Conference on Future Internet of Things and Cloud (FiCloud-2014), 2014.
57. B. P. Rao, P. Saluia, N. Sharma, A. Mittal, and S. V. Sharma, "Cloud Computing for Internet of Things & Sensing Based Applications," in Sensing technology (ICST), 2012 Sixth International Conference on, IEEE, 2012, pp. 374–380.
58. C. Stergiou, K. E. Psannis, B. G. Kim, and B. Gupta, "Secure Integration of IoT and Cloud Computing," Future Generation Computer Systems, vol. 78, pp. 964–975, 2018.

23 Cloud Computing, IoT and Machine Learning Techniques for Detection and Classification of Tomato Plants Diseases Due to Pests

Tejinder Deep Singh and Ramesh Bharti

23.1 INTRODUCTION

In recent years, agriculture has had a significant impact on the world economy. The growing stress on the agricultural system is a result of population growth and urbanization, which cause a steady decline in the total area under cultivation. Due to ongoing agricultural reform, modern technology and the expansion of agriculture are closely linked. In addition to using natural resources, understanding and managing information resources are now part of the sustainable development of modern agriculture [1]. Due to ecological environment degradation over the past few years, insect pests and crop diseases frequently undergo widespread outbreaks. Both the quantity and quality of agricultural goods are directly impacted by regular outbreaks of crop illnesses and insect pests, which also result in financial losses. To avoid needless losses, research into agricultural illnesses and insect pests is essential. The technical name for tomatoes is *Solanum lycopersicum*. Nine out of ten farmers produce tomatoes in fields, and they can thrive on practically any well-drained soil. Many gardeners also produce tomatoes in their backyards so they may utilize them in their cooking areas and enjoy wonderful food [2].

Farmers and gardeners occasionally struggle to track the growth of their plants properly. Tomato diseases may not be seen on plants occasionally, or they develop unsightly, disease-filled black spots at the bottom. Finding affected areas of the leaf is the initial step in identifying a tomato plant disease. Then, detecting abnormalities like brown or black spots and openings in plants helps to look at insects. On the same farm, tomatoes, potatoes, and closely related vegetables cannot be grown more than once every three years [3]. It is best to begin growing any grass or crop before cultivating tomatoes to preserve soil fertility. Two groups of tomato issues can be distinguished: Five diseases originate from insects, while the remaining 16 are brought on by bacteria,

DOI: 10.1201/9781032656694-23

fungi, or improper agricultural practices. Bacterial wilt is a serious disease caused by the bacteria *Ralstonia solanacearum*. The bone in question has a long lifespan in soil and can enter the roots through wounds created naturally during the emergence of secondary roots, wounds created artificially during cultivation or transplanting, or even by insects. Hot temperatures and excessive moisture are favorable to the development of diseases. The water-conducting membrane of plants becomes slimy as a result of bacteria that quickly develop there. The green foliage may persist, but this affects the plant's vascular system. When examined in cross-section, a diseased stem becomes brown and has yellowish stuff pouring out of it (Figure 23.1) [4].

Accurately and quickly predicting plant diseases and pests is the initial step in managing them. Consequently, future risk evaluations and management plans will be feasible. To predict identical diseases and insect pests, statistical analysis and the use of numerous case studies are also crucial [5]. Historically, many pest control processes were completed by hand, and agricultural professionals utilized experience to determine pests through time-consuming and recurrent assessment, estimation, and statistical analysis. Modern information technology is desperately required to support it because fictitious experiences and technological inconsistencies make it tough to correctly classify diseases and pests and lead to deviations and omissions in the way information is processed [6].

Figure 23.2 depicts the general procedure for using conventional image recognition processing to recognize plant diseases. The most promising method for closing the gap is automatic plant picture identification, which is attracting a lot of interest in the fields of botany and computing [7]. In a variety of applications, including plant recognition, facial recognition, and others, an image represents the most important information. Numerous studies concentrate on the categorization, segmentation, and quality evaluation of plant leaves. A few elementary image processing steps are required for plant leaf recognition in order to recognize and classify plant leaves. This technique includes steps of image capture, preprocessing, feature extraction, and classification [8].

FIGURE 23.1   Images a and b show spotted leaves due to caterpillar larvae, and images c and d show molded leaves due to fungus in *Solanum lycopersicum*.

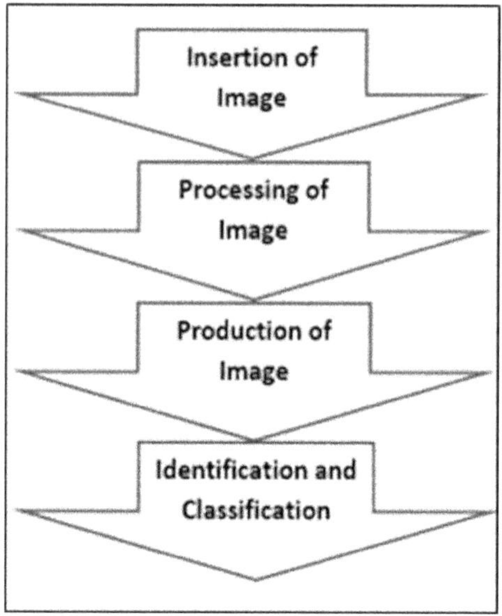

FIGURE 23.2 Traditional image processing techniques.

Tomato leaf diseases are identified by ML classification approaches. The number of samples taken and the classification techniques used determine its precision. The main classification techniques are supervised, unsupervised, semi-supervised, and reinforcement learning. In supervised learning, a variety of inputs and outputs are offered, and during the training phase, relationships between them are identified [9]. Forecasting intended results and building a model that captures dependencies and relationships among incoming data are the main goals. Even with unsupervised learning, the model that is produced captures relationships, but there is no connection between output and inputs [10]. The learning process groups together patterns that are similar. Three types of apple diseases were discovered using an improved SVM with 93% accuracy. Ashwani Kumar Dubey et al. [11] applied K-means clustering to divide lesion areas. They merged the global color histogram, coherence vector, and local binary pattern to identify the texture and color attributes of apple spots. Intelligent systems based on computer vision are used to increase productivity in agriculture, accounting for a significant portion of agricultural product maintenance. Artificial intelligence (AI) includes machine learning, which enables systems to learn automatically and get better over time without having to write code explicitly. Since ML methods have various uses, the primary issue remains feature engineering.

Deep learning (DL) has recently entered the agriculture field as it has demonstrated promising results in a variety of applications [12]. Medical experts can forecast plant anomalies with the aid of DL approaches based on automated computer-aided diagnostic systems, enabling them to choose the best course of action. DL systems function admirably today due to technological advancements, and the system's output helps experts make the right decisions. Deep NN's input layer, several hidden layers, and outcome layer make up its fundamental design.

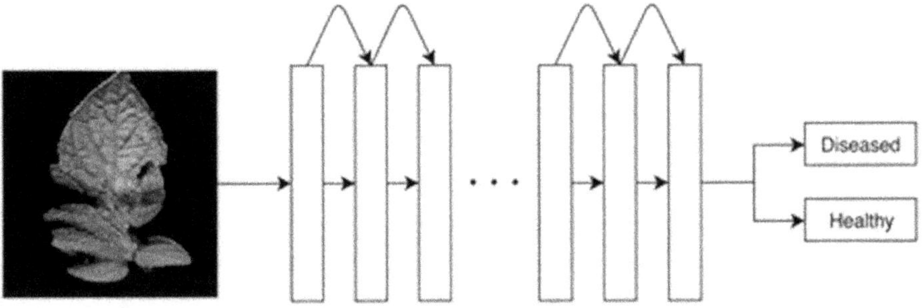

FIGURE 23.3   DL classifier for image classification [14].

When input data is sent to DNNs, the output is computed sequentially across layers of the network. CNN proved to be best at classifying images and made astounding progress. DL was successfully applied to a variety of tasks, including scene analysis, text analytics, object detection, etc. The first step in CNN is to take a picture of a crop that piques our interest and then feed that picture into a model that has been created for additional evaluation and output (Figure 23.3) [13].

23.2   TOMATO LEAF DISEASE DETECTION USING MACHINE LEARNING

This section discusses traditional ML algorithms for identifying tomato diseases. Crop leaves are typically regarded as the primary source of identification for tomato plant diseases, and symptoms of serious diseases may appear on leaves. Researchers from all around the world are drawn to the current study area that identifies tomato disease using photographs of affected leaves because of potential benefits and successful outcomes. Numerous studies offered to identify tomato leaf disease reported various models, approaches, and attributes. In order to wrap up prior research in this field, a literature review was carried out. Figure 23.4 shows a diagram of several ML models that identify and categorize plant diseases.

Sabrol and Satish [16] categorized five distinct diseases: Late blight, septoria, bacterial spot, bacterial canker, leaf curl in tomato leaves, and stem images. Twenty-four shape, color, and texture attributes were considered to categorize images using an 83-node decision tree. A 10-fold cross-validation with 97.3% testing accuracy was observed.

Hlaing and Zaw [17] introduced a feature vector derivation method based on statistical models for identifying six tomato diseases. SIFT feature vector's dimensionality was decreased using Generalized Extreme Value distribution, which reduced the algorithm's computational demands. The system's 10-fold cross-validated performance was reported to have an accuracy rate of 84.7%. In 56.8 seconds, the classifier was trained. The suggested method also managed to predict 12000 instances per second.

Harakannanavar et al. [18] considered tomato leaves with disorders. Farmers can quickly identify diseases based on their early symptoms. Tomato leaves are

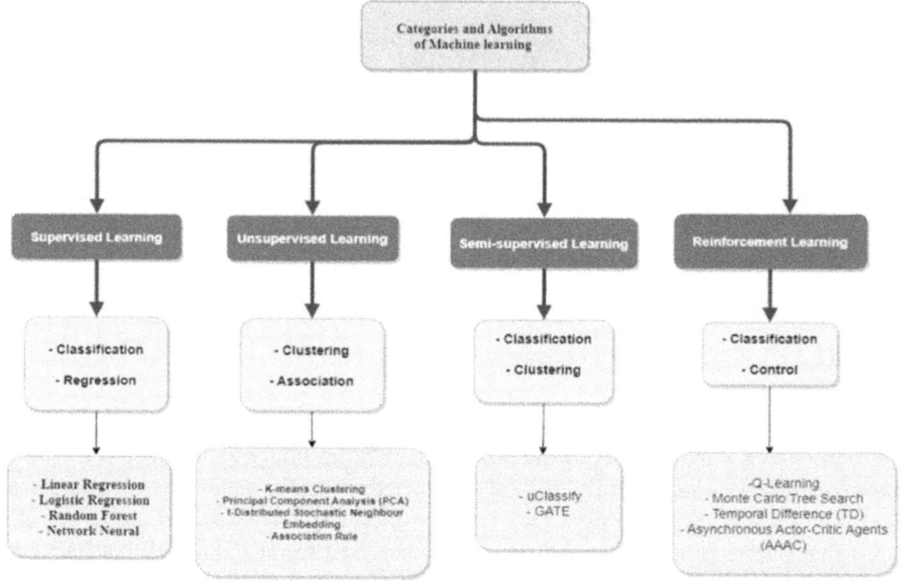

FIGURE 23.4 ML classifiers for categorizing tomato leaf diseases [15].

downsized to 256 × 256 pixels before histogram equalization. K-means clustering is used to segment the dataset into Voronoi cells. Utilizing contour tracing, the leaf boundary is extracted. More leaf features are obtained by applying Discrete Wavelet Transform, PCA, and GLCM. On tomato instances, SVM, KNN, and CNN achieved 88%, 97%, and 99.6% accuracy.

Patil and Manohar [19] used an enhanced radial basis function NN classifier for tomato leaf disease detection. Modified sunflower optimization improves the proposed ERBFNN. Gaussian filter is used first to eliminate noise from images, and the Contrast-limited adaptive technique is used to equalize the histogram. The diseased part of the input picture is segmented using color features obtained from every leaf picture and provided to the partition phase. Accuracy, Jaccard and Dice's coefficient, precision, recall, F-score, and mean intersection over union are used to assess the effectiveness of ERBFNN. Its effectiveness is evaluated compared to region growing, fuzzy c-means, and radial basis functions. Results show that the 98.92% accurate ERBFNN surpassed RBFNN, FCM, and RG.

Bhatia et al. [20] used a sensor-based Tomato Powdery Mildew Disease dataset with an Extreme Learning Machine. The authors used Synthetic Minority Oversampling, Random Oversampling, Random Undersampling, and Importance Sampling to balance the TPMD database. Accuracy and area under the curve have been used to analyze ELM. Results show that CA and AUC rise for resampling techniques employed here. With 89.19% accuracy and 88.57% AUC, ELM outperformed the IMPS approach.

Govardhan and Veena [21] used Random Forest to recognize and categorize tomato diseases. Features are extracted from preprocessed pictures. System diagnoses early and late blight, Spider Mite, Yellow Leaf Curl Virus, and Target Spot. Tomato leaf diseases can be challenging to identify because symptoms often resemble nutritional deficiencies, such as mosaic virus. They vary depending on the plant's age at contamination time. Findings demonstrate that all testing pictures provide the best image classification with 95% of the system's effectiveness.

Qasrawi et al. [22] used 3000 photos of five tomato illnesses—Alternaria solani, Botrytis cinerea, Panonychus citri, Phytophthora infestans, and Tuta absoluta. The photos were taken in Tubas, Jenin, and Tukarem, three West Bank communities. ML classifiers employed image embedding and hierarchical clustering. They used neural networks, random forests, naive Bayes, SVM, decision trees, and logistic regression for forecasting and categorization. The accuracy of the proposed model was examined in relation to a database of tomato plant diseases compiled by specialists in plant pathogens. The accuracy of the clustering model for seven diseases was 70%, compared to accuracy rates of 70.3% and 68.9% for NN and LR, respectively. The suggested method is accurate for clustering, identifying, and categorizing tomato plant diseases. Applying ML techniques to agriculture could benefit Palestinian farmers by improving their ability to control tomato diseases.

Mokhtar et al. [23] proposed SVM for identifying two different tomato leaf diseases. Both training and testing phases used datasets that included 200 pictures of leaves having Powdery mildew and early blight. Cauchy, Invmult, and Laplacian kernels are used in SVM. The proposed approach was chosen using grid search, and its performance was assessed using N-fold cross-validation. Using Cauchy and Laplacian kernels at 100% and 98% accuracy, respectively, compared to the Invmult kernels at 78% accuracy, outcomes appear to demonstrate the high accuracy provided by the recommended strategy.

Dhaya [24] focused on fusarium wilt disease that commonly affects plants. Fusarium oxysporum disease primarily affects tomato, sweet potato, tobacco, legume, and cucurbit plants and is frequently brought on by soil. The suggested algorithm builds models by forecasting disease twice for improved precision. The public dataset contains 87k pictures, 60% of which are of affected leaves and 40% of which are of healthy plant leaves. With a large dataset, the suggested hybrid naive Bayes and image recognition algorithm accurately detected disease with 96% accuracy.

Xian and Ngadiran [25] used an Extreme Learning Machine with single-layer feed-forward NN to categorize plant diseases by analyzing leaf pictures. The picture is preprocessed using HSV color space, and attributes are extracted using Haralick textures. ELM classifier is fitted with features to train and test the model. The dataset considered is a subset of the PlantVillage record. When compared to SVM and decision tree, outcomes from ELM achieve the highest accuracy of 84.94%.

Raza et al. [26] used depth information and combined thermal and visible light pictures to identify plants with Oidium neolycopersici. Features can be extracted from pictures by employing local and global statistics to more accurately identify diseased plants. Plants that later developed a disease due to natural transmission but were not initially infected with fungus at the beginning of the experiment can be recognized using the feature set (Table 23.1).

TABLE 23.1

Tomato Leaf Diseases Detection Using Machine Learning

Reference	Technique	Dataset	Accuracy (%)
Muthukannan and Latha [27]	Fuzzy Logic	Private	95
Sabrol and Satish [16]	Decision Tree	Private	97.3
Sabrol and Kumar [28]	MultiLayer Feed-Forward BPNN	Private	87.20
Sabrol and Kumar [29]	Decision Tree	Private	78
Hlaing and Zaw [17]	Statistical-model	PlantVillage	84.7
Lu et al. [30]	Moth-Flame Optimization and rough set	Tomato leaves	86
Hlaing and Zaw [31]	SVM-Quadratic	PlantVillage	85.10
Annabel and Muthulakshmi [32]	SVM Random Forest	Private	82.60
	MDC		87.60
	Random Forest		94.10
Das et al. [33]	SVM	Kaggle (Public)	87.60
	LR		70.05
	RF		67.30
Basavaiah and Anthony [34]	Decision Tree	PlantVillage	90

23.3 DEEP LEARNING ARCHITECTURE USED ON TOMATO PLANTS

Deep learning is a multilayered automated learning system developed from ANNs and used to create information categorized into unsupervised, supervised, and reinforcement learning. DL models have been used in several applications that have solved a variety of picture recognition issues and improved research outcomes in fields like automatic plant disease identification, natural language processing, and medical diagnosis. DL is a novel method for identifying plant species and leaf diseases. Figure 23.5 shows a diagram for implementing DL algorithms for the identification and categorization of tomato leaf disease. The research on using DL models to identify tomato illnesses from leaf photos is covered in this section.

Zamora-Izquierdo et al. [36] proposed a Cyber-Physical Systems (CPS), which interacts with crop gadgets to collect data and perform real-time atomic control actions. The edge plane of the platform is in charge of surveillance and managing main PA tasks near the access network to increase system reliability against network access failures. Finally, the cloud platform collects current and past records and hosts data analytics modules in a FIWARE deployment. IoT protocols like Message Queue Telemetry Transport (MQTT) or Constrained Application Protocol (CoAP) are used to communicate with CPS, while Next Generation Service Interface (NGSI) is employed for southbound and northbound access to the cloud.

Khan and Meera [37] recommended a DL technique that automates the diagnosis of tomato leaf ailments. To categorize image datasets based on the apparent effects of diseases on plant leaves, CNN is trained from scratch. They trained a deep

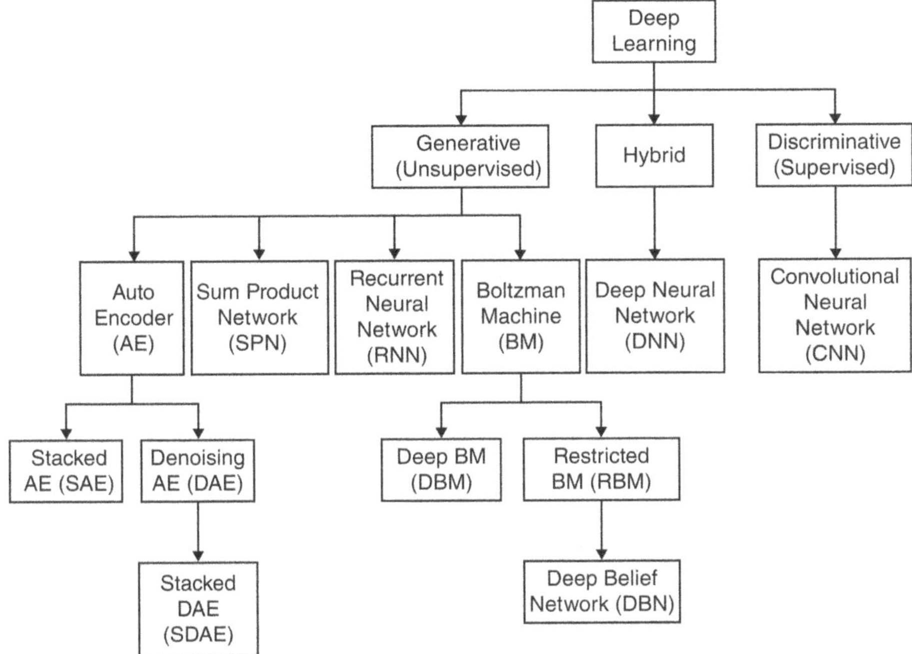

FIGURE 23.5 DL algorithms for categorizing tomato leaf diseases [35].

CNN to recognize early and late blight fungal diseases that affect tomatoes utilizing a database made up of leaf pictures collected from various sources. With 97.25 % accuracy, a DNN sequential model is created, proving the viability of suggested system testing on a dataset that combines real-world data with pictures downloaded from the Internet and basic data from PlantVillage.

Zhang et al. [38] presented a 3-channel CNN to detect tomato leaf disease. They employed three convolutional feature extraction models with a fused dense network. With 15.817 images and eight classes of controlled environments, the PlantVillage dataset allowed the Multi-class Classifier to achieve an accuracy of 91.15%. With Layers Custom as a predominate technique, simulation is done in Matlab with a Top View of images. One of the three RGB color channels was given to every CNN. The suggested model outshined Sparse Representation-based Classification, Global-Local Singular Value Decomposition, SVM, and Image Processing Technology by a wide margin.

Brahimi et al. [39] contrasted the efficacy of SVM and RF classifiers to AlexNet and GoogleNet CNNs that were trained from scratch and utilized transfer learning to identify nine tomato leaf diseases. The dataset included 14,828 images and was categorized using a CNN with 99.18% accuracy. CNN models outperformed SVM and RF classifiers and also pretrained CNNs outperformed CNNs trained from scratch.

Elhassouny and Smarandache [40] showed that deep CNNs can be implemented for identifying plant illnesses on mobile devices. Images from the PlantVillage dataset were used to train MobileNets to recognize ten typical tomato leaf diseases. The model's accuracy was only slightly impacted by the selected optimization method. Additionally, despite the cost of a longer training period, greater accuracy could be attained by reducing the learning rate. According to the learning rate selected, accuracy ranged from 85.9% to 90.3%. Considering the prevalence of smart devices, potential of farmers to get the most from this innovation will grow with creation of CNN designs which can be used on mobile devices.

Kaur et al. [41] For localization of items, 1610 tomato leaf images representing various classes were taken from the PlantVillage database. Modified Mask R-CNN is suggested for recognition of tomato leaf diseases. The proposed model includes a Region CNN with a light head in an effort to reduce computation and storage costs. The effectiveness of computing metrics is enhanced by changing ratios of anchor in the RPN network as well as feature extraction topology. It is contrasted to existing models to determine whether the suggested approach is practically better. Proposed model produced accuracy, F-measure, and precision of 0.88, 0.912, and 0.98, respectively. The lesion detection period is decreased by two times compared to current classifiers as the model's capability is enhanced with few parameters.

Anandhakrishnan and Jaisakthi [42] invented computerized software for identifying tomato leaf diseases by employing deep CNN. Eighteen thousand one hundred sixty images were used from the PlantVillage dataset and divided 60% images for training and 40% for testing. A 98.40% accuracy was attained for the testing set using the suggested DCNN model.

Anim-Ayeko et al. [43] suggests ResNet-9 that farmers can use to identify blight disease in potatoes and tomatoes. Three thousand nine hundred ninety training samples were used from "PlantVillage Dataset". The classifier was trained with hyperparameter values and tested on 1,331 images, following augmentation of the training set and hyperparameter optimization process. 99.25% accuracy, 99.67% precision, 99.33% recall, and 99.33% F-measure were all attained. Saliency maps were used to explain the suggested model's assumptions and provide justifications for its predictions, allowing for a thorough understanding of the model. ResNet-9 considered leaf shape, diseased, and green areas for predictions. Findings may help develop and apply CNN models for classifying tomato and potato leaf images. This modeling framework will be useful for future research, which will test a number of additional variables to identify leaf infections at earlier times for protecting crops.

Li et al. [44] suggests a method called LMBRNet for identifying tomato leaves diseases classifiers that use complementary grouped dilated residual feature extraction blocks. To extract distinctive data on tomato illnesses and tomato leaf receptive fields, four subsidaries with convolutional kernels of various dimensions were created. Residual connection was used to address network degradation and gradient disappearance issues. Public datasets RS, SIW, and PlantVillage-corn were used to test LMBRNet. All three datasets showed high accuracy, with scores of 82.32%, 88.37%, and 97.25% on RS, SIW, and PlantVillage-corn, demonstrating LMBRNet's strong generalizability. When applied to specific datasets, LMBRNet outperforms MobileNetV3S and MobileNetV3L. Its (4.1M) metrics are greater than MobileNetV3S (2.9M) and lesser than MobileNetV3L (5.4M), suggesting that, despite having a limited no. of parameters, it has better generalization.

Islam et al. [45] devised a methodology for categorizing tomato leaf diseases depending on the output of multiple concurrent CNN with various configurations. The network performs significantly better with using LeakyReLU-Swish, Elu-Swish, and ClippedReLU-Swish activation layers and Batch Normalization-Instance Normalization layer, accomplishing accuracy of 99.0%, 97.5%, and 98.0% with training, validation, and testing instances.

Sunil et al. [46] put forth Multilevel Feature Fusion Network for the categorization of tomato leaf diseases. To categorize leaves images, ResNet50, MFFN, and Adaptive Attention Mechanism are used. The suggested technique accomplished 99.88% training, 99.9% validation, and 99.8% test accuracy. It performed better than the currently used methods appropriate for the dataset. The research offers information on specific pesticides that are in accordance with the kind of leaf disease.

Sharma et al. [47] introduces deeper lightweight CNN for leaf disease detection across various crops. Passage layer and a series of collective blocks are added in the suggested model to obtain attributes. These advantages in feature reuse and propagation help to solve the vanishing gradient issue. Point-wise and detachable convolutional blocks are also used to lower various learning metrics. Citrus, cucumber, grapes, and tomato are publicly accessible datasets used to validate DLMC-Net's performance. Accuracy, error, sensitivity, specificity, F-measure, and Mathews correlation coefficient are used to compare outcomes of a suggested classifier against seven existing models. The recommended approach has outshined all models with 93.56%, 92.34%, 99.5%, and 96.5% accuracy for citrus, cucumber, grapes, and tomato, respectively. It is 2nd best classifier compared to the 6.4 million trainable metrics needed.

Agarwal et al. [48] created a CNN model to find diseases in tomatoes. Three convolutional with maximum pooling layers, with distinct filters, provided a suggested solution. We used data from PlantVillage's dataset of tomato leaves for simulation. There are nine diseased class labels and one healthy label in the data that contains images. Data augmentation was used to balance images inside the class. The suggested approach's precision for classes ranges from 76% to 100%, and accuracy is 91.2%. The suggested model requires 1.5 MB memory, whereas pretrained classifiers require 100 MB storage, showing CNN's benefits over others.

Paymode and Malode [49] employed CNN techniques for detecting Multi-Crops Leaf Disease. DL model extracted features from images to classify unhealthy and healthy leaves. The VGG classifier is utilized to measure performance better. For model training and testing, crops leaves pictures dataset is considered. Accuracy, sensitivity, precision, and F-measure were computed and tracked. Created model more accurately categorizes leaves with disease. Suggested research has 95.71% accuracy for tomatoes and 98.40% accuracy for grapes. The suggested technique directly supports boosting agricultural food production.

Astani et al. [50] seeks to quickly and accurately identify 13 tomato diseases in farms and labs through 260 ensemble techniques created with feature extraction. The accuracy and precision of the suggested method were assessed under laboratory and outdoor conditions with various challenges using two databases, PlantVillage and Taiwan tomato leaves. Best ensemble classifier identifies diseases with 95.98% accuracy based on background clutter, multiple leaves, brightness changes, disease similarity, and shadow conditions. A comparison among 260 suggested ensemble models in recommended methodology and various DL models was conducted. The proposed approach performed better than most advanced DL model.

Abd Algani **et al.** [51] presented the ACO-CNN model in which ACO was used to examine the efficacy of disease detection in plant leaves. CNN is used to remove color, texture, and leaf arrangement from given images. Parameters for measuring effectiveness that were used to analyze the suggested methodology showed that the recommended approach outperforms existing ones with the highest accuracy. The stages of disease detection include image acquisition, separation, removing noise, and classification.

Ngugi et al. [52] suggested removing noise from leaves photos captured using mobile devices by employing CNNs. A leaf would be surrounded by other leaves, stalks, and mulch. These supplementary traits will be removed by the segmentation network, leaving only the target leaf. A dataset was produced to train and test suggested systems. It is made up of 1,408 tomato leaf images and ground truth masks that go with them that were captured under challenging field settings. They present cutting-edge leaf image segmentation outcomes with mean weighted intersection over union over 0.96 and boundary F1 scores over 0.91. Particularly, our suggested segmentation network, KijaniNet, outshines all rivals with 0.9439 boundary F1-measure and mean weighted intersection over the union of 0.9766. The suggested method outperforms rival background subtraction computations while requiring neither user input nor imposing restrictions on the target leaf's orientation or illumination. Additionally, compared to other methods, all CNN models can segment a 256×256 RGB picture in less than 0.12 seconds when running on GPU and less than 2.1 seconds on CPU.

Wspanialy and Moussa [53] trained a disease model to recognize previously unidentified tomato leaf diseases, and its severity forecasting can yield outcomes that are comparable to human assessment. Leaves shape has predictive ability for disease detection. Future data gathering initiatives can increase the diversity of datasets by reducing accidental bias in backgrounds and providing a better understanding of PlantVillage information set drawbacks and its image-capturing methods. One could conceptualize a diseased-healthy binary classifier as a healthy leaf predictor. In addition to being able to identify diseases that haven't been trained on before, such a classifier can assist in reducing the effort needed to create an extensive data set for identifying general diseases. This study also presented a brand-new dataset of proportional disease severity with annotations and an associated classifier. Ordinal categories work better for diseases brought by viruses and insects, while proportional area measures work best for estimating the severity of bacterial and fungal diseases. The findings can be put into practice by incorporating classifiers into an automatic examining system, lowering expenses and error and boosting accuracy.

Baser et al. [54] uses TomConv model, which employs CNN to classify tomato leaves into ten distinct groups. The publicly accessible dataset PlantVillage, which contains more than 16000 pictures of healthy and diseased leaves, was employed for this research project. The suggested model is the most straightforward of all current cutting-edge classifiers. Images of tomato leaves underwent preprocessing to reduce their size to 150×150 pixels. The model consists of a max pooling layer after a four-layered CNN. It divides the corpus into training and validation datasets in 80:20 proportion, trains with 105 iterations for pictures of tomato leaves, and achieves a 98.19% accuracy. The proposed model is contrasted with existing ones based on various factors, including a number of classes, layers, and accuracy. Outcomes are encouraging because they outperform all current cutting-edge models.

Fuentes et al. [55] compared the performance of various DL networks for disease categorization in tomatoes. Faster R-CNN, R-FCN, and SSD were some of these. For classification, the Front Viewpoint is applied to RGB images. Keras framework employs TensorFlow language, with data augmentation being the main methodology. They discovered a technique for local and global class annotation and data augmentation as their foundation. Outcomes were more accurate, and there were fewer false positives.

Guo et al. [56] implemented a tomato leaf illness detection system on Android mobile using the AlexNet model for identifying diseases accurately in their early, middle, and late phases.

Fuentes et al. [57] suggested deep architecture with a targeted feature extraction to identify plant illnesses. It was created to spot diseases in every picture. It was confirmed on the tomato plant diseases dataset and pests that they independently gathered and successfully applied to challenging real-world scenarios.

Natarajan et al. [58] presented an automated system for spotting diseases on cultivated land. To identify and categorize tomato disease in plants, Faster R-CNN with feature extraction is used. The suggested model was trained and tested using a tomato dataset that includes roughly 1090 pictures of early and late stages of tomato diseases. Even in complex plant surroundings, the suggested system ResNet50 effectively predicts early blight, leaf curl, septoria, and bacterial leafspot.

Monika et al. [59] conducted research between 2019 and 2020. Aphids Myzus persicae and Macrosiphum euphorbiae, serpentine leaf miner Liriomyziatrifolii, fruit borer Helicoverpa armigera, whitefly Trialeurodes vaporariorum and aphids Myzuspersicae and Macrosiphum euphorbiae were all observed during surveys. Low-hills were home to aphid species M. persicae, while mid-hills were found to have tomato infestations by M. euphorbiae. More tomato pest infestations were noted in Himachal Pradesh's mid-hills than in low-hills.

Fuentes et al. [60] presented the DL method for identifying diseases and pests in tomatoes utilizing real-time pictures taken from cameras. They focus on three families of detectors: Single Shot Multibox Detector, Region-based Fully CNN, and Faster Region-based CNN. They combined VGG net and Residual Network with every meta-architecture. They also suggest a technique for local and global class annotation as well as data augmentation to improve precision and decrease false positives during training. Tomato Diseases and Pests Database contains pictures of many inter-class variations, like the infection portion in leaf, which is utilized for training and testing models from beginning to end. The suggested method is capable of identifying nine distinct diseases and pests and is able to handle complex situations that may arise in the plant's surroundings.

Ashok et al. [61] suggested a methodology that links input image pixel intensity and contrasts it with trained image instances using CNN for feature extraction. All adjusting metrics of leaf parts are optimized by lowering error relative to the training set.

This study suggests using image processing methods on the basis of segmentation and clustering to detect tomato diseases. Leaf picture is classified using an image classifier that uses ANNs, fuzzy logic, and hybrid algorithms to distinguish between disease-affected and healthy leaves. The accuracy of the suggested method is 98%.

Nagamani et al. [62] examined ML tools like Fuzzy SVM, CNN, and Region-based CNN to detect tomato leaf disease. Tomato leaves pictures with six diseases and healthy samples were used to confirm findings. Images are trained using picture scaling, color thresholding, flood filling, gradient local ternary pattern, and Zernike moments' attributes. Analysis and comparison of Fuzzy SVM, CNN with R-CNN aims to identify the most precise classifier for plant disease detection. When contrasted with other methods, R-CNN achieves 96.735 % accuracy.

Bhandari et al. [63] In addition to healthy leaves, this study seeks to visually recognize nine infectious diseases in tomato leaves, including early and late blight, leaf mold, 2-spotted spider mite, Septoria leaf and bacterial spot, and mosaic virus. Without segmenting the dataset, they used EfficientNetB5, with ten cross-folds. The classifier produced 99.84%, 98.28%, and 99.07% accuracy for training, validation, and testing data. It is suggested that Gradient-weighted class activation mapping and locally interpretable model-related discussions be used for integration into agricultural practice as well as for forecasting effectiveness.

Tian et al. [64] proposed a DL classifier based on private and public datasets of tomato leaf images for identifying diseases. VGG16, InceptionV3, and Resnet50 were trained and evaluated. To identify nine different diseases and healthy tomato leaves, they installed a trained model into an Android application called TomatoGuard. TomatoGuard performed better than APP Plantix with 99% testing accuracy.

Martins Crispi **et al.** [65] created an ML classifier to assess the severity of tomato leafminer fly attacks. The database included pictures of pest signs on tomato leaves that were taken in the field. Acquired pictures were manually annotated into three groups, including background, leaves, and leafminer fly symptoms. For a multi-class categorization, metrics like accuracy, precision, and recall contrasted three classifiers and four distinct backbones. U-Net classifier with Inceptionv3 produced the best outcomes, according to a comparison of segmentation outcomes. The best classifier for estimating symptom severity had lower RMSE values and was FPN with ResNet34 and DenseNet121 backbones.

Bouni et al. [66] Finding and controlling plant illnesses, which have certain effects on food production, population health, plant quality, and productivity, is a significant challenge in agriculture. It requires time to accurately identify different plant diseases with the naked eye. Managing tomato diseases necessitates constant attention throughout the crop lifecycle and accounts for a sizable portion of the overall cost of production. They used automation and pretrained DNN to classify tomato diseases. Plant disease monitoring can be done using digital image processing. DL has significantly outperformed more traditional methods in digital image processing. This article uses transfer learning and a deep CNN to predict tomato diseases. AlexNet, ResNet, VGG-16, and DenseNet form the core of the CNN classifier. Adam and RmsProp approach analyze the relative performance of networks, showing that DenseNet with RmsPropmethodology obtains significant outputs having 99.9% accuracy.

Tarek et al. [67] assessed ResNet50, InceptionV3, AlexNet, MobileNetV1, MobileNetV2, and MobileNetV3, that were trained on ImageNet dataset. MobileNetV3 Small and MobileNetV3 Large have 98.99% and 99.81% precision, respectively. To assess each model's effectiveness, the detection duration on tomato leaf pictures was calculated on the workstation where all models had been employed. To construct an IoT system for detecting tomato diseases, classifiers were also installed on Raspberry Pi 4. On the workstation and Raspberry Pi 4, MobileNetV3 Small has shown 66 ms and 251 ms latency, respectively. MobileNetV3 Large had 348 ms latency on Raspberry Pi 4 and 50 ms on the workstation (Table 23.2).

TABLE 23.2

Tomato Leaves Disease Recognition Using DL Approaches

Reference	Approach Used	Dataset	Performance Parameter
Kim et al. [55]	Faster R-CNN, ResNet	5000 field pictures 9 pests diseases classes	mAP = 0.8306 Accuracy = 85.98%
Boukhalfa et al. [39]	AlexNet, GoogLeNet	PlantVillage	Accuracy = 99.185% Precision = 98.529% Recall = 98.532% F-measure = 98.518%
Smarandache et al. [40]	MobileNets	PlantVillage	Accuracy = 90.3%
Fuentes et al. [68]	Faster R-CNN, VGG16 feature extraction	8927 self-gathered pictures, 9 anomalies and one backgroundclass	mAP = 96%
Durmus et al. [69]	AlexNet, SqueezeNet	PlantVillage	Accuracy= 94.3%
Toda and Okura [70]	Modified Inceptionv3	PlantVillage	Accuracy = 97.1%
Huang et al. [38]	Custom three channel CNN	PlantVillage (15,817 Images/ 8 Class Controlled Environment)	Accuracy = 91.16%
Rangarajan et al. [71]	VGG16 and AlexNet	PlantVillage, 6 tomato diseases	Accuracy = 96.2%
Sharma et al. [72]	S-CNN	PlantVillage, 17,000 Real pictures 10 Classes	Accuracy = 98.6%
Khamparia et al. [73]	ResNet	PlantVillage Subset	Accuracy = 99.4%
Wang et al. [74]	Mask R-CNN and ResNet101	1,430 Images 11 Classes	Accuracy = 99.64%
Ashqar and Abu-Naser [75]	LeNet	PlantVillage 9,000 pictures 6 Classes Control Environment	Accuracy = 99.84%
Suryawati et al. [76]	VGG	PlantVillage 18,160 instances 10 Classes	Accuracy = 95.24%

(Continued)

TABLE 23.2 *(Continued)*
Tomato Leaves Disease Recognition Using DL Approaches

Reference	Approach Used	Dataset	Performance Parameter
de Luna et al. [77]	AlexNet	4,923 pictures 4 Classes	Accuracy = 95.75%
Sardogan et al. [78]	LeNet classifier and LVQ	PlantVillage 500 photos 5 Classes Control Environment	Accuracy = 86%
Zhang et al. [79]	ResNet and SGD	5,550 PlantVillage photos, 8 Classes	Accuracy = 96.51%
Tm et al. [80]	LeNet	PlantVillage18,160 pictures,10 Classes	Accuracy = 94.85%
Fuentes et al. [81]	Refinement Filter Bank Based ResNet	8,927 real instances 10 Classes	Accuracy = 96%
Zeng et al. [82]	HoResNet Based ResNet	PlantVillage 10,478 pictures 6 Classes Control Environment	Accuracy = 91.79%
Rangarajan et al. [83]	AlexNet VGG16	PlantVillage 13,262 pictures 6 Classes	Accuracy = 97.49% Accuracy = 97.23%
Durmus et al. [69]	AlexNet SqueezeNet	54,306 photos 38 Classes Control Environment	Accuracy = 95.65% Accuracy = 94.3%
Atabay [84]	ResNet	PlantVillage 19,742 instances 10 Classes	Accuracy = 97.57%
Yamamoto et al. [85]	AlexNet	18,149 pictures 9 Classes Control Environment	Accuracy = 78%

23.4 CONCLUSION

The agricultural industry has suffered many difficulties recently. This article offers a study of recent findings in AI-based disease identification of tomato leaves. This paper examines several ML and DL approaches applied to categorizing tomato leaf diseases. Various related research papers have been examined and analyzed in this study, focusing on the dataset, preprocessing methods utilized, algorithms, and accuracy. According to the literature review, the DL model performs better for identifying tomato diseases using images of leaves than other traditional methods, including machine learning and neural networks. Early diagnosis of tomato plant disease lowers expenses by avoiding needless pesticide treatment of plants. For the early diagnosis of tomato leaf disease, a new emerging technology called deep learning using hyperspectral imaging is advised. As tomato plant illnesses spread to neighboring plants, their severity has increased over time. This has made it possible to use specialized DL classifiers to predict and categorize tomato leaf diseases throughout their entire life cycle. Fusion of low and high-level CNN classifier attributes can be utilized to speed up convergence and improve prediction accuracy. Incorporating drones and agriculture robots in the future to automatically capture photographs of plant leaves and classify disease-affected plants.

REFERENCES

1. Pallagani, V., V. Khandelwal, B. Chandra, V. Udutalapally, D. Das, and S. P. Mohanty. "DCrop: A deep-learning-based framework for accurate prediction of diseases of crops in smart agriculture." In *Proceedings of the 2019 IEEE International Symposium on Smart Electronic Systems (iSES) (Formerly iNiS)*, Rourkela, India, 2019.
2. Abdullahi, H. S., R. Sheriff, and F. Mahieddine. "Convolution neural network in precision agriculture for plant image recognition and classification." In *Proceedings of the 2017 Seventh International Conference on Innovative Computing Technology (INTECH)*, Luton, UK, 16–18 August 2017, pp. 1–3.
3. Altalak, M., M. A. Uddin, A. Alajami, and A. Rijg. "Smart agriculture applications using deep learning technologies: A survey." *MDPI* 12 (2022): 5919.
4. Ahmed, A. A., and G. H. Reddy. "A mobile-based system for detecting plant leaf diseases using deep learning." *AgriEngineering* 3 (2021): 478–493.
5. Sladojevic, S., M. Arsenovic, A. Anderla, D. Culibrk, and D. Stefanovic. "Deep neural networks based recognition of plant diseases by leaf image classification." *Computational Intelligence and Neuroscience* 5 (16) (2016): 3289801.
6. Arivazhagan, S., R. N. Shebiah, S. Ananthi, and S. V. Varthini. "Detection of unhealthy region of plant leaves and classification of plant leaf diseases using texture features." *Agricultural Engineering International: CIGR Journal* 15 (2013): 211–217.
7. Khan, S., and M. Narvekar. "Novel fusion of color balancing and superpixel based approach for detection of tomato plant diseases in natural complex environment." *The Journal of King Saud University Computer and Information Sciences* 34 (6, PartB) (2020): 3506–3516.
8. Kumar, A., and M. Vani. "Image based tomato leaf disease detection." In *Proceedings of the 2019 10th International Conference on Computing, Communication and Networking Technologies (ICCCNT)*, Kanpur, India, 30 December 2019, pp. 1–6.
9. Patel, S. M., M. B. Jain, S. S. Pai, and S. D. Korde. "Smart agriculture using IoT and machine learning." *IRJET* 1 (2021): 47–59.
10. Gupta, A., D. Nagda, P. Nikhare, and A. Sandbhor. "Smart crops prediction using IoT and machine learning." *IJERT* 9 (2021): 18–21.
11. Ashwani Kumar Dubey, Y., R. Ratan, and A. Rocha. "Computer vision based analysis and detection of defects in fruits causes due to nutrients deficiency." *Cluster Computing* 23 (2020): 1817–1826.
12. Zhao, S., Y. Peng, J. Liu, and S. Wu. "Tomato leaf disease diagnosis based on improved convolution neural network by attention module." *Agriculture* 11 (2021): 651.
13. Bhujel, A., N. E. Kim, E. Arulmozhi, J. K. Basak, and H. T. Kim. "A lightweight attention-based convolutional neural networks for tomato leaf disease classification." *Agriculture* 12 (2022): 228.
14. Jackulin, C., and S. Murugavalli. "A comprehensive review on detection of plant disease using machine learning and deep learning approaches." *Measurement: Sensors* 24 (2022): 100441.
15. Taye, M. M. "Understanding of machine learning with deep learning: Architectures, workflow, applications and future directions." *Computers* 12, no. 5 (2023): 91.
16. Sabrol, H., and S. Kumar. "Tomato plant disease classification in digital images using classification tree." In *2016 International Conference on Communication and Signal Processing (ICCSP)*, 2016, pp. 1242–1246.

17. Hlaing, C. S., and S. M. M. Zaw. "Model-based statistical features for mobile phone image of tomato plant disease classification." In *2017 18th International Conference on Parallel and Distributed Computing, Applications and Technologies (PDCAT)*, 2017, pp. 223–229.
18. Harakannanavar, S. S., J. M. Rudagi, V. I. Puranikmath, A. Siddiqua, and R. Pramodhini. "Plant leaf disease detection using computer vision and machine learning algorithms." *Global Transitions Proceedings* 3, no. 1 (2022): 305–310.
19. Patil, M. A., and M. Manohar. "Enhanced radial basis function neural network for tomato plant disease leaf image segmentation." *Ecological Informatics* 70 (2022): 101752.
20. Bhatia, A., A. Chug, and A. P. Singh. "Application of extreme learning machine in plant disease prediction for highly imbalanced dataset." *Journal of Statistics and Management Systems* 23, no. 6 (2020): 1059–1068.
21. Govardhan, M., and M. B. Veena. "Diagnosis of tomato plant diseases using random forest." In *2019 Global Conference for Advancement in Technology (GCAT)*, IEEE, 2019, pp. 1–5.
22. Qasrawi, R., R. Z. MalakAmro, M. Sawafteh, and S. V. Polo. "Machine learning techniques for tomato plant diseases clustering, prediction and classification." In *2021 International Conference on Promising Electronic Technologies (ICPET)*, IEEE, 2021, pp. 40–45.
23. Mokhtar, U., M. A. S. Ali, A. E. Hassenian, and H. Hefny. "Tomato leaves diseases detection approach based on support vector machines." In *2015 11th International computer engineering conference (ICENCO)*, IEEE, 2015, pp. 246–250.
24. Dhaya, R. "Flawless identification of *Fusarium oxysporum* in tomato plant leaves by machine learning algorithm." *Journal of Innovative Image Processing (JIIP)* 2, no. 04 (2020): 194–201.
25. Xian, T. S., and R. Ngadiran. "Plant diseases classification using machine learning." *Journal of Physics: Conference Series* 1962, no. 1 (2021): 012024. IOP Publishing.
26. Raza, S.-E.-A., G. Prince, J. P. Clarkson, and N. M. Rajpoot. "Automatic detection of diseased tomato plants using thermal and stereo visible light images." *PLoS One* 10, no. 4 (2015): e0123262.
27. Muthukannan, K., and P. Latha. "Fuzzy inference system based unhealthy region classification in plant leaf image." *International Journal of Computer and Information Engineering* 8, no. 11 (2015): 2103–2107.
28. Sabrol, H., and S. Kumar. "Fuzzy and neural network based tomato plant disease classification using natural outdoor images." *Indian Journal of Science and Technology* 9, no. 44 (2016a): 1–8.
29. Sabrol, H., and S. Kumar. "Intensity based feature extraction for tomato plant disease recognition by classification using decision tree." *The International Journal of Computer Science and Information Security* 14, no. 9 (2016b): 622–626.
30. Lu, J., M. Zhou, Y. Gao, and H. Jiang. "Using hyperspectral imaging to discriminate yellow leaf curl disease in tomato leaves." *Precision Agriculture* 19, no. 3 (2018): 379–394.
31. Hlaing, C. S., and S. M. M. Zaw. "Tomato plant diseases classification using statistical texture feature and color feature." In *2018 IEEE/ACIS 17th International Conference on Computer and Information Science (ICIS)*, IEEE, 2018, pp. 439–444.
32. Annabel, L. S. P., and V. Muthulakshmi. "AI-Powered Image-Based Tomato Leaf Disease Detection." In *2019 Third International conference on I-SMAC (IoT in Social, Mobile, Analytics and Cloud)(I-SMAC)*, IEEE, 2019, pp. 506–511.
33. Das, D., M. Singh, S. S. Mohanty, and S. Chakravarty. "Leaf Disease Detection using Support Vector Machine." In *2020 International Conference on Communication and Signal Processing (ICCSP)*, IEEE, 2020, pp. 1036–1040.

34. Basavaiah, J., and A. A. Anthony. "Tomato leaf disease classification using multiple feature extraction techniques." *Wireless Personal Communications* 115, no. 1 (2020): 633–651.

35. Selvaganapathy, S. G., M. Nivaashini, and H. P. Natarajan. "Deep belief network based detection and categorization of malicious URLs." *Information Security Journal: A Global Perspective* 27, no. 3 (2018): 145–161.

36. Zamora-Izquierdo, M. A., J. Santa, J. A. Martínez, V. Martínez, and A. F. Skarmeta. "Smart farming IoT platform based on edge and cloud computing." *Biosystems Engineering (Elsevier)* 177 (2019): 4–17.

37. Khan, S., and N. Meera. "Disorder detection in tomato plant using deep learning." In *Advanced Computing Technologies and Applications: Proceedings of 2nd International Conference on Advanced Computing Technologies and Applications—ICACTA 2020*, Springer Singapore, 2020, pp. 187–197.

38. Zhang, S., W. Huang, and C. Zhang. "Three-channel convolutional neural networks for vegetable leaf disease recognition." *Cognitive Systems Research* 53 (2019): 31–41.

39. Brahimi, M., K. Boukhalfa, and A. Moussaoui. "Deep learning for tomato diseases: Classification and symptoms visualization." *Applied Artificial Intelligence* 31 (2017): 299–315.

40. Elhassouny, A., and F. Smarandache. "Smart mobile application to recognize tomato leaf diseases using Convolutional Neural Networks." In *Proceedings of 2019 International Conference of Computer Science and Renewable Energies*, ICCSRE, 2019, pp. 1–4.

41. Kaur, P., S. Harnal, V. Gautam, M. P. Singh, and S. P. Singh. "An approach for characterization of infected area in tomato leaf disease based on deep learning and object detection technique." *Engineering Applications of Artificial Intelligence* 115 (2022): 105210.

42. Anandhakrishnan, T., and S. M. Jaisakthi. "Deep convolutional neural networks for image based tomato leaf disease detection." *Sustainable Chemistry and Pharmacy* 30 (2022): 100793.

43. Anim-Ayeko, A. O., C. Schillaci, and A. Lipani. "Automatic blight disease detection in potato (*Solanum tuberosum* L.) And tomato (*Solanum lycopersicum,* L. 1753) plants using deep learning." *Smart Agricultural Technology* 4 (2023): 100178.

44. Li, M., G. Zhou, A. Chen, L. Li, and Y. Hu. "Identification of tomato leaf diseases based on LMBRNet." *Engineering Applications of Artificial Intelligence* 123 (2023): 106195.

45. Islam, M. P., K. Hatou, T. Aihara, S. Seno, S. Kirino, and S. Okamoto. "Performance prediction of tomato leaf disease by a series of parallel convolutional neural networks." *Smart Agricultural Technology* 2 (2022): 100054.

46. Sunil, C. K., C. D. Jaidhar, and N. Patil. "Tomato plant disease classification using multilevel feature fusion with adaptive channel spatial and pixel attention mechanism." *Expert Systems with Applications* 228 (2023): 120381.

47. Sharma, V., A. K. Tripathi, and H. Mittal. "DLMC-net: Deeper lightweight multi-class classification model for plant leaf disease detection." *Ecological Informatics* 75 (2023): 102025.

48. Agarwal, M., A. Singh, S. Arjaria, A. Sinha, and S. Gupta. "ToLeD: Tomato leaf disease detection using convolution neural network." *Procedia Computer Science* 167 (2020): 293–301.

49. Paymode, A. S., and V. B. Malode. "Transfer learning for multi-crop leaf disease image classification using convolutional neural network VGG." *Artificial Intelligence in Agriculture* 6 (2022): 23–33.

50. Astani, M., M. Hasheminejad, and M. Vaghefi. "A diverse ensemble classifier for tomato disease recognition." *Computers and Electronics in Agriculture* 198 (2022): 107054.

51. Abd Algani, Y., O. J. M. Caro, L. M. R. Bravo, C. Kaur, M. S. A. Ansari, and B. Kiran Bala. "Leaf disease identification and classification using optimized deep learning." *Measurement: Sensors* 25 (2023): 100643.

52. Ngugi, L. C., M. Abdelwahab, and M. Abo-Zahhad. "Tomato leaf segmentation algorithms for mobile phone applications using deep learning." *Computers and Electronics in Agriculture* 178 (2020): 105788.

53. Wspanialy, P., and M. Moussa. "A detection and severity estimation system for generic diseases of tomato greenhouse plants." *Computers and Electronics in Agriculture* 178 (2020): 105701.

54. Baser, P., J. K. R. Saini, and K. Kotecha. "TomConv: An improved CNN model for diagnosis of diseases in tomato plant leaves." *Procedia Computer Science* 218 (2023): 1825–1833.

55. Fuentes, A., S. Yoon, S. C. Kim, and D. S. Park. "A robust deep-learning based detector for real-time tomato plant diseases and pests recognition." *Sensors* 17, no. 9 (2017): 2022.

56. Guo, X., T. Fan, and X. Shu. "Tomato leaf diseases recognition based on improved multi-scale AlexNet (in Chinese)." *Transactions of the Chinese Society of Agricultural Engineering* 35, no. 14 (2019): 162–169.

57. Fuentes, A., S. Yoon, and D. S. Park. Deep learning-based techniques for plant diseases recognition in real-field scenarios. In *Advanced Concepts for Intelligent Vision Systems - 20th International Conference*, 2020. pp. 3–14.

58. Natarajan, V. A., M. M. Babitha, and M. S. Kumar. "Detection of disease in tomato plant using deep learning techniques." *International Journal of Modern Agriculture* 9, no. 4 (2020): 525–540.

59. Monika, S. M., P. L. Sharma, and I. K. Kasi. "Incidence of major insect pest infesting tomato in low and mid hills of Himachal Pradesh." (2022).

60. Fuentes, A., S. Yoon, S. C. Kim, and D. S. Park. "A robust deep-learning-based detector for real-time tomato plant diseases and pests recognition." *Sensors* 17, no. 9 (2017): 2022.

61. Ashok, S., G. Kishore, V. Rajesh, S. Suchitra, S. G. G. Sophia, and B. Pavithra. "Tomato leaf disease detection using deep learning techniques." In *2020 5th International Conference on Communication and Electronics Systems (ICCES)*, IEEE, 2020, pp. 979–983.

62. Nagamani, H. S., and H. Sarojadevi. "Tomato leaf disease detection using deep learning techniques." *International Journal of Advanced Computer Science and Applications* 13, no. 1 (2022).

63. Bhandari, M., T. B. Shahi, A. Neupane, and K. B. Walsh. "BotanicX-AI: Identification of tomato leaf diseases using an explanation-driven deep-learning model." *Journal of Imaging* 9, no. 2 (2023): 53.

64. Tian, K., J. Zeng, T. Song, Z. Li, A. Evans, and J. Li. "Tomato leaf diseases recognition based on deep convolutional neural networks." *Journal of Agricultural Engineering* 54, no. 1 (2023).

65. Martins Crispi, G., D. S. M. Valente, D. Marçal de Queiroz, A. Momin, E. Fernandes-Filho, and M. CoutinhoPicanço. "Using deep neural networks to evaluate leafminer fly attacks on tomato plants." *AgriEngineering* 5, no. 1 (2023): 273–286.

66. Bouni, M., B. Hssina, K. Douzi, and S. Douzi. "Impact of pretrained deep neural networks for tomato leaf disease prediction." *Journal of Electrical and Computer Engineering* 2023 (2023).

67. Tarek, H, S. Eisa, H. Aly, and M. Abul-Soud. "Optimized deep learning algorithms for tomato leaf disease detection with hardware deployment." *Electronics* 11, no. 1 (2022): 140.

68. Fuentes, A. F., S. Yoon, J. Lee, and D. S. Park. "High-performance deep neural network-based tomato plant diseases and pests diagnosis system with refinement filter bank." *Frontiers in Plant Science* 9 (2018): 1–15.

69. Durmus, H., E. O. Gunes, and M. Kirci. "Disease detection on the leaves of the tomato plants by using deep learning." In *2017 6th Int Conf Agro-Geoinformatics*, 2017, pp. 1–5. doi: 10.1109/Agro-Geoinformatics.2017.8047016

70. Toda, Y., and F. Okura. "How convolutional neural networks diagnose plant disease." *Plant Phenom* 2019 (2019): 1–14.

71. Rangarajan, A. K., R. Purushothaman, and A. Ramesh. "Tomato crop disease classification using pre-trained deep learning algorithm." *Procedia Computer Science* (2018).

72. Sharma, P., Y. P. S. Berwal, and W. Ghai. "Performance analysis of deep learning CNN models for disease detection in plants using image segmentation." *Information Processing in Agriculture* (2019): 566–574.

73. Khamparia, A., G. Saini, D. Gupta, A. Khanna, S. Tiwari, and V. H. C. de Albuquerque. "Seasonal crops disease prediction and classification using deep convolutional encoder network." *Circuits Systems and Signal Processing* (2019).

74. Wang, Q., F. Qi, M. Sun, J. Qu, and J. Xue. "Identification of tomato disease types and detection of infected areas based on deep convolutional neural networks and object detection techniques." *Computational Intelligence and Neuroscience* 2019 (2019): 15.

75. Ashqar, B. A. M., and S. S. Abu-Naser. "Image-based tomato leaves diseases detection using deep learning." *International Journal of Academic Engineering Research* 2 (2019): 10–16.

76. Suryawati, E., R. Sustika, R. S. Yuwana, A. Subekti, and H. F. Pardede. "Deep structured convolutional neural network for tomato diseases detection." In *2018 International Conference on Advanced Computer Science and Information Systems (ICACSIS)*, 2018, pp. 385–390.

77. de Luna, R. G., E. P. Dadios, and A. A. Bandala. "Automated image capturing system for deep learning-based tomato plant leaf disease detection and recognition." In *TENCON 2018–2018 IEEE Region 10 Conference*, 2018, pp. 1414–1419.

78. Sardogan, M., A. Tuncer, and Y. Ozen. "Plant leaf disease detection and classification based on CNN with LVQ algorithm." In *2018 3rd International Conference on Computer Science and Engineering (UBMK)*, 2018, pp. 382–385.

79. Zhang, K., Q. Wu, A. Liu, and X. Meng. "Can deep learning identify tomato leaf disease?" *Advance in Multimedia* (2018): 6710865:1–6710865:10.

80. Tm, P., A. Pranathi, K. SaiAshritha, N. B. Chittaragi, and S. G. Koolagudi. "Tomato leaf disease detection using convolutional neural networks." In *2018 Eleventh International Conference on Contemporary Computing (IC3)*, 2018, pp. 1–5.

81. Fuentes, A. F., S. Yoon, J. Lee, and D. S. Park. "High-performance deep neural network based tomato plant diseases and pests diagnosis system with refinement filter bank." *Frontiers in Plant Science* 9 (2018): 1162.

82. Zeng, W., M. Li, J. Zhang, L. Chen, S. Fang, and J. Wang. "High-order residual convolutional neural network for robust crop disease recognition." In *Proceedings of the 2Nd International Conference on Computer Science and Application Engineering, CSAE '18*, ACM, New York, NY, USA, 2018, pp. 1–5, Article number 101. doi:10.1145/3207677.3277952

83. Rangarajan, A. K., R. Purushothaman, and A. Ramesh. "Tomato crop disease classification using pre-trained deep learning algorithm." *Procedia Computer Science* 133 (2018): 1040–1047.

84. Atabay, H. A. "Deep residual learning for tomato plant leaf disease identification." *Journal of Theoretical and Applied Information Technology* 95 (2017): 866.

85. Yamamoto, K., T. Togami, and N. Yamaguchi. "Super-resolution of plant disease images for the acceleration of image-based phenotyping and vigor diagnosis in agriculture." *Sensors* 17 (2017).

The Impact of Cloud and IoT Integration in the Education Industry

*Tejinder Deep Singh and
Kanwaldeep Kaur Kanwal*

24.1 INTRODUCTION

The current educational paradigm places a strong emphasis on traditional features such as textbooks, exams, grades, and evaluations while frequently ignoring creative and innovative learning approaches. Teachers strictly stick to set syllabi, pupils limit their studies to these prescribed portions, and the process finishes after exams. More imaginative and profound thought, on the other hand, has catalyzed a dramatic shift. Cloud computing has developed as an attractive and unique notion within educational institutions. This approach provides students and administrative staff with quick and cost-effective access to a variety of application platforms and resources via on-demand websites. The utilization of radio frequency identification (RFID) tags enhances practical learning as it enables science students to record observations from their classroom instead of needing to be physically present in the field. Educational settings have incorporated computer technologies to elevate the learning experience and achieve better results in modern digital learning environments of the twenty-first century, particularly in the realm of the Internet of Things (IoT) computing. Professors, lecturers, instructors, and teachers can use computer technologies to construct and operate courses via a digital platform, providing interactive lectures without requiring real classroom participation. Educators and university lecturers can provide multimedia presentations inside an immersive and globally accessible synchronous digital milieu within the scope of interactive educational pedagogy. This synchronous method not only saves time and energy but also improves overall efficiency in operating digital classrooms.

Moreover, multimedia presentations offer an engaging experience, captivating the audience through auditory, visual, and evaluative dimensions, especially when considering the audio, video elements, 3-D simulation environment, and visual enhancements that amplify the impact of lectures. The integration of computers into the education of the twenty-first century will eradicate monotony from the dynamic process of teaching and learning. The technology of computers incorporated within educational institutions not only catalyzes technological progress but also shapes the landscape of digital pedagogy [1]. Recent times have witnessed remarkable strides in computer technology, encompassing all aspects of human managerial duties and embracing a comprehensive range of media for interactive involvement, thus

DOI: 10.1201/9781032656694-24

holistically addressing contemporary and pervasive pedagogical accomplishments in the realm of teaching and learning.

Presently, the forefront of research and innovation is dominated by computer development, swiftly progressing within the realm of education through inventive modernization. This evolution centers on the augmented accessibility of contemporary computing tools like smartphones, (PCs) personal computers, and (PDAs) personal digital assistants, each holding the potential to profoundly increase knowledge transmission [2]. The growth of science and technology has revealed the fundamental elements of the digital era, particularly in terms of information handling, often referred to as information communication technologies (ICTs). Computer advancement has had a dramatic impact on the lives of students throughout Nigeria, Africa, and the entire global spectrum, reaching even the most remote corners of our globe. Undoubtedly, computers, mobile technologies, and the internet have substantially dictated the experiences of ordinary graduate and undergraduate students within the academic landscape of the twenty-first century [3].

The researchers conducted a worldwide analysis of computer technology applications in education, specifically honing in on the 21st-century design of e-learning within IoT computing. Across the globe, numerous educational establishments have adopted smart e-learning platforms, transcending geographical barriers and offering learning opportunities to e-learners regardless of location. The COVID-19 pandemic, which mandated the shutdown of academic institutions, undeniably underscored the vital role of e-learning development. This paradigm showcased multiple advantages, including enhanced learning and knowledge dissemination, improved curriculum access, cost savings, and widespread utilization of mobile computing for comprehensive intellectual interaction and digital learning [4]. E-learning platforms in the 21st century, exemplified by tools like Zoom, exhibit high intelligence and adaptability. They offer interactive and versatile features, creating a structured yet adaptable learning setting that caters to digital learners. This environment ensures seamless and simultaneous information sharing [5]. Modern and forward-thinking educational technologies designed for the twenty-first century, such as e-learning, blended learning, distance learning, mobile learning (m-learning), virtual learning, and e-libraries, enable members of the academic community, such as professors, lecturers, instructors, teachers, researchers, and students, to easily develop and maintain essential digital skills [6].

For information exchange, e-learning education systems rely largely on internet infrastructure. Unfortunately, the internet and its infrastructure have evolved to promote illegal acts. Because e-learning systems are distributed, open, and networked, information within them continually crosses internet protocol spaces, exposing them to security concerns [7]. The proliferation of communication and information technology, mixed with the prevalence of cybercrime, poses a significant obstacle to post-COVID-19 e-learning investment. This study uses technology monitoring to examine ICT penetration, cybercrime trends, and preventive approaches for reducing cyber hazards in Nigeria's educational landscape. Furthermore, the chapter delves into improving information communication security, mobile cloud computing, and technology infrastructures in e-learning, as well as the possibility of reinforcing platform security through policy-based computing.

24.2   BACKGROUND OF EDUCATION

The fusion of personal computing devices (desktops, laptops, gadgets), alongside Platform-as-a-Service, Infrastructure-as-a-Service, and Software-as-a-Service, collaborates to provide essential functionalities like e-tutoring, self-assessment, and communication tools such as the Zoom, chat applications, forums, and video calling apps, all contributing to the facilitation of remote learning. Mobile e-learning environments emphasize technological mobility, while ubiquitous learning environments underscore the social interdependence within various activities and connections between entities in their environmental framework. Consequently, the concept of ubiquitous learning synergizes mobile learning and pervasive learning, engendering heightened mobility and social connectivity. Mobile and pervasive computing technologies, mobiles, and computer devices that are embedded like global positioning systems (GPS) and RFID tags consisting of radio transponders, receiver-transmitters, sensors, pads, badges, and wireless sensor networks, support ubiquitous learning (U-learning) environments [8]. Context-aware U-learning is a subtype of ubiquitous computer learning that uses mobile devices, wireless communication, and sensor technologies in educational pursuits [9]. These technologies primarily collect environmental data and give customized education based on the contextual awareness of the e-learner. This awareness includes the learner's point of view, environmental conditions, system design, and the U-learning environment, providing a holistic view of student involvement, including the learners' understanding of the learning process's constraints.

Initiatives are currently underway to build versatile U-learning settings for various applications to connect Radio Frequency ID tags within the area of e-learning context [10]. This endeavor entails incorporating context awareness into the e-learning framework, resulting in an adaptable U-learning digital arena. The digital landscape of U-learning has evolved to supplement traditional learning models by combining multimedia devices (audio, video, simulation) and synchronous chat-based interactive pedagogy [11]. Synchronous U-learning allows classroom students to explore the vast digital universe by interacting with mobile devices and computing technology via various embedded digital instruments.

24.3   WHY USE CLOUD IN EDUCATION?

By moving to the cloud, almost three by four institutes reduced the cost of their applications. Of the educational institutions, 35% have stored a minimum of 1 terabyte of data on cloud platforms. Based on data, 43% of higher education establishments have opted for or are in the process of considering the implementation of cloud computing solutions.

24.3.1   Diminished Costs

Leveraging cloud-based services can aid institutions in reducing costs and expediting the integration of novel technologies to meet changing educational demands. Students gain access to office applications without buying, installing, or maintaining the latest versions on their personal computers. Additionally, educators benefit from a pay-as-you-go model for selected applications.

Easy Access: The lesson organizes laboratories, evaluates assignments, keeps track of notes, and even transfers computerized materials like PowerPoint presentations – essentially, any digital resources utilized in the training process are efficiently moved over.

Security: Particular person's information, material, photographs, and text - whatever we put on the cloud usually needs verification through ID and password and is, thus, virtually unavailable to anyone.

Shareability: Cloud computing opens up a world of new possible results for students, particularly those underserved by traditional training frameworks. With cloud computing, one may reach a broader range of students.

No costly programming required: Software-as-a-Service (SaaS) is a major focus of cloud-based registration. Many product initiatives are now available for free or on a convenience membership basis, significantly lowering the cost of crucial applications for students.

Easy Update: Make changes to a lesson and then need to reverse them? Don't be concerned. Cloud computing will save many corrections and variants of a record so that you may track the evolution of a thing consecutively.

Through various means like these, cloud computing reduces costs and ensures that every student can avail top-notch training and resources. Whether you are a

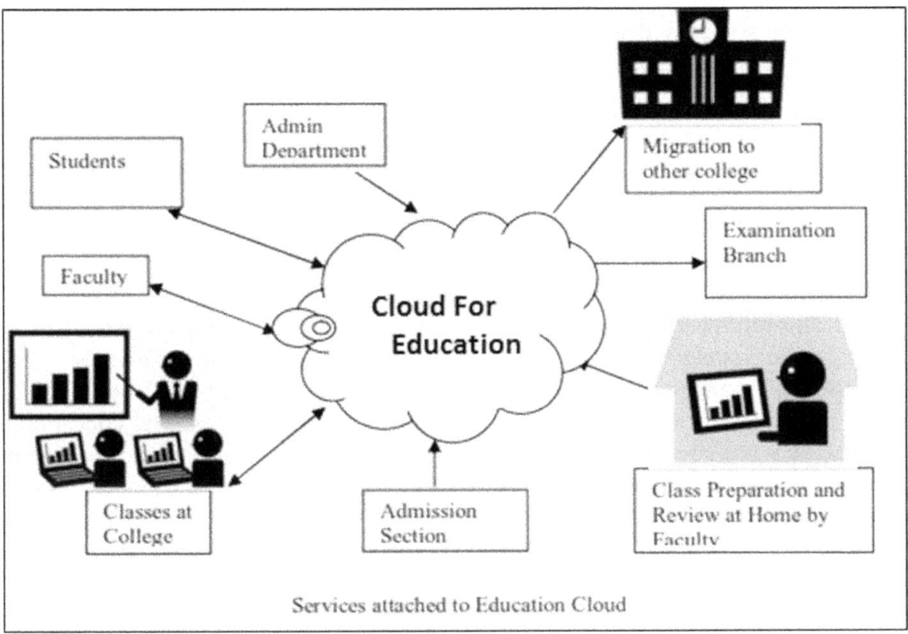

FIGURE 24.1   Education services to the cloud [12].

chairperson, educator, learner, or student's caretaker, the present moment offers a favorable opportunity to explore the potential benefits of cloud-based applications for you, your children, and your educational institution.

24.4   IMPLEMENTING THE INTEGRATION OF CLOUD AND IoT IN EDUCATION SECTOR: A NEED

The trajectories of these two realms, IoT and Cloud, have evolved autonomously. Nevertheless, a multitude of shared benefits arising from their integration, as delineated in literature, offer a glimpse into the future. On one hand, the IoT can leverage the cloud's nearly boundless capacity and resources to overcome its technical limitations. Specifically, cloud computing presents a viable solution for managing internet-based services, along with orchestrating the interaction and utilization of various entities or data applications. Conversely, cloud computing stands to gain from its alliance with the IoT, broadening its scope to engage with real-world entities in a more widespread and dynamic manner, thus furnishing novel services across diverse real-world scenarios. The four foundational pillars of IoT underscore the need for an educational framework that equips a fresh generation of digital citizens to comprehend the foundational technology of IoT, its societal implications through widespread adoption, and the appropriate application of the knowledge assimilated. Higher education programs must ensure that upcoming engineers possess the expertise to design and develop technological systems that capture the spirit of our new-found openness and engagement.

In the field of computer science, the task at hand entails developing novel and scalable educational models capable of accommodating large numbers of global students, enticing prospective students with a wide range of interests, and delivering an inventive curriculum that reflects the profound changes in computing technology. The integration of IoT and cloud computing has the potential to overcome the aforementioned educational system issues. Individuals can use cloud computing to manage and retrieve data over the internet. In a higher education setting, the cloud is a resource available to various entities, including students, instructors, administrative staff, the Examination Branch, and the Admission Branch. It provides these primary users with global access and assigns unique logins based on each user's specialized duties.

Educators can upload classroom tutorials, assignments, and evaluations to a cloud server. This repository is made available to students over the internet, allowing them to access the complete range of educational content provided by their teachers. This accessibility extends to both home and college settings, operating 24 hours a day, seven days a week, and spanning a variety of electronic devices. The educational framework is prepared to discover areas of challenge where errors are likely to occur through a study of students' learning records. This feedback mechanism provides teachers with insights that allow them to improve their teaching materials and practices. Aside from in-class use, students can use online educational materials for both preparation and review at home. Because servers and educational resources are shared among institutions, the use of cloud computing systems results in reduced operational costs. Many of these services and technologies can be smoothly transferred to cloud platforms, which are directly accessible via the internet.

24.5   THE PILLARS OF IoT IN EDUCATION

The use of IoT technology in the field of education is currently in its initial stages. However, specific institutions are leading the path by showcasing how IoT can be effectively employed to educate both youth and the broader public. This segment delves into the influence of each fundamental aspect on education and explores the necessities for fostering, developing, and expanding some of the strategies presently under consideration or being put into action.

> **People:** In the contemporary era, the majority of individuals access the internet through a variety of devices and social platforms. Predicting future internet access methods remains uncertain, but it is evident that people will achieve a heightened level of connectivity. This connectivity holds significant implications for the education sector, necessitating an understanding of internet usage patterns to enhance learning and knowledge application. The forthcoming key asset will be rapid mastery of skills. Successful learners will maintain relevance and lead the way. Alvin Toffler's insight underscores the ability to continually learn and adapt. Forming crucial connections, individuals will not only tap into expert insights but also link with peers who share passions, fostering vibrant knowledge-sharing communities. Streaming and live video will become the norm for global information exchange, exemplified by platforms like OEDB, COURSERA, EDX, and MOOCs, offering free access to world-class education and fostering diverse networks.
>
> **Process:** The process holds immense significance in the collaborative generation of value within the interconnected realm of IoT, encompassing individuals, data, and objects. Valuable connections are established when precise information is delivered to the right individual, appropriately timed, and presented. Ensuring that young individuals access tailored learning opportunities enhances educational efficiency, expedites mastery, and fuels inspiration. Such prospects also amplify student retention and apps of newfound data, which are pivotal for upcoming professional and societal accomplishments. Feedback on a student's progress proves invaluable; a 10th-class geography student, for instance, can gauge real-time ranking against peers. Traditional exams assessing and comparing performance might gradually be replaced by an ongoing, personalized measurement method, delivering continuous guidance for improved comprehension and aptitude.
>
> **Data:** As internet-connected objects progress, their intelligence will expand, providing increasingly pertinent data. Rather than merely presenting raw information, interconnected entities will soon convey advanced insights to machines, computers, and humans, facilitating swifter decision-making and in-depth analysis.

The educational impact is significant. Students could tag objects for research purposes, collect data, and improve research accuracy by evaluating them with other technologies. They might have access to study data, track environmental programs,

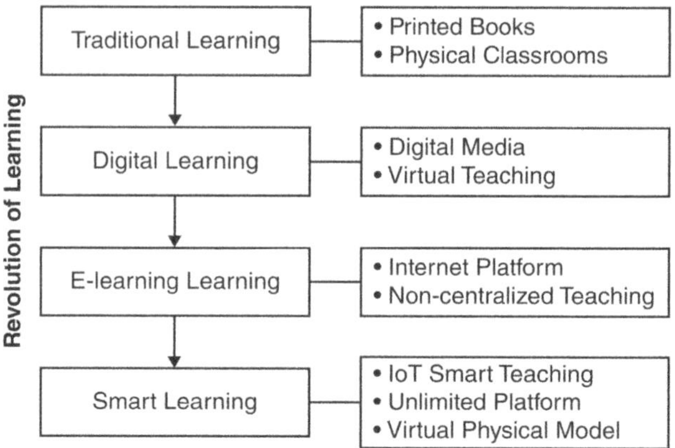

FIGURE 24.2   Revolution of learning [13].

see wildlife via webcams, and collect data via sensors attached to their clothing. The dependability of this data has a significant impact on students' engagement and learning.

24.6   LITERATURE REVIEW

Shams Tabrej Siddiqui et al. [14] said that IoT devices encompass a range of items such as intelligent lighting, ovens, and machinery for industrial data collection. Initially, early iterations of IoT devices transmitted data to remote servers in the cloud for analysis. However, this approach led to congestion in the data pipeline due to the massive volume of gigabytes being transferred. To address this, edge computing emerged, enabling IoT devices to process data locally rather than send it to distant servers. This entails processing data at the periphery of your local network rather than transporting it away. The versatility of this technology renders it valuable in educational settings. Distinct criteria govern IoT edge computing as opposed to non-IoT edge computing. Given the limited data processing and storage capabilities of IoT devices, it becomes necessary for processing to occur externally, beyond the device itself. This edge environment serves as a platform for efficiently managing a multitude of IoT devices and their associated data, leading to potential cost reductions. Particularly due to the demands of real-time applications, the adoption of IoT edge computing architecture has found prominence in the realm of education. This study seeks to establish an IoT edge computing framework tailored for the education sector. The primary objective is to assist in scrutinizing the architecture of educational systems. The framework will spotlight challenges and considerations tied to IoT implementation. Moreover, the research delves into the utilization of IoT edge computing within education, evaluates its limitations, and puts forth recommendations. Additionally, the study appraises the performance of IoT edge computing in the context of educational systems.

Anna Felicia et al. [15] presented that IoT implementations have found extensive usage across multiple sectors, but their integration into education has been somewhat less pronounced compared to other industries. Nevertheless, this doesn't negate the evident benefits and influence that IoT incorporation can bring to the field of education. In this paper, we meticulously examine a range of established IoT applications within the education domain over recent years, aiming to provide readers with a comprehensive overview. Our investigation reveals that the deployment of IoT in the education sector encompasses diverse facets, spanning from intelligent learning environments to interactive educational toys. These applications are explored across all educational levels, from early childhood education to university settings. Drawing from the present trajectory of technological advancement, we also offer insights into potential future developments of IoT in education. To unify existing research, which predominantly centers around enhancing learning outcomes, we introduce the concept of the "Internet of Education Things" (IoEdT). This term refers to a collection of technologies aimed at enhancing the learning environment and educational processes. These enhancements could range from creating more intelligent campuses to monitoring the educational journey closely.

Dewar Rico-Bautista et al. [9] presented that higher education institutions view emerging technologies as an opportunity to enhance both their academic and administrative procedures. Nevertheless, the mere presence and utilization of these novel technologies within these establishments do not automatically ensure the desired positive impact on their operations. Consequently, numerous technology adoption endeavors falter, resulting in setbacks and reluctance to embrace change. This failure is primarily attributed to an excessive focus on the technology itself, disregarding the practical realities of the institution. A remedy to circumvent such predicaments and redirect efforts involves aligning the adoption process with the institution's specific context. As artificial intelligence (AI), cloud computing, IoT, and Big Data technologies gain prominence, the availability of tools capable of gauging institutional receptiveness becomes crucial. The primary objective is to gauge the degree of incorporation within HEIs, concentrating on the needs and perspectives of their users. Various models have been formulated to elucidate the factors driving people's acceptance and utilization of technology. Frameworks such as TAM, UTAUT, UTAUT2, DOI, TPB, TRA, and others have emerged as valuable tools for investigating technology adoption patterns. This categorization of models leads to the conclusion that technology integration should harmonize and merge seamlessly with the institution's existing processes. Such alignment necessitates increased engagement with top-level management to ensure a successful and effective implementation.

Yu-Hsin Hung et al. [2] discuss that the adoption of cloud computing across these diverse sectors has garnered the attention of scholars and experts. Between 2010 and 2019, this study delved into the technological patterns shaping innovative advancements in cloud computing, analyzing a total of 3697 studies on cloud computing. The findings underscore that the advancement of cloud computing is predominantly motivated by the pursuit of intelligence and automation. This exploration resulted in the identification of three core research categories: critical review, system design, and systematic analysis. Cloud computing services, including XaaS (Anything as a Service), EaaS, LaaS, and MaaS, exhibit links with noteworthy themes. This investigation harnessed machine-learning algorithms to examine data from the domains of education, logistics, and manufacturing. The outcomes yielded remarkably accurate results, surpassing a 90% accuracy rate and an impressive AUC (Area Under the Curve) score. Employing laptops, desktops, tablets, electronic gadgets, and smartphones, this study configured and validated machine-learning models using third-party cloud platforms, which offer limitless scalability and adaptability for data analysis. This empowerment enables users to generate more precise predictions and make well-informed decisions that align with the specific demands of their businesses.

Muhammad Kashif Saeed et al. [16] emphasized the potential future IoT significance, which plays a multifaceted role in enhancing modern human life. It has a significant role in propelling the development of smart cities, sophisticated surroundings, safety, intellect, industrial management, automation, farming, healthcare, and especially higher education. This technology is harnessed to elevate convenience for individuals, signifying a notable progression within the education sector. Its ability to gather data, enhance learning experiences, and foster improved learning environments signifies a remarkable evolution. The reach of IoT is rapidly expanding across diverse sectors, particularly within the realms of smart cities and higher education institutions. This makes a more extensive adoption of this technology highly advantageous. This article offers a concise overview of the IoT and its potential to enhance higher education. The concepts of smart campuses, intelligent classrooms, and advanced laboratories are explored within this study. The authors delineate the benefits it brings to students, educators, and universities, amplifying learning speed and higher education processes. The article also includes a comparative analysis between traditional educational institutions and those empowered by IoT. It elucidates the superiority of the latter in contrast to conventional classroom education. Addressed within the paper are the challenges associated with its implementation in universities. Furthermore, the authors outline their plans to develop a framework to deploy these services in future endeavors.

Muhammad Zubair Asghar et al. [17] Using AI and the IoT, this study dives into the current landscape of ideological and political classroom training. The project aims to improve classroom reform by adding real-life experiences in order to overcome the limitations of traditional techniques. To educate political and ideological concepts, a modern and sophisticated methodology is designed. The IoT is the foundation of this cognitive and political pedagogical system, spanning the perceptual, network, and application layers. Real-time data from classroom activities is collected via IoT and wireless networks, sent to a central data center, and used for applications, data analysis, and modeling. The usefulness of this approach in transforming ideological and political education is confirmed through simulations and classroom experiments.

Parul Agarwal et al. [3] The impact of abstract technology on various aspects of life, including education, has been profound. In recent years, significant transformations have occurred in the traditional classroom dynamic. Education relied on chalkboard teachings in the past, but technology has brought about substantial improvements. Today's classrooms include not only teachers, students, and books but also robots, e-books, and laptops. Modern education transcends boundaries through e-lectures and tutorials, offering easy access to vast amounts of data. The shift from "Teacher-centric" to "Student-centric" is evident. This paper concentrates on the potential of Blockchain and IoT applications in education, exploring their advantages and potential solutions through an extensive literature review. Challenges are acknowledged, and these technologies hold promising prospects to revolutionize education in the future, particularly Blockchain, despite its early stages.

Biju Bajracharya et al. [18] emphasized the rapid expansion of manufacturing low-power, compact, cost-effective computing technology, coupled with its widespread accessibility, enabling its integration into any physical object, linked to the internet, forming what is known as the IoT. IoT possesses the capability to engage with various entities, irrespective of time or location, interconnecting individuals, processes, data, and objects. This integration streamlines daily tasks, unlocking avenues for growth and innovation. While pervading multiple sectors, IoT's influence is gaining prominence in education, promising enhanced teaching and learning experiences. This study delves into IoT technology and its potential educational applications and evaluates its advantages and constraints, envisioning a smart education system or intelligent learning environment fostered by the IoT.

Sofia Kusuma et al. [19] presented that integrating IoT and extensive data analysis in e-learning presents considerable prospects and hurdles for educational institutions and learners. Current trends highlight how the rise of pervasive computing, social media, evolving IoT capabilities, and technologies such as cloud computing, big data, and analytics are not just elevating fundamental teaching and research elements but are also cultivating a novel digital culture and nurturing an IoT-centric society. This article primarily aims to examine the influence of IoT and big data analytics on e-learning and explore various approaches in the domain of digital education.

Ugochukwu O. Matthew et al. [20] discussed how integrating technologies like big data, cloud computing, IoT, and AI will shape the educational landscape, reflecting a highly automated society and boosting enterprise productivity through a transformative global educational reform. This study explored enhancing managerial performance through investments in contextual and self-adaptive electronic learning systems, addressing manageability challenges. It underscores the significance of technology in education, emphasizing teachers' digital adaptation to prepare students for success in the evolving digital realm. The report emphasizes the fusion of enterprise platform flexibility and technological innovation to ensure the sustainability of global education in the 21st century.

Jing Wang et al. [21] discussed recent progress in science, informatics, globalization, astronautics manufacturing, robotics, and AI has led to significant transformations in higher education. Effectively managing resources to enhance educational quality in an interactive and efficient environment poses a noteworthy challenge for both educators and students. This article introduces the implementation of an Internet of Things-powered Interactive System (IoT-IS) for Smart Learning aimed at evaluating the performance of teachers and students within the SEL platform. The study explores psychometric techniques aligned with advanced educational tools to enhance effective instruction according to higher education standards. Moreover, a dynamic higher education system incorporates an attention score approach through active learning, utilizing facial expression analysis from online classroom footage to assess student engagement. Experimental outcomes underscore the superiority of this approach, achieving a student performance ratio of 98.5%, accuracy of 95.3%, efficiency of 96.7%, reliability of 93.2%, and a probability ratio of 94.5% compared to existing methods.

Velankanni Alex [22] presented an article that aids in the advancement of leadership styles among higher education administrators through research. It offers insights into addressing challenges within higher education and effecting necessary changes. It examines leadership philosophies that promote such transformations. A valuable aspect is the comparative analysis of global educational systems, allowing our higher education system to adopt suitable approaches aligned with our culture and context. Review papers provide valuable guidance to management, administrators, and school principals, enabling the implementation of new methodologies and enhancement of leadership philosophies. By assimilating the core principles and leadership models presented, educational leadership and management strategies can be further elevated.

Kanagasabapathi et al. [23] focused on every employee who aspires to advance in their field. Employees must improve their knowledge, skills, and attitudes (KSA) to align with their future positions if they want to succeed in their careers. The primary goal of the current study is to investigate the effects of professional development methods through education continuity, training, and e-learning on executives' quality of work life in Chennai. The current investigation is an exploratory one. A self-developed questionnaire was used to obtain primary data utilizing the survey method of data collection. Various executives from different vehicle manufacturing auto companies located in and around Chennai City participated in the poll. To evaluate the primary data, descriptive methods of stats methods, confirmatory analysis of factors, and structural equation modeling were used. The study's findings show that career development techniques such as training, continuing education, and e-learning considerably favor the executives' quality of work in Chennai's automobile sectors.

R. Naganathan et al. [11] discussed that Corporate English, which is urgently needed, requires students to have strong business vocabulary, idioms, and phrasal verbs (V-IP) during placement. As a result, the pre-final year must be calibrated with this V-IP. The study's goal is to use edutainment to encourage kids to learn without feeling compelled to. Pictionary is, hence, an image game used in Google Classroom to teach V-IP to pupils. Furthermore, this study provides empirical evidence showing that the experimental group's use of the tool produced a substantial difference from the control group's use of the direct technique. With the use of the Google Classroom application and the Pictionary game, this paper's main goal is to teach students business vocabulary so they can improve their language skills and adopt a flipped classroom methodology.

Arasu Raja [24] highlighted that continuous workforce learning is essential across industries, including public sector enterprises competing with their private counterparts. E-learning systems, an emerging technology in India, facilitate ongoing education via the internet and electronic networks. This study aims to assess how e-learning system quality impacts organizational performance. Quota sampling was used to select population samples, and primary research data underwent percentage analysis and structural equation modeling. Results indicated high-quality e-learning systems significantly enhance performance in Chennai's public sector banks. Customized e-learning system design, development, and implementation can thus elevate staff learning, potentially leading to substantial positive growth in these banks' performance, as indicated by this survey's outcomes.

Hesti Kusumaningrum et al. [25] discussed the study to assess the efficacy of the distance education program offered by MORA's Education and Training Center for Administrative Staff, which has been operational for over five years. As a result, an evaluation study is required to ascertain the program's efficacy. It uses a qualitative evaluative research design. The Kirkpatrick Four-Level, which consists of Reactions, Learning, Behavior, and Results, is referred to in the assessment research paradigm. Data were gathered through surveys, document analyses, interviews, and observational studies. The final analysis of this investigation reveals the following conclusions: The distribution of certificates was regarded as unsatisfactory at the reaction level, with the indicator facilities, instructor competency, and administrative service all given good ratings. A good category is given to the degree of learning with the indicator of the instructor's management skills, the study plan's flexibility, and the exam time's flexibility. Participants' test results met expectations in terms of content. According to the participants' degree of conduct, self-learning with electronic modules is preferable to participating in online chats. The final level's results demonstrate that participants' knowledge and abilities have increased as a result of participating in distance education. This study offers suggestions for future work to enhance the effectiveness of online learning.

S K Pulist et al. [26] focused on the Right to Information Act of India, enacted on 15 June 2005, replacing the Freedom of Information Act of 2002, which marked a significant shift toward transparent governance and citizen engagement. Functioning as a tool for addressing grievances, the Act has been utilized by citizens and NGOs to expose instances of corruption. Even the Indira Gandhi National Open University (IGNOU), funded by the federal government, falls under the Act's purview. Investigating IGNOU students' perception of the Act, the study found that they consider it vital for accessing government information with a basic grasp of its rules. Responsible usage was acknowledged, though a few misused it. Implementation challenges included infrastructure, information management, and officer training, emphasizing the need for improved student support and knowledge dissemination pathways.

V. Selvakumar et al. [27] presented that embracing educational technology is inevitable for advancing teaching methods in this era of technology. This study aims to enhance the environment of self-financed engineering institutions in Chennai through e-learning techniques. The research employed a descriptive approach to surveying 150 educators from various colleges in and around Chennai, utilizing a self-administered questionnaire with four sections: personal details, e-learning course information, e-learning practices assessment, and institutional climate evaluation. The analysis utilized frequency assessment, descriptive statistics, and structural equation modeling. Results reveal that mean scores for e-learning practices and institutional climate surpass moderate levels, with all hypothetical relationships from the research model statistically significant at the 1% level. E-learning significantly impacts individuals and institutionally, indicating a strong positive connection between e-learning practices and teaching faculty's perception of the institutional environment, showcasing the substantial positive effect of e-learning tools on faculty members' institutional atmosphere perception in self-financing engineering institutions.

Siti Hajar Halili [28] discussed that leveraging advancements in technology within educational contexts can enhance and invigorate the learning experience, fostering increased student engagement with course materials. Education 4.0 represents a convergence of digital integration into our daily lives, a trend further reinforced by Malaysia's pursuit of higher education reform with a focus on embracing the fourth industrial revolution. Achieving this goal involves reshaping the utilization of technological innovations in teaching and learning. Notable technological progressions under Industry 4.0 encompass 3D printing, augmented reality, virtual reality, cloud computing, holograms, biometrics, multi-touch LCD screens, the IoT, AI, big data, and QR codes, all tailored for educational purposes. This essay delves into the expanding role of technology in education.

Dyandra Maheswari et al. [29] said there are many different ways to go about the educational process. As technology has developed, educational techniques have become more effective than conventional learning techniques. One application that makes use of modern technology to support the educational system is remote learning. Despite these advancements, the current remote learning approach has drawbacks, such as a lack of immersion and involvement. Like IoT, Metaverse is a recent technology advancement in the field of education. Learning can be engaging and immersive through the Metaverse, which integrates the real and virtual worlds. In order to improve learning, this paper investigates IoT integration in the Metaverse. This study used a systematic literature review methodology to find out more about the research concerns and make inferences about them. The results of this study show that IoT and the Metaverse are both advantageous to the educational system and that incorporating IoT into the Metaverse can assist with data mapping to provide a more accurate digital twin outcome and construct an immersive ecosystem for a more productive learning environment.

24.7 CONCLUSION

E-learning in the twenty-first century offers a number of opportunities for expanding education as well as learning, exceeding the limits of traditional classrooms. The question of security and sustainable investment is currently the most challenging situation facing 21st-century e-learning education. To prevent security breaches, paying close attention to security problems, particularly those relating to available, personal, similar, and confidential in the context of M-learning, E-learning, and U-learning. E-learning education is a cutting-edge concept that enables faculty and student contact in a collaborative learning environment and connects teaching and learning to foster intellectual dialogue. communication between the involved parties to give users a sense of security and comfort against espionage. The impact of intelligent networks linking those who are separated from education is enormous. This article highlights the combination of IoT and cloud and their potential influence on raising learner engagement and motivation, accelerating time to mastery, and making education more relevant. However, dependable connectivity and ongoing access must be ensured if the benefits of linking people, data, processors, and objects are to be realized. In order for IoT to be embedded, regulators and faculty must be well versed in possible threats as well as how to utilize them.

REFERENCES

1. Chen, X., Zou, D., Cheng, G., & Xie, H. Detecting latent topics and trends in educational technologies over four decades using structural topic modeling: A retrospective of all volumes of computer & education. Computers & Education. (2020):103855.
2. Hung, Y. -H. Investigating how the cloud computing transforms the development of industries. IEEE Access. (2019);7:181505–181517, doi: 10.1109/ACCESS.2019.2958973
3. Agarwal, P., Sheikh, M. I., & Obaid, A. J. Blockchain and IoT technology in transformation of education sector. (2021);17(12):4–18.
4. Matthew, U. O., Kazaure, J. S., & Haruna, K. Multimedia Information system (MIS) for knowledge generation and ICT policy framework in education: Innovative Sustainable educational investment. International Journal of Information Communication Technologies and Human Development (IJICTHD). (2020);12(3):28–58.
5. Blau, I., Shamir-Inbal, T., & Avdiel, O. How does the pedagogical design of a technology-enhanced collaborative academic course promote digital literacies, self-regulation, and perceived learning of students? The Internet and Higher Education. (2020);45:100722.
6. Boticki, I., Uzelac, N., Dlab, M. H., & Hoić-Božić, N. Making synchronous CSCL work: A widget-based learning system with group work support. Educational Media International. (2020);57(3):187–207.
7. Al-Fraihat, D., Joy, M., & Sinclair, J. Evaluating e-learning systems success: An empirical study. Computers in Human Behavior. (2020);102:67–86.
8. Huda, A. The impact of integrated blended learning technology in higher education studies. (2018), doi: 10.13140/RG.2.2.30205.74726
9. Rico-Bautista, D., Maestre-Gongora, G., & Guerrero, C. D. Smart university: IoT adoption model. Paper presented at the 2020 Fourth World Conference on Smart Trends in Systems, Security and Sustainability (WorldS4), London, UK, (pp. 821–826). (2020), doi: 10.1109/WorldS450073.2020.9210369
10. Barbosa, J. L. V. Ubiquitous computing: Applications and research opportunities. Paper presented at the 2015 IEEE International Conference on Computational Intelligence and Computing Research (ICCIC). (2015).
11. Naganathan, R., & GangaLakshmi, C. E-learning using Pictionary and google classroom-a flipped classroom methodology. The Online Journal of Distance Education and e-Learning. (2019);7(1):15.
12. Margianti, E., & Mutiara, A. (2015). Application of cloud computing in education. doi: 10.13140/RG.2.1.3506.0247
13. Madni, S. H. H., Ali, J., Husnain, H. A., Masum, M. H., Mustafa, S., Shuja, J., Maray, M., & Hosseini, S. Factors influencing the adoption of IoT for e-learning in higher educational institutes in developing countries. Frontiers in Psychology. (2022);13:915596.
14. Siddiqui, S. T., Khan, M. R., Khan, Z., Rana, N., Khan, H., & Alam, M. I. Significance of internet-of-things edge and fog computing in education sector. Paper presented at the 2023 International Conference on Smart Computing and Application (ICSCA), Hail, Saudi Arabia, (pp. 1–6). (2023), doi: 10.1109/ICSCA57840.2023.10087582
15. Felicia, A., Wong, W. K., Loh, W. N., & Juwono, F. H. Increasing role of IoT in education sector: A review of Internet of educational things (IoEdT). Paper presented at the 2021 International Conference on Green Energy, Computing and Sustainable Technology (GECOST) (pp. 1–6). IEEE. (2021, July).
16. Saeed, M. K., Munir, A., Shah, K., Hassan, M., Khan, J., & Nawaz, B. Usage of internet of things (IoT) technology in the higher education sector. Journal of Engineering Science and Technology. (2021);16(5):4181–4191.
17. Asghar, M. Z., Lajis, A., Alam, M. M., Rahmat, M. K., Nasir, H. M., Ahmad, H., Al-Rakhami, M. S., Al-Amri, A., & Albogamy, F. R. A deep neural network model for the detection and classification of emotions from textual content. Complexity. (2022);2022:1–12.

18. Bajracharya, B., Blackford, C., & Chelladurai, J. Prospects of internet of things in education system. Prospects. (2018);6(1):1–7.

19. Kusuma, S., & Viswanath, D. K. IOT and big data analytics in e-learning: A technological perspective and review. International Journal of Engineering & Technology. (2018);7(1):164–167.

20. Matthew, U. O., Kazaure, J. S., & Okafor, N. U. Contemporary development in e-learning education, cloud computing technology & internet of things. EAI Endorsed Transactions on Cloud Systems. (2021);7(20):e3.

21. Wang, J., & Yu, Z. Smart educational learning strategy with the internet of things in higher education system. International Journal on Artificial Intelligence Tools. (2022);31(05):2140101.

22. Alex, V. A comparative study on the challenges in developing leadership style in higher education in five selected countries. The Online Journal of Distance Education and e-Learning. (2019);7(1):1–6.

23. Thangavelu, C., & Kanagasabapathi, J. R. Career development practices through training, continuing education and e-learning and its impact on quality of work life of executives at automobile industries in Chennai City. The Online Journal of Distance Education and e-Learning. (2019);7(1):7.

24. Raja, A. Impact of quality of e-learning systems on performance of public sectors banks in Chennai City. The Online Journal of Distance Education and e-Learning. (2019);7(1):23.

25. Kusumaningrum, H., Sayhrial, Z., & Nuriah, T. Implementing Kirkpatrick model for evaluating distance education in government institution. The Online Journal of Distance Education and e-Learning. (2019);7(1):33.

26. Pulist, S. K. Perception of ODL students towards right to information act 2005: A case of IGNOU, India. The Online Journal of Distance Education and e-Learning. (2019);7(1):41.

27. Selvakumar, M. V., & Maran, K. Role of e-learning practices for teaching faculty on enhancing institutional climate at self-finance engineering colleges at Chennai city. The Online Journal of Distance Education and e-Learning. (2019);7(1):51.

28. Halili, S. H. Technological advancements in education 4.0. The Online Journal of Distance Education and e-Learning. (2019);7(1):63–69.

29. Maheswari, D., Ndruru, F. B. F., Rejeki, D. S., Moniaga, J. V., & Jabar, B. A. Systematic literature review on the usage of IoT in the metaverse to support the education system. Paper presented at the 2022 5th International Conference on Information and Communications Technology (ICOIACT) (pp. 307–310). IEEE. (2022, August).

25 Integration of Artificial Intelligence and Internet of Things in the Banking Sector
A Convergence of Frontiers

Anita Tanwar, Shivani Malhan, Nidhi Punj, and Arvinder Kaur

25.1 INTRODUCTION

Cloud computing has evolved as a game-changing technology transforming how organizations and individuals access and use computing resources. As technology advances, the conventional model of owning and managing on-premises hardware and software is losing way to cloud computing's more flexible, efficient, and cost-effective approach. Cloud computing, at its heart, allows customers to utilize a wide range of computing services and resources, such as data storage, processing power, databases, and applications for software, via the internet. Instead of relying simply on local servers or desktop computers, users can access the massive computational power and storage capacity of cloud service providers' far-away data centers. This paradigm shift from traditional infrastructure to virtualized resources has created a plethora of opportunities for both companies and people to streamline operations, improve collaboration, and drive innovation [1].

Cloud computing is distinguished by fundamental characteristics that distinguish it from traditional IT architecture. These include resource pooling, on demand self-service for customers, widespread network access, quick flexibility, and quantifiable service. Users can utilize these capabilities to provision and manage resources independently, access cloud services from a range of devices and locations, maximize resource use, scale resources up and down as needed, and pay for only the resources they use. A wide variety of services tailored to particular needs are offered via cloud computing. Software as a Service (SaaS) delivers programs over the internet, Infrastructure as a Service (IaaS) offers virtualized computing resources, and Framework as a Service offers a framework for development and deployment. Each service model gives consumers the freedom to pick the level of control they want, whether it's controlling their infrastructure, concentrating only on application development, or leveraging ready-to-use software applications [2].

DOI: 10.1201/9781032656694-25

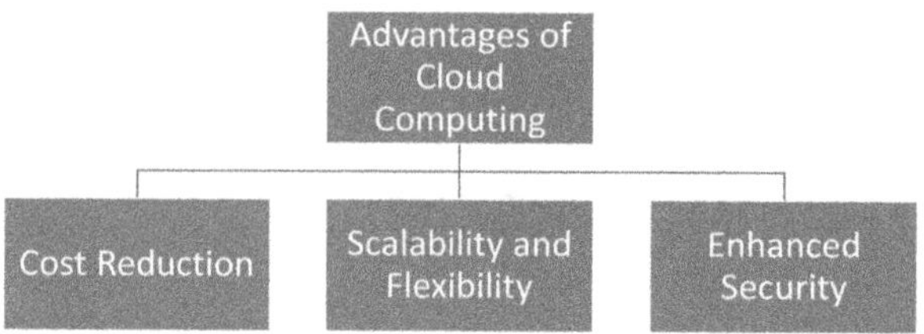

FIGURE 25.1   Advantages of cloud computing.

The advantages, as mentioned in Figure 25.1, of cloud computing are numerous, and they have had a significant impact on enterprises of all sizes and across industries. The main benefit is cost reductions because cloud computing reduces the requirement for substantial up-front expenditures in technology and infrastructure, enabling businesses to use a pay-as-you-go model. The ability to scale resources up or down on demand allows firms to quickly respond to changing needs and market demands, avoiding overprovisioning and underutilization. Cloud computing also promotes cooperation and mobility by enabling users to access resources and apps from any location with an internet connection. This has facilitated remote work, increased productivity, and allowed global teams to collaborate more efficiently. Furthermore, reliable cloud providers give high levels of dependability and availability, which are supported by sophisticated data centers and redundancy measures. These providers use strict security procedures to secure data and resources, often outperforming individual enterprises' security capabilities. As a result, cloud computing has evolved into a reliable and secure platform for storing and processing sensitive information [3].

However, the growing popularity of cloud computing has raised concerns about data security, privacy, and compliance. Data storage in third-party data centers and potential vendor lock-in are two of the obstacles businesses must overcome to gain the benefits of cloud computing while limiting risks. Cloud computing continues to affect how businesses and consumers use technology in this quickly expanding digital world. The route to cloud adoption is a strategic imperative that involves careful planning, rigorous security measures, and compliance with regulatory obligations. Organizations that embrace cloud computing and capitalize on its revolutionary powers position themselves to succeed in a more dynamic and interconnected environment [4]. The financial landscape is experiencing a radical upheaval as a result of the convergence of two major innovations: Cloud computing and AI.

The integration of these technologies has given rise to the concept of "smart banking," which is changing financial organizations' traditional processes and offerings. This chapter examines the disruptive impact of cloud computing and AI on the financial industry, highlighting the benefits, problems, and future prospects of this dynamic synergy. As technology advances, new options for optimizing processes, providing individualized consumer experiences, and making data-driven decisions emerge. Cloud computing allows financial organizations to shift away from traditional on-premises infrastructure while still providing cost-effective and scalable solutions. Meanwhile, artificial intelligence (AI) applications such as tailored customer experiences, fraud detection, credit risk assessment, and investment counseling are revolutionizing how financial institutions connect with clients and manage risks [4]. One paper examines how cloud computing has revolutionized IT, emphasizing access to resources and infrastructure. SaaS, PaaS, and IaaS cloud architectures are divided into several cloud types in the study. With a focus on key players in cloud computing, it analyzes trends and gaps in current research. The work offers useful insights for cloud providers, users, and academics. The main topics of this essay are the growth of cloud computing, its definition, its traits, and the implementation of cloud services. Big data integration into cloud computing has positive development and application potential. The TOE framework has been used to analyze cloud computing adoption in Lebanese organizations. Their research uncovered benefits such as cost reductions and scalability, as well as drawbacks such as security problems. Despite its small sample size and inherent temporal restrictions, the work sheds light on regional cloud adoption dynamics and inspires more research [5].

Throughout this chapter, we will look at the advantages of cloud computing and AI and the problems that financial institutions encounter when implementing these technologies. We will also look at the challenges and future potential of smart banking. Financial institutions may position themselves to stay ahead in an increasingly competitive and technologically driven sector by recognizing the effects and possibilities of cloud computing and AI.

25.2 REVIEW OF LITERATURE

The term "cloud computing" refers to a model that enables users to access a variety of technology services as well as assets over the internet, according to numerous research. Cloud computing differs from conventional IT architecture in several ways, including immediate self-service kiosks, widespread network access, resource sharing, quick adaptability, and measurable service [6, 7]. These features enable customers to set up and handle resources on their own, access cloud-based services from different devices or locations, maximize resource use, and only pay for real usage. Three basic service models in cloud computing have been identified by studies [7]. IaaS offers virtualized computer resources, giving consumers control and management over their infrastructure. Platform as a Service (PaaS) provides a foundation for creation and execution, allowing users to concentrate entirely on application development. SaaS provides an approach to delivering programs via the World Wide Web, allowing customers to access ready-to-use software. Each service arrangement provides management and adaptability to meet the needs of various users. Some research [8, 9] has highlighted the many benefits of cloud computing, including its potential for cost savings. Cloud computing helps companies switch to a pay-as-you-go business model by removing the need for significant up-front hardware and infrastructure investments. A corporation can respond swiftly to shifting customer and market demands while avoiding over- and under-provisioning if it has the ability to scale its resources in response to demand. Cloud computing also enables improved organizational cooperation, agility, and creativity. A concept known as "cloud computing" enables quick, unlimited, and instantaneous network access to a shared repository of computational resources. The importance of cloud architecture in information systems is highlighted, and various aspects are explained in this article. The paper also discusses the difficulties and opportunities that the cloud-based model presents, and it concludes by outlining the key SWOT factors for information delivery techniques. Cloud computing is a pay-as-you-go concept that provides virtualized resources and services for computing in a scalable network environment. The growing acceptance of cloud computing has resulted in its incorporation by a number of worldwide IT corporations and government agencies, frequently for crucial application and data hosting [9].

Several studies [10, 11] have found various cloud security issues. Security breaches, illicit use, and information loss are among the most serious concerns. Cloud-based collaborative infrastructure offers possible dangers, including the "noisy neighbor" impact, in which one tenant's behaviors affect others. Furthermore, adhering to data protection requirements may prove difficult because data may be stored in numerous jurisdictions.

Researchers [11] emphasize the importance of implementing adequate safety precautions and best practices in cloud systems. Encrypting data at rest and in transit is crucial for safeguarding sensitive information. Multi-factor authentication, access control, and identity management all help to prevent unauthorized access. Regular safety and monitoring checks are crucial for swiftly identifying and minimizing risks. The confidentiality and privacy of data are crucial in the cloud. Differential privacy is one strategy suggested by studies [12, 13] to safeguard individual data while allowing for relevant analysis. To strike a compromise between data utility and secrecy, safeguarding privacy data exchange systems and anonymity approaches were additionally investigated.

It is vital to build credibility between providers of cloud services and users. Integrity in cloud computing operations, especially unambiguous SLAs (service level agreements) or reporting of security-related incidents, has been highlighted by studies [13, 14]. Third-party certification and audits by outsiders contribute to trust in cloud companies' security policies.

25.3 UNDERSTANDING CLOUD COMPUTING IN BANKING

25.3.1 What Is Cloud Computing?

Cloud computing shows a paradigm shift in the delivery and management of computing resources. Instead of maintaining on-premises infrastructure, financial institutions store, process, and manage data, applications, and services on distant servers hosted on the internet. This transition enables banks to adopt a more agile, scalable, and cost-effective operating model. Cloud computing is a transformative paradigm that has fundamentally reshaped the manner in which people, corporations, and organizations acquire and employ computing resources. Historically, the field of computing included the possession and upkeep of tangible hardware and software infrastructure, a practice that entailed significant costs, consumed substantial time, and lacked adaptability. On the other hand, cloud computing alters this scenario by offering a virtualized setting in which anyone may access a communal collection of computer resources, including processing capabilities, storage, and networking, over the internet. The on demand self-service capacity is considered a core element of cloud computing. Users have the ability to provide and manage resources independently without the need for direct involvement from service providers. This implies that both people and enterprises have the ability to rapidly adjust their operations to accommodate their requirements, therefore avoiding the time-consuming processes often connected with establishing conventional infrastructure. In addition, cloud computing provides a high degree of expansive network accessibility, allowing users to conveniently access apps, data, and services across various devices equipped with internet connectivity, including laptops, tablets, smartphones, and desktop PCs. The enhanced accessibility of this technology enables users to overcome geographical limitations and facilitates seamless work and collaboration across several places [14].

One of the key advantages of cloud computing is the concept of resource pooling. Cloud providers consolidate computational resources from a diverse array of physical computers and assign them flexibly to cater to different customers' needs. The use of resource optimization leads to enhanced efficiency and less wastage. The notion of quick elasticity is an additional characteristic that distinguishes cloud computing. This implies that resources have the capability to be rapidly adjusted in size, either increased or decreased, to effectively manage varying workloads. Cloud computing guarantees customers with the necessary resources at the exact moment they require them, whether it involves managing heightened traffic during peak hours or downsizing during less busy periods. The economic element of cloud computing is supported by the measured service model. Customers are invoiced according to their real-time use, enabling them to circumvent substantial initial financial outlays. The utilization of a pay-as-you-go strategy results in potential cost reductions, particularly for enterprises that experience fluctuating computing requirements across different periods [15].

Cloud computing comprises three main service models. IaaS is a cloud computing model that gives virtualized computing resources to users. PaaS is a cloud computing service that provides users with an environment for developing and deploying applications. SaaS is a cloud computing model that enables users to access fully working software applications via the internet. Furthermore, cloud computing encompasses several deployment models. Other entities make Public clouds available and accessible to the wider population. Private clouds are exclusively allocated to a singular company, providing enhanced governance and heightened levels of data protection. Hybrid clouds are formed by the integration of public and private clouds, while multi-cloud strategies entail the use of services from several providers to enhance functionality and minimize potential dangers [1].

25.3.2 Advantages of Cloud Computing in Banking [13]

Adoption of cloud computing in the financial sector has various benefits (given in Figure 25.1):

a. Cost Reduction: By eliminating the need for large up-front expenditures in hardware and software, cloud computing allows banks to pay only for the resources they use, resulting in significant cost savings.

b. Scalability and Flexibility: Financial organizations may quickly scale their computer resources up or down based on fluctuating demand, assuring optimal performance during peak periods while achieving savings during slack periods.

c. Enhanced Security: To protect significant financial data from breaches and cyber crimes, reputable cloud service providers deploy comprehensive security measures, frequently beyond the capability of individual banks.

25.3.3 AI REVOLUTIONIZING BANKING SERVICES [14]

Financial institutions may use data-driven insights and automated processes and provide users with individualized experiences. Smart AI banking facilities are mentioned below (Given in Figure 25.2).

25.3.3.1 Personalized Customer Experience

The creation of tailored client experiences via chatbots and virtual assistants is one of the revolutionary implications of AI in the banking sector. AI-powered solutions have the capability to engage with clients in real time, effectively responding to their inquiries, delivering support, and making personalized suggestions. Through the examination of consumer data and the comprehension of behavioral patterns, AI systems have the capability to effectively interact with users in a manner that is tailored to their individual preferences. This personalized approach serves to augment levels of customer satisfaction and foster a sense of loyalty among customers. Chatbots, specifically, provide round-the-clock accessibility and prompt reaction durations, guaranteeing that clients obtain rapid assistance whenever required. The aforementioned degree of customer interaction not only enhances the overall customer experience but also serves to maximize operational efficiency within the banking sector.

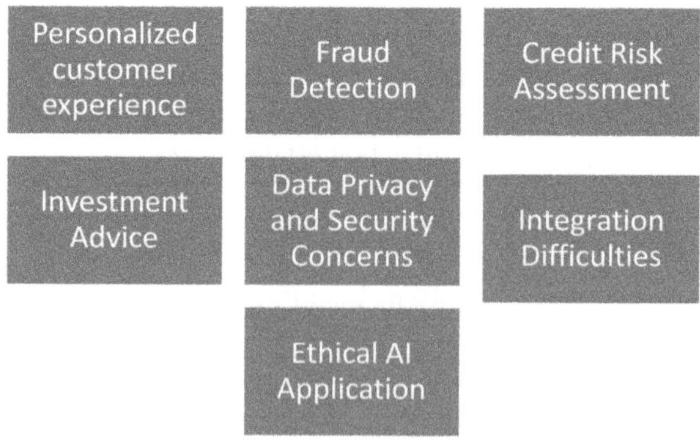

FIGURE 25.2   Smart Banking AI applications.

25.3.3.2 Fraud Detection

The integration of AI inside the banking sector is of paramount importance in fraud detection and prevention. AI systems have a high level of proficiency in efficiently analyzing large volumes of data in real-time, with the purpose of detecting typical patterns and abnormalities that might potentially signify instances of fraudulent behavior. Through continually acquiring fresh data, these systems can adapt to ever-changing fraud methods and improve their accuracy as time progresses. The implementation of a proactive strategy for fraud detection serves the dual purpose of protecting clients' financial assets and upholding the overall integrity and reliability of the banking industry.

25.3.3.3 Credit Risk Assessment

The utilization of machine learning algorithms has brought about a significant transformation in the domain of credit risk assessment within the banking sector. The algorithms employed in this context analyze several datasets, including previous transactions, credit ratings, and economic indicators, to enhance the precision of lending determinations. The utilization of AI-driven credit risk assessment empowers banks to provide loans to a wider spectrum of consumers while mitigating risks by accurately forecasting the probability of loan defaults or delinquencies. The aforementioned phenomenon yields favorable outcomes for both borrowers, granting them the opportunity to get credit. For banks, it enables them to optimize their lending portfolios and mitigate possible losses.

25.3.3.4 Investment Advice

The utilization of AI's data-driven insights has also been included in financial advice services. AI-powered systems utilize advanced algorithms to examine a diverse range of financial data, market movements, news, and customer preferences in order to provide tailored investment suggestions. The suggestions above take into account various risk variables and prospective returns and are in accordance with the investor's financial objectives. AI plays a significant role in enhancing wealth management by empowering customers to make well-informed investment decisions that align with their long-term financial goals.

25.3.3.5 Data Privacy and Security Concerns

The transfer of sensitive financial data to cloud-based platforms presents significant issues in terms of data privacy and security. When collaborating with cloud providers, it is imperative for banks to give utmost importance to the implementation of strong security protocols, encryption systems, and compliance requirements. Maintaining consumer confidence and adherence to regulatory compliance necessitates implementing measures to safeguard data from unauthorized access, breaches, and cyberattacks. It is imperative to collaborate with cloud service providers that offer sophisticated security measures and ongoing monitoring to effectively address these concerns.

25.3.3.6 Integration Difficulties

The process of incorporating cloud-based technology into pre-existing legacy systems is a multifaceted and arduous task for financial institutions. The achievement of seamless integration is of utmost importance in order to prevent any delays in services and guarantee the uninterrupted functioning of essential financial processes. The successful implementation of new cloud solutions in an organization typically necessitates careful and detailed planning, thorough assessments of compatibility, and the creation of suitable middleware to facilitate the integration of these solutions with existing legacy infrastructure. The achievement of a good integration process is characterized by the seamless and consistent flow of data between various systems, hence upholding a superior degree of operational efficiency.

25.3.3.7   Ethical AI Application

AI has a multitude of advantages, yet its use within the finance industry might give rise to ethical considerations. It is imperative for financial institutions to exercise utmost vigilance in effectively resolving instances of algorithmic bias and prejudice that may emerge as a result of skewed training data. The necessity of transparent and explainable AI models lies in their ability to uphold fairness in lending procedures and decision-making. The implementation of ethical principles for deploying AI and the frequent evaluation of AI systems' conduct are crucial measures in upholding customer confidence, complying with regulatory requirements, and mitigating the inadvertent influence of biases on significant financial judgments.

25.3.4   Smart Banking in the Future [16, 17]

25.3.4.1   Hyper-Personalization

The ongoing progress of smart banking will lead to adopting an enhanced degree of hyper-personalization. In this evolutionary process, AI systems will persistently examine individual consumers' distinct preferences, behaviors, and requirements in real-time. The comprehension of this dynamic enables banks to create highly customized experiences that not only augment consumer delight but also cultivate enduring loyalty.

25.3.4.2   AI Expansion that Is Ongoing

The banking sector is now seeing a continuous and dynamic increase in the implementation of AI technology. The trajectory of this phenomenon is closely connected to the ongoing advancements in natural language processing, sentiment analysis, and the complexity of deep machine learning algorithms. These technological breakthroughs facilitate the further integration of AI in the banking sector as financial organizations leverage these developing capabilities to generate more intelligent, adaptable, and customized services. With the maturation of AI technologies, their influence in creating the future of banking is becoming increasingly prominent. This convergence of innovation and customer-centric experiences is expected to have a big impact on the financial sector.

25.3.4.3   Fraud Prevention and Risk Management

In the domain of risk reduction and fraud detection, the predictive capabilities of AI will be of utmost importance. AI enables financial institutions to effectively anticipate potential risks and enhance their security measures, therefore safeguarding both their own interests and those of their clients. Real-time data analysis plays a pivotal role in this undertaking, allowing prompt detection of possible hazards and the adoption of proactive actions to safeguard the security and welfare of financial institutions and their esteemed clientele.

25.3.5 AI's IMPACT ON THE FINANCIAL SECTOR: REVOLUTIONIZING DIGITAL BANKING

From the beginning in the 1950s, AI has dramatically advanced. In the banking sector, technology has evolved from automating routine processes like entering data and calculations to being a game-changer. Conversational AI is being quickly implemented by banks to deliver excellent client experiences, optimize operations, and bolster security measures due to the availability of huge quantities of information and recent technological developments [18].

Giving customers a remarkable customer experience: Conversational AI has been at the cutting edge of this development as AI has come a long way over the years. Customers can have conversational interactions with their bank accounts using this kind of technology since it makes use of natural language processing (NLP). By processing text or speech input, it enables machines to comprehend human language and perform operations like checking balances, transferring money, paying bills, and providing individualized financial advice based on analysis of personal financial data. By using the strength of AI, chatbots that operate are able to learn from prior consumer interactions, thereby enhancing their accuracy and responsiveness. Conversational AI may give consumers prompt, personalized responses by utilizing machine learning models, which employ neural networks similar to those in our brains to accurately analyze complicated information [19]. This makes banking more easy and more available. Banks must, however, make certain that the technology is applied safely and in accordance with privacy laws. It is critical for banks to take advantage of conversational AI's advantages while reducing any potential hazards related to these developments as they become more common in the internet-based banking industry. Enhance operational effectiveness: AI is revolutionizing the banking sector by increasing operational effectiveness and reducing costs. Banks are now able to streamline their processes for making decisions, and analyze massive data sets to forecast outcomes, find patterns, and spot abnormalities thanks to deep learning, a popular AI technique. Banks can meet the unique demands and preferences of their clients thanks to customized products and services. Additionally, banks can evaluate and comprehend consumer requests thanks to the use of NLP technology, which results in customized responses that improve the customer experience. AI-powered chatbots are becoming more and more common in banking because they simplify repetitive operations, shorten wait times, and offer clients quick service. Furthermore, Robotic Process Automation (RPA) is being used in a number of key business operations to automate manual and repetitive work, freeing up people to concentrate on more intelligent and complicated duties. Finally, AI is being used to prevent and detect fraud by analyzing patterns of transactions to spot fraudulent activity and minimize losses [20].

Strengthening data security and compliance: The incorporation of AI in the banking industry to improve data safety and compliance with regulations is an exciting trend. The sector is being changed by the use of emerging innovations such as neural networks, NLP, automated forecasting, and blockchain. By examining data on transaction patterns, banks may now utilize algorithms that use machine learning to detect fraudulent actions. Meanwhile, NLP analyzes unprocessed information to find possible dangers and security issues. By evaluating consumer data and behavior to forecast future trends and potential security issues, predictive analytics provides preventive measures to stop clients from defaulting on loans or credit payments. Last but not least, the application of blockchain technology adds an additional degree of security by guaranteeing transaction integrity and prohibiting unwanted access to confidential data via a decentralized ledger system. We can all agree that AI is at the vanguard of the banking revolution. Forward-looking banks are currently integrating AI into their business processes and are seeing great outcomes. These are a few instances showing how the incorporation of AI improves banking efficiency and provides more comprehensive knowledge than any time before, resulting in higher satisfaction among customers. JPMorgan Chase, a financial institution, is a prime instance of a corporation that uses AI at an enormous scale in its online banking system. The bank utilizes machine learning techniques to examine huge volumes of records in order to correctly identify probable fraud situations. The technology they use assists them in reducing the number of false positives and providing more security for their clients when they utilize online resources. Another instance is Bank of America, which created its digital assistant Erica, which enables voice-enabled operations through mobile devices or home appliances like Amazon's Echo. With the use of input from customers, Erica can assist customers in keeping track of their spending, paying their bills, and receiving tailored financial advice. Wells Fargo utilizes chatbots that are driven by the use of NLP to enhance customer service. These chatbots enable consumers to ask inquiries regarding the status of their accounts, transactions, and various other relevant issues whenever they want, without requiring help from live representatives [21].

Financial organizations that use AI technology may reap many benefits, but they will also face new problems. When introducing AI, financial organizations often encounter the following issues [22]:

The training and decision-making processes of AI systems rely substantially on the availability of high-quality data. Large volumes of data are common in the financial sector, yet this data is typically scattered across many systems, unreliable, or missing entirely. Maintaining reliable, consistent, and easily accessible data can be difficult. Financial information is very private and must comply with stringent privacy and security regulations like GDPR and HIPAA. When using AI, it's important to think about how to protect people's personal information. It's a delicate balancing act to make sure AI systems follow all the rules while still providing useful insights. As an example, many people dismiss deep learning and other forms of AI as "black boxes" due to the difficulty of understanding the reasoning behind their decisions. Financial organizations have a responsibility to stakeholders, regulators, and customers to explain the rationale behind AI-driven choices. Trust can only be established via open communication and clear meaning.

Compliance with Regulations The financial sector is tightly regulated to combat crime such as fraud, money laundering, and terrorist financing. Due to the ever-changing nature of AI technology and laws, ensuring compliance with them may be difficult.

AI models used in finance might mistake and lose efficiency over time as a result of fluctuations in the market. Financial losses and regulatory difficulties may be avoided with the use of well-established model risk management procedures for keeping an eye on, validating, and updating AI models.

Fairness and Bias: Discriminatory results can be produced by AI systems that have inherited biases from their training data. Unfair loan practices and consumer treatment can emerge from biased choices in financial organizations. One of the biggest obstacles to implementing AI is making sure it's fair and not biased in any way.

Problem: a shortage of experts in the fields of machine learning, data science, and specific application domains, which are all necessary for the successful creation and rollout of AI systems. It may be difficult for financial institutions to attract and retain workers with the expertise needed to develop and maintain AI models.

Management of Change: Introducing new AI technology frequently necessitates major adjustments to established procedures, workflows, and job descriptions. Successful adoption can be hampered by factors such as resistance to change, a lack of employee buy-in, and the requirement for training.

While AI has great potential, it is difficult to ensure that it can expand to meet the demands of ever-increasing data volumes and processing loads. Systems should be built to scale up without compromising functionality or speed.

Return on investment (ROI) and expenses AI implementation can be time-consuming and costly due to the need to spend on hardware, software, and support staff. To determine if the effort is worthwhile, financial institutions must weigh the expenses against the potential return on investment.

Many banks and other financial organizations rely on complicated legacy systems that were never intended to integrate with today's AI tools. AI integration may be difficult and requires careful planning to minimize interruptions in these systems.

High levels of uncertainty and volatility are common in the financial markets. It's possible that AI algorithms that have been trained on historical data could fail to adapt to the dynamic nature of the market and provide unexpected results.

IT, data science, compliance, and risk management, among others, must work together strategically to address these difficulties. The framework in which AI may be securely and successfully used in financial institutions is shaped in large part by regulatory bodies.

25.3.6 The Impact of Azure in Increasing AI Usage in Banking: Changing the Next Generation of Banking [23, 24]

With the help of AI and machine learning (ML) methods, the banking and finance industry is going through an exciting revolution. On-premise solutions are prohibitively costly because the setup calls for a lot of computing and storage capacity. It's a good thing that the use of cloud computing has become so revolutionary. Banks are able to hold and manage huge quantities of data fast and effectively thanks to nearly limitless data processing and storage capacity. As a bank's requirements for AI rise, so can its use of cloud resources thanks to the unparalleled elasticity of its cloud-based systems. The highlight, though? The pay-as-you-go price structure offered by cloud-based services makes them economical because it does not require large initial investments in infrastructure or technology. Additionally, security remains of utmost importance in the banking industry. For the greatest degree of protection, cloud service providers offer strong features like encryption, authentication, and access control. With the power of the cloud at their disposal, banks are confidently employing algorithms based on ML and AI to obtain insights, enhance processes, and provide unmatched experiences for customers while upholding the highest standards of security. Thanks to Microsoft's cloud-based computing system, Azure artificial intelligence, a collection of AI tools for programmers and data researchers, has been instrumental in changing the BFSI sector. The usage of AI has been made possible by Azure, which has created countless opportunities. Through straightforward calls to the API, users can quickly access superior AI models for voice, language acquisition, making choices, and vision. In addition, cutting-edge tools like Jupyter Notebooks for Java & Visual Studio Code (VS Code) enable talented individuals to build their own models for ML. Additionally, the possibility for development has increased thanks to the accessibility of open-source tools like PyTorch and TensorFlow that offer limitless possibilities. Whether a bank wants to use ethical AI techniques or a company wants to use data to drive choices, Azure AI delivers the adaptability and innovative studies that make everything happen.

The following is a list of Azure AI services and tools that facilitate the creation and implementation artificial intelligence solutions [25]:

ML in Azure: Azure ML is a service offered in the cloud that lets programmers and data scientists create, test, and use large-scale ML models. Deep learning, automated neural networks, and model management are just a few of the advantages it provides.

Cognitive Solutions: Pre-built AI models are available through Azure Cognition Services, which can be quickly incorporated into apps to carry out functions like sentiment analysis, image and audio recognition, and language interpretation.

Databricks for Azure: This analytics system, which is based on Apache Spark, makes it quick, simple, and collaborative for engineers and data scientists to work together on AI and big data initiatives. It offers an integrated analytics framework for business intelligence, AI, and data preparation.

Analytics in Azure Synapse: Big data and data warehousing are combined in this analytics solution. This tool allows users to run complex analytics queries on enormous datasets while creating an environment for linking data, management, and preparations.

AI in Azure: Language comprehension, image recognition, making choices, and identifying anomalies are among the AI services included in this collection. These tools are made to make it easy and quick for programmers to create intelligent applications.

25.3.7 THE IMPLEMENTATION OF AI IN THE BANKING SECTOR POSES SEVERAL CHALLENGES [26, 27]

There is great potential for AI to improve banking in terms of both efficiency and the quality of service provided to customers. But there are a lot of issues that need to be thought out and planned for.

When introducing AI, assuring the quality and integration of data is a major problem for financial institutions. A massive amount of information, kept in a wide variety of forms and systems, is produced by banking activities every day. Consolidating, cleaning, and harmonizing this data is essential for efficient AI implementation. Inaccurate AI results can be the result of low-quality data, undermining the very benefits AI is meant to provide. Data privacy and security are of even greater importance in the banking industry due to the sensitive nature of customers' personal information. Compliance with data protection rules like GDPR and HIPAA is essential since AI applications process and analyze sensitive personal and financial data. Finding a happy medium between using client data for personalized marketing and protecting that data from breaches is an ongoing problem.

While AI's prediction powers are impressive, they are not completely safe from the biases inherent in past data. This can lead to biased judgments with serious moral consequences. Maintaining AI fairness calls for constant vigilance, algorithmic improvements, and the creation of rules to avoid biased results. Beyond biases, the social impact of AI on financial inclusion and economic fairness are also important ethical considerations in the field of artificial intelligence. One of the many challenging aspects of using AI in banking is maintaining regulatory compliance. There is a complex network of rules that the industry must follow, and AI models must be transparent and auditable in order to be trusted. For purposes of explaining AI judgments to regulators, customers, and internal stakeholders, it is crucial that these decisions be interpretable. Maintaining AI's complexity while achieving transparency is a fine balancing act. In order to fully realize the benefits of AI, financial institutions must have access to a qualified workforce that can design, build, and support AI systems. However, the market suffers from an AI talent gap, making competition for jobs extremely tough. Financial institutions will need to come up with creative solutions to the problem of attracting and retaining top AI professionals by providing competitive pay, interesting work, and room for professional development [28].

There may be some initial pushback from personnel used to the status quo when AI technology is introduced into the banking industry's workflows. Successfully implementing organizational change calls for widespread education, open lines of communication, and teamwork. Workers should be made aware of how AI may help them do their jobs better. The importance of scalability grows with the widespread use of AI models. There are considerable technological obstacles to ensuring that AI systems can accurately and efficiently manage the volume and complexity of real-world data. If scalability concerns aren't resolved, performance and the quality of the client experience might suffer. Any plan to adopt AI must take cost into account. It may be necessary to spend a sizable sum on initial setup costs, infrastructure, personnel training, and upkeep. Financial institutions need to do detailed cost-benefit studies to show how AI helps them and their clients. Overall, the banking industry's road toward incorporating AI has enormous promise but also enormous hurdles. A comprehensive strategy involving technological expertise, ethics, regulatory compliance, personnel acquisition, and effective change management is required to successfully navigate these obstacles. Successful banks will be able to capitalize on AI-driven innovation while mitigating risk and prioritizing their customers' best interests [29, 30].

25.4 CONCLUSION

A new era of banking called "smart banking" has begun as a result of the fusion of cloud computing and AI. By enabling financial institutions to provide unmatched consumer experiences, streamline their business processes, and confidently make data-driven choices, this potent confluence has radically changed the financial landscape. Financial institutions must adopt these game-changing technologies if they want to succeed in the quickly developing field of smart banking.

The emergence of smart banking and the revolutionary effects of cloud computing and AI have changed how financial institutions function. As a result of its cost-effectiveness, scalability, and strong security, cloud computing has become a crucial component of contemporary banking operations. In the meantime, AI applications have transformed financial services by enabling tailored customer experiences, more proficient fraud detection, accurate credit risk assessment, and cutting-edge investment advice services.

25.5 FUTURE IMPLICATIONS OF THE RESEARCH

Adopting cloud computing and AI, however, comes with risks and problems. In addition to the challenges of effortlessly integrating these innovations into current systems, the security and confidentiality of information need to be carefully considered. Financial organizations must adopt and uphold ethical AI procedures because they are most important when using AI.

Banking organizations must carefully plan their shift to smart banking, work efficiently with cloud service providers, and prioritize ethical AI principles if they are to successfully negotiate these hurdles. Only through this will they be able to fully utilize these technologies and protect the information and confidence of those they serve.

Future smart banking developments are expected to be more fascinating. As AI tools analyze real-time client data to provide customized experiences that increase satisfaction and brand loyalty, hyper-personalization will become the norm. Additionally, the predictive power of AI will transform the management of risks and prevention of fraud, guaranteeing that financial companies are well-equipped to efficiently confront new threats.

As a result, the financial sector is being transformed by smart banking, which is powered by cloud computing and AI. Financial institutions may prosper and stay at the vanguard of this technology transformation by adopting these innovations wholeheartedly, confronting issues head-on, and preserving ethical standards. Smart banking's continued development will result in even more significant breakthroughs that improve consumer experiences and propel the financial sector to success.

REFERENCES

1. Angwin, J., Larson, J., Mattu, S., & Kirchner, L. (2016). Machine bias: There's software used across the country to predict future criminals. And it's biased against blacks. ProPublica.
2. Dangwal, A., Kukreti, M., Angurala, M., Sarangal, R., Mehta, M., & Chauhan, P. (2023, March). A review on the role of artificial intelligence on tourism. In 2023 10th International Conference on Computing for Sustainable Global Development (INDIACom) (pp. 164–168). IEEE.
3. Ahmad, A., Qasim, R., Aljohani, N. R., & Baig, F. (2019). A survey of cloud computing security management. Journal of Computing and Security, 11(2), 101–117.
4. Bertsekas, D. P., & Gallagher, R. G. (2017). Data Networks (2nd ed.). Prentice Hall.
5. Buyya, R., Srirama, S. N., & Casale, G. (2018). Cloud Computing: Principles and Paradigms (2nd ed.). Wiley.
6. Chang, L. M., Yeh, C. T., & Lin, H. C. (2017). Security analysis of cloud computing based on cloud computing security threats. Security and Communication Networks, 10(11), 2495–2500.
7. Davenport, T. H., & Ronanki, R. (2018). Artificial intelligence for the real world. Harvard Business Review, 96(1), 108–116.
8. Armbrust, M., Fox, A., Griffith, R., Joseph, A. D., Katz, R., Konwinski, A., & Zaharia, M. (2010). A view of cloud computing. Communications of the ACM, 53(4), 50–58.
9. Buyya, R., Yeo, C. S., Venugopal, S., Broberg, J., & Brandic, I. (2009). Cloud computing and emerging IT platforms: Vision, hype, and reality for delivering computing as the 5th utility. Future Generation Computer Systems, 25(6), 599–616.
10. Mell, P., & Grance, T. (2011). The NIST definition of cloud computing. National Institute of Standards and Technology, 53(6), 50.
11. Rimal, B. P., Jukan, A., & Katsaros, D. (2009). Architectural requirements for cloud computing systems: An enterprise cloud approach. In 2009 15th IEEE International Conference on Parallel and Distributed Systems (pp. 243–250). IEEE.
12. Brenner, S., Garbers, B., & Kapitza, R. (2014). Adaptive and scalable high availability for infrastructure clouds. In Proceedings of the 14th IFIP WG 6.1. International Conference on Distributed Applications and Interoperable Systems (pp. 16–30, vol 8460). Springer, New York.
13. Rittinghouse, J. W., & Ransome, J. F. (2016). Cloud Computing: Implementation, Management, and Security. CRC Press.
14. Mather, T., Kumaraswamy, S., & Latif, S. (2009). Cloud Security and Privacy: An Enterprise Perspective on Risks and Compliance. O'Reilly Media, Inc..
15. Rahman, M. S., Khan, S. U., & Kowalkiewicz, M. (2018). A systematic review of cloud computing tools and technologies. Computers & Electrical Engineering, 70, 329–344.

16. Xu, H., Fu, X., & Yao, D. D. (2017). A survey of network anomaly detection techniques. Journal of Network and Computer Applications, 60, 19–31.
17. Deng, R. H., Chen, L., Lei, B., & Thuraisingham, B. (2016). A survey of privacy-preserving data sharing and mining in cloud computing. Security and Communication Networks, 9(18), 5135–5154.
18. Jensen, M., & Schwenk, J. (2019). Differential privacy in cloud-based applications. IEEE Cloud Computing, 6(2), 12–21.
19. Sun, H., Zeadally, S., & Wang, J. (2019). Cloud service level agreements: A survey. IEEE Access, 7, 129718–129732.
20. Sood, K. K., Sood, P., & Sharma, S. K. (2019). Trust-based security analysis of cloud service-level agreements. Future Generation Computer Systems, 96, 298–307.
21. Diakopoulos, N. (2016). Accountability in algorithmic decision making. Digital Journalism, 4(6), 614–626.
22. Feng, X., & Hu, J. (2019). Hyper-personalization in financial services: A conceptual framework. Frontiers in Business Research in China, 13(1), 1–23.
23. Grabis, J., & Wang, S. (2019). Fintech: The impacts of AI, big data, cloud computing, blockchain, and cybersecurity. Journal of Business and Economic Policy, 6(1), 8–15.
24. Jin, X., & Yoon, J. (2019). AI-driven customer experience: The future of personalization in banking. Journal of Financial Services Marketing, 24(1), 29–39.
25. Mone, C. T., & Munir, R. (2018). Artificial intelligence and big data analytics in the finance industry: Implications for organizational decision-making. In Innovations in Financial and Business Analytics: Implications for the Future (pp. 1–24). IGI Global.
26. Rajabi, S., Mousakhani, M., Parizi, R. M., & Baradaraneh, Y. (2021). AI for risk management in banks. In The Road to Autonomous Finance (pp. 13–33). Springer.
27. Tiwari, R., & Bhatt, D. K. (2020). Cloud computing implementation in banks: A systematic literature review and future research directions. International Journal of Bank Marketing, 38(5), 1163–1195.
28. Rana, A., Bansal, R., & Gupta, M. (2022). Big data: A disruptive innovation in the insurance sector. In Big Data Analytics in the Insurance Market (pp. 165–183). Emerald Publishing Limited.
29. Zhong, S., Huang, S., & Zhang, J. (2018). A review of cloud computing: Design challenges in architecture and security. Journal of Cloud Computing, 7(1), 1–26.
30. Vinoth, S., Vemula, H. L., Haralayya, B., Mamgain, P., Hasan, M. F., & Naved, M. (2022). Application of cloud computing in banking and e-commerce and related security threats. Materials Today: Proceedings, 51, 2172–2175.

26 From Data to Decisions
Cloud, IoT, and AI Integration

Prerna and Sanjay Sharma

26.1 INTRODUCTION: BACKGROUND AND DRIVING FORCES

Cloud computing is the linchpin of IoT, enabling scalable, secure, and real-time operations. Its role extends beyond mere data storage and processing, encompassing resource optimization, accessibility, and security enhancements [1]. This synergy between cloud computing and IoT represents a transformative force that underpins the digital transformation of industries and societies, shaping a future characterized by efficiency, innovation, and connectivity.

Cloud-IoT-AI integration represents the convergence and synergy of three transformative technological domains: Cloud computing, the Internet of Things (IoT), and artificial intelligence (AI). This integration seeks to harness the collective power of these domains to unlock new possibilities, insights, and efficiencies in various industries and applications. Let's delve into the core elements of cloud-IoT-AI integration:

A. Cloud Computing:
  i.   Scalable Infrastructure: Cloud computing provides a scalable and flexible data storage, processing, and computation infrastructure. It allows organizations to allocate computing resources based on demand dynamically.
  ii. Accessibility: Cloud services are accessible from anywhere with an internet connection, enabling remote data management and processing [2].
  iii. Resource Optimization: Cloud platforms offer tools for optimizing resource usage, cost control, and efficient scaling of applications.

B. IoT:
  i.   Connected Devices: IoT refers to the network of interconnected physical devices, sensors, and everyday objects capable of collecting and exchanging data over the Internet.
  ii. Data Generation: IoT devices generate vast amounts of data, including sensor readings, environmental data, and user interactions, creating a continuous stream of information [3].
  iii. Real-time Monitoring: IoT enables real-time monitoring of assets, environments, and processes, allowing immediate responses and decision-making.

DOI: 10.1201/9781032656694-26

C. AI:

    i.   Data Analysis: AI encompasses a range of techniques, including machine learning and deep learning, which can analyze large datasets to discover patterns, make predictions, and automate decision-making.

   ii.   Pattern Recognition: AI algorithms excel at recognizing complex patterns and anomalies in data, which is particularly valuable when dealing with the diverse and dynamic data generated by IoT devices.

  iii.   Automation: AI can automate tasks, reducing human intervention and improving operational efficiency [4].

Cloud IoT Integration: Cloud platforms serve as the backbone for IoT applications, providing the infrastructure needed to collect, store, and process data generated by IoT devices. IoT devices are equipped with sensors and communication modules that send data to the cloud in real-time, where it can be analyzed and acted upon. Cloud IoT integration ensures that data from IoT devices is securely and reliably transmitted to the cloud, enabling centralized data management and analysis [5].

AI in Cloud IoT Integration: AI algorithms are applied to the data collected from IoT devices within the cloud infrastructure. AI-driven analytics can identify data patterns, anomalies, and trends, allowing organizations to gain valuable insights [6]. AI models can make real-time predictions, automate decision-making processes, and trigger actions or alerts based on the data analysis.

Edge Computing in Cloud-IoT-AI Integration: Edge computing extends this integration by introducing local data processing and AI capabilities at the network's edge, closer to the IoT devices. This approach reduces latency, as data can be processed locally, and only relevant information is transmitted to the cloud for further analysis [7]. Edge devices can make rapid decisions without relying solely on cloud-based AI models, which is essential for applications requiring real-time responses.

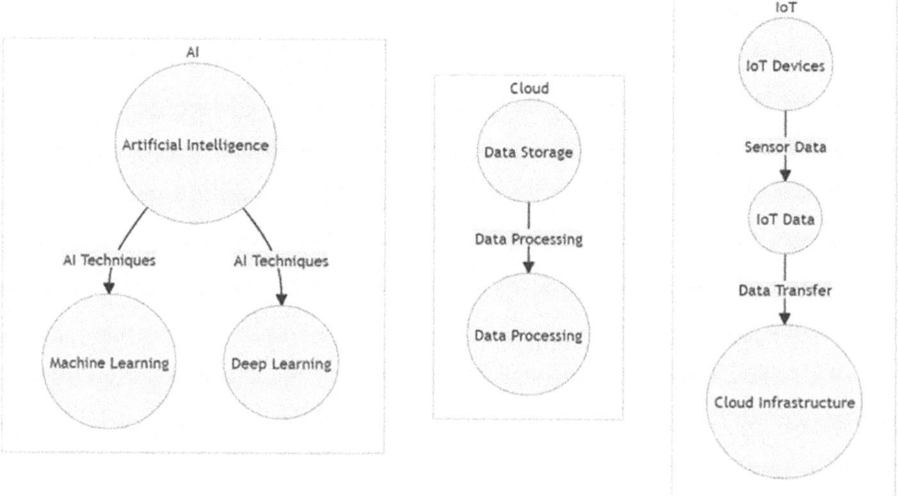

FIGURE 26.1 AI-cloud-IoT integration.

26.2 CLOUD COMPUTING IN IoT: ENABLING SCALABLE, SECURE, AND REAL-TIME OPERATIONS

The synergy between cloud computing and the IoT has given rise to a transformative technological paradigm that redefines the way we interact with the digital world. At the core of this convergence lies cloud computing, a ubiquitous and versatile technology that facilitates the efficient operation and management of IoT ecosystems.

26.2.1 ROLE OF CLOUD COMPUTING IN IoT

Cloud computing is the linchpin of IoT ecosystems, providing a robust and multi-faceted foundation for their operation. Its role can be delineated through several key attributes (refer Table 26.1):

- **Data Centralization:** IoT devices, often resource-constrained, generate copious volumes of data. Cloud computing is the central repository for this data, ensuring it is collected, processed, and stored efficiently [8].
- **Scalability:** IoT deployments can vary widely in size and complexity. Cloud computing offers the ability to scale resources dynamically to accommodate fluctuations in data volume and processing demands, ensuring that IoT systems remain responsive and adaptive.
- **Accessibility:** By leveraging cloud services, IoT data and applications become accessible from anywhere with an internet connection. This accessibility fosters remote monitoring, control, and management of IoT devices and systems [1].
- **Resource Offloading:** Cloud infrastructure offloads computational and storage burdens from IoT devices, enabling them to conserve energy and extend

---

TABLE 26.1

Key Attributes of Cloud-IoT-AI Integration

Row	Integration Components	Data-Driven Transformation	Edge Computing in IoT	Ethical and Security Considerations	Societal Impact and Collaboration
1.	Cloud Computing	Shift toward data-driven	Significance of edge	Challenges in convergence	Societal impact on smart cities
2.	IoT	Unlocking data potential	Real-time decision-making	Ethical considerations	Interdisciplinary collaboration
3.	AI	Massive IoT data influx	Reduced latency	Data security	Precision agriculture impact
4.	Edge Computing	Advanced analytics	Local data processing	Responsible AI use	Addressing pressing challenges
5.	Societal Impact	Synergies among domains	Real-time operations	Transparency in AI decisions	Industry reshaping opportunity

their operational lifespan. This is especially crucial in battery-powered IoT sensors.

- **Data Analytics:** Cloud-based data analytics and machine learning algorithms can extract actionable insights from IoT data, facilitating predictive maintenance, anomaly detection, and pattern recognition.

26.2.2 THE NEED FOR SCALABLE AND SECURE INFRASTRUCTURE

Scalability and security are paramount considerations in the IoT landscape, and cloud computing addresses these needs effectively:

- **Scalability:** IoT ecosystems are characterized by their dynamic nature, subject to rapid expansion and contraction. Scalable cloud infrastructure ensures IoT deployments can grow or shrink seamlessly, adapting to evolving requirements.
- **Resource Allocation:** Cloud services enable efficient resource allocation, ensuring that computing resources are provisioned on demand. This optimizes resource utilization and minimizes operational costs [2].
- **Security:** Security is an existential concern in IoT, where vulnerabilities can lead to dire consequences. Cloud providers invest heavily in security measures, offering robust authentication, encryption, access controls, and threat detection, bolstering the security of IoT data and communications.
- **Data Privacy:** Compliant with stringent data privacy regulations, cloud infrastructure ensures the protection of sensitive IoT data, mitigating the risk of data breaches and unauthorized access.
- **Real-time Data Processing and Storage:**

Real-time operations are imperative in IoT, and cloud computing plays a pivotal role in enabling real-time data processing and storage:

- **Low-Latency Processing:** Cloud platforms have high-performance computing resources facilitating low-latency data processing. This is vital for time-sensitive IoT applications like autonomous vehicles and industrial automation [2].
- **Data Streams:** Real-time data streams from IoT devices can be ingested and processed in the cloud. This enables immediate responses to critical events, such as triggering alerts or initiating automated actions.

26.3 CLOUD IoT INTEGRATION: TECHNICAL ASPECTS, DATA TRANSFER PROTOCOLS, AND REAL-TIME DATA SYNCHRONIZATION

The convergence of cloud computing and the IoT represents a pivotal moment in the evolution of technology. This integration marries the ubiquity and computational power of the cloud with the sensory capabilities of IoT devices, thereby enabling the seamless flow of data between the physical and digital realms [9].

- **Technical Aspects of Cloud IoT Integration:**
  Cloud IoT integration encompasses various technical components and processes crucial for the harmonious operation of the combined ecosystem:
  - **IoT Devices and Sensors:** IoT devices, embedded with sensors, collect data from the physical world. These devices are typically equipped with communication modules to transmit data to the cloud.
  - **Cloud Infrastructure:** Cloud providers offer scalable infrastructure that includes data centers, servers, and storage facilities. This infrastructure serves as the centralized hub for collecting, processing, and storing IoT data.
  - **Connectivity:** IoT devices rely on various communication technologies such as Wi-Fi, cellular networks, LPWAN (Low-Power Wide Area Network), and MQTT (Message Queuing Telemetry Transport) to transmit data to the cloud [10].
  - **Data Ingestion:** Cloud services facilitate the ingestion of IoT data, which involves receiving, validating, and storing data from multiple devices in a structured format.
  - **Data Processing:** Once data is ingested, cloud-based data processing pipelines execute data transformations, analytics, and machine learning algorithms to derive insights and initiate actions.

26.4 DATA TRANSFER PROTOCOLS AND STANDARDS

Data transfer in cloud IoT integration relies on established protocols and standards to ensure interoperability, security, and efficiency:

- **MQTT (Message Queuing Telemetry Transport):** MQTT is a lightweight, efficient, and widely adopted publish-subscribe protocol used for real-time communication between IoT devices and the cloud. Its low overhead makes it suitable for constrained IoT devices [10].
- **HTTP/HTTPS:** The Hypertext Transfer Protocol (HTTP) and its secure variant (HTTPS) are used for data transfer over the Internet. These protocols are commonly used for RESTful APIs and web services.
- **CoAP (Constrained Application Protocol):** CoAP is designed for resource-constrained IoT devices and provides a lightweight alternative to HTTP for data transfer.
- **AMQP (Advanced Message Queuing Protocol):** AMQP is a messaging protocol that supports message queuing and asynchronous communication, making it suitable for IoT applications requiring reliable and ordered message delivery.
- **IoT Standards:** IoT standards such as OPC UA (Unified Architecture), DDS (Data Distribution Service), and O-MI (Open Messaging Interface) provide standardized communication and data exchange mechanisms for IoT devices [11].

26.5   CHALLENGES AND SOLUTIONS IN REAL-TIME DATA SYNCHRONIZATION

Real-time data synchronization between IoT devices and the cloud introduces complex challenges:

- **Latency:** Latency in data transfer can hinder real-time applications. Edge computing, where data is processed locally before being sent to the cloud, reduces latency.
- **Data Volume:** The sheer volume of data generated by IoT devices can overwhelm networks and cloud infrastructure. Compression and data filtering can mitigate this challenge [12].
- **Data Security:** Securing data in transit is paramount. The use of encryption and secure communication protocols safeguards data during transmission.
- **Interoperability:** Ensuring compatibility between diverse IoT devices and cloud platforms requires adherence to standardized protocols and open interfaces.
- **Reliability:** Real-time applications demand reliable data transfer. Implementing redundant communication paths and error-handling mechanisms enhances reliability.
- **Scalability:** Scalability challenges can arise as the number of IoT devices grows. Cloud providers offer auto-scaling capabilities to address this issue.

  Cloud IoT integration relies on a multifaceted technical architecture encompassing IoT devices, cloud infrastructure, connectivity, data transfer protocols, and standards [13]. Though fraught with challenges, real-time data synchronization is vital for applications that require immediate responses and decision-making. By navigating these complexities and leveraging established protocols, the integration of cloud and IoT unlocks transformative capabilities that drive innovation, efficiency, and connectivity across diverse industries.
- **Historical Data Storage:** Cloud storage solutions enable long-term retention of historical IoT data, facilitating trend analysis, compliance reporting, and retrospective investigations.

26.6   BENEFITS OF CLOUD-IoT-AI INTEGRATION

- Data-Driven Insights: Integration enables organizations to extract valuable insights from IoT data, leading to data-driven decision-making [14].
- Efficiency: Automation powered by AI reduces manual intervention, streamlining processes and reducing operational costs.
- Real-time Response: The combination of cloud, IoT, and AI allows for real-time monitoring and immediate responses to events and anomalies.
- Scalability: Cloud resources can be easily scaled to accommodate growing IoT deployments and increasing data volumes.
- Innovation: Integration drives innovation in various domains, from smart cities and healthcare to manufacturing and agriculture.

26.7 IoT DATA AND AI

In the rapidly evolving landscape of the IoT, data constitutes the lifeblood that flows through the digital veins of interconnected devices and sensors. Harnessing the full potential of this voluminous and diverse data requires applying advanced analytical techniques, and herein lies the profound significance of data analytics in IoT. Moreover, integrating AI into this ecosystem elevates the process from mere analysis to intelligent decision-making and transformative insights.

26.7.1 Importance of Data Analytics in IoT

IoT, often likened to a vast sensor network, generates an unprecedented stream of data that encompasses diverse formats and dimensions. This data is a treasure trove of information waiting to be unearthed, provided it is effectively captured, processed, and analyzed. Data analytics in IoT is not merely an option but a necessity, and its importance reverberates across several dimensions:

- **Operational Efficiency:** Data analytics optimizes operations by providing real-time insights into device performance, allowing for predictive maintenance and resource allocation. This efficiency extends to industries like manufacturing, where downtime is minimized through proactive fault detection.
- **Cost Reduction:** Efficient resource utilization, informed by data analytics, leads to significant cost reductions, as organizations can make informed decisions regarding energy consumption, supply chain logistics, and asset utilization [15].
- **Enhanced User Experience:** In consumer IoT, data analytics facilitates personalization and improves user experiences through tailored recommendations and responsive services.
- **Data-Driven Decision-Making:** Data analytics empowers organizations to make data-driven decisions, resulting in informed strategies, risk mitigation, and performance improvements.

26.8 VARIOUS AI TECHNIQUES IN IoT

With its diverse set of techniques, AI augments the analytical capabilities of IoT by adding a layer of intelligence capable of extracting meaningful insights from the deluge of data. Several AI techniques come to the forefront in this context:

- **Machine Learning (ML):** ML algorithms, particularly supervised and unsupervised learning, find applications in IoT for tasks such as anomaly detection, predictive maintenance, and classification. They adapt to changing data patterns, making them suitable for dynamic IoT environments [16].
- **Deep Learning:** Deep neural networks excel in processing unstructured data, such as images and audio, making them indispensable for IoT applications like surveillance, autonomous vehicles, and voice recognition systems.

- **Natural Language Processing (NLP):** NLP techniques enable IoT devices to understand and respond to human language, thus enhancing user interaction and enabling voice-controlled IoT ecosystems.

26.9   HOW AI DERIVES INSIGHTS FROM IoT DATA

AI's prowess lies in its ability to discern patterns, anomalies, and correlations within IoT data, transforming raw information into actionable insights. The process of deriving insights involves several key stages:

- **Data Preprocessing:** Before analysis, IoT data is preprocessed to clean noise, handle missing values, and normalize data for consistency. This ensures the quality and reliability of subsequent insights [17].
- **Feature Engineering:** AI models leverage feature engineering techniques to extract relevant information and create meaningful input variables for analysis. This step is crucial for accurate predictions and classifications.
- **Model Training:** AI models like neural networks are trained on historical IoT data. During training, these models learn patterns, relationships, and dependencies within the data.

Inference and Prediction: Once trained, AI models can make real-time inferences and predictions based on new incoming IoT data. For example, they can predict equipment failures, recommend actions, or identify abnormal behavior.

In essence, the marriage of IoT data and AI techniques transforms IoT systems from passive data collectors into intelligent decision-makers capable of adapting to changing conditions and optimizing operations. This symbiosis can revolutionize industries, drive innovation, and enhance the quality of life through a more connected and intelligent world [17].

26.10   EDGE COMPUTING AND AI: REDEFINING DATA PROCESSING AT THE NETWORK'S EDGE

The fusion of edge computing and AI stands at the forefront of technological innovation, heralding a paradigm shift in data processing and decision-making. Edge computing, characterized by decentralized data processing near the source of data generation, and AI, which empowers machines to learn, reason, and act intelligently, converge to reshape industries and redefine how data is harnessed and transformed. This scholarly exploration delves into the definitions, significance, operational mechanisms, and illustrative use cases of edge computing integrated with AI, underscoring their transformative potential [18].

26.10.1   DEFINING EDGE COMPUTING AND ITS SIGNIFICANCE

Edge computing signifies the dissemination of computational capabilities from centralized cloud data centers to the network's periphery, or "edge." It enables data processing and analytics closer to the data source, reducing latency and enhancing real-time responsiveness. Its significance in the convergence with AI can be summarized as follows:

- **Latency Reduction:** Edge computing circumvents the inherent latency of cloud-based processing by performing computations locally. This is vital for applications demanding instantaneous decision-making, such as autonomous vehicles and industrial automation [19].
- **Bandwidth Optimization:** By processing data at the edge, superfluous data transmissions to the cloud are minimized, conserving bandwidth and reducing operational costs, particularly in scenarios with bandwidth constraints.
- **Data Privacy and Security:** Edge computing enhances data privacy and security by keeping sensitive data within a localized environment. This mitigates risks associated with data transit and storage in the cloud.
- **Scalability:** Edge computing platforms can scale horizontally to accommodate a growing number of edge devices and handle increasing data volumes efficiently.
- **Resilience:** Decentralization and processing distribution improve system resilience, ensuring continued operation even in network interruptions or cloud service outages.

26.10.2 OPERATIONAL MECHANISM OF EDGE DEVICES WITH AI

Edge devices with AI capabilities operate through a well-defined process encompassing the following stages:

- **Data Collection:** Edge devices equipped with sensors and data acquisition components collect data from their local environment. This data can include sensory inputs, images, audio, or other relevant information [20].
- **Local Data Processing:** On-board AI models or edge computing platforms process the collected data locally. These AI models may include machine learning algorithms, deep neural networks, or specialized AI chips (e.g., GPUs or TPUs).
- **Inference and Decision-Making:** Edge AI models make real-time inferences and decisions based on the data after processing. For instance, the AI model might analyze sensor data in an autonomous drone to adjust flight parameters.
- **Action or Data Transmission:** Depending on the inference, the edge device can initiate an action, such as controlling a machine, issuing an alert, or transmitting processed data to a centralized system or cloud for further analysis [21].

26.10.3 EXAMPLES OF INDUSTRIES BENEFITING FROM EDGE AI

- **Manufacturing:** In smart factories, edge AI is used for quality control, predictive maintenance, and process optimization. Cameras equipped with AI can detect defects in real time, reducing production waste.

- **Healthcare:** Edge AI enables real-time analysis of patient data from wearable devices, facilitating early disease detection and personalized treatment recommendations.
- **Autonomous Vehicles:** Edge AI powers self-driving cars by processing sensor data (e.g., lidar, cameras) locally for rapid decision-making, ensuring passenger safety.
- **Agriculture:** Smart agricultural equipment equipped with edge AI can optimize crop management, detect crop diseases, and automate irrigation, increasing yields and resource efficiency.
- **Retail:** In retail, edge AI is used for real-time inventory management, customer behavior analysis, and personalized shopping experiences.
- **Energy:** Edge AI in energy grids monitors power consumption patterns, detects anomalies, and automatically adjusts energy distribution for optimal efficiency and reliability [22].

The convergence of edge computing and AI heralds a transformative era in data processing and decision-making. Edge computing's significance lies in its ability to reduce latency, optimize bandwidth, enhance security, and ensure real-time responsiveness. Edge devices with AI capabilities operate through data collection, local processing, inference, and action, enabling intelligent decision-making at the network's edge [23]. Industries across the spectrum, from manufacturing and healthcare to autonomous vehicles and agriculture, are reaping the benefits of this convergence, ushering in a new era of efficiency, safety, and innovation.

26.11 CASE STUDIES AND PRACTICAL EXAMPLES: UNVEILING THE TRANSFORMATIVE POWER OF CLOUD-IoT-AI INTEGRATION

In the annals of technological innovation, the triumvirate of cloud computing, the IoT, and AI emerges as a revolutionary convergence, promising to reshape industries and societies. As we embark on this scholarly exploration, we unveil real-world scenarios across diverse industries, demonstrating the tangible benefits of cloud-IoT-AI integration. These case studies illuminate how organizations have harnessed this synergy to achieve operational efficiency and cost reduction while paving the way for innovation and optimization.

- **Smart Healthcare:**

    In the realm of healthcare, cloud-IoT-AI integration is orchestrating a paradigm shift. Medical wearables, such as smartwatches and health-monitoring devices, continuously gather patient data, from vital signs to activity levels. This data, transmitted to the cloud, undergoes AI-driven analysis for early disease detection and personalized healthcare recommendations. The benefits are twofold:
    - **Improved Patient Outcomes:** Early detection of health anomalies and predictive insights enable timely interventions, improving patient outcomes and reducing hospital readmissions.

- **Cost Reduction:** Preventive care and remote monitoring reduce the burden on healthcare systems, lowering hospital stays and emergency intervention costs.
- **Manufacturing Excellence:**
  Manufacturing has embraced cloud-IoT-AI integration to optimize operations and enhance productivity. In smart factories, IoT sensors monitor equipment performance, detecting anomalies and predicting maintenance needs. These insights are processed in real time through edge computing or cloud-based AI. The advantages are conspicuous:
  - **Reduced Downtime:** Predictive maintenance minimizes equipment downtime, ensuring continuous production and significant cost savings.
  - **Quality Enhancement:** AI-powered quality control systems identify defects with high precision, reducing waste and enhancing product quality.
- **Connected Transportation:**
  The transportation industry is transforming with the fusion of the cloud, IoT, and AI. Autonomous vehicles rely on real-time sensor data processed at the edge, with cloud-based AI providing route optimization and decision-making support. The outcomes are transformative:
  - **Safety:** Cloud-IoT-AI integration enhances road safety by detecting and responding to hazards with split-second precision, reducing accidents and associated costs.
  - **Efficiency:** Smart traffic management optimizes traffic flow, reducing congestion and fuel consumption and leading to lower operational costs.
- **Agricultural Revolution:**
  In agriculture, integrating cloud, IoT, and AI fosters precision farming. IoT sensors monitor soil conditions, crop health, and weather, while cloud-based AI models provide actionable insights. The results are striking:
  - **Increased Yields:** Data-driven insights enable optimized planting, irrigation, and pest control, leading to increased crop yields and reduced resource wastage.
  - **Sustainability:** Cloud-IoT-AI integration promotes sustainable farming practices, reducing environmental impact and long-term operational costs.
- **Retail Personalization:**
  Retailers leverage Cloud-IoT-AI integration to enhance customer experiences. IoT devices track shopper behavior, and cloud-based AI analyzes this data to offer personalized product recommendations. The benefits include:
  - **Higher Sales:** Personalized recommendations increase sales and customer loyalty, compensating for IoT and AI technology investment.
  - **Inventory Optimization:** Real-time inventory management reduces overstocking and stockouts, lowering carrying costs and losses.

The tangible benefits of cloud-IoT-AI integration resonate across industries, from healthcare and manufacturing to transportation, agriculture, and retail. These case studies underscore the potential for operational efficiency and cost reduction, highlighting how organizations use this convergence to innovate, optimize, and remain competitive in an increasingly connected world. As one traverses this transformation landscape, these real-world examples stand as beacons illuminating the boundless possibilities of cloud-IoT-AI integration.

26.12 CHALLENGES AND ETHICAL CONSIDERATIONS IN INTEGRATING CLOUD IoT AND AI

The integration of cloud computing, the IoT, and AI heralds a new era of technological advancement. However, with this integration come significant challenges and ethical considerations that demand attention.

- **Privacy and Data Security Concerns:**
  - **Data Proliferation:** IoT devices generate vast amounts of personal and sensitive data. The challenge lies in ensuring that this data is handled with utmost care and doesn't fall into the wrong hands.
  - **Data Encryption:** Strong encryption mechanisms must be employed for data both in transit and at rest to safeguard it from unauthorized access or breaches.
  - **Consent and Control:** Users' consent should be sought for data collection and processing. Transparent data control mechanisms must be provided, allowing users to have control over their data.
  - **Data Localization:** Regulations governing data localization can pose challenges. Ensuring compliance with varying data residency requirements is crucial [24].

26.13 RESPONSIBLE AI USE AND TRANSPARENCY

- **Bias and Fairness:** AI models can inadvertently perpetuate biases present in training data. Ethical considerations demand ongoing scrutiny to mitigate these biases and ensure fairness.
- **Explainability:** AI models often operate as black boxes. Ensuring transparency by implementing explainable AI techniques is vital to understanding decisions.
- **Accountability:** Assigning accountability for AI-driven decisions is challenging. Clear lines of responsibility must be established to address issues or errors.
- **Regulatory Compliance:** Compliance with evolving regulations, such as GDPR in Europe or CCPA in California, requires continuous monitoring and adjustments in AI practices [25].

26.14 BEST PRACTICES FOR MITIGATING ETHICAL CHALLENGES

- **Data Governance:** Develop robust policies outlining data collection, storage, and sharing practices while ensuring compliance with relevant regulations.

- **Privacy by Design:** Incorporate privacy and security measures from the inception of IoT devices and AI models, adhering to principles like Privacy by Design.
- **Ethical AI Training:** Prioritize ethical considerations during AI model training by scrutinizing training data for bias and implementing fairness-aware algorithms.
- **Explainable AI:** Use solvable AI models that offer transparency into decision-making processes, allowing users to understand and trust AI-driven outcomes.
- **User Consent:** Obtain explicit user consent for data collection and processing, clearly articulating the purposes and sharing practices and offering options for data deletion.
- **Ethics Committees:** Establish ethics committees or advisory boards to review and provide guidance on the ethical implications of AI and IoT initiatives.
- **Continuous Monitoring:** Implement constant monitoring and auditing processes to ensure compliance with ethical standards and regulatory requirements.

Integrating cloud IoT and AI holds immense potential for innovation, efficiency, and societal advancement. However, it must be approached sincerely about its ethical considerations and challenges. Privacy, data security, responsible AI use, and transparency are non-negotiable aspects of this integration. Organizations and policymakers must collaborate to ensure that ethical best practices are adopted and evolve with technological advancements. By addressing these ethical considerations, one can harness the full potential of cloud IoT and AI integration while upholding the principles of privacy, security, fairness, and accountability.

26.15 SOCIETAL IMPACT AND FUTURE TRENDS OF INTEGRATING CLOUD IoT AND AI

The integration of cloud computing, the IoT, and AI is a technological convergence that transcends mere innovation; it is a transformative force poised to reshape societies and industries.

- **Potential Societal Impact:**
  - **Enhanced Quality of Life:** Cloud-IoT-AI integration fosters the creation of smart cities, where interconnected devices optimize traffic flow, reduce energy consumption, and enhance public services. This leads to improved urban living conditions and a higher quality of life.
  - **Healthcare Revolution:** Remote monitoring of patients through IoT devices and AI-driven diagnostics enable early disease detection and personalized healthcare. This not only saves lives but also reduces the burden on healthcare systems.
  - **Precision Agriculture:** In agriculture, cloud-IoT-AI integration transforms traditional farming into precision agriculture. Smart sensors

and AI-driven insights enhance crop yields, conserve resources, and address food security challenges.

- **Environmental Sustainability:** AI-driven predictive analytics and IoT sensors help manage and reduce environmental impact. From monitoring air quality to optimizing energy consumption, this convergence supports sustainable practices.
- **Accessibility and Inclusivity:** Cloud-IoT-AI can provide solutions for persons with disabilities. Smart homes and AI-powered assistive technologies improve accessibility and inclusivity for all [26].
- **Addressing Global Challenges:**
  - **Smart Cities:** Cloud-IoT-AI integration plays a pivotal role in addressing the challenges of rapid urbanization. It enables efficient traffic management, waste reduction, energy conservation, and the creation of sustainable urban environments.
  - **Precision Agriculture:** In a world facing growing food demands and climate change, precision agriculture ensures efficient resource use, minimizes waste, and boosts agricultural productivity.
  - **Healthcare Access:** IoT-enabled telemedicine and AI-powered diagnostics extend healthcare access to remote or underserved regions, bridging healthcare gaps and improving global health outcomes.
  - **Environmental Conservation:** IoT sensors and AI analytics help monitor ecosystems, detect natural disasters, and manage resources, contributing to global efforts to combat climate change and protect biodiversity.

26.16 FUTURE TRENDS AND DEVELOPMENTS

- **Edge AI:** Edge computing combined with AI will become more prevalent, enabling real-time decision-making at the device level, reducing latency, and ensuring privacy.
- **AI Ethics:** Ethical considerations in AI will gain prominence. Regulatory frameworks and guidelines will emerge to address issues like bias, fairness, and transparency.
- **5G Integration:** The rollout of 5G networks will accelerate the adoption of cloud-IoT-AI integration, providing the high-speed, low-latency connectivity required for real-time applications.
- **IoT Ecosystem Expansion:** IoT ecosystems will expand to include a broader range of devices, from household appliances to industrial equipment, further connecting the physical world to the digital realm.
- **AI Augmentation:** Human-AI collaboration will intensify. AI will augment human capabilities in decision-making, creative tasks, and problem-solving across various domains.
- **Cross-Industry Integration:** Industries will increasingly leverage insights from one another. For example, healthcare solutions may find applications in smart cities and vice versa.
- **Quantum Computing:** As quantum computing matures, it will unlock new possibilities for AI algorithms, enabling the processing of complex data sets at previously deemed unattainable speeds.

Cloud-IoT-AI integration is poised to shape our societies and address pressing global challenges. Its potential impact on various aspects of life, from healthcare and agriculture to environmental sustainability and accessibility, is profound. As this convergence continues to evolve, future trends suggest greater connectivity, ethical considerations, and the augmentation of human capabilities through AI [27]. One can chart a course toward a more connected, efficient, and inclusive world by harnessing the transformative power of cloud-IoT-AI integration.

26.16.1 Interdisciplinary Collaboration: Fostering Innovation in Integrating Cloud IoT and AI

Integrating cloud computing, the IoT, and AI represents a technological frontier that demands the convergence of expertise across diverse domains.

- **Importance of Collaboration:**
  - **Cross-Domain Expertise:** The successful integration of cloud-IoT-AI necessitates expertise spanning cloud infrastructure, IoT device development, data analytics, machine learning, and more. Collaboration facilitates the exchange of knowledge and insights across these domains.
  - **Complex Problem-Solving:** The challenges posed by this convergence are multifaceted, demanding interdisciplinary perspectives to solve complex problems effectively. Collaboration brings together experts who can address these challenges comprehensively.
  - **Holistic Solutions:** Interdisciplinary collaboration fosters the development of holistic solutions. Combining cloud resources, IoT data, and AI algorithms leads to comprehensive, intelligent systems that offer greater value than the sum of their parts.
- **Synergies Created by Integration:**
  - **Data Enrichment:** IoT devices generate massive datasets. Cloud-based AI can analyze this data to derive meaningful insights, enhancing decision-making in real-time applications like autonomous vehicles and industrial automation.
  - **Edge Intelligence:** The synergy between edge computing (IoT) and cloud-based AI enables real-time decision-making, reducing latency and optimizing resource usage in applications like smart grids and healthcare monitoring.
  - **Scalability:** Cloud infrastructure scales dynamically, accommodating IoT deployments of varying sizes. AI models can be trained and deployed in the cloud, supporting IoT devices regardless of their processing capabilities.
  - **Predictive Maintenance:** Integrating IoT sensors with cloud-based AI models allows for predictive maintenance in manufacturing, where anomalies in machinery can be detected early, reducing downtime and maintenance costs.
- **Encouragement for Innovation:**
  - **Research Initiatives:** Encourage readers to engage in research initiatives exploring cloud-IoT-AI integration's uncharted territories.

Interdisciplinary research teams can investigate novel use cases and solutions.

- **Industry Collaborations:** Advocate for partnerships between academia and industry to bring cutting-edge research into practical applications. Collaborative projects can drive innovation and accelerate adoption.
- **Education and Training:** Promote education and training programs that prepare individuals to work at the intersection of cloud computing, IoT, and AI. These programs empower future innovators to bridge the knowledge gaps.
- **Open Collaboration:** Encourage open collaboration through forums, conferences, and knowledge-sharing platforms. Open-source projects can facilitate the development of shared resources and best practices [28].

26.17 CONCLUSION: UNLEASHING THE TRANSFORMATIVE POWER OF CLOUD-IoT-AI INTEGRATION

Cloud-IoT-AI integration represents a frontier of technological innovation with profound implications for industries and societies. Interdisciplinary collaboration is not merely advantageous but essential in unlocking the full potential of this convergence. The synergies created by uniting experts from cloud computing, IoT, and AI domains lead to holistic solutions that comprehensively address complex challenges. By actively engaging in collaborative efforts and contributing to innovation in this field, readers can play a pivotal role in shaping a future characterized by connectivity, efficiency, and transformative technological advancements. The convergence of cloud computing, the IoT, and AI heralds a technological revolution of unprecedented proportions.

- **Key Takeaways:**
  - **Synergy of Convergence:** Cloud-IoT-AI integration is more than the sum of its parts. The synergy of these domains unlocks transformative potential across industries, from healthcare and manufacturing to transportation and agriculture.
  - **Societal Impact:** The convergence brings forth profound societal impact, enhancing quality of life, revolutionizing healthcare, promoting environmental sustainability, and addressing global challenges such as urbanization, food security, and accessibility.
  - **Ethical Considerations:** The integration comes with ethical considerations, including privacy, data security, bias mitigation, transparency, and responsible AI use. Adhering to ethical best practices is crucial for harnessing the convergence's benefits while minimizing risks.
  - **Interdisciplinary Collaboration:** Collaboration among cloud computing, IoT, and AI experts is pivotal. It facilitates cross-domain knowledge exchange, complex problem-solving, and the development of holistic solutions that drive innovation.
  - **Future Trends:** Anticipated trends include the rise of edge AI, increased focus on AI ethics, integration with 5G networks, expansion

of IoT ecosystems, AI augmentation of human capabilities, and the advent of quantum computing.

- **The Transformative Potential:** The transformative potential of cloud-IoT-AI integration cannot be overstated. It has the power to redefine industries, revolutionize processes, and enhance the well-being of societies worldwide. In healthcare, it means timely disease detection and personalized treatment. In manufacturing, it signifies optimized production and reduced downtime. In smart cities, it promises efficient resource utilization and improved urban living. In precision agriculture, it ensures sustainable farming practices and increased crop yields. These are but a glimpse of the possibilities this convergence holds.

- **Inspiration for Exploration and Contribution:** The realm of cloud-IoT-AI integration offers fertile ground for exploration and contribution. Those who seek to advance knowledge and innovation should embrace the spirit of inquiry and innovation. Engaging in interdisciplinary collaborations that bridge knowledge gaps and create synergies is paramount. Initiating research endeavors that reveal novel use cases and solutions is essential. Leveraging open collaboration platforms to disseminate knowledge and best practices is encouraged. Collaborating with industry partners to translate cutting-edge research into practical applications is a strategic approach. Individuals who aspire to be catalysts for change should assume the role of architects of progress. They can shape the future through their contributions to this dynamic convergence. By doing so, they position themselves at the vanguard of technological advancement and assume a pivotal role in shaping a world characterized by connectivity, efficiency, and transformative technological achievements. In the ever-evolving landscape of technology, the convergence of cloud computing, IoT, and AI serves as a testament to human ingenuity and the relentless pursuit of progress. Scholars, innovators, and visionaries should wholeheartedly embrace this convergence enthusiastically. Through their collective efforts, they have the potential to unleash its full transformative capacity and pave the path toward a brighter, more interconnected, and more inclusive future.

REFERENCES

1. K. Paranjape, M. Schinkel, and P. Nanayakkara, "Short keynote paper: Mainstreaming personalized healthcare-transforming healthcare through a new era of artificial intelligence," *IEEE Journal of Biomedical and Health Informatics*, vol. 24, no. 7, pp. 1860–1863, 2020. doi:10.1109/jbhi.2020.2970807
2. S. Khanam, S. Tanweer, and S. Khalid, "Artificial intelligence surpassing human intelligence: Factual or hoax," *The Computer Journal*, vol. 64, no. 12, pp. 1832–1839, 2020. doi:10.1093/comjnl/bxz156
3. L. Jiao *et al.*, "Brain-inspired remote sensing interpretation: A comprehensive survey," *IEEE Journal of Selected Topics in Applied Earth Observations and Remote Sensing*, vol. 16, pp. 2992–3033, 2023. doi:10.1109/jstars.2023.3247455

4. B. Hu, Z.-H. Guan, G. Chen, and C. L. Chen, "Neuroscience and network dynamics toward brain-inspired intelligence," *IEEE Transactions on Cybernetics*, vol. 52, no. 10, pp. 10214–10227, 2022. doi:10.1109/tcyb.2021.3071110

5. X. Ren, and Y. Chen, "How can artificial intelligence help with space missions - A case study: Computational intelligence-assisted design of space tether for payload orbital transfer under uncertainties," *IEEE Access*, vol. 7, pp. 161449–161458, 2019. doi:10.1109/access.2019.2951136

6. S. Saravanan *et al.*, "Explainable artificial intelligence (EXAI) models for early prediction of Parkinson's disease based on spiral and wave drawings," *IEEE Access*, vol. 11, pp. 68366–68378, 2023. doi:10.1109/access.2023.3291406

7. M. Nazar, M. M. Alam, E. Yafi, and M. M. Su'ud, "A systematic review of human–computer interaction and explainable artificial intelligence in healthcare with artificial intelligence techniques," *IEEE Access*, vol. 9, pp. 153316–153348, 2021. doi:10.1109/access.2021.3127881

8. X. Xu *et al.*, "Artificial intelligence for edge service optimization in internet of vehicles: A survey," *Tsinghua Science and Technology*, vol. 27, no. 2, pp. 270–287, 2022. doi:10.26599/tst.2020.9010025

9. G. Y. Odongo, R. Musabe, D. Hanyurwimfura, and A. D. Bakari, "An efficient lora-enabled smart fault detection and monitoring platform for the power distribution system using self-powered IoT devices," *IEEE Access*, vol. 10, pp. 73403–73420, 2022. doi:10.1109/access.2022.3189002

10. R. Ortigueira, J. A. Fraire, A. Becerra, T. Ferrer, and S. Cespedes, "Ress-IoT: A scalable energy-efficient MAC protocol for direct-to-satellite IoT," *IEEE Access*, vol. 9, pp. 164440–164453, 2021. doi:10.1109/access.2021.3134246

11. K. Fizza, P. P. Jayaraman, A. Banerjee, N. Auluck, and R. Ranjan, "Iot-QWatch: A novel framework to support the development of quality-aware autonomic IoT applications," *IEEE Internet of Things Journal*, vol. 10, no. 20, pp. 17666–17679, 2023. doi:10.1109/jiot.2023.3278411

12. S. N. Swamy, and S. R. Kota, "An empirical study on system level aspects of internet of things (IoT)," *IEEE Access*, vol. 8, pp. 188082–188134, 2020. doi:10.1109/access.2020.3029847

13. N. Neshenko, E. Bou-Harb, J. Crichigno, G. Kaddoum, and N. Ghani, "Demystifying IOT security: An exhaustive survey on IoT vulnerabilities and a first empirical look on internet-scale IOT exploitations," *IEEE Communications Surveys & Tutorials*, vol. 21, no. 3, pp. 2702–2733, 2019. doi:10.1109/comst.2019.2910750

14. Y. Liu, W. Yu, W. Rahayu, and T. Dillon, "An evaluative study on IoT ecosystem for smart predictive maintenance (IOT-SPM) in manufacturing: Multiview requirements and data quality," *IEEE Internet of Things Journal*, vol. 10, no. 13, pp. 11160–11184, 2023. doi:10.1109/jiot.2023.3246100

15. J. Hwang, L. Nkenyereye, N. Sung, J. Kim, and J. Song, "IoT service slicing and task offloading for edge computing," *IEEE Internet of Things Journal*, vol. 8, no. 14, pp. 11526–11547, 2021. doi:10.1109/jiot.2021.3052498

16. Y. Li *et al.*, "Toward location-enabled IOT (Le-IoT): IoT positioning techniques, error sources, and error mitigation," *IEEE Internet of Things Journal*, vol. 8, no. 6, pp. 4035–4062, 2021. doi:10.1109/jiot.2020.3019199

17. F. Meneghello, M. Calore, D. Zucchetto, M. Polese, and A. Zanella, "IoT: Internet of threats? A survey of practical security vulnerabilities in real IOT devices," *IEEE Internet of Things Journal*, vol. 6, no. 5, pp. 8182–8201, 2019. doi:10.1109/jiot.2019.2935189

18. S. Khan *et al.*, "Towards an applicability of current network forensics for cloud networks: A SWOT analysis," *IEEE Access*, vol. 4, pp. 9800–9820, 2016. doi:10.1109/access.2016.2631543

19. D. Yao, C. Yu, L. T. Yang, and H. Jin, "Using crowdsourcing to provide QoS for mobile cloud computing," *IEEE Transactions on Cloud Computing*, vol. 7, no. 2, pp. 344–356, 2019. doi:10.1109/tcc.2015.2513390

20. N. Sultan, "Knowledge management in the age of cloud computing and web 2.0: Experiencing the power of disruptive innovations," *IEEE Engineering Management Review*, vol. 41, no. 4, pp. 98–108, 2013. doi:10.1109/emr.2013.2288159

21. K. Lee, "Comments on 'Secure data sharing in cloud computing using revocable-storage identity-based encryption," *IEEE Transactions on Cloud Computing*, vol. 8, no. 4, pp. 1299–1300, 2020. doi:10.1109/tcc.2020.2973623

22. B. Zhou, A. V. Dastjerdi, R. N. Calheiros, S. N. Srirama, and R. Buyya, "MCloud: A context-aware offloading framework for heterogeneous Mobile cloud," *IEEE Transactions on Services Computing*, vol. 10, no. 5, pp. 797–810, 2017. doi:10.1109/tsc.2015.2511002

23. F. Fowley, C. Pahl, P. Jamshidi, D. Fang, and X. Liu, "A classification and comparison framework for cloud service brokerage architectures," *IEEE Transactions on Cloud Computing*, vol. 6, no. 2, pp. 358–371, 2018. doi:10.1109/tcc.2016.2537333

24. X. Ren *et al.*, "Ai-bazaar: A cloud-edge computing power trading framework for ubiquitous AI services," *IEEE Transactions on Cloud Computing*, pp. 1–13, 2022. doi:10.1109/tcc.2022.3201544

25. M. T. Islam, S. Karunasekera, and R. Buyya, "Performance and cost-efficient spark job scheduling based on deep reinforcement learning in cloud computing environments," *IEEE Transactions on Parallel and Distributed Systems*, vol. 33, no. 7, pp. 1695–1710, 2022. doi:10.1109/tpds.2021.3124670

26. I. Petri *et al.*, "Market models for federated clouds," *IEEE Transactions on Cloud Computing*, vol. 3, no. 3, pp. 398–410, 2015. doi:10.1109/tcc.2015.2415792

27. S. N. Shirazi, A. Gouglidis, A. Farshad, and D. Hutchison, "The extended cloud: Review and analysis of Mobile edge computing and fog from a security and resilience perspective," *IEEE Journal on Selected Areas in Communications*, vol. 35, no. 11, pp. 2586–2595, 2017. doi:10.1109/jsac.2017.2760478

28. S. Chaisiri, B.-S. Lee, and D. Niyato, "Optimization of resource provisioning cost in cloud computing," *IEEE Transactions on Services Computing*, vol. 5, no. 2, pp. 164–177, 2012. doi:10.1109/tsc.2011.7

27 Comprehensive Review of Recent Trends, Challenges, Applications, and Case Studies in Fog Computing

Tisha Aggarwal, Yogesh Lohumi, Duragaprasad Gangodkar, and Prakash Srivastava

27.1 INTRODUCTION

The term "fog computing" was originally coined by the industry to describe the primary constructive idea behind it: fog occupies a position between the ground and the data centers, where user devices are situated, as shown in Figure 27.1. A term frequently employed in conjunction is "edge computing," which denotes tasks positioned at the network's edge in contrast to the cloud. It should be noted that the term "edge" can refer to several architectural levels. In a technical context, "edge" often pertains to nodes in a construction site and is situated on premises with the user, such as being integrated into a network gateway or a machine controller [1]. Fog computing can be carried out collectively on either one fog computing node or several nodes. This approach provides redundancy and enhances scalability and flexibility, allowing for the incorporation of additional fog nodes when increased computing power is required. Fog computing and cloud computing share many ideas since conducting activities using mechanisms like virtualization and containerization is possible. Research into fog computing has experienced exponential growth, driven by numerous studies published across various domains. The examination of existing papers enables the identification and categorization of challenges associated with fog computing, thus illuminating the key issues that impede the widespread adoption and deployment of fog computing solutions. Evaluating the methodologies and findings garnered in diverse studies assists in recognizing common themes and determining the reliability of the findings presented in the reviewed papers [2–4]

Internet of Things (IoT) application development presents a number of challenges for integrated cloud computing (CC), including performance, security, latency, and network interruptions. Fog computing has arisen in response to these worries, bringing CC closer to the IoT space [5]. Fog computing's primary function is to make it easier to analyze and store the data that IoT devices generate at the edge. Instead of sending the data to a cloud server, these processes would be carried out locally at

DOI: 10.1201/9781032656694-27

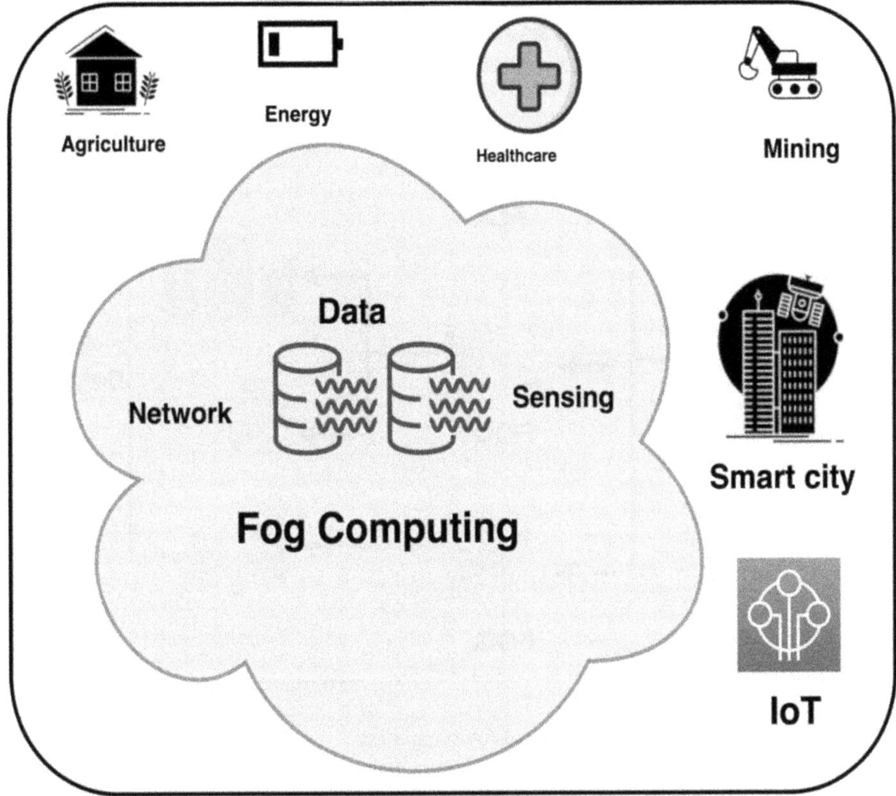

FIGURE 27.1 Integration of fog computing and IoT applications.

the fog node. Fog computing delivers services with higher quality and faster reaction times than typical CC. Consequently, a potential approach for allowing the IoT to offer effective and highly secure services to its plethora of clients is fog computing. By positioning them closer to devices, at the network edge, or in accordance with Service Level Agreements (SLAs), it enables service management and resource allocation outside of the CC [3, 6]. Fog computing architectures represent the structural designs and frameworks used to deploy and manage fog computing infrastructure, as shown in Figure 27.2. The architecture aims to evenly deploy computing assets and services near the network edge, enabling efficient data processing, low latency, and improved performance [7, 8].

27.2 FOG COMPUTING ARCHITECTURE AND TECHNOLOGIES

27.2.1 HIERARCHICAL ARCHITECTURE

Fog computing resources are structured in a hierarchical architecture, consisting of multiple layers organized in a tiered manner. Each layer serves specific functions and purposes. Cloud, fog, and edge are typically the three subcategories of the layers.

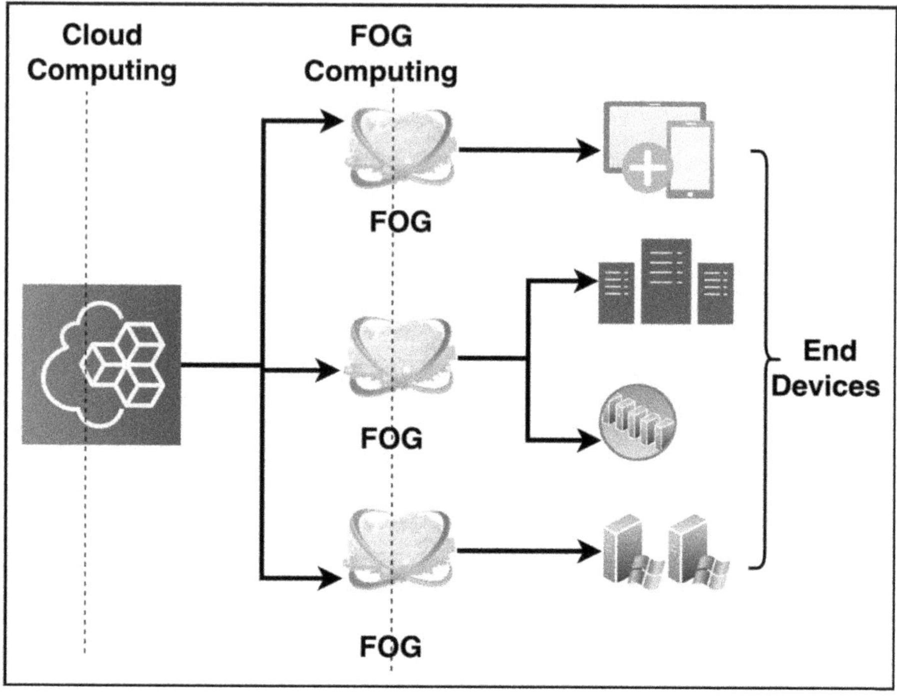

FIGURE 27.2   Generalized structure of fog computing.

The cloud layer addresses the incorporated cloud foundation, which handles weighty computational assignments and capacity. Localized processing and data aggregation are provided by intermediate fog nodes in the fog layer located close to edge devices. The edge devices that generate data and perform initial processing make up the edge layer. Examples of these devices include IoT devices or sensors. In a hierarchical system, data, and computation flow from the edge to the fog layer and then the cloud layer [9, 10].

27.2.2 FLAT ARCHITECTURE

In fog computing, a flat architecture is characterized by a network of widely dispersed fog nodes interconnected without a strict hierarchy [11]. Mist hubs are deployed in various locations within the organization. Mist computing is a highly decentralized layer where processing occurs directly on end devices, such as sensors, ensuring ultra-low latency. Each mist hub can communicate directly with other mist hubs, edge devices, and the cloud. Mist computing sits between fog and cloud computing, bringing processing closer to the network's edge. Fog nodes work together in this architecture to divide processing and storage tasks efficiently. Level structures advance flat adaptability, adaptation to non-critical failure, and low-inertness correspondence. Each haze hub at a level of engineering has the capacity to offer types of assistance and execute applications freely [12].

27.2.3 Hybrid Architecture

A hybrid architecture uses the advantages of both flat and hierarchical architecture. It incorporates different layers, such as the various leveled designs, yet underscores shared correspondence and joint effort, such as level engineering. The mixed model attempts to create some kind of harmony between incorporated control and decentralized activities. It could involve a mix of fog nodes with varying capabilities, from lightweight edge devices to fog servers that are more powerful. The half-and-half design gives asset distribution and navigation adaptability, permitting the ideal utilization of accessible assets across the mist figuring framework [13]. Edge devices are the network's endpoints, typically IoT devices or sensors close to the data source or at the edge of the network. In fog computing environments, these devices are the entry point for data collection and processing and generate a significant amount of data. Fog nodes are intermediate computing entities between the edge devices and the cloud or data center. Data filtering, aggregation, and pre-processing of running applications or services closer to the data source are all responsibilities of fog nodes. By offloading computational tasks from edge devices and performing distributed processing, they are crucial in reducing latency, network congestion, and bandwidth utilization. Edge investigation alludes to the method involved in performing information examinations and determining experiences at the edge of the organization. By leveraging analytics capabilities at the edge, fog computing systems can make real-time decisions and respond quickly to dynamic situations without relying heavily on cloud connectivity. Fog computing relies on various communication technologies to facilitate seamless connectivity and data transfer between cloud, fog nodes, and edge devices [14].

Cellular networks and wireless communication technologies, like Wi-Fi, Bluetooth, Zigbee, etc., enable reliable and effective data transport in areas with fog computing [13].

Fog computing environments require efficient and reliable communication protocols and standards to facilitate seamless data transmission, interoperability, and collaboration between fog nodes, edge devices, and the cloud.

27.3 ADVANCEMENTS IN FOG COMPUTING

Recent research and development advancements in fog computing have advanced significantly in various areas, including architecture design, resource management, security, data analytics, and application-specific implementations [15]. Researchers have suggested innovative architectural approaches to address the special needs of fog computing. This includes adaptive designs that dynamically change the composition of the fog nodes based on demand and resource availability, as well as hybrid architectures that combine hierarchical and flat models to capitalize on their respective advantages [16].

A growing trend is edge-centric designs, which emphasize the fusion of edge and fog computing to provide a holistic distributed computing paradigm. Advances in resource management techniques focus on resource supply, load balancing, and effective job allocation in fog computing environments [17]. In order to optimize task offloading choices, researchers have investigated dynamic resource allocation algorithms that consider network circumstances, device capabilities, and energy efficiency [11].

27.3.1 Security and Privacy in Fog Computing

Security and privacy have been key problems. Recent studies have suggested secure data transmission methods, authentication procedures, and access control frameworks for fog computing environments as solutions to these problems [18]. In order to preserve sensitive data while enabling efficient data processing and analytics at the edge, data anonymization and encryption have been investigated [19].

27.3.1.1 Data Analytics

By enabling real-time decision-making and data analytics at the edge, fog computing lowers latency and bandwidth needs. Recent developments have concentrated on creating effective edge analytics frameworks and algorithms [20]. In order to allow intelligent and real-time analytics in fog computing settings, methods including distributed stream processing, federated learning, and edge-based machine learning have been investigated. For thorough analysis and decision support, researchers have also looked at data fusion and aggregation techniques to combine data from various edge devices and fog nodes [21].

27.3.2 Implementations for Particular Applications

Fog computing has been employed in a number of industries, including smart cities, healthcare, transportation, and industrial automation. Recent studies have demonstrated application-specific fog computing implementations, adjusting the architecture and algorithms to suit the needs of certain domains. Examples include intelligent traffic management systems, healthcare monitoring and diagnosis platforms, and energy-efficient industrial control systems leveraging fog computing capabilities.

27.3.2.1 Integration with Emerging Technologies

Fog computing has been integrated with emerging technologies to enhance its capabilities. Integration with edge artificial intelligence (AI) and machine learning (ML) enables more advanced analytics and edge-based decision-making. Researchers have explored the use of blockchain technology to enhance data security, trust, and accountability in fog computing environments. Inventive haze processing models, such as fog-to-cloud reconciliation and edge-driven structures, have arisen to address the developing necessities of edge registering and the rising intricacy of appropriated frameworks.

27.3.3 Fog-to-Cloud Integration

Fog-to-cloud integration architectures aim to seamlessly integrate cloud and fog computing resources, taking advantage of each paradigm.

In this architecture, fog nodes are set up at the network's edge to handle real-time data processing, analytics, and control tasks. Fog nodes work as bridges between the cloud and edge devices, offloading processing tasks from edge devices and providing localized services.

Fog nodes aggregate and filter data, performing initial analytics and decision-making. They can then selectively forward relevant data or summarized information to the cloud for further research or long-term storage. Fog-to-cloud integration architectures ensure efficient utilization of resources, reduce latency, and minimize network bandwidth consumption by leveraging fog computing capabilities while benefiting from the scalability and storage capacities of the cloud [22].

27.3.4 EDGE-CENTRIC ARCHITECTURES

Edge-driven models center around driving computational capacities and knowledge straightforwardly to the brink of gadgets, improving edge registering abilities.

In this engineering, edge gadgets are enabled with handling, stockpiling, and examination abilities, permitting them to locally perform progressed undertakings. Edge devices are capable of decision-making, ML, and real-time data processing, eliminating the need for constant connection with the cloud or fog nodes. This architecture promotes low-latency processing, enhanced privacy and security, and reduced network bandwidth requirements by minimizing data transmission to remote nodes. Edge-centric architectures are particularly beneficial in scenarios with strict latency requirements or limited cloud connectivity, such as remote or disconnected environments [22].

27.4 FOG COMPUTING APPLICATIONS

Fog computing has found a variety of purposes across various industries, enabling efficient data processing, intelligent decision-making, and real-time analytics at the network edge [23].

27.4.1 SMART CITIES

Fog computing contributes significantly to the creation of smart cities by providing real-time data processing and analysis for a range of applications. Traffic management systems use fog computing to handle data from sensors, cameras, and linked cars. This enables real-time traffic monitoring, congestion management, and improved signal control. Environmental monitoring systems utilize fog computing to process data from distributed sensors, facilitating real-time air quality monitoring, weather prediction, and disaster management. Smart lighting systems leverage fog computing to enable intelligent lighting control, optimize energy consumption, and provide additional services such as security monitoring and parking management.

27.4.2 MODERN WEB OF THINGS (IoT)

Haze figuring upgrades IoT applications by empowering ongoing examination and control in modern conditions. Prescient support frameworks use haze figuring to handle sensor information from hardware and gear, empowering ongoing checking, peculiarity identification, and proactive upkeep planning. Modern robotization frameworks influence haze figuring to empower neighborhood direction and control, diminishing dormancy and guaranteeing convenient reactions in basic cycles. Resource tracking and executive frameworks use edge computing to manage location and sensor data, enabling real-time monitoring, inventory optimization, and supply chain visibility [24, 25].

27.4.3 Transportation and Planned Operations

Haze figuring upgrades transportation and strategies tasks by empowering continuous information handling, examination, and astute independent direction. Armada, the board frameworks influence haze figuring to handle ongoing information from vehicles, empowering course enhancement, eco-friendliness examination, and driver conduct observing. Store network the executive frameworks supply chain management and haze figuring with cloud computing everywhere in this section to handle information from RFID labels, sensors, and associated gadgets, working with ongoing stock administration, shipment following, and request anticipating. Haze processing is fundamental in improving Web of Things IoT applications across different areas, including savvy urban communities, medical services, transportation, and modern robotization. It broadens the capacities of distributed computing by bringing calculation, stockpiling, and systems administration assets nearer to the edge of the organization, where IoT gadgets and sensors are found. This vicinity to the edge empowers a few advantages and headways in IoT applications [26].

27.4.4 Brilliant Urban Communities

Low Idleness: Haze figuring diminishes the dormancy in information handling and direction by carrying the registering assets nearer to the IoT gadgets. This empowers ongoing observation and quicker reaction to gridlock, natural checking, and crises.

27.4.5 Transfer Speed Enhancement

By handling and investigating information at the edge, haze figuring decreases how much information that should be communicated to the cloud. This advancement saves transmission capacity and diminishes network clogs.

27.4.6 Confined Administrations

Haze hubs in savvy urban communities can offer limited types of assistance, for example, shrewd road lighting, squandering the executives, and stopping the board. These administrations can work independently without depending vigorously on the cloud, guaranteeing ceaseless activity even in case of organizational disturbances.

27.4.7 Modern Robotization: Continuous Observing and Contro

Haze figuring empowers ongoing checking and control of modern cycles by bringing process assets nearer to the assembling floor. This diminishes dormancy and supports opportune decision-production for basic activities. By breaking down sensor information locally, mist processing can recognize examples and abnormalities in modern hardware. This considers prescient upkeep, identifying possible disappointments before they happen, and decreasing personal time. Fog computing enables the deployment of localized safety systems that can react quickly to hazardous conditions. For example, in industrial settings, fog nodes can detect the presence of toxic gases or abnormal operating conditions, triggering immediate alarms or shutdowns [27].

27.4.8 Case Studies that Highlight Successful Fog Computing Deployments

Cisco's Smart Connected Digital Platform is shown in Figure 27.3. Cisco implemented fog computing in Barcelona as part of their Smart Connected Digital Platform. Fog nodes were deployed throughout the city to process data from various IoT gadgets, including parking sensors, garbage management systems, and smart street lighting. The fog nodes carried out edge decision-making and real-time analytics, enabling efficient traffic management, improved waste collection, and enhanced public safety. The haze figuring approach decreased the information traffic shipped off the cloud, limited idleness, and offered continuous types of assistance in any event, during network disturbances [28].

27.4.8.1 GE's Splendid Assembling Suite

GE executed haze registering in their Splendid Assembling Suite, an answer for modern computerization. Haze hubs were conveyed on the plant floor to gather and

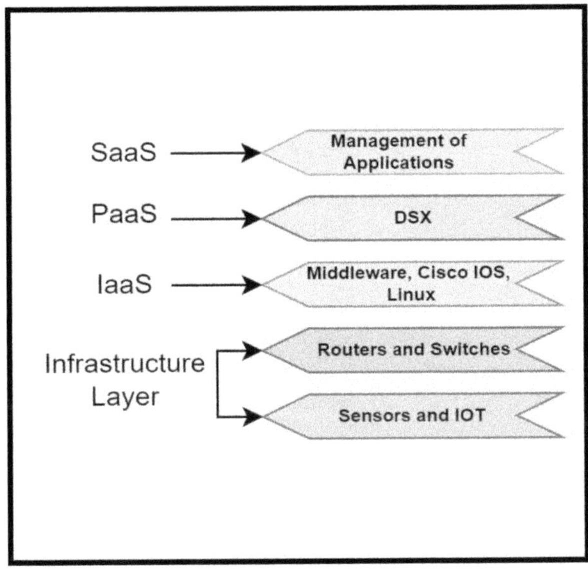

FIGURE 27.3 Technical architecture of Cisco's fog computing platform.

deal with information from sensors and modern machines. The haze hubs performed constant investigation and AI calculations locally, empowering prescient support, continuous observing, and improved creation processes. The haze figuring engineering diminished the idleness in information handling, upgraded functional effectiveness, and limited margin time by proactively distinguishing and tending to hardware issues.

27.4.8.2 China Portable's Brilliant Transportation Arrangement

China Versatile sent haze figuring in their shrewd transportation arrangement, zeroing in on rush hour gridlock the executives and well-being. Haze hubs were introduced in traffic light control cupboards and side-of-the-road units to deal with information from traffic cameras, sensors, and associated vehicles. The fog nodes performed real-time video analytics, vehicle detection, and traffic flow optimization locally. This approach reduced the reliance on CC, minimized network congestion, and enabled faster response times for traffic management and incident detection [29].

27.5 CHALLENGES IN FOG COMPUTING

For fog computing systems to be successfully implemented, a number of issues must be resolved, as shown in Figure 27.4

27.6 CHALLENGES IN ARCHITECTURE AND DESIGN

27.6.1 HETEROGENEITY

Most often, fog computing systems use a variety of hardware, software, and communication methods. Integrating and managing heterogeneous devices and platforms pose challenges in terms of interoperability, standardization, and scalability.

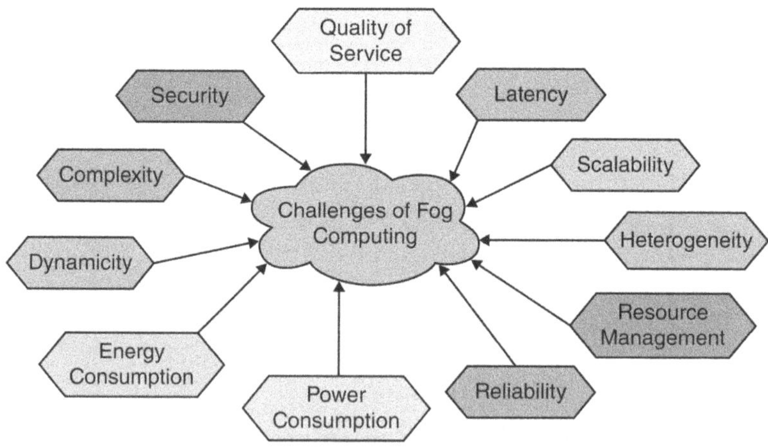

FIGURE 27.4 Challenges of fog computing.

27.6.2 RESOURCE CONSTRAINTS

Fog nodes often have limited processing power, storage capacity, and energy resources. Designing efficient algorithms and resource management techniques to optimize resource utilization and ensure effective task allocation is challenging.

27.6.3 DYNAMIC NETWORK TOPOLOGY

The network topology in fog computing settings can be dynamic due to the mobility of devices and changing connectivity. In such dynamic circumstances, providing smooth communication and coordination among fog nodes and IoT devices is challenging.

27.7 DATA MANAGEMENT CHALLENGES

27.7.1 DATA VOLUME AND SCALABILITY

The enormous volume of data produced by IoT devices in fog computing systems poses difficulties in storing, processing, and analyzing that data. Efficient data management techniques, such as data filtering, aggregation, and compression, need to be employed to handle the data volume and ensure scalability.

27.7.2 DATA SECURITY AND PRIVACY

Fog computing systems process and store sensitive data at the edge. Ensuring data security, privacy, and protection against unauthorized access or breaches is a significant challenge. Appropriate encryption, authentication, and access control mechanisms must be implemented.

27.7.3 DATA CONSISTENCY AND SYNCHRONIZATION

In distributed fog computing systems, ensuring data consistency and synchronization across multiple fog nodes and the cloud is challenging. Addressing issues related to data replication, versioning, and conflict resolution becomes crucial.

27.7.4 NETWORK CONGESTION AND BANDWIDTH LIMITATIONS

Fog computing systems rely on network communication for data transfer between cloud servers, fog nodes, and IoT devices. Network congestion and limited bandwidth can impact the system's performance and latency. Efficient data routing, traffic management, and prioritization mechanisms are required.

27.7.5 NETWORK RELIABILITY AND CONNECTIVITY

Fog nodes and IoT devices may operate in environments with unreliable or intermittent network connectivity. Ensuring reliable communication, fault tolerance, and seamless handoffs between different network interfaces (e.g., Wi-Fi, cellular) is challenging.

27.7.6 LATENCY AND RESPONSE TIME

Fog computing aims to reduce latency by processing data at the edge. However, achieving low latency and real-time response across a distributed fog infrastructure is challenging, particularly when dealing with resource-constrained devices and complex data processing tasks.

27.7.7 ORCHESTRATION AND TASK OFFLOADING

Efficient task offloading and workload management techniques are required to determine which tasks should be processed locally on fog nodes and which should be offloaded to the cloud. Dynamic decision-making based on factors like resource availability, latency requirements, and energy efficiency is challenging.

27.7.8 SYSTEM MONITORING AND MANAGEMENT

Monitoring and managing a distributed fog computing system involving numerous devices and nodes can be complex. Effective monitoring, fault detection, software updates, and system configuration management mechanisms are necessary to ensure system reliability and performance.

27.7.9 SCALABILITY AND LOAD BALANCING

As IoT devices and fog nodes increase, ensuring scalability and load balancing across the system becomes challenging. Efficient load distribution and resource allocation strategies need to be employed to handle the increasing workload [30].

27.8 ISSUES AND CHALLENGES

Some important aspects to note when it comes to open-source issues and challenges

27.8.1 HETEROGENEITY

Fog computing environments surround various operating systems, devices, and communication technologies. This diversity poses challenges for open-source projects in terms of developing software that runs on different hardware platforms and operating systems and accommodates various network protocols.

27.8.2 LACK OF STANDARDIZATION

Lack of standardized platforms and frameworks is one of the primary challenges in fog computing. Open-source projects usually differ in architecture, communication protocols, and data models. This makes it difficult to integrate different computing solutions and hinders interoperability.

27.8.3 Security and Privacy

Fog computing devices often operate in untrusted and physically vulnerable environments, thus introducing new security and privacy concerns. Developing open-source solutions that can efficiently utilize these constrained resources simultaneously and providing reliable and scalable services is a significant challenge.

27.8.4 Resource Constraints

Fog devices have constrained computing resources, such as processor speed, memory, and storage size. So, developing open-source solutions that can efficiently utilize these constrained resources with reliable and scalable services is a challenge.

27.8.5 Manageability and Orchestration

Managing and orchestrating many fog devices distributed across different locations is typical. Open-source solutions should provide mechanisms for device discovery, configuration, monitoring, and orchestration of services in a fog computing environment.

27.8.6 Integration with Cloud and Edge Computing

Typically, fog computing is viewed as a development of both cloud and edge computing. Open-source projects should focus on seamless integration with existing cloud and edge computing platforms to leverage their capabilities and provide a comprehensive computing environment.

27.9 EMERGING TRENDS IN FOG COMPUTING

Despite some of the inadequacies and problems noted in the literature, fog computing emerges as a crucial architectural component for ubiquitous computing. Several emerging trends and technologies are shaping the future of fog computing, enabling its continued growth and enhancing its capabilities. Here are a few prominent patterns to investigate:

27.9.1 Edge Man-Made Intelligence and AI

The combination of computerized reasoning (man-made intelligence) and AI (ML) at the edge is a critical pattern in haze registering. Edge artificial intelligence empowers ongoing information handling, examination, and direction, diminishing the requirement for broad information transmission to the cloud. This pattern permits haze hubs to perform progressed investigation, prescient displaying, and insightful robotization at the edge, improving the responsiveness and effectiveness of mist registering frameworks.

27.9.2 5G and Organization Cutting

Sending 5G organizations brings quicker information transmission speeds, lower dormancy, and expanded network limit. Haze figuring, joined with 5G, empowers more hearty and productive IoT arrangements. Network cutting, a 5G component, permits the formation of virtual organizations with explicit qualities to take special care of different mist registering applications. This pattern upgrades the versatility, dependability, and nature of administration in mist figuring conditions.

27.9.3 CONTAINERIZATION AND ORGANIZATION

Containerization advances, like Docker and Kubernetes, are becoming famous in haze registering. Holders give lightweight and convenient conditions to conveying applications across haze hubs, empowering simpler administration, versatility, and asset assignment. Compartment coordination stages work on the sending, scaling, and organizing of containerized applications in haze registering conditions.

27.9.4 BLOCKCHAIN FOR HAZE FIGURING

The reconciliation of blockchain innovation with mist registering is an arising pattern. Blockchain gives decentralized and secure information to the board, as well as verification and agreement components, which can improve the trust, security, and protection of mist figuring frameworks. It empowers secure information sharing, unchanging review trails, and savvy contracts in mist conditions.

27.9.5 UNIFIED LEARNING

Combined learning is a protection safeguarding AI strategy that permits disseminated edge gadgets to cooperatively prepare models without sharing crude information. It is particularly suitable for fog computing, as it enables edge devices to contribute their local data for model training while maintaining data privacy. Federated learning empowers fog nodes to learn from a vast distributed dataset without transferring sensitive data to the cloud.

27.9.6 FOG-TO-CLOUD CONTINUUM

The integration and seamless coordination between fog computing and CC form a fog-to-cloud continuum. Fog nodes and cloud resources are seen as complementary components, enabling flexible workload distribution and dynamic resource allocation. Fog nodes preprocess and filter data at the edge, while the cloud provides high-level analytics, long-term storage, and global coordination. This continuum optimizes resource utilization, reduces latency, and supports scalable and efficient fog computing deployments [31].

27.10 OPEN RESEARCH DIRECTIONS

It's important to highlight the ongoing and potential future areas of exploration and progress in the context of open-source fog computing:-

27.10.1 INTEROPERABILITY AND STANDARDIZATION

Research may concentrate on creating open protocols and standards for fog computing to enhance compatibility between different fog computing platforms and devices. This includes defining interfaces, communication protocols, and standardized data representations to allow smooth integration and collaboration among various fog computing solutions.

27.10.2 EDGE AI AND MACHINE LEARNING

With the rise in the trend of edge AI and ML applications, research can be done in open-source frameworks and algorithms that give efficient and scalable machine learning at the fog level. This includes developing lightweight and distributed machine learning models, federated learning approaches, and edge AI frameworks that leverage fog resources for data processing and inference.

27.10.3 QUALITY OF SERVICE AND REAL-TIME PROCESSING

Fog computing aims to provide low-latency and real-time services. Researchers can explore open-source solutions for quality of service (QoS) management, real-time data processing frameworks, stream processing engines, and event-driven architectures that ensure timely and reliable service delivery in fog environments.

27.10.4 COMMUNITY COLLABORATION AND CONTRIBUTIONS

Encouraging community collaboration and contributions is important for the success of open-source fog computing projects. Research can explore methodologies and frameworks to improve community engagement, promote inclusive participation, and facilitate knowledge sharing and collaboration among developers and researchers.

27.11 CONCLUSION AND FUTURE SCOPE

Our study suggests that fog computing can be helpful for a sizable count of computing jobs across various application use cases and deployment scenarios. We distributed a set of comparable computing jobs and demonstrated how they may be completed on network nodes. The reviewed papers show that there is an eventuality for calculation in all network situations. The necessity to drop computing tasks arises from the fact that sensor technology is often not powerful enough to do such computations alone. On the other hand, cloud computation is usually not an appropriate outcome for such comparable offloading due to limitations regarding reliability, privacy & security concerns, or regulations. Data can be filtered by fog computing operations to help lessen the strain on the network or to protect privacy. Additionally, with their capability to act close to the users, fog computing tasks contribute a crucial component, making systems more dependable or reliable. However, to make the gains productive, it is important to elevate attention from the specific use cases for further complete infrastructures, as stated above. This analysis and discussion, which details the variety of deployment options, serves as a signpost in that direction. Further research is needed to develop standardized protocols, interfaces, and frameworks that promote interoperability among diverse devices, platforms, and communication technologies in fog computing. Explore novel approaches for edge intelligence and machine learning in fog computing. Research can focus on developing lightweight AI models, distributed learning algorithms, federated learning techniques, and edge analytics frameworks to enable intelligent decision-making, real-time insights, and advanced data analysis at the edge.

Explore application-specific use cases and domains where fog computing can bring significant benefits. Research on domain-specific requirements, architectural design patterns, and optimization techniques will provide valuable insights into tailoring fog computing solutions for specific applications such as industrial automation, smart cities, transportation, and healthcare.

REFERENCES

1. F. A. Kraemer, A. E. Braten, N. Tamkittikhun, and D. Palma, "Fog computing in healthcare–A review and discussion," *IEEE Access*, vol. 5, pp. 9206–9222, 2017.
2. V. K. Quy, N. Van Hau, D. Van Anh, and L. A. Ngoc, "Smart healthcare IoT applications based on fog computing: Architecture, applications and challenges," *Complex Intell. Syst.*, vol. 8, no. 5, pp. 3805–3815, 2022.
3. A. A. Laghari, A. K. Jumani, and R. A. Laghari, "Review and state of art of fog computing," *Arch. Comput. Methods Eng.*, vol. 28, no. 5, pp. 3631–3643, 2021.
4. M. S. Sofla, M. H. Kashani, E. Mahdipour, and R. F. Mirzaee, "Towards effective offloading mechanisms in fog computing," *Multimed. Tools Appl.*, vol. 81, pp. 1997–2042, 2022.
5. K. Ma, A. Bagula, C. Nyirenda, and O. Ajayi, "An IoT-based fog computing model," *Sensors (Basel)*, vol. 19, no. 12, p. 2783, 2019.
6. H. Sabireen, and V. Neelanarayanan, "A review on fog computing: Architecture, fog with IoT, algorithms and research challenges," *ICT Express*, vol. 7, no. 2, pp. 162–176, 2021.
7. S. S. Gill, and N. Kumar, "A comprehensive review on fog computing: Architecture, applications, security, and future research directions," *J. Ambient Intell. Humaniz. Comput.*, vol. 11, no. 4, pp. 3895–3926, 2020.
8. Y. Liu, J. E. Fieldsend, and G. Min, "A framework of fog computing: Architecture, challenges, and optimization," *IEEE Access*, vol. 5, pp. 25445–25454, 2017.
9. D. A. Chekired, L. Khoukhi, and H. T. Mouftah, "Industrial IoT data scheduling based on hierarchical fog computing: A key for enabling smart factory," *IEEE Trans. Industr. Inform.*, vol. 14, no. 10, pp. 4590–4602, 2018.
10. B. Tang, Z. Chen, G. Hefferman, T. Wei, H. He, and Q. Yang, "A hierarchical distributed fog computing architecture for big data analysis in smart cities," in *Proceedings of the ASE BigData & SocialInformatics 2015*, 2015.
11. V. Karagiannis, and S. Schulte, "Comparison of alternative architectures in fog computing," in *2020 IEEE 4th International Conference on Fog and Edge Computing (ICFEC)*, 2020.
12. A. Nagaraj, *Introduction to Sensors in IoT and Cloud Computing Applications.* Bentham Science Publishers, 2021.
13. V. Karagiannis, and S. Schulte, "Distributed algorithms based on proximity for self-organizing fog computing systems," *Pervasive Mob. Comput.*, vol. 71, p. 101316, 2021.
14. B. V. Natesha, and M. R. Guddeti, "Meta-heuristic based hybrid service placement strategies for two-level fog computing architecture," *J. Netw. Syst. Manag.*, vol. 30, no. 3, 2022.

15. R. Roman, J. Lopez, and M. Mambo, "Mobile edge computing, fog et al.: A survey and analysis of security threats and challenges," *Future Gener. Comput. Syst.*, vol. 78, pp. 680–698, 2018.

16. K.-K. R. Choo, R. Lu, L. Chen, and X. Yi, "A foggy research future: Advances and future opportunities in fog computing research," *Future Gener. Comput. Syst.*, vol. 78, pp. 677–679, 2018.

17. D. Jo, and G. J. Kim, "IoT+ AR: Pervasive and augmented environments for 'Digi-log' shopping experience," *Hum.-Centric Comput. Inf. Sci.*, vol. 9, no. 1, pp. 1–17, 2019.

18. S. Sicari, A. Rizzardi, and A. Coen-Porisini, "Insights into security and privacy towards fog computing evolution," *Comput. Secur.*, vol. 120, p. 102822, 2022.

19. M. Abdel-Basset, H. Hawash, N. Moustafa, I. Razzak, and M. Abd Elfattah, "Privacy-preserved learning from non-iid data in fog-assisted IoT: A federated learning approach," *Digit. Commun. Netw*, vol. 10, pp. 404–415, 2022.

20. B. M. Alencar, J. P. Canário, R. Lobão Neto, C. Prazeres, A. Bifet, and R. A. Rios, "Fog-DeepStream: A new approach combining LSTM and concept drift for data stream analytics on fog computing," *Internet Things*, vol. 22, p. 100731, 2023.

21. T.-A. N. Abdali, R. Hassan, A. H. M. Aman, and Q. N. Nguyen, "Fog computing advancement: Concept, architecture, applications, advantages, and open issues," *IEEE Access*, vol. 9, pp. 75961–75980, 2021.

22. S. A. AlQahtani, "An evaluation of e-health service performance through the integration of 5G IoT, fog, and cloud computing," *Sensors (Basel)*, vol. 23, no. 11, pp. 1–22, 2023.

23. R. K. Naha *et al.*, "Fog computing: Survey of trends, architectures, requirements, and research directions," *IEEE Access*, vol. 6, pp. 47980–48009, 2018.

24. G. S. S. Chalapathi, V. Chamola, A. Vaish, and R. Buyya, "Industrial internet of things (IIoT) applications of edge and fog computing: A review and future directions," in *Fog/Edge Computing For Security, Privacy, and Applications*, Cham: Springer International Publishing, 2021, pp. 293–325.

25. V. V. Rohinidevi, P. K. Srivastava, N. Dubey, S. Tiwari, A. Tiwari, and Sweeti, "A Taxonomy towards fog computing resource allocation," in *2022 2nd International Conference on Innovative Sustainable Computational Technologies (CISCT)*, 2022, pp. 1–5.

26. D. Korzun, A. Varfolomeyev, A. Shabaev, and V. Kuznetsov, "On dependability of smart applications within edge-centric and fog computing paradigms," in *2018 IEEE 9th International Conference on Dependable Systems, Services and Technologies (DESSERT)*, 2018.

27. M. Ijaz, G. Li, L. Lin, O. Cheikhrouhou, H. Hamam, and A. Noor, "Integration and applications of fog computing and cloud computing based on the internet of things for provision of healthcare services at home," *Electronics (Basel)*, vol. 10, no. 9, p. 1077, 2021.

28. K. Tange, M. De Donno, X. Fafoutis, and N. Dragoni, "A systematic survey of industrial internet of things security: Requirements and fog computing opportunities," *IEEE Commun. Surv. Tutor.*, vol. 22, no. 4, pp. 2489–2520, 2020.

29. J. Satyakrishna, and R. K. Sagar, "Analysis of smart city transportation using IoT," in *2018 2nd International Conference on Inventive Systems and Control (ICISC)*, 2018.

30. S. Khan, S. Parkinson, and Y. Qin, "Fog computing security: A review of current applications and security solutions," *J. Cloud Comput. Adv. Syst. Appl.*, vol. 6, no. 1, pp. 1–22, 2017.

31. O. Ali, M. K. Ishak, M. K. L. Bhatti, I. Khan, and K.-I. Kim, "A comprehensive review of internet of things: Technology stack, middlewares, and Fog/edge computing interface," *Sensors (Basel)*, vol. 22, no. 3, p. 995, 2022.

Index

Note: *Italicized* and **bold** page numbers refer to figures and tables respectively.

A

AAL, *see* Ambient assisted living (AAL)
Accessibility, cloud computing and, 8
Accuracy, sensors, 47
Acoustic communication, 259
    advancements, 264
    applications of, 251
    challenges in, 251
    integration with Cloud-IoT, 251
    principles of, 251
Acoustic monitoring, 144
Active sensors, 49
Actuators, 50, *50*, 236, 355
    types of, 50–51, 356–357
Adaptive learning, 242–243
Advanced Messaging Queuing Protocol (AMQP), 54, 465
Agriculture, 157–168, *see also* Agriculture, Cloud IoT in
    agricultural supply chain optimization, 174
    automation in, 63, *64*
    cloud computing in, 157–158
    IoT in, 159, 360
    literature review of, 159–161
    smart, 385
Agriculture, Cloud IoT in, 161–164
    advantages of, 162–163
    applications of, 171–175
    autonomous machinery, 164, 175
    burst time of cloudlets, 166, **167**
    collaborative platforms, 163
    cost-effectiveness, 162
    crop monitoring, 164, 174
    data accessibility, 162
    data backup, 163
    data-driven decision-making, 164
    data management, 162
    data storage, 162
    environmental monitoring, 164, 174–175
    flexibility, 162
    framework, 165–166
    literature survey, 171–173
    livestock monitoring and management, 163, 174
    performance evaluation, 165–166, *166*, **166**
    precision agriculture, 163, 173–174
    predictive analytics, 164
    real-time data analysis and decision-making, 162–163
    robotics, 164
    Round Robin scheduling algorithm, 164–165, *165*
    scalability, 162
    security, 163
    smart irrigation, 163, 174
    supply chain management, 163–164
    sustainability, 164, 174–175
Agriculture Mobile Crowd Sensing (AMCS), 160–161
AI, *see* Artificial intelligence (AI)
Air-powered actuators, 51
Alibaba Cloud IoT platform, 160
Amazon, 16, 22, 70
    Amazon Go, 70, 77
    Echo, 454
Amazon Web Services (AWS), 4, 338
Ambient assisted living (AAL), 122, 129–130
AMCS, *see* Agriculture Mobile Crowd Sensing (AMCS)
AMQP, *see* Advanced Messaging Queuing Protocol (AMQP)
Analog sensor, 49
Analytics, 46
ANNs, *see* Artificial neural networks (ANNs)
Anti-poaching technologies, 145
Application Layer, IoT architecture, 56, 57–58, 384
Application Service Providers (ASPs), 335
AR, *see* Augmented reality (AR)
Architecture, of cloud computing, 8, *9*, 157–158, 324–343
Arduino, 160
ARPANET, 44, 45, 335
Artificial intelligence (AI), 330, 344–351
    AI-driven decision support, 110–111
    in education sector, 300–301
    global impact of, 346–348
    impact on healthcare, 83
    for intelligent systems, 365–378
    in IoT devices, 361
    in IoT healthcare, 132–133, **133**
    personalization, 299
    tiny, 345
    in underwater communication, 216–217
    in wildlife conservation, 146, 150
Artificial neural networks (ANNs), 83, 84

For Product Safety Concerns and Information please contact our EU
representative GPSR@taylorandfrancis.com
Taylor & Francis Verlag GmbH, Kaufingerstraße 24, 80331 München, Germany

www.ingramcontent.com/pod-product-compliance
Ingram Content Group UK Ltd.
Pitfield, Milton Keynes, MK11 3LW, UK
UKHW022319100726
473146UK00009B/540